TRANSITION, TURBULENCE AND COMBUSTION MODELLING

ERCOFTAC SERIES

VOLUME 6

Series Editors

Aims and Scope of the Series

ERCOFTAC (European Research Community on Flow, Turbulence and Combustion) was founded as an international association with scientific objectives in 1988. ERCOFTAC strongly promotes joint efforts of European research institutes and industries that are active in the field of flow, turbulence and combustion, in order to enhance the exchange of technical and scientific information on fundamental and applied research and design. Each year, ERCOFTAC organizes several meetings in the form of workshops, conferences and summerschools, where ERCOFTAC members and other researchers meet and exchange information.

The ERCOFTAC Series will publish the proceedings of ERCOFTAC meetings, which cover all aspects of fluid mechanics. The series will comprise proceedings of conferences and workshops, and of textbooks presenting the material taught at summerschools.

The series covers the entire domain of fluid mechanics, which includes physical modelling, computational fluid dynamics including grid generation and turbulence modelling, measuring-techniques, flow visualization as applied to industrial flows, aerodynamics, combustion, geophysical and environmental flows, hydraulics, multi-phase flows, non-Newtonian flows, astrophysical flows, laminar, turbulent and transitional flows.

The titles published in this series are listed at the end of this volume.

Transition, Turbulence and Combustion Modelling

Lecture Notes from
the 2nd ERCOFTAC Summerschool
held in Stockholm, 10-16 June, 1998

Edited by

A. HANIFI

The Aeronautical Research Institute of Sweden,
Bromma, Sweden

P. H. ALFREDSSON

Department of Mechanics,
Royal Institute of Technology,
Stockholm, Sweden

A.V. JOHANSSON

Department of Mechanics,
Royal Institute of Technology,
Stockholm, Sweden

and

D. S. HENNINGSON

Department of Mechanics,
Royal Institute of Technology,
Stockholm, Sweden
and
The Aeronautical Research Institute of Sweden,
Bromma, Sweden

KLUWER ACADEMIC PUBLISHERS

DORDRECHT / BOSTON / LONDON

A C.I.P. Catalogue record for this book is available from the Library of Congress.

ISBN 0-7923-5989-5

Published by Kluwer Academic Publishers,
P.O. Box 17, 3300 AA Dordrecht, The Netherlands.

Sold and distributed in North, Central and South America
by Kluwer Academic Publishers,
101 Philip Drive, Norwell, MA 02061, U.S.A.

In all other countries, sold and distributed
by Kluwer Academic Publishers,
P.O. Box 322, 3300 AH Dordrecht, The Netherlands.

Printed on acid-free paper

CONTENTS

About 60 participants attended the summerschool at KTH in June 1998. In the front (from left) lecturers Alfredsson, Henningson, Johansson, Friedrich, Riley, Lele, Hanifi and Burden.

Preface

During mid-June 1998 the second ERCOFTAC Summer School on Turbulence and Transition Modelling was held at KTH (The Royal Institute of Technology in Stockholm, Sweden). The summer school was primarily intended as an introduction to the field of turbulence and transition with objective to spread knowledge of modelling tools to current and prospective users from industry as well as academia. The summer school consisted of two parts: a tutorial part (June 10-11) and a lectures part (June 12-16). During the weekend a visits to the fluid mechanics laboratory and a boat trip to Sandhamn in the Stockholm archipelago were organized.

The summer school was attended by 60 participants (excluding the lecturers) from 9 countries; Sweden (42), Germany (4), Netherlands (3), Switzerland (3), France (2), Italy (2), Poland (2), Japan (1) and Israel (1). Of these 50 belong to universities, 2 to research institutes and 8 came from industry. Of the Swedish participants 17 were from the Mechanics department at KTH.

During the tutorials a brief summary of basic physics and characteristic feature of turbulence and transition were given and basic methodologies of the modelling of turbulence and transition were introduced. The tutorials were given by researchers at the organizing institution:

- Prof. P.H. Alfredsson and Prof. D.S. Henningson: Tutorials on Stability and Transition.
- Dr. A. Hanifi: Tutorials on Advanced Stability theory.
- Dr. A.D. Burden: Tutorials on Combustion.
- Prof. A.V. Johansson: Tutorials on Turbulence.

Following the tutorials examples of more advanced modeling approaches were given by invited lecturers:

- Dr. Daniel Arnal (ONERA/CERT, France): Transition prediction in industrial applications.
- Dr. Uwe Dallmann (DLR Göttingen, Germany): Transition physics and modelling in compressible flows. [1]
- Prof. Rainer Friedrich (TU Munich, Germany): Modelling of turbulence in compressible flows.
- Prof. Sanjiva Lele (Stanford Univ., USA): DNS techniques in incompressible and compressible flows, including the modelling of turbulent noise generation.
- Dr. Olivier Metais (Grenoble, France): LES techniques in incompressible and compressible flows.

[1] This lecture is not included in the book

- Prof. Jim Riley (Univ. of Washington, USA): Modelling of turbulent combustion.

Lecture notes that had been prepared by each lecturer in advance, comprising about 400 pages, were handed out to the participants. The lecture notes have been revised and this book is the outcome of this revised and edited material. The editors believe that it will serve two purposes, first it will through the chapters 1, 2 and 4, which are of a tutorial character, give an introduction to the equations governing compressible viscous flows, and give a description of problems related to modelling of stability, transition and turbulence phenomena. The reader is assumed to have a background in boundary layer and stability theory as well as turbulence according to a first basic university course in these subjects. A good reference, which will be referred to in the following is Hallbäck *et al.* (1995), which is the book produced after the summerschool at KTH in 1995. The second purpose of the book is that through chapters 3 and 5-8 give state-of-the-art descriptions of various practical and research methods used at present.

The editors and organizers want to thank all those involved for a successful summer school. Financial support to the summer school was obtained from ERCOFTAC, The Swedish Royal Academy of Science and the Department of Mechanics, KTH.

Stockholm, May 1999
Ardeshir Hanifi
Henrik Alfredsson
Arne Johansson
Dan Henningson

Chapter 1

INTRODUCTION

P.H. ALFREDSSON AND A.D. BURDEN
Department of Mechanics, KTH
SE-10044 Stockholm, Sweden

Compressible fluid flow is an important branch of fluid dynamics but it is seldom given extensive coverage in university courses. This is especially true of the viscous aspects, *i.e.* modelling of laminar and turbulent boundary layer flows as well as the important area of transition to turbulence. It is clear that one of the major applications for such knowledge is the aerodynamics of high speed aircraft, both sub and supersonic, as well as spacecraft. However large effects of compressibility on boundary layer flows also occur in other areas of engineering, for instance in various gas turbine applications. Combustion is another area where varying density and composition make physical modelling a difficult task.

This chapter will first give the equations of motion, energy and state and also discuss pertinent gas (air) properties. It also includes a description of the theory for laminar compressible boundary layers to elucidate some of the differences between incompressible and compressible boundary layers. Finally it gives a short introduction to properties of gas mixtures and combustion fundamentals. The reader who needs a textbook on compressible flow in general and further references is refered to *e.g.* Anderson [1]. A comprehensive account of boundary layer theory is given by Schlichting [6], whereas a review on compressible turbulent shear layers is found in Spina, Smits & Robinson [8]. Examples of standard books on combustion are those of Williams [11] and Glassman [3].

1.1. Equations for compressible flows

Compared to incompressible flow, compressible flows present additional complications since the material properties can no longer be assumed to be constant when variations in both pressure and temperature are large. Apart from the Reynolds number the following parameters or conditions will affect the development of for example a boundary layer:

1. Mach number.
2. Prandtl number.
3. The temperature dependence of the gas viscosity and heat conductivity.
4. The boundary condition for the temperature field, *e.g.* adiabatic or constant temperature wall.

In the compressible case both the density and temperature may vary as functions of $\vec{x}$ so we need two more equations in order to be able to solve, for example, the boundary layer flow and we also need material data for the gas. The two new equations are the perfect gas law (or some other equation of state for the gas such as van der Waal's) and the energy equation.

The basic equations are derived by applying the fundamental principles of mass conservation, Newton's 2nd Law, and the First Law of Thermodynamics to an arbitrary fluid particle. Mathematically, a fluid particle is contained in, and defines, a material volume, $V(t)$, which follows the flow so that each point on its surface, $S(t)$, moves with the flow velocity, $\vec{u}$, at that point. In the derivations we need to work out derivatives with respect to time, t, of time-dependent volume integrals and for this we will use the mathematical identity,

$$\frac{\mathrm{d}}{\mathrm{d}t}\int_{V(t)} \Phi(\vec{x},t)\,\mathrm{d}V \;=\; \int_{V(t)} \left\{\frac{\partial \Phi}{\partial t} \;+\; \nabla\cdot(\Phi\vec{u})\right\}\mathrm{d}V\,, \tag{1.1}$$

which is sometimes refered to as Reynolds' transport equation.

Mass conservation

The mass of a fluid particle is, of course, constant,

$$\frac{\mathrm{d}}{\mathrm{d}t}\int_{V(t)} \rho(\vec{x},t)\,\mathrm{d}V \;=\; 0\,,$$

where ρ is the mass density of the gas. Using now the mathematical identity in equation (1.1), the right-hand side of equation (1.1), with $\Phi = \rho$, must vanish for all $V(t)$, so the integrand itself has to vanish,

$$\frac{\partial \rho}{\partial t} \;+\; \nabla\cdot(\rho\vec{u}) \;=\; 0\,. \tag{1.2}$$

This is the equation expressing conservation of mass in a compressible fluid. When ρ is constant it reduces to the familiar continuity condition, $\nabla\cdot\vec{u} = 0$, of incompressible flow.

Before proceeding to the equation of motion and the energy budget we will develop the mathematical identity in equation (1.1) with the help of equation (1.2) for ρ. Writing $\Phi = \rho\phi$ in equation (1.1), and applying equation (1.2) to the integrand on the right-hand side, we find,

$$\frac{\partial}{\partial t}(\rho\phi) + \nabla\cdot(\rho\phi\vec{u}) = \rho\frac{\mathrm{D}\phi}{\mathrm{D}t}, \tag{1.3}$$

where,

$$\frac{\mathrm{D}\phi}{\mathrm{D}t} = \frac{\partial\phi}{\partial t} + (\vec{u}\cdot\nabla)\,\phi, \tag{1.4}$$

is the material time derivative, following the flow, so that equation (1.1) can be written in the form,

$$\frac{\mathrm{d}}{\mathrm{d}t}\int_{V(t)}\rho\,\phi\,\mathrm{d}V = \int_{V(t)}\rho\,\frac{\mathrm{D}\phi}{\mathrm{D}t}\,\mathrm{d}V\,. \tag{1.5}$$

Equation (1.3) contains two equivalent and alternative forms for the left-hand sides of typical conservation equations.

Equation of motion

The equation of motion is derived by applying Newton's 2nd Law to an arbitrary fluid particle. The surrounding fluid exerts a force on the surface of the fluid particle which is given in terms of the stress tensor, σ_{ij}. In Cartesian coordinates, the component of this force in the x_i-direction is,

$$\oint \sigma_{ij}n_j\,\mathrm{d}S = \int\frac{\partial}{\partial x_j}\sigma_{ij}\,\mathrm{d}V\,,$$

where $S(t)$ is the surface boundary to $V(t)$ and $\vec{n}$ is the unit surface normal which points outwards from $V(t)$ (since we used Gauss theorem). Now, in the absence of body forces, Newton's 2nd Law for the fluid particle takes the form,

$$\frac{\mathrm{d}}{\mathrm{d}t}\int_{V(t)}\rho u_i\,\mathrm{d}V = \int_{V(t)}\frac{\partial}{\partial x_j}\sigma_{ij}\,\mathrm{d}V\,.$$

Using equation (1.5), the volume integrals over the arbitrary volume $V(t)$ can be discarded to yield,

$$\rho\frac{\mathrm{D}u_i}{\mathrm{D}t} = \frac{\partial}{\partial x_j}\sigma_{ij}\,. \tag{1.6}$$

This equation is sometimes refered to as Cauchy's general equation of motion for a fluid. (The left-hand side can be rewritten using equation (1.3).)

In a *Newtonian fluid*, the stress tensor is given by,

$$\sigma_{ij} = -p\,\delta_{ij} + 2\mu\left(s_{ij} - \tfrac{1}{3}s_{kk}\delta_{ij}\right) + \mu_v s_{kk}\delta_{ij}\,, \tag{1.7}$$

where,

$$s_{ij} = \frac{1}{2}\left(\frac{\partial u_i}{\partial x_j} + \frac{\partial u_i}{\partial x_j}\right),$$

is the rate-of-strain tensor and μ_v is the bulk viscosity, which resists compression and expansion in the same way that μ resists strain. According to Stokes' hypothesis $\mu_v = 0$ and μ_v can in fact be shown to be zero for monoatomic gases. For air $\mu_v \approx 0.6\mu$ and, in most cases, it can safely be neglected in the atmosphere on earth. The bulk viscosity is often ignored in combustion calculations. For carbon dioxide $\mu_v \approx 1000\mu$ so μ_v may well be important in the atmospheres of other planets, for example when calculating the flow around a spacecraft which enters such an atmosphere. Both Venus and Mars have atmospheres which consist of more than 95% CO_2.

Insertion of the expression in equation (1.7) for the stress tensor into the general equation of motion, equation (1.6), yields,

$$\rho\frac{\mathrm{D}u_i}{\mathrm{D}t} = -\frac{\partial p}{\partial x_i} + \frac{\partial}{\partial x_j}\left[2\mu\left(s_{ij} - \tfrac{1}{3}s_{kk}\delta_{ij}\right)\right], \tag{1.8}$$

when the bulk viscosity is neglected. This is the Navier-Stokes equation for compressible flow. It contains several terms but in the boundary layer approximation it is possible to simplify the equation and write it in a compact vector form,

$$\rho\left[\frac{\partial \vec{u}}{\partial t} + (\vec{u}\cdot\nabla)\,\vec{u}\right] = -\nabla p + \nabla\cdot(\mu\nabla\vec{u})\,,$$

though even in this form some of the terms are negligible in the boundary layer approximation.

The energy equations

According to the First Law of Thermodynamics, the change in total energy, *i.e.* inner energy plus macroscopic kinetic energy, of a fluid particle is given by the amount of heat added to the particle plus the amount of work carried out on the particle by, *e.g.*, the surrounding fluid,

$$\mathrm{d}\left(e + \tfrac{1}{2}|\vec{u}|^2\right) = \delta q + \delta w\,.$$

Here e is the specific inner energy, $\frac{1}{2}|\vec{u}|^2$ is the specific kinetic energy of the macroscopic flow, δq is the addition of heat per unit mass and δw is the amount of work carried out on the fluid per unit mass.

Heat transfer (conduction and radiation)

The transfer of heat to the fluid particle can be due to both heat conduction and thermal radiation. In either case, the rate of transfer of heat to the fluid particle can be expressed in terms of the heat flux vector, $\vec{q}$, by,

$$\dot{Q} = \oint_{S(t)} -\vec{q}\cdot\vec{n}\mathrm{d}S = \int_{V(t)} -\nabla\cdot\vec{q}\,\mathrm{d}V\,,$$

where $S(t)$ is the surface boundary to $V(t)$ and $\vec{n}$ is the unit surface normal which points outwards from $V(t)$ (hence the minus sign).

In this section we will only discuss heat conduction although heat transfer by thermal radiation may be important both at high Mach numbers and in combustion. In a simple gas, or in a mixture of gases with constant composition, the heat flux can usually be modelled by the gradient-diffusion expression,

$$\vec{q} = -k\nabla T\,, \tag{1.9}$$

which is often refered to as Fourier's law.

Internal work

The surrounding fluid exerts a force $\sigma_{ij}n_j\mathrm{d}S$ on a surface element of the fluid particle. Consequently, the rate at which the surrounding fluid carries out work on the fluid particle is,

$$\dot{W} = \oint_{S(t)} (\sigma_{ij}n_j)\,u_i\,\mathrm{d}S = \int_{V(t)} \frac{\partial}{\partial x_j}(u_i\sigma_{ij})\,\mathrm{d}V \tag{1.10}$$

using Gauss theorem again.

Total energy

The energy budget of the fluid particle is thus,

$$\frac{\mathrm{d}}{\mathrm{d}t}\int_{V(t)} \rho\left(e + \tfrac{1}{2}u^2\right)\mathrm{d}V = \int_{V(t)} \rho\frac{\mathrm{D}}{\mathrm{D}t}\left(e + \tfrac{1}{2}u^2\right)\mathrm{d}V = \dot{Q} + \dot{W}\,.$$

Using equation (1.5), we can now discard the volume integrals over the arbitrary volume $V(t)$ and write,

$$\rho\frac{\mathrm{D}}{\mathrm{D}t}\left(e + \tfrac{1}{2}u^2\right) = -\frac{\partial}{\partial x_j}(q_j - u_i\sigma_{ij})\,. \tag{1.11}$$

The right-hand side of this equation has the form, $-$div(flux), and is thus a pure transport term. Since total energy is absolutely conserved there are no sources or sinks in its budget.

Inner energy and dissipation

An equation for the kinetic energy of the macroscopic flow can be derived in a straightforward way by taking the scalar product of the velocity, $\vec{u}$, with the equation of motion, equation (1.6);

$$\rho \frac{\mathrm{D}}{\mathrm{D}t}\left(\tfrac{1}{2}|\vec{u}|^2\right) = u_i \frac{\partial}{\partial x_j}(\sigma_{ij}) = -\frac{\partial}{\partial x_j}(-u_i\sigma_{ij}) - \sigma_{ij}\frac{\partial u_i}{\partial x_j}\,.$$

The right-hand side of this equation has been rewritten so that the flux of total energy due to the stress can be identified from equation (1.11).

Subtraction of the equation governing the kinetic energy from the equation for the total energy yields now the equation governing the inner energy,

$$\rho \frac{\mathrm{D}e}{\mathrm{D}t} = -\frac{\partial q_j}{\partial x_j} + \sigma_{ij}\frac{\partial u_i}{\partial x_j}\,.$$

Clearly, the term, $\sigma_{ij}\frac{\partial u_i}{\partial x_j}$, in the right-hand sides of the two equations above describes transfer of energy from kinetic energy to inner energy. For a Newtonian fluid it takes the form,

$$\sigma_{ij}\frac{\partial u_i}{\partial x_j} = -p\frac{\partial u_i}{\partial x_i} + \mu\left[\frac{1}{2}\left(\frac{\partial u_i}{\partial x_j} + \frac{\partial u_j}{\partial x_i}\right)^2 - \frac{2}{3}\left(\frac{\partial u_i}{\partial x_i}\right)^2\right] + \mu_v\left(\frac{\partial u_i}{\partial x_i}\right)^2\,.$$

In this expression,

$$-p\frac{\partial u_i}{\partial x_i} = -p\nabla\cdot\vec{u}\,,$$

is the rate at which work is carried out in compressing the fluid, and,

$$\Phi = \tfrac{1}{2}\mu\left(\frac{\partial u_i}{\partial x_j} + \frac{\partial u_j}{\partial x_i}\right)^2 + \left(\mu_v - \tfrac{2}{3}\mu\right)\left(\frac{\partial u_i}{\partial x_i}\right)^2,$$

is the rate of viscous disspation of kinetic or mechanical energy to inner energy (or heat).

Final form of the energy equation

Collecting everything together we can now write,

$$\rho\frac{\mathrm{D}e}{\mathrm{D}t} = -\nabla(-k\nabla T) - p\nabla\cdot\vec{u} + \Phi\,, \qquad (1.12)$$

for a Newtonian gas with constant composition.

It is generally useful in the analysis of compressible flow to introduce the enthalpy,

$$h = e + \frac{1}{\rho} p = e + pv ,$$

where $v = 1/\rho$ is the specific volume. Using equation (1.3),

$$\rho \frac{\mathrm{D}h}{\mathrm{D}t} = \frac{\partial}{\partial t}(\rho h) + \nabla \cdot (\rho h \vec{u}) = \rho \frac{\mathrm{D}e}{\mathrm{D}t} + \frac{\partial p}{\partial t} + \nabla \cdot (p \vec{u}) .$$

The equation governing the specific enthalpy is now found to be,

$$\rho \frac{\mathrm{D}h}{\mathrm{D}t} = \frac{\mathrm{D}p}{\mathrm{D}t} - \nabla \cdot (-k \nabla T) + \Phi , \tag{1.13}$$

for a Newtonian gas with constant composition and the equation governing the temperature, T, is

$$\rho c_p \frac{\mathrm{D}T}{\mathrm{D}t} = \frac{\mathrm{D}p}{\mathrm{D}t} - \nabla \cdot (-k \nabla T) + \Phi . \tag{1.14}$$

1.1.1. GAS (AIR) MATERIAL PARAMETERS

In this section we will present some of the properties for the gas mixture that we usually denote as air. In many compressible flows of practical interest one can, without any large error, view the gas as perfect (or ideal). In the limit of a perfect gas intermolecular forces can be neglected so that the gas molecules move independently of each other until they hit another molecule in a billiard-ball like collision. Such a perfect gas has the following equation of state

$$p = \rho R T . \tag{1.15}$$

R is the specific gas constant, which for a particular gas can be calculated as

$$R = \frac{\mathcal{R}}{\mathcal{M}}$$

where $\mathcal{R}$ is the universal gas constant and $\mathcal{M}$ is the gas (average) molecular weight. The universal gas constant is

$$\mathcal{R} = 8314 \frac{\mathrm{J}}{\mathrm{kmol} \cdot \mathrm{K}}$$

Air consists mainly of di-atomic molecules of nitrogen (78 %) and oxygen (21 %). In addition to these consituents there are small amounts of carbon dioxide, argon etc. The weights of the the N_2 and O_2 molecules are 28.0

kg/kmol and 32.0 kg/kmol, giving air the average molecule weight of approximately 29.0 kg/kmol and a corresponding gas constant of $R_{\text{air}} = 287$ J/kgK.

For temperatures above 2000 K the N_2 molecules start to dissociate and for temperatures above 4000 K the O_2 start to dissociate. In this temperature range the molecular weight will also change and thereby also the specific gas constant (which hence no longer is a constant).

There are several other material parameters that have to be considered when dealing with compressible flows. Two important such parameters are the specific heats at constant pressure (c_p) and at constant volume (c_v), which in an ideal gas satisfy $c_p - c_v = R$. The ratio between the two specific heats is usually denoted by

$$\gamma = \frac{c_p}{c_v}$$

An ideal gas for which R, c_p, c_v and thus γ are constant is usually called a thermally and calorically perfect gas. Statistical mechanics predicts that for each degree of freedom of the molecule the average energy is $\frac{1}{2}kT$ (where k is the Boltzmann constant). This gives the specific internal energy as

$$e = \frac{n}{2}RT$$

where n is the number of degrees of freedom and the specific enthalpy

$$h = (\frac{n}{2} + 1)RT$$

which in turn gives

$$\gamma = \frac{n+2}{n}$$

For a monoatomic gas (like He, Ar, Ne etc.) n=3, *i.e.* the number of degrees of freedom corresponds to the three independent directions of space. For a diatomic molecule as for instance N_2 or O_2 the translational degrees of freedom are complemented by two rotational degrees of freedom[1]. For these two cases, γ becomes $\frac{5}{3}$ and $\frac{7}{5}$, respectively. For higher temperatures vibrations about the centre of mass for the two molecules start to become important, though the energy of this state only gradually increases with temperature. However, at sufficiently high temperatures the energy of this state is the same as for the other energy modes and the number of degrees of freedom thereby increases by 2 (one each for the kinetic and potential energy) and γ becomes equal to $\frac{9}{7}$.

[1]Note that there are only two degrees of freedom since rotation around the axis binding the two molecules cannot be detected

For compressible boundary layers the variation of viscosity and thermal conductivity with temperature must be taken into account. The variation of density is also important and can be determined from equation (1.15). A commonly used relation for the dynamic viscosity is that of Sutherland

$$\frac{\mu}{\mu_0} = \frac{T_0 + \mathcal{T}}{T + \mathcal{T}} \left(\frac{T}{T_0} \right)^{3/2}$$

where the reference temperature T_0=273.1 K for which μ_0=1.716$\cdot 10^{-5}$ kg/ms and the constant $\mathcal{T}$=100.6 K. A simpler expression is the formula of Chapman which gives a linear variation of the viscosity with temperature, *i.e.*

$$\frac{\mu}{\mu_0} = \frac{T}{T_0}$$

If the temperature range of interest is not too large this expression can be used with good accuracy.

The Prandtl number is defined as

$$Pr = \frac{\mu c_p}{k}$$

where k is the thermal conductivity. It is usually convenient to express the variation in k through the Prandtl number, the viscosity and c_p, however for air in the temperature range 300-2000 K $Pr = 0.71 \pm 0.02$, *i.e.* it is almost constant.

In figure 1.1 material properties for air in the temperature range 100-1000 K are shown. The gas properties shown are the dynamic viscosity (μ), the thermal conductivity (k), the Prandtl number (Pr) and the ratio between the specific heats (γ).

1.2. Laminar boundary layers

1.2.1. THE INCOMPRESSIBLE BOUNDARY LAYER

Consider the flow of a viscous fluid along a flat surface, where we denote the streamwise direction x, the direction normal to the plate y (see figure 1.2) and the corresponding velocity components by u and v. If the surface is rigid and impermeable the boundary condition states that the flow velocity should be tangential to the surface. The no-slip boundary condition furthermore forces the fluid velocity to be zero at the wall, *i.e.*

$$u = 0 \quad , \quad v = 0 \quad \text{at} \quad y = 0.$$

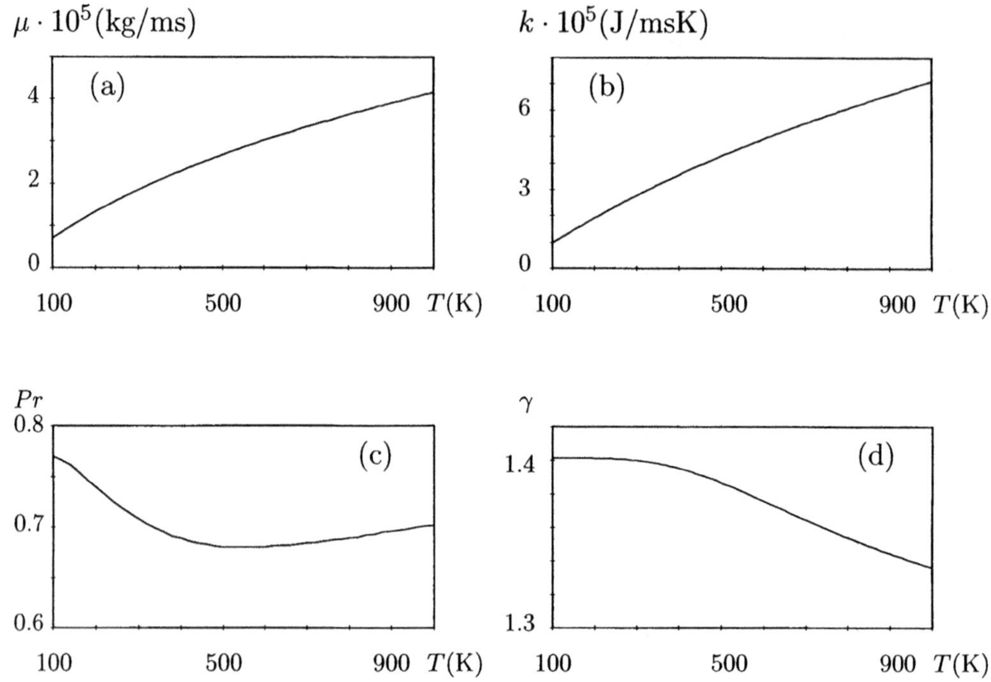

Figure 1.1. Variation for air with temperature of (a) dynamic viscosity, (b) thermal conductivity, (c) Prandtl number, (d) ratio of specific heats (γ).

Prandtl developed the so called boundary layer theory in which the viscous effects are limited to a thin region close to the surface, the boundary layer, whereas the flow outside this region can be viewed as inviscid. The theory is valid for flows for which the Reynolds number is high (or equivalently the non-dimensional distance from the leading edge should be large).

It is possible to use simple physical arguments to analyse how the boundary layer thickness, δ, varies with the relevant parameters. Assume that viscosity diffuses information of the boundary condition at the wall outwards into the flow. A typical viscous diffusion time scale would be

$$t_{\text{viscous}} \sim \frac{\delta^2}{\nu}$$

where ν is the kinematic viscosity of the fluid. Similarily a typical timescale for convecting a disturbance from the leading edge downstream a distance L with the external free stream velocity U_e would be

$$t_{\text{convective}} \sim \frac{L}{U_e}$$

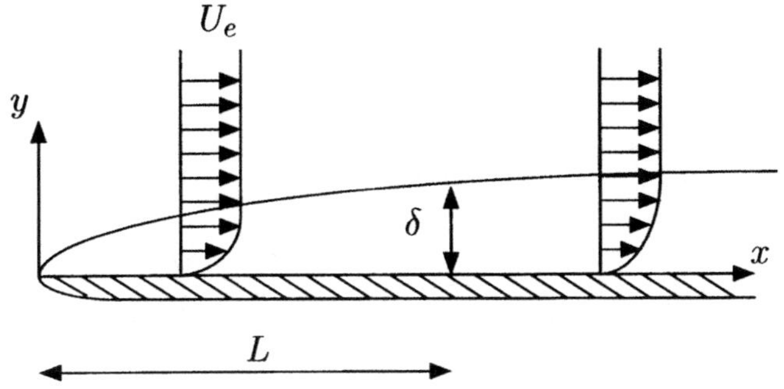

Figure 1.2. Boundary layer development along a flat plate.

At the edge of the boundary layer these two time scales would be of the same order, *i.e.*

$$\frac{\delta^2}{\nu} \sim \frac{L}{U_e}$$

which can be rewritten

$$\frac{\delta}{L} \sim \sqrt{\frac{\nu}{U_e L}} = Re_L^{-\frac{1}{2}} \tag{1.16}$$

where Re_L is the Reynolds number based on the distance from the leading edge of the flat plate. The assumption that there exists a boundary layer is that the boundary layer is thin, *i.e.* $\delta << L$, which implies that the Reynolds number must be high. It is also seen from equation (1.16) that the boundary layer thickness grows as

$$\delta \sim \sqrt{L}$$

The Blasius boundary layer

We now consider the boundary layer along a flat plate in the absence of a pressure gradient, *i.e.* the free stream velocity is constant along the plate. The Navier-Stokes equation and the continuity equation for a fluid with constant properties can be written

$$\frac{D\vec{u}}{Dt} = -\frac{1}{\rho}\nabla p + \nu \nabla^2 \vec{u} \tag{1.17}$$

$$\nabla \cdot \vec{u} = 0 \tag{1.18}$$

We now assume that the flow is steady and two-dimensional, *i.e.* there is no time dependence, nor variation in the spanwise direction (z). By projecting the NS-equation onto two directions, *viz.* the streamwise and the direction normal to the surface, we obtain

$$u\frac{\partial u}{\partial x}+v\frac{\partial u}{\partial y}=-\frac{1}{\rho}\frac{\partial p}{\partial x}+\nu(\frac{\partial^2}{\partial x^2}+\frac{\partial^2}{\partial y^2})u \tag{1.19}$$

$$u\frac{\partial v}{\partial x}+v\frac{\partial v}{\partial y}=-\frac{1}{\rho}\frac{\partial p}{\partial y}+\nu(\frac{\partial^2}{\partial x^2}+\frac{\partial^2}{\partial y^2})v \tag{1.20}$$

$$\frac{\partial u}{\partial x}+\frac{\partial v}{\partial y}=0 \tag{1.21}$$

From the continuity equation (1.21) we can estimate the magnitude of the v-component by assuming that a typical length in the x-direction is L, a typical length in the y-direction is δ and the u-component is of the order of the free stream velocity U_e. We then obtain

$$\frac{U_e}{L}\sim\frac{v}{\delta}$$

which gives

$$v\sim\frac{\delta}{L}U_e \quad \text{or} \quad v\sim Re_x^{-\frac{1}{2}}U_e \quad \text{where} \quad Re_x=\frac{U_e x}{\nu}$$

The terms in equation (1.20) can now be estimated in a similar way

$$\mathcal{O}(\frac{uv}{L})+\mathcal{O}(\frac{v^2}{\delta})=-\frac{1}{\rho}\frac{\partial p}{\partial y}+\mathcal{O}(\nu\frac{v}{L^2})+\mathcal{O}(\nu\frac{v}{\delta^2})$$

By introducing the estimate $v\sim\frac{\delta}{L}U_e$ we find that

$$\frac{1}{\rho}\frac{\partial p}{\partial y}=\mathcal{O}(\frac{\delta}{L}\frac{U_e^2}{L})$$

which tends towards zero as $Re\rightarrow\infty$. This gives the result that the pressure is effectively constant through the boundary layer in the boundary layer approximation. From an experimental point of view this means that if the pressure is measured by static pressure holes at the surface the pressure is the same as if the pressure was measured in the inviscid free stream. Finally the terms in equation (1.19) are estimated as

$$\mathcal{O}(\frac{u^2}{L})+\mathcal{O}(\frac{vu}{\delta})=-\frac{1}{\rho}\frac{\partial p}{\partial x}+\mathcal{O}(\nu\frac{u}{L^2})+\mathcal{O}(\nu\frac{u}{\delta^2})$$

By introducing the estimates of u and v it is clear that the two terms on the left hand side are of equal magnitude whereas, of the two viscous terms, the rightmost term is the largest. Since equation (1.20) showed that the pressure was constant across the boundary layer we can write

$$\frac{\partial p}{\partial x} = \frac{\mathrm{d}p_e}{\mathrm{d}x}$$

For a flow where U_e =const. Bernoullis equation for the exterior flow yields $\frac{\mathrm{d}p}{\mathrm{d}x} = 0$ and equation (1.19) reduces to

$$u\frac{\partial u}{\partial x} + v\frac{\partial u}{\partial y} = \nu\frac{\partial^2 u}{\partial y^2} \tag{1.22}$$

which together with the continuity equation forms a system of equations which determines the two unknowns u and v.

To solve the system of equations appropriate boundary conditions have to be applied which in this case are

$$u = 0 \quad , \quad v = 0 \quad \text{at} \quad y = 0$$

$$u \to U_e \quad \text{for} \quad \frac{y}{\delta} \to \infty$$

We have also implicitly given the boundary conditions at the beginning of the plate by stating that the approaching flow should be parallel to the plate and uniform in the normal direction.

Similarity solution

For equation (1.22) it is possible to find a similarity solution which reduces the partial differential equation to an ordinary differential equation. We do this by introducing the stream fuction ψ defined by

$$u = \frac{\partial \psi}{\partial y} \quad , \quad v = -\frac{\partial \psi}{\partial x}$$

which can be seen to identically satisfy the continuity equation. Using this definition equation (1.22) can be written as

$$\frac{\partial \psi}{\partial y}\frac{\partial^2 \psi}{\partial y \partial x} - \frac{\partial \psi}{\partial x}\frac{\partial^2 \psi}{\partial y^2} = \nu\frac{\partial^3 \psi}{\partial y^3} \tag{1.23}$$

We introduce a similarity coordinate

$$\eta = \sqrt{\frac{U_e}{\nu x}}y \sim \frac{y}{\delta}$$

and assume that the streamwise velocity can be written as

$$u = U_e g(\eta) = \frac{\partial \psi}{\partial y} \tag{1.24}$$

In physical terms this means that we assume that the shape of the velocity profile is given uniquely by the function g and the growth of the boundary layer is taken into account in the definition of η where the y-coordinate is scaled with a lengthscale which is of the order of the boundary layer thickness. Equation (1.24) also implies that the streamfunction can be written as

$$\psi = U_e \int^{y} g(\eta) \mathrm{d}y = U_e \sqrt{\frac{\nu x}{U_e}} \int^{\eta} g(\eta) \mathrm{d}\eta = \sqrt{\nu x U_e} f(\eta)$$

Using this expression for ψ we can immediately calculate the expressions for u and v as

$$u = \frac{\partial \psi}{\partial y} = U_e f'(\eta)$$

$$v = -\frac{\partial \psi}{\partial x} = \sqrt{\frac{\nu U_e}{x}} \frac{1}{2} [\eta f'(\eta) - f(\eta)]$$

With these expressions equation (1.23) yields

$$f''' + \frac{1}{2} f f'' = 0$$

This equation is of third order so three boundary conditions are needed as discussed above and these become

$$u(y=0) = 0 \quad \Rightarrow \quad f'(\eta = 0) = 0$$

$$v(y=0) = 0 \quad \Rightarrow \quad f(\eta = 0) = 0$$

$$u(y \to \infty) = U_e \quad \Rightarrow \quad f'(\eta \to \infty) = 1$$

The equation can be solved with for instance a shooting method and the corresponding u and v velocity profiles are shown in figure 1.3. It is interesting to note that close to the wall the U profile is almost linear whereas further out it has a fairly strong curvature. The boundary layer thickness defined as the position where the velocity has reached $0.99U_e$ is located at approximately $\eta = 5$. The v-velocity also shows the expected behaviour close to the wall ($v \sim \eta^2$) and reaches its maximum value outside the boundary layer. Note that the transverse velocity is constant throughout the inviscid exterior flow. Physically the exterior flow is continuously displaced outwards, away from the plate, by the growth of the boundary layer.

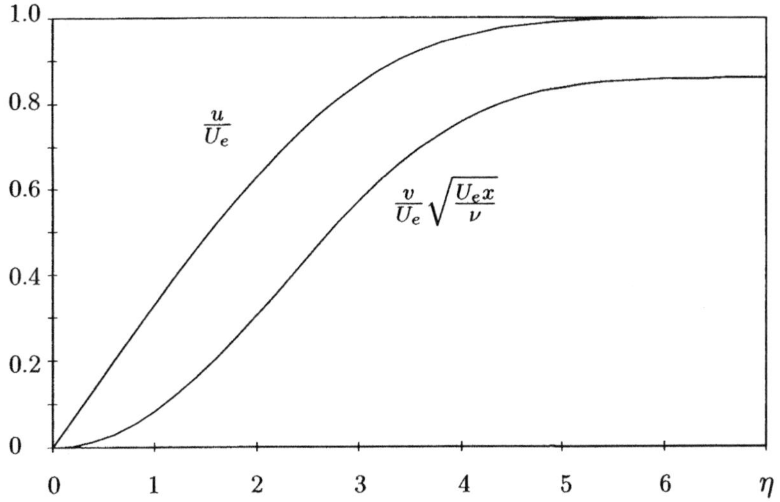

Figure 1.3. u and v velocity profiles of Blasius boundary layer.

It is also possible to find similarity solutions for boundary layer flows in which the free stream velocity varies in the downstream direction as some power of x. These profiles are called Falkner-Skan profiles but will not be discussed further here (see *e.g.* Schlichting, 1979).

Boundary layer parameters

The boundary layer thickness was defined above as the position where the flow velocity was 99% of the free stream velocity. This is a somewhat arbitrary definition of the boundary layer thickness and it may be quite hard to determine it accurately both in numerical calculations and experiments. Instead the so called displacement thickness

$$\delta^* = \int_0^\infty \left(1 - \frac{u}{U_e}\right) dy$$

and the momentum-loss thickness

$$\theta = \int_0^\infty \frac{u}{U_e}\left(1 - \frac{u}{U_e}\right) dy$$

have clear physical interpretations and are also possible to determine quite accurately both from numerical and experimental data. The displacement thickness can be interpreted as the distance streamlines are displaced normal to the wall due to the growth of the boundary layer (remember that the distance between streamlines is related to the mass flux). The displacement

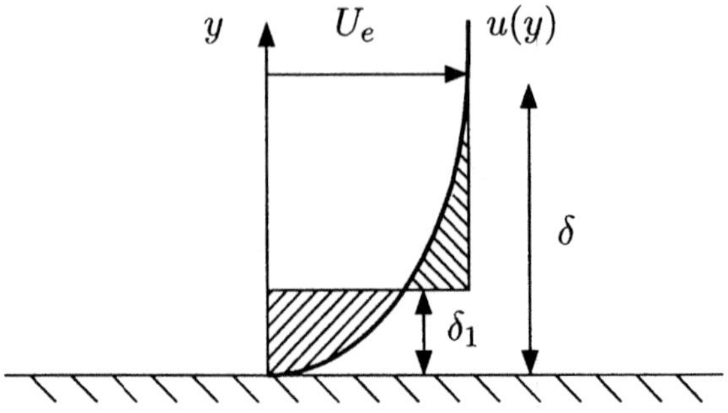

Figure 1.4. Geometric interpretation of displacement thickness.

thickness can also be given a simple geometric interpretation as is shown in figure 1.4.

The momentum-loss thickness on the other hand is directly related to the skin friction drag (D)

$$D = \rho U_e^2 \theta = \int_0^x \tau_w \mathrm{d}x$$

Another quantity of significant interest is the local skin friction coefficient which is defined as

$$c_f = \frac{\tau_w}{\frac{1}{2}\rho U_e^2}$$

where τ_w is the wall shear stress defined as

$$\tau_w = \left(\mu \frac{\mathrm{d}u}{\mathrm{d}y}\right)_{y=0}$$

and $\mu(=\rho\nu)$ is the dynamic viscosity. It is also possible to find a relation between the local skin friction coefficient and the momentum-loss thickness as

$$\frac{1}{2}c_f = \frac{\mathrm{d}\theta}{\mathrm{d}x}$$

In the case of the Blasius boundary layer the skin friction coefficient can easily be calculated as

$$c_f = 2Re_x^{-\frac{1}{2}} f''(0) = 0.664 Re_x^{-\frac{1}{2}}$$

where $f''(0)$=0.332 is given by the similarity solution. This shows that the skin friction coefficient decreases with the Reynolds number, *i.e.* the skin friction is highest close to the leading edge and decreases along the plate.

Another boundary layer parameter is the so called shape factor defined as

$$H = \frac{\delta^*}{\theta}$$

which for the Blasius boundary layer has the value 2.59.

Limitations of the boundary layer theory

The boundary layer theory is valid as long as typical variations of the flow in the streamwise direction are small compared to variations normal to the surface. This is however not true in the immediate vicinity of the leading edge and the boundary layer solution is hence not valid there. In order to investiagte the flow around the stagnation region other methods have to be employed but these will not be discussed here.

1.2.2. BOUNDARY LAYER EQUATIONS FOR COMPRESSIBLE FLOW

The boundary layer equations can be obtained in a similar manner as for the incompressible case and become for a stationary, two-dimensional boundary layer

$$\frac{\partial(\rho u)}{\partial x} + \frac{\partial(\rho v)}{\partial y} = 0 \tag{1.25}$$

$$\rho\left(u\frac{\partial u}{\partial x} + v\frac{\partial u}{\partial y}\right) = -\frac{\mathrm{d}p}{\mathrm{d}x} + \frac{\partial}{\partial y}\left(\mu\frac{\partial u}{\partial y}\right) \tag{1.26}$$

$$\frac{\partial p}{\partial y} = 0 \tag{1.27}$$

$$\rho c_p\left(u\frac{\partial T}{\partial x} + v\frac{\partial T}{\partial y}\right) = u\frac{\mathrm{d}p}{\mathrm{d}x} + \frac{\partial}{\partial y}\left(k\frac{\partial T}{\partial y}\right) + \mu\left(\frac{\partial u}{\partial y}\right)^2 \tag{1.28}$$

together with the perfect gas law

$$p = \rho RT \tag{1.29}$$

Boundary conditions

The above equations have to be complemented with suitable boundary conditions; for the velocity these will be the same as for the incompressible case whereas new boundary conditions have to be set for the temperature field. One of these boundary conditions is simply that the temperature should approach the exterior temperature outside the boundary layer, *i.e.*

$$T \to T_e \quad \text{for} \quad \frac{y}{\delta} \to \infty \, .$$

The boundary condition at the wall, however, depends on the type of wall. Two specific cases can be distinguished; the constant-temperature wall and the adiabatic wall. In the case of a constant-temperature wall the boundary condition becomes

$$T = T_w \quad \text{at} \quad y = 0 \, ,$$

whereas for the adiabatic or insulated wall the boundary condition can be written

$$\frac{\partial T}{\partial y} = 0 \quad \text{at} \quad y = 0 \, .$$

In practice the boundary condition is usually neither of these two but is given by a certain heat transfer to the body, but the two wall boundary conditions given above can be seen as two limiting cases.

Integral equation

Just as in the incompressible case, it is possible to formulate a useful integral relation for the boundary layer without explicitly knowing the shape of the boundary layer profile. This is often called the von Karman momentum integral equation. The starting point is the momentum equation

$$\rho \left(u \frac{\partial u}{\partial x} + v \frac{\partial u}{\partial y} \right) = - \frac{\mathrm{d}p}{\mathrm{d}x} + \frac{\partial}{\partial y} \left(\mu \frac{\partial u}{\partial y} \right) \tag{1.30}$$

where

$$\frac{\mathrm{d}p}{\mathrm{d}x} = -\rho_e U_e \frac{\mathrm{d}U_e}{\mathrm{d}x} \tag{1.31}$$

which is obtained from the Euler equation. By integrating equation (1.30) from the wall to the edge of the boundary layer we obtain

$$\int_0^\delta \left[\rho \left(u \frac{\partial u}{\partial x} + v \frac{\partial u}{\partial y} \right) - \rho_e U_e \frac{\mathrm{d}U_e}{\mathrm{d}x} \right] \mathrm{d}y \; + \; \left(\mu \frac{\partial u}{\partial y} \right)_{y=0} = 0 \tag{1.32}$$

The rightmost term can be identified to be the wall shear stress, τ_w. The continuity equation

$$\frac{\partial(\rho u)}{\partial x} + \frac{\partial(\rho v)}{\partial y} = 0$$

is also integrated, but from the wall to y in order to obtain an expression for ρv such that

$$\rho v = -\int_0^y \frac{\partial(\rho u)}{\partial x} \mathrm{d}y$$

which can then be inserted into equation (1.32) giving

$$\tau_w = \int_0^\delta \left[\rho_e U_e \frac{\mathrm{d}U_e}{\mathrm{d}x} - \rho u \frac{\partial u}{\partial x} + \frac{\partial u}{\partial y} \int_0^y \frac{\partial(\rho u)}{\partial x} \mathrm{d}y'\right] \mathrm{d}y$$

By partial integration the last term in the integral on the RHS can be rewritten

$$\int_0^\delta \frac{\partial u}{\partial y} \left(\int_0^y \frac{\partial(\rho u)}{\partial x} \mathrm{d}y'\right) \mathrm{d}y = \int_0^\delta (U_e - u) \frac{\partial(\rho u)}{\partial x} \mathrm{d}y$$

so that the equation can finally be written as

$$\tau_w = \rho_e U_e^2 \frac{1}{U_e} \frac{\mathrm{d}U_e}{\mathrm{d}x} \delta_1 + \frac{\mathrm{d}}{\mathrm{d}x}(\rho_e U_e^2 \delta_2). \tag{1.33}$$

Here the displacement thickness for the compressible case is defined as

$$\delta_1 = \int_0^\delta \left(1 - \frac{\rho u}{\rho_e U_e}\right) \mathrm{d}y \tag{1.34}$$

and the momentum loss thickness is defined as

$$\delta_2 = \int_0^\delta \frac{\rho u}{\rho_e U_e} \left(1 - \frac{u}{U_e}\right) \mathrm{d}y\ . \tag{1.35}$$

By dividing equation (1.33) by $\rho_e U_e^2$ we get

$$\frac{\tau_w}{\rho_e U_e^2} = \frac{\mathrm{d}\delta_2}{\mathrm{d}x} + \frac{1}{U_e} \frac{\mathrm{d}U_e}{\mathrm{d}x}(\delta_1 + 2\delta_2) + \delta_2 \frac{1}{\rho_e} \frac{\mathrm{d}\rho_e}{\mathrm{d}x}$$

It is possible to rewrite the term $\frac{1}{\rho_e}\frac{\mathrm{d}\rho_e}{\mathrm{d}x}$ by using the definition of the speed of sound

$$a^2 = (\frac{\partial p}{\partial \rho})_s$$

to obtain

$$\frac{\mathrm{d}p_e}{\mathrm{d}x} = (\frac{\partial p}{\partial \rho})_s \frac{\mathrm{d}\rho_e}{\mathrm{d}x} = a^2 \frac{\mathrm{d}\rho_e}{\mathrm{d}x}$$

Combining this with equation 1.31 we can write

$$\frac{1}{\rho_e}\frac{\mathrm{d}\rho_e}{\mathrm{d}x} = -M_e^2 \frac{1}{U_e}\frac{\mathrm{d}U_e}{\mathrm{d}x}$$

Finally the integral equation can be written as

$$\frac{1}{2}c_f = \frac{\mathrm{d}\delta_2}{\mathrm{d}x} + \frac{1}{U_e}\frac{\mathrm{d}U_e}{\mathrm{d}x}(\delta_1 + 2\delta_2 - \delta_2 M_e^2)$$

Note that this expression is also valid in the incompressible case for which $M_e = 0$.

Physical interpretation of δ_1 and δ_2

Consider the flow sketched in figure 1.5. The streamline shown in the figure is the one that passes through the edge of the boundary layer at the point where the velocity profile has been sketched. In front of the body the distance between this streamline and the streamline following the wall is $\delta - \delta_1$ since the definition of the displacement thickness is that it is the distance that the streamlines are displaced in the outer flow as compared to the inviscid case. The mass flow between the two streamlines per unit width in front of the body should be equal to the mass flow at the position where the boundary layer thickness is δ, *i.e.*

$$\rho_e U_e(\delta - \delta_1) = \int_0^\delta u\rho \mathrm{d}y$$

which can be rewritten as

$$\rho_e U_e \delta_1 = \rho_e U_e \int_0^\delta \mathrm{d}y - \int_0^\delta \rho u \mathrm{d}y$$

which immediately gives the definition of the displacement thickness, equation (1.34).

Similarily we can find an expression for the momentum loss between the plate origin and the position where the boundary layer thickness is δ.

$$\text{Momentum loss} = \rho_e U_e^2(\delta - \delta_1) - \int_0^\delta \rho u^2 \mathrm{d}y$$

By using the definition of δ_1 the RHS can be rewritten and shown to be equal to $\rho_e U_e^2 \delta_2$ with δ_2 defined in equation (1.35).

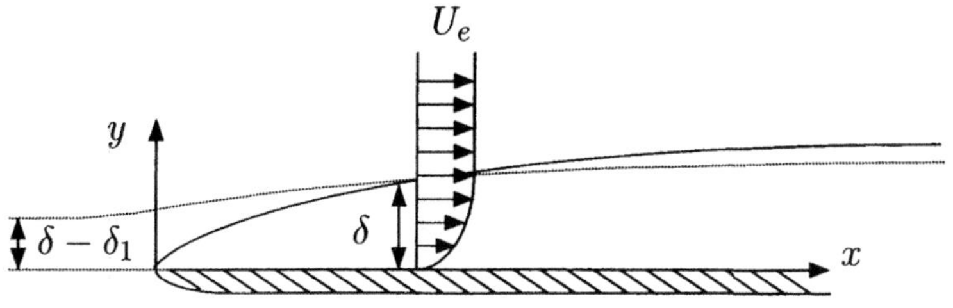

Figure 1.5. Analysis of displacement thickness.

Boundary layer solutions for Pr=1

For a gas with moderate temperature variations c_p can be assumed to be constant. With this assumption, and for the special case of Pr =1 ($Pr = \frac{\mu c_p}{k}$), it is possible to solve the boundary layer equations analytically by assuming that the temperature field can be written as

$$T = T(u)$$

i.e. the temperature field is directly related to the velocity distribution in the boundary layer. This is possible when the diffusivities of vorticity and heat are the same, *i.e.* Pr=1, which means that the growth of both the velocity and temperature boundary layers is the same.

With the assumption $T = T(u)$ the energy equation 1.28 can be written

$$\rho c_p \frac{\mathrm{d}T}{\mathrm{d}u}\left(u\frac{\partial u}{\partial x} + v\frac{\partial u}{\partial y}\right) = u\frac{\mathrm{d}p}{\mathrm{d}x} + \frac{\partial}{\partial y}\left(k\frac{\mathrm{d}T}{\mathrm{d}u}\frac{\partial u}{\partial y}\right) + \mu\left(\frac{\partial u}{\partial y}\right)^2 \quad (1.36)$$

Multiplying the momentum equation by $c_p\frac{\mathrm{d}T}{\mathrm{d}u}$ and subtracting it from equation (1.36) we obtain

$$u\frac{\mathrm{d}p}{\mathrm{d}x} + \frac{\partial}{\partial y}\left(k\frac{\mathrm{d}T}{\mathrm{d}u}\frac{\partial u}{\partial y}\right) + \mu\left(\frac{\partial u}{\partial y}\right)^2 - c_p\frac{\mathrm{d}T}{\mathrm{d}u}\left[-\frac{\mathrm{d}p}{\mathrm{d}x} + \frac{\partial}{\partial y}\left(\mu\frac{\partial u}{\partial y}\right)\right] = 0 \quad (1.37)$$

and after rearranging the terms this can be written as

$$\left(k\frac{\mathrm{d}^2T}{\mathrm{d}u^2} + \mu\right)\left(\frac{\partial u}{\partial y}\right)^2 + \frac{\mathrm{d}p}{\mathrm{d}x}\left(c_p\frac{\mathrm{d}T}{\mathrm{d}u} + u\right) - c_p\frac{Pr-1}{Pr}\frac{\mathrm{d}T}{\mathrm{d}u}\frac{\partial}{\partial y}\left(\mu\frac{\partial u}{\partial y}\right) = 0$$

A solution to this equation can be found by requiring that all three terms are zero simultaneously, which can be fulfilled if

$$\frac{\mathrm{d}^2T}{\mathrm{d}u^2} = -\frac{\mu}{k} \quad , \quad \frac{\mathrm{d}p}{\mathrm{d}x} = 0 \quad \text{and} \quad Pr = 1 \tag{1.38}$$

It is also possible to find a solution for $\frac{\mathrm{d}p}{\mathrm{d}x} \neq 0$ if instead $c_p\frac{\mathrm{d}T}{\mathrm{d}u} = -u$. This last requirement will in fact turn out to correspond to the wall being adiabatic as will be shown below.

We can easily find a relation between temperature and velocity from equation (1.38) by integrating twice, giving

$$T(u) = -\frac{\mu}{k}\frac{u^2}{2} + Au + B = -\frac{1}{c_p}\frac{u^2}{2} + Au + B \tag{1.39}$$

In this case $\frac{\mu}{k} = \frac{1}{c_p}$ since $Pr = 1$. The constants A and B have to be obtained from the boundary conditions.

Adiabatic wall

For the adiabatic wall, heat transfer to the wall should be zero so that the temperature gradient vanishes at the wall, *i.e.*

$$\frac{\partial T}{\partial y} = 0 \quad \text{at} \quad y = 0$$

If $T = T(u)$ we can rewrite

$$\frac{\partial T}{\partial y} = \frac{\mathrm{d}T}{\mathrm{d}u}\frac{\partial u}{\partial y} = 0$$

As $\frac{\partial u}{\partial y} > 0$ this means that $\frac{\mathrm{d}T}{\mathrm{d}u} = 0$ at the wall where also $u = 0$. This immediately gives $A = 0$. In the outer part of the boundary layer the temperature should approach the temperature in the outer inviscid flow

$$T \to T_e \quad \text{for} \quad \frac{y}{\delta} \to \infty$$

and with this condition we obtain

$$B = T_e + \frac{1}{c_p}\frac{U_e^2}{2}$$

and

$$T = T_e + \frac{1}{2c_p}(U_e^2 - u^2)$$

TABLE 1.1. Adiabatic wall temperature rise for various Mach numbers. $T_e = 300\text{K}$.

M_e	$\Delta T/T_e$	$\Delta T(K)$
1	0.2	60
3	1.8	540
5	5.0	1500

The adiabatic wall temperature becomes

$$T_{w,\text{adiabatic}} = T_e + \frac{1}{2c_p}U_e^2 = T_e\left(1 + \frac{\gamma - 1}{2}M_e^2\right) .$$

The physical interpretation of the rise in wall temperature is that it is due to friction. Note that this temperature rise is exactly the same as one would have at a stagnation point in an adiabatic compressible flow at the same Mach number. Table 1.1 gives the temperature rise for a few different Mach numbers.

In the case of $\frac{\mathrm{d}p}{\mathrm{d}x} \neq 0$ a requirement was that $c_p\frac{\mathrm{d}T}{\mathrm{d}u} = -u$. Calculating $\frac{\mathrm{d}T}{\mathrm{d}u}$ from equation (1.39) gives

$$\frac{\mathrm{d}T}{\mathrm{d}u} = -\frac{2}{c_p}u + A$$

which shows that $A = 0$. This is equivalent to $\frac{dT}{du} = 0$ at the wall, *i.e.* this solution is only valid for an adiabatic wall.

The velocity and corresponding temperature distributions for a typical case are illustrated in figure 1.6.

Constant wall temperature

In the case of constant wall temperature $T(y = 0) = T_w$ we obtain

$$\frac{T - T_w}{T_e} = \left(1 - \frac{T_w}{T_e}\right)\frac{u}{U_e} + \frac{\gamma - 1}{2}M_e^2\frac{u}{U_e}\left(1 - \frac{u}{U_e}\right)$$

For this case we can have heat transfer either to or from the wall depending on the temperature difference between the wall and the free stream. The heat transfer can easily be determined by calculating the value of the temperature derivative at the wall.

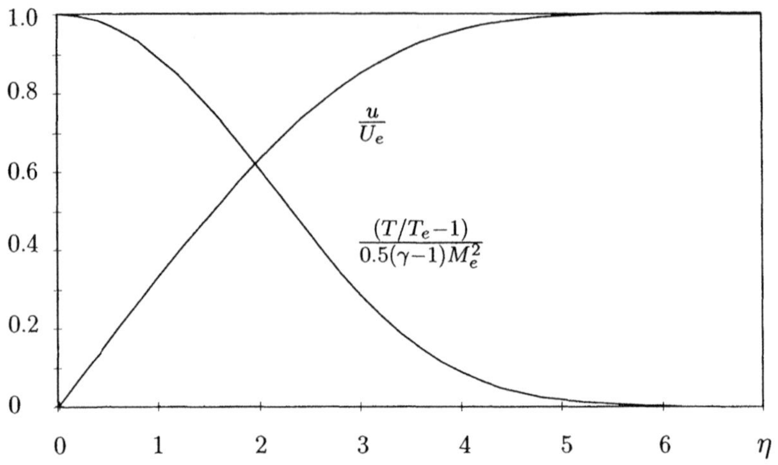

Figure 1.6. Typical velocity and temperature distributions corresponding to $Pr = 1$ and with adiabatic wall.

The Howarth-Dorodnitzyn transformation

In this section we will describe a transformation of the compressible boundary layer equations which reduces the x-momentum equation to the Blasius boundary layer equation by transforming the coordinates in such a way that the unknown variation in the density across the boundary layer is hidden in a tranformed variable. The temperature equation is also tranformed and can be written as a linear ordinary differential equation with variable coefficients (containing the solution to the Blasius equation). By solving the temperature equation it is hence possible to obtain the temperature distribution and then determine the density distribution inside the boundary layer (since the pressure is assumed to be constant across the boundary layer there is a direct relation between temperature and density through the equation of state).

The starting point is to introduce a transformed normal coordinate $\tilde{y} = \tilde{y}(x, y)$ such that

$$\tilde{y} = \int_0^y \frac{\rho(x,y)}{\rho_e} dy$$

In analogy with the solution of the incompressible boundary layer equations we introduce a stream function, Ψ, which satisfies the continuity equation and is defined as

$$\frac{\rho}{\rho_e} u = \frac{\partial \Psi}{\partial y} \quad ; \quad \frac{\rho}{\rho_e} v = -\frac{\partial \Psi}{\partial x}$$

In the transformed coordinates we write the stream function and the temperature as

$$\Psi(x,y) = \tilde{\Psi}(x,\tilde{y})$$
$$T(x,y) = \tilde{T}(x,\tilde{y})$$

It is now possible to rewrite the boundary layer equations in the new variables and, noting that

$$\frac{\partial \tilde{y}}{\partial y} = \frac{\rho}{\rho_e},$$

we find that u and v can be written

$$u = \frac{\partial \tilde{\Psi}}{\partial \tilde{y}}$$

$$v = -\frac{\rho_e}{\rho}\left(\frac{\partial \tilde{\Psi}}{\partial x} + u\frac{\partial \tilde{y}}{\partial x}\right)$$

If these expressions are introduced into the x-momentum equation (1.26) we obtain

$$\rho_e\left(\frac{\partial \tilde{\Psi}}{\partial \tilde{y}}\frac{\partial^2 \tilde{\Psi}}{\partial \tilde{y}\partial x} - \frac{\partial \tilde{\Psi}}{\partial x}\frac{\partial^2 \tilde{\Psi}}{\partial \tilde{y}^2}\right) = \frac{\partial}{\partial \tilde{y}}\left(\frac{\mu\rho}{\rho_e}\frac{\partial^2 \tilde{\Psi}}{\partial \tilde{y}^2}\right) \tag{1.40}$$

We assume that the viscosity of the gas varies linearly with temperature such that $\mu/T = \mu_e/T_e =$ const. Since the pressure is constant across the boundary layer we can write

$$\rho T = \rho_e T_e$$

giving that

$$\mu\rho = \mu_e\frac{T}{T_e}\rho_e\frac{T_e}{T} = \mu_e\rho_e$$

We introduce a stretched similarity coordinate and the corresponding self-similar streamfunction such that

$$\eta = \tilde{y}\sqrt{\frac{\rho_e U_e}{\mu_e x}}$$

$$\tilde{\Psi} = \sqrt{\frac{\mu_e U_e x}{\rho_e}} f(\eta)$$

$$u = U_e f'(\eta)$$

This reduces equation (1.40) to the Blasius equation

$$f''' + \frac{1}{2} f f'' = 0$$

where the boundary conditions are the same as for the incompressible case.

To transform the energy equation (1.28), with $\frac{\mathrm{d}p}{\mathrm{d}x} = 0$, we obtain

$$\frac{\partial \tilde{\Psi}}{\partial \tilde{y}} \frac{\partial \tilde{T}}{\partial x} - \frac{\partial \tilde{\Psi}}{\partial x} \frac{\partial \tilde{T}}{\partial \tilde{y}} = \frac{\partial}{\partial \tilde{y}} \left(\frac{k\rho}{c_p \rho_e^2} \frac{\partial \tilde{T}}{\partial \tilde{y}} \right) + \frac{\mu\rho}{c_p \rho_e^2} \left(\frac{\partial^2 \tilde{\Psi}}{\partial \tilde{y}^2} \right)^2$$

Let us now introduce a transformed non-dimensional temperature function

$$\frac{T}{T_e} = \tilde{\Theta}(\eta).$$

and using the relation

$$\frac{k\rho}{c_p \rho_e^2} = \frac{\mu\rho}{Pr\, \rho_e^2} = \frac{\mu_e \rho_e}{Pr\, \rho_e^2} = \frac{\mu_e}{Pr\, \rho_e}$$

for the first term on the RHS, the energy equation is found to take the form

$$\tilde{\Theta}'' + \frac{Pr}{2} f \tilde{\Theta}' + Pr \frac{U_e^2}{c_p T_e} f''^2 = 0$$

One should note that this equation has variable coefficients but is linear in Θ. The last term can be rewritten in terms of the Mach number so that

$$\tilde{\Theta}'' + \frac{Pr}{2} f \tilde{\Theta}' + Pr(\gamma - 1) M_e^2 f''^2 = 0$$

For the case of $Pr = 1$ it is possible to find a simple analytical solution, by separating Θ into a 'homogeneous' and a 'particular' part,

$$\Theta = \Theta_h + \Theta_p$$

First study the homogeneous equation

$$\tilde{\Theta}_h'' + \frac{1}{2} f \tilde{\Theta}_h' = 0$$

and rewrite it

$$\frac{\tilde{\Theta}_h''}{\tilde{\Theta}_h'} = -\frac{f}{2}$$

By using the Blasius equation we can rewrite the RHS

$$\frac{\tilde{\Theta}_h''}{\tilde{\Theta}_h'} = \frac{f'''}{f''}$$

which can be integrated to give the 'homogeneous' solution

$$\tilde{\Theta}_h' = Af'' \Rightarrow \tilde{\Theta}_h = Af' + B$$

where A and B are integration constants. A particular solution can also be found and is given by

$$\tilde{\Theta}_p = \frac{1}{2}(\gamma - 1)M_e^2(f' - f'^2)$$

so that the complete solution is given as

$$\Theta = \Theta_h + \Theta_p = Af' + B + \frac{1}{2}(\gamma - 1)M_e^2(f' - f'^2)$$

The constants A and B can now be determined from the boundary conditions

$$\Theta = \Theta_w \quad \text{at} \quad \eta = 0 \quad \text{and} \quad f'(0) = 0$$
$$\Theta = 1 \quad \text{at} \quad \eta \to \infty \quad \text{and} \quad f'(\infty) = 1$$

giving $B = \Theta_w$ and $A = 1 - \Theta_w$ and the following solution

$$\Theta = \Theta_w + (1 - \Theta_w)f' + \frac{1}{2}(\gamma - 1)M_e^2(f' - f'^2) \tag{1.41}$$

It has been shown that for Prandtl numbers which are slightly different from one, an approximation of the temperature field can be written

$$\Theta = \Theta_w + (1 - \Theta_w)f' + \sqrt{Pr}\frac{1}{2}(\gamma - 1)M_e^2(f' - f'^2) \tag{1.42}$$

From the above equations it is now possible to calculate the velocity and temperature distributions. Figure 1.7 shows these distributions for Mach numbers in the range from 0 to 5. The temperature boundary condition has been taken to be adiabatic. It is clearly shown how the boundary layer thickness increases with M_e.

Calculation of boundary layer thickness

The displacement thickness is defined as

$$\delta_1 = \int_0^\infty \left(1 - \frac{\rho u}{\rho_e U_e}\right) \mathrm{d}y$$

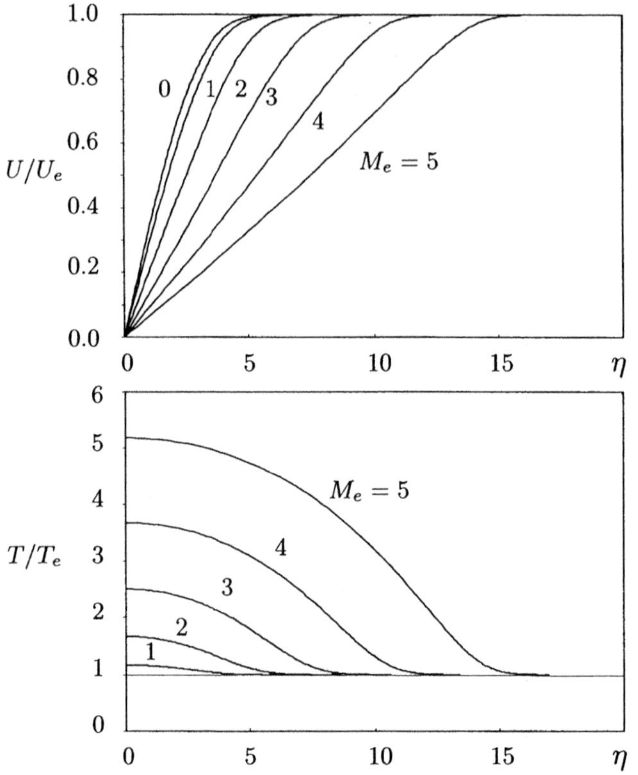

Figure 1.7. Velocity and temperature distribution for 6 different Mach numbers (M_e=0,1,2,. . . ,5) for adiabatic wall conditions. The dynamic viscosity is assumed to vary linearly with temperature. Pr=0.7.

By introducing the tranformed coordinate, $\mathrm{d}\tilde{y} = \frac{\rho}{\rho_e}\mathrm{d}y$, we obtain

$$\delta_1 = \int_0^\infty \left(\frac{\rho_e}{\rho} - \frac{u}{U_e}\right) \mathrm{d}\tilde{y}$$

which can be rewritten, assuming a perfect gas ($\frac{\rho_e}{\rho} = \frac{T}{T_e} = \Theta$),

$$\delta_1 = \int_0^\infty \left(\frac{T}{T_e} - \frac{u}{U_e}\right) \mathrm{d}\tilde{y}$$

By introducing the similarity coordinate η we obtain

$$\delta_1 = \frac{x}{\sqrt{Re}} \int_0^\infty (\Theta - f')\mathrm{d}\eta$$

and then with the temperature distribution from equation (1.41) we obtain

$$\delta_1 = \frac{x}{\sqrt{Re}} \int_0^\infty \left[\Theta_w + (1-\Theta_w)f' + \frac{\gamma - 1}{2} M_e^2 f'(1-f') - f'\right] \mathrm{d}\eta$$

By partial integration of the term involving the Mach number and using the Blasius equation we finally obtain

$$\delta_1 = \frac{x}{\sqrt{Re}}\left\{\Theta_w \lim_{\eta\to\infty}[\eta - f(\eta)] + \frac{\gamma-1}{2}M_e^2 2f''(0)\right\}$$

or

$$\delta_1 = \frac{x}{\sqrt{Re}}\left(1.721\Theta_w + 0.664\frac{\gamma-1}{2}M_e^2\right)$$

For the incompressible case, $\Theta_w = 1$ and $M_e = 0$. For compressible flow ($M_e > 0$) typically $\Theta_w > 1$ which means that for a given Re the compressible boundary layer is thicker than its incompressible counterpart.

The momentum-loss thickness can also be calculated in a similar manner and becomes

$$\delta_2 = \frac{x}{\sqrt{Re}}f''(0)$$

whereas the wall shear stress can be calculated as

$$\tau_w = \left(\mu\frac{\partial u}{\partial y}\right)_{y=0} = \mu_w\frac{\rho_w}{\rho_e}\left(\frac{\partial u}{\partial \tilde{y}}\right)_{\tilde{y}=0} = \mu_w\frac{\rho_w}{\rho_e}U_e\sqrt{\frac{U_e}{\nu_e x}}f''(0)$$

This gives finally (assuming $\mu \sim T$)

$$\tau_w = \frac{1}{2}\rho_e U_e^2\frac{1}{\sqrt{Re}}f''(0)$$

which is the same expression as in the incompressible case!

Temperature conditions at the wall

From the temperature distribution, equation (1.42), we can easily determine the conditions at the wall for the two cases of an adiabatic wall and a wall of constant temperature.

For an adiabatic wall the temperature gradient at the wall is zero, *i.e.* $\Theta'(0) = 0$. The temperature gradient can be obtained from equation (1.42),

$$\tilde{\Theta}' = (1-\Theta_w)f'' + \sqrt{Pr}\frac{\gamma-1}{2}M_e^2(f'' - 2f'f'')$$

Introducing the boundary conditions ($f'(0) = 0$ and $\Theta'(0) = 0$) we obtain the wall temperature as

$$\Theta_w = 1 + \sqrt{Pr}\frac{\gamma-1}{2}M_e^2 = \Theta_r$$

where Θ_r is called the recovery temperature. It is interesting to compare the increase in temperature at the wall with the temperature increase obtained for an isentropic velocity decrease to zero velocity $(U_e^2/2c_p)$. This ratio is called the recovery factor, r, and is

$$r = \frac{T_w - T_e}{U_e^2/2c_p} = \sqrt{Pr}$$

For a constant temperature wall, equation (1.42) shows that the temperature derivative at the wall is

$$T' = f''(0)(T_r - T_w)$$

Flow over conical bodies

The Mangler transformation transforms the axisymmetric boundary layer equations into the two-dimensional boundary layer equations and the results from this case can hence be used for the flow on conical bodies. We define a coordinate system with the coordinate directions $\hat{s}$ in the streamwise direction and $\hat{n}$ in the direction normal to the surface. The local radius of the body is given by $r_0(s)$. Under the assumption that the boundary layer thickness is much smaller than the local body radius ($\delta << r_0$) the boundary layer equations become

$$\frac{1}{r_0}\frac{\partial}{\partial s}(r_0 \rho u) + \frac{\partial}{\partial n}(\rho v) = 0$$

$$\rho\left(u\frac{\partial u}{\partial s} + v\frac{\partial u}{\partial n}\right) = -\frac{\mathrm{d}p}{\mathrm{d}s} + \frac{\partial}{\partial n}\left(\mu\frac{\partial u}{\partial \eta}\right)$$

$$\rho c_p\left(u\frac{\partial T}{\partial s} + v\frac{\partial T}{\partial n}\right) = u\frac{\mathrm{d}p}{\mathrm{d}s} + \frac{\partial}{\partial n}\left(\frac{\mu c_p}{Pr}\frac{\partial T}{\partial n}\right) + \mu\left(\frac{\partial u}{\partial n}\right)^2$$

The boundary layer assumption also implies that the pressure is constant across the boundary layer.

As can be seen the boundary layer equations are the same as for the flat plate boundary layer except for the continuity equation. The latter can be derived by viewing a small fluid element (see figure 1.8) situated at a distance of r from the symmetry line. The mass flowing into the volume through surface a is $(\rho u)_a r\Delta\phi\Delta n$ and out through surface b, $(\rho u)_b r\Delta\phi\Delta n$. The net flux of mass in the s-direction is hence $\Delta(\rho ur)\Delta\phi\Delta n$. In the same manner the net flux of mass across surface c and d can be estimated as $\Delta(\rho vr)\Delta\phi\Delta s$. There will be no mass flux through the other two sides since the flow is axi-symmetric. The net flux to the volume should be zero giving

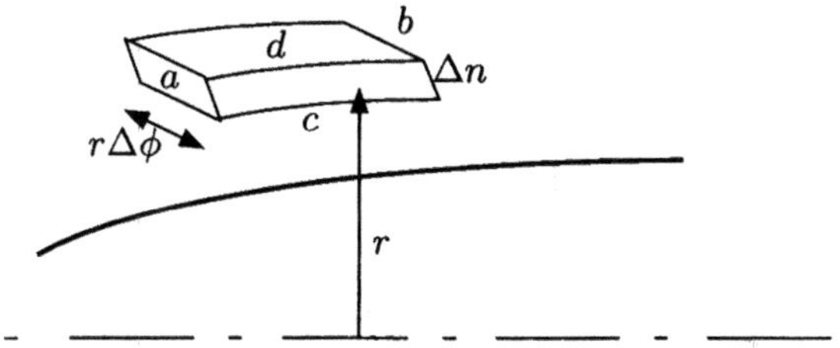

Figure 1.8. Definition sketch of flow element in axisymmetric flow.

$$\Delta(\rho ur)\Delta\phi\Delta n + \Delta(\rho vr)\Delta\phi\Delta s = 0$$

which, after division with $\Delta n\Delta s\Delta\phi$, gives

$$\frac{\Delta(\rho ur)}{\Delta s} + \frac{\Delta(\rho vr)}{\Delta n} = 0$$

By letting $\Delta s \to 0$ and $\Delta n \to 0$ we can write

$$\frac{\partial(\rho ur)}{\partial s} + \frac{\partial(\rho vr)}{\partial n} = 0$$

The second term could be written

$$\frac{\partial(\rho vr)}{\partial n} = \rho v + r\frac{\partial(\rho v)}{\partial n}$$

since $\frac{\partial r}{\partial n} \approx 1$. In the boundary layer approximation we can show that the first term on the RHS is much smaller than the second term

$$\frac{\rho v}{r\frac{\partial \rho v}{\partial n}} \sim \frac{\rho v}{r\frac{\rho v}{\delta}} = \frac{\delta}{r} \ll 1$$

The Mangler transformation

In order to transform the axisymmetric boundary layer equations we introduce new coordinates

$$\hat{s} = \int_0^s \frac{r_0^2}{R^2} ds \tag{1.43}$$

$$\hat{n} = \frac{r_0}{R} n \tag{1.44}$$

where R is an arbitrary reference length, and the transformed velocity components

$$u = \hat{u}$$

$$v = \frac{r_0}{R}\hat{v} - \frac{r_0'}{r_0} n u$$

where $r_0' = \frac{dr_0}{ds}$. The transformed variables should now be put into the continuity, momentum and energy equations although we carry out the derivation in detail only for the continuity equation. First we calculate

$$\frac{\partial}{\partial s} = \frac{\partial \hat{s}}{\partial s}\frac{\partial}{\partial \hat{s}} + \frac{\partial \hat{n}}{\partial s}\frac{\partial}{\partial \hat{n}} = \frac{r_0^2}{R^2}\frac{\partial}{\partial \hat{s}} + \frac{r_0'}{R} n \frac{\partial}{\partial \hat{n}}$$

$$\frac{\partial}{\partial n} = \frac{\partial \hat{s}}{\partial n}\frac{\partial}{\partial \hat{s}} + \frac{\partial \hat{n}}{\partial n}\frac{\partial}{\partial \hat{n}} = 0 + \frac{r_0}{R}\frac{\partial}{\partial \hat{n}}$$

In the boundary layer approximation we can write $r \approx r_0 + \mathcal{O}(\delta)$ and the continuity equation becomes

$$\frac{1}{r_0}\frac{\partial(\rho u r_0)}{\partial s} + \frac{\partial(\rho v)}{\partial n} = 0$$

which can be rewritten

$$\frac{r_0'}{r_0}\rho u + \frac{\partial(\rho u)}{\partial s} + \frac{\partial(\rho v)}{\partial n} = 0$$

By introducing the transformed variables we obtain the following equation

$$\frac{r_0'}{r_0}\rho\hat{u} + \frac{r_0^2}{R^2}\frac{\partial(\rho\hat{u})}{\partial \hat{s}} + \frac{r_0'}{R}\frac{R}{r_0}\hat{n}\frac{\partial(\rho\hat{u})}{\partial \hat{n}} + \frac{r_0}{R}\frac{\partial}{\partial \hat{n}}\left(\rho\frac{r_0}{R}\hat{v} - \rho\frac{r_0'}{r_0}\frac{R}{r_0}\hat{n}\hat{u}\right) = 0$$

It can easily be shown that the sum of terms 1, 3 and 5 in this expression equals zero and that the remaining two terms reduce to

$$\frac{\partial(\rho\hat{u})}{\partial \hat{s}} + \frac{\partial(\rho\hat{v})}{\partial \hat{n}} = 0$$

which has the same form as the 2D version of the continuity equation.

In the new coordinates the axisymmetric version of the boundary layer equations for streamwise momentum and energy have the following form

$$\rho\hat{u}\frac{\partial \hat{u}}{\partial \hat{s}} + \rho\hat{v}\frac{\partial \hat{u}}{\partial \hat{n}} = \rho_e U_e \frac{dU_e}{d\hat{s}} + \frac{\partial}{\partial \hat{n}}\left(\mu\frac{\partial \hat{u}}{\partial \hat{n}}\right)$$

and

$$\rho c_p \left(\hat{u}\frac{\partial T}{\partial \hat{s}} + \hat{v}\frac{\partial T}{\partial \hat{n}} \right) = -\hat{u}\rho_e U_e \frac{dU_e}{d\hat{s}} + \frac{\partial}{\partial \hat{n}} \left(\frac{\mu c_p}{Pr} \frac{\partial T}{\partial \hat{n}} \right) + \mu \left(\frac{\partial \hat{u}}{\partial \hat{n}} \right)^2$$

It is seen that both the momentum and the energy equations have the same form as in the two-dimensional case and the boundary conditions will also transform to the same form.

The pointed cone at zero angle of attack

As an example we will study the pointed cone to determine how the boundary layer thickness changes along the cone as compared to flat plate boundary layer. We use the transformations of the coordinates (equations 1.43 and 1.44) where for the pointed cone $r_0 = c \cdot s$ where c is a constant. This gives

$$\hat{s} = \int_0^s \frac{c^2}{R^2} s^2 ds = \frac{c^2}{R^2}\frac{s^3}{3} = \frac{r_0^2}{R^2}\frac{s}{3}$$

or

$$\frac{r_0^2}{R^2} = 3\frac{\hat{s}}{s}$$

and for the normal coordinate

$$\hat{n} = \sqrt{3}\sqrt{\frac{\hat{s}}{s}} n \Rightarrow \frac{\hat{n}}{\sqrt{\hat{s}}} = \sqrt{3}\frac{n}{\sqrt{s}}$$

But as $\frac{n}{\sqrt{s}} \sim \eta$ and $\frac{\hat{n}}{\sqrt{\hat{s}}} \sim \hat{\eta}$ we can conclude that $\hat{\eta} = \sqrt{3}\eta$ or $\hat{\delta} = \sqrt{3}\delta$. This means that for a given distance from the leading edge of the plate or from the apex of the cone, the boundary layer on the plate will be $\sqrt{3}$ thicker than that on a cone. To obtain the same boundary layer thickness the distance along the cone must be three times larger than the distance along the plate.

Heat transfer

An important aspect of compressible boundary layer flows is the heat transfer between the boundary layer and the wall. Usually the heat transfer is given in terms of a non-dimensional number, the so called Nusselt number (Nu), defined as

$$Nu = \frac{\alpha \Delta T}{k\frac{\Delta T}{L}} = \frac{\alpha L}{k} \tag{1.45}$$

where α is the actual heat transfer coefficient (W/m^{2}K) and L is a typical reference length. By writing the middle step in equation (1.45) the interpretation of the Nusselt number becomes clear: It is the ratio between the actual heat transfer for the present situation and the heat transfer that would be obtained if only heat conduction was acting. If Nu=1 this would indicate that heat transfer was only from heat conduction. However the value depends on the choice of the length scale L which may be taken to be an integral size of the body which is related to, but not equal to, the length scale appropriate for evaluating the heat conduction term.

Under certain conditions, a direct relation exists between the heat transfer and the wall shear stress at a surface. This analogy is called the Reynolds analogy and is based on the fact that both momentum transfer and heat transfer are due to molecular diffusion. If the analogy holds then the ratio between the heat transfer and the skin friction should be constant. This ratio can be expressed as

$$\frac{\text{heat flux}}{\text{momentum flux}} = \frac{q}{\tau_w} = \frac{-(k\frac{\partial T}{\partial y})_{y=y_w}}{(\mu\frac{\partial u}{\partial y})_{y=y_w}}$$

If we rewrite the variables in the RHS in non-dimensional form we obtain

$$\frac{q}{\tau_w} = \frac{k\Delta T}{\mu U_e}\frac{(-k\frac{\partial \theta}{\partial y})_{y=y_w}}{(\mu\frac{\partial u}{\partial y})_{y=y_w}}$$

where $\theta = T/\Delta T$ and y is made non-dimensional with an arbitrary length scale. We now introduce the Prandtl number $Pr = \mu c_p/k$, the skin friction coefficient $c_f = \tau_w/\frac{1}{2}\rho U_e^2$, and the Nusselt number $Nu_x = qx/k\Delta T$ which gives

$$\frac{2}{c_f}Nu_x\frac{k\Delta T}{x\rho U_e^2} = \frac{c_p\Delta T}{PrU_\infty}\frac{(-\frac{\partial \theta}{\partial y})_{y=y_w}}{(\frac{\partial u}{\partial y})_{y=y_w}}$$

where the length scale, x, is taken to be the distance from the leading edge of the plate. We can finally rewrite the equation by introducing the Reynolds number $Re_x = \rho U_e x/\mu$ so that

$$\frac{2}{c_f}\frac{Nu_x}{Re_x Pr} = \frac{1}{Pr}\frac{(-\frac{\partial \theta}{\partial y})_{y=y_w}}{(\frac{\partial u}{\partial y})_{y=y_w}}$$

The non-dimensional group on the LHS is usually denoted the Stanton number $St = Nu_x/Re_x Pr$. If the ratio between the temperature and velocity gradients is independent of x then the heat flux is proportional to the skin friction. In fact for $Pr = 1$ this ratio does become equal to one and for this special case we can write

$$Nu_x = \frac{c_f}{2} Re_x$$

A useful non-dimensional number can usually be expressed as a ratio between two important parameters or variables, although the interpretation of the Stanton number is not at first sight obvious. However the interpretation can be made by rewriting the Stanton number in terms of its physical variables which gives

$$St = \frac{q}{\rho c_p U_e \Delta T}$$

Now the denominator can be interpreted as the extra heat flux per unit area for a fluid with velocity U_e and an excess temperature of ΔT. So the interpretation of the Stanton number is the ratio between the heat flux at the wall and the heat convected through the boundary layer.

1.3. Combustion

The control and exploitation of fire and combustion have played a decisive role in the history of mankind and in particular in the development of modern technological society. Our early ancestors were able to use fire for cooking, to protect themselves from predators, and as an aid in the making of tools. The use of fire has been a prerequisite for the development of many of the features which nowadays distinguish mankind from the rest of the animals. The use of fire in preparing food may even have played a role in the development of the physiological prerequisites for speech, which in turn is a basis for nearly all social structures, not least the human activity known as science.

In a great many cultures the importance of fire is expressed in a legend that it was stolen from the gods, or at least from the heavens near the sun. Often it was stolen from a divine blacksmith, one of the earliest mechanical engineers. Indeed, fire, or combustion, has always played a central role in mechanical engineering. During the industrial revolution, steam engines, with fire at their hearts, released the factories from the necessity of being built in places where a good head of water was available to drive the water wheels. The new sources of power paved the way for the establishment of the mechanical workshop industry in what became known as the Black Country of England. Later on, black 'London' smog was replaced by brown 'Los Angeles' smog, produced from the exhaust fumes from internal combustion engines in vehicles. Nowadays, any odour from exhaust fumes experienced on the streets of Stockholm is more likely to come from a bus run on ethanol than from a conventionally fueled car or lorry engine. Combustion technology is just as important as ever for vehicle propulsion and energy conversion, and is still one of the major sources of pollution both of our immediate environment and of the earth's global ecological system — see, *e.g.*, the review by Prather & Logan [5].

The rest of this section provides an elementary introduction to the study of combustion. It has been written for readers with a knowledge of fluid mechanics at university level, a knowledge of chemistry at secondary-school level, and, consequently, a non-professional interest in combustion, rather than specific formal knowledge. The next subsection presents the equations describing mixtures of gases as opposed to simple gases. The following subsection extends the analysis of plane shock fronts to include heat release due to chemical reactions and thus introduces the concepts of detonations and flame fronts. Before proceeding to a closer analysis of a flame front, combustion chemistry and reactions in spatially homogeneous mixtures are presented in a subsection each. The internal structure of a flame front is presented from a thermal point of view in the last subsection.

Under turbulent conditions combustion is thought of as taking place either at a reaction front, such as a flame front, or in a fine-scale structure in which fuel, oxidant and heat are mixed. These two extremes, fronts and small mixed zones, are both known to be extinguished by high rates of strain, or stretch, but stretch extinction is not presented in this chapter. Local extinction, due to turbulence, plays a significant role in several branches of combustion engineering and the basic understanding of this phenomenon is an area of ongoing research — see, *e.g.*, Poinsot *et al.* [4] and Vervisch & Poinsot [9].

1.3.1. GAS MIXTURES WITH VARYING COMPOSITION

The concentrations of the component or constituent gases in a mixture can be expressed in terms of either mass densities, ρ_α, or molar densities, n_α, where α denotes a particular chemical species. (In this section n denotes the molar density of a gas rather than a unit surface normal.) The total densities of the mixture are given by simple sums over all the various species,

$$\rho = \sum_\alpha \rho_\alpha \qquad \text{and} \qquad n = \sum_\alpha n_\alpha .$$

It is more convenient to describe the composition of the mixture in terms of mass or molar fractions defined by,

$$Y_\alpha = \frac{\rho_\alpha}{\rho} \qquad \text{and} \qquad X_\alpha = \frac{n_\alpha}{n} .$$

For example, for Dry Air:

$$Y_{N_2} = 0.76 \qquad Y_{O_2} = 0.23 \qquad Y_{Ar} = 0.01$$

$$X_{N_2} = 0.78 \qquad X_{O_2} = 0.21 \qquad X_{Ar} = 0.01$$

The mass and molar fractions are very nearly equal since air is dominated by nitrogen, N_2, and the molar mass of the second component, O_2, is very close to that of nitrogen.

Under conditions under which each constituent gas behaves separately like a perfect gas, the mixture as a whole can be expected to behave like a perfect gas,

$$p = n\mathcal{R}T = \rho RT,$$

with specific gas constant,

$$R = \sum_\alpha Y_\alpha R_\alpha = \frac{\mathcal{R}}{\mathcal{M}_\mathrm{m}},$$

where the mean molar mass of the mixture is given by,

$$\frac{1}{\mathcal{M}_\mathrm{m}} = \sum_\alpha \frac{Y_\alpha}{\mathcal{M}_\alpha} = \frac{n}{\rho} \qquad \text{and} \qquad \mathcal{M}_\mathrm{m} = \frac{\rho}{n} = \sum_\alpha X_\alpha \mathcal{M}_\alpha .$$

For example, for Dry Air, $\mathcal{M}_\mathrm{m} = 29\,\mathrm{kg/kmol}$ and $R = 287\,\mathrm{J/(kg\,K)}$.

The mass density of a particular species is governed by,

$$\frac{\partial \rho_\alpha}{\partial t} + \nabla \cdot (\rho_\alpha \vec{u}) = -\nabla \cdot \left(\rho_\alpha \vec{V}_\alpha\right) + w_\alpha, \tag{1.46}$$

where $\vec{V}_\alpha$ is the diffusion velocity of the species α and w_α is the rate at which it is produced in chemical reactions. The diffusion velocities are defined relative to $\vec{u}$, the mass-weighted mean velocity of the mixture, and must therefore satisfy,

$$\sum_\alpha \rho_\alpha \vec{V}_\alpha = 0.$$

Since chemical reactions conserve mass, *i.e.* $\sum w_\alpha = 0$, the sum over all species, *i.e.* α, of equation (1.46) yields the mass conservation equation, (1.2).

Equation (1.46) is often written in the form,

$$\frac{\partial}{\partial t}(\rho Y_\alpha) + \nabla \cdot (\rho Y_\alpha \vec{u}) = \rho \frac{\mathrm{D}Y_\alpha}{\mathrm{D}t} = -\nabla \cdot \left(\rho Y_\alpha \vec{V}_\alpha\right) + w_\alpha, \tag{1.47}$$

which has essentially the same structure as most other equations, such as the equation for the inner energy and that for the enthalpy. In equations

(1.46) and (1.47) the diffusion velocity, $\vec{V}_\alpha$, has to be expressed in terms of ρ_α *etc.*

Based on the Chapman-Cowling-Enskog solution of Boltzmann's equation, the Stefan-Maxwell equations give the following implicit expression for the diffusion velocities,

$$\begin{aligned} \nabla X_\alpha &= \sum_\beta \frac{X_\alpha X_\beta}{\mathcal{D}_{\alpha\beta}} \left(\vec{V}_\beta - \vec{V}_\alpha \right) + (Y_\alpha - X_\alpha) \frac{1}{p} \nabla p \\ &+ \sum_\beta \frac{X_\alpha X_\beta}{\mathcal{D}_{\alpha\beta}} \left(\frac{\mathcal{D}_{T\beta}}{\rho_\beta} - \frac{\mathcal{D}_{T\alpha}}{\rho_\alpha} \right) \frac{1}{T} \nabla T \, , \end{aligned} \tag{1.48}$$

in the absence of body forces. In these equations $\mathcal{D}_{\alpha\beta}$ is the binary diffusion coefficient for a pair of species. The contribution from the pressure gradient, ∇p, can be significant if there are significant concentrations of molecules with widely differing molecular masses but in general this contribution is negligible. The contribution from the temperature gradient is an example of Onsager's principle of reciprocity. It drives relatively light molecules to hot regions and relatively heavy molecules to cold regions. This effect can be significant, for example, in the combustion of hydrogen in air where it can account for a major part of the transport of hydrogen. All the same, in general it can be ignored. Neglecting the terms in ∇p and ∇T, equation (1.48) relates the concentration gradients, ∇X_α, to linear sums of diffusion velocities. Now, equation (1.48) can be inverted to yield an explicit expression for the diffusion velocities, $\vec{V}_\alpha$, in terms of concentration gradients,

$$\rho_\alpha \vec{V}_\alpha = -\rho \mathcal{D}_\alpha \nabla Y_\alpha \, , \tag{1.49}$$

i.e. Fick's law of diffusion, under either of the following two conditions:

1) If the mixture is dominated by one particular species, as for example nitrogen in air, since then $\mathcal{D}_\alpha \approx \mathcal{D}_{\alpha \mathrm{N}_2}$ can be an acceptable approximation.
2) If $\mathcal{D}_{\alpha\beta} \approx \mathcal{D}_0$ is an acceptable approximation for all pairs α and β (since this tends to imply $\mathcal{M}_\alpha \approx \mathcal{M}_\beta$ and $X_\alpha \approx Y_\alpha$).

In straightforward analogy with the Prandtl number, the Schmidt number of a species in the mixture is defined by,

$$Sc_\alpha = \frac{\mu}{\rho \mathcal{D}_\alpha} \, ,$$

and the Lewis number of a species in the mixture is defined by,

$$Le_\alpha = \frac{k}{\rho c_p \mathcal{D}_\alpha} \, .$$

Lewis numbers are particularly important in combustion.

In Fick's law of diffusion, equation (1.49), as well as in Fourier's law for the conduction of heat, equation (1.9), the transport flux is given by a gradient-diffusion expression,

$$\vec{j} = -\mathcal{D}\nabla\phi \qquad i.e. \qquad j_l = -\mathcal{D}\frac{\partial\phi}{\partial x_l}, \tag{1.50}$$

in which the flux vector, $\vec{j}$, is directed from maxima in the scalar field, ϕ, towards minima by the gradient, $-\nabla\phi$. The minus sign ensures that the transport goes in the right direction, down the slope from maxima to minima, while the strength of the transport is determined by the magnitude of the gradient, $|\nabla\phi|$, together with the diffusion coefficient or diffusivity, $\mathcal{D}$.

The fundamental expression for the *enthalpy* of each pure species is,

$$h_\alpha = h_{\alpha,\mathrm{ref}} + \int_{T_{\mathrm{ref}}}^{T} c_p(T')\,\mathrm{d}T',$$

where $h_{\alpha,\mathrm{ref}}$ includes the enthalpy of formation of the species. The enthalpy density of the mixture is obtained by summing the enthalpy densities of the individual components,

$$\rho h = \sum_\alpha \rho_\alpha h_\alpha,$$

so that,

$$h = \sum_\alpha Y_\alpha h_\alpha = \int_{T_{\mathrm{ref}}}^{T} c_p(T')\,\mathrm{d}T' + \sum_\alpha Y_\alpha h_{\alpha,\mathrm{ref}}, \tag{1.51}$$

where

$$c_p = \sum_\alpha Y_\alpha c_{p,\alpha}.$$

This is the specific heat at constant pressure of the mixture which can vary with composition as well as with the thermodynamical quantities.

The enthalpy of the mixture satisfies the conservation equation, (1.13), with the exception that in a mixture, since the individual species carry their enthalpy with them when they diffuse through the mixture, the molecular transport flux of enthalpy is now,

$$\vec{q} = -k\nabla T + \sum_\alpha \rho_\alpha h_\alpha \vec{V}_\alpha, \tag{1.52}$$

rather than merely $-k\nabla T$ as in equation (1.9). In equation (1.52), a further contribution from the diffusion of species has been neglected since this Dufour effect is even weaker than the ∇T contribution in equation (1.48).

One of the important uses of the equations presented here is the theoretical and numerical study of combustion. Combustion, *i.e.* burning, is usually associated with something getting hot, *i.e.* rises in temperature. In the rest of this section, heat release, and its effect on the temperature will be identified in the theoretical formalism. The 'heat' must be in the $\int c_p \mathrm{d}T$ contribution to the enthalpy. (Note that the distinction between this contribution and the $\sum Y_\alpha h_{\alpha,\mathrm{ref}}$ contribution is slightly arbitrary since it depends on the choice of T_{ref}.)

Define, now,

$$h_T = \int_{T_{\mathrm{ref}}}^{T} c_p(T')\,\mathrm{d}T' = h - \sum_\alpha Y_\alpha h_{\alpha,\mathrm{ref}}\,. \tag{1.53}$$

This is principally an intermediary quantity between enthalpy, h, and temperature, T, but we can refer to it as thermal enthalpy or just 'heat'. Multiplying the equations for the mass fractions, (1.47), by $h_{\alpha,\mathrm{ref}}$ and subtracting the sum of them from the enthalpy equation, (1.13), we obtain,

$$\rho\frac{\mathrm{D}h_T}{\mathrm{D}t} = \frac{\mathrm{D}p}{\mathrm{D}t} - \nabla\cdot\left(-k\nabla T + \sum_\alpha Y_\alpha h_{\alpha,T}\vec{V}_\alpha\right) + \Phi - \sum_\alpha h_{\alpha,\mathrm{ref}} w_\alpha\,. \tag{1.54}$$

In this equation, $h_{\alpha,T} = h_\alpha - h_{\alpha,\mathrm{ref}}$, and $-\sum h_{\alpha,\mathrm{ref}} w_\alpha$ is the rate of heat release in chemical reactions, *i.e.* the rate at which enthalpy of formation is converted to heat. The heat release will be positive, *i.e.* the sum will be negative, since species such as fuel, with a high value of $h_{\mathrm{Fu,ref}}$, are being consumed, so w_{Fu} is negative, while species such as combustion products, with a low value of $h_{\mathrm{Pr,ref}}$ are being produced, so w_{Pr} is positive.

The transport equation for the temperature can be derived from $h = \sum Y_\alpha h_\alpha$, assuming $\mathrm{d}h_\alpha = c_{p,\alpha}\mathrm{d}T$ and using equation (1.13) for h, with equation (1.52) for $\vec{q}$, and equation (1.47) for Y_α;

$$\rho c_p\frac{\mathrm{D}T}{\mathrm{D}t} = \frac{\mathrm{D}p}{\mathrm{D}t} - \nabla\cdot(-k\nabla T) - \sum_\alpha \rho_\alpha \vec{V}_\alpha\cdot\nabla h_\alpha + \Phi - \sum_\alpha h_\alpha w_\alpha\,. \tag{1.55}$$

The third term, $\sum_\alpha \rho_\alpha \vec{V}_\alpha\cdot\nabla h_\alpha$, is simply due to the mass-weighted definition of the mixture's velocity, $\vec{u}$. It is negligible when the specific heats of the constituent gases are sufficiently similar. The dissipation, Φ, is negligible at low enough Mach numbers. The last term in equation (1.55) gives a mathematical expression to what everybody already knew: burning makes things hot; heat release raises the temperature.

1.3.2. SHOCKS, DETONATIONS AND DEFLAGRATIONS

Most readers will be acquainted with the Rankine-Hugoniot curve presenting possible jump conditions over a shock wave in the p-v-diagram of thermodynamics. In this section, this analysis is extended to situations in which chemical reactions release heat in or close to plane fronts. The Rankine-Hugoniot curve, together with the Rayleigh line, makes a clear distinction between detonation fronts, corresponding to shock waves, and deflagrations or flame fronts which propagate at speeds below the speed of sound.

From the start the analysis proceeds as in the non-reacting case. Values of the mass density, pressure, velocity, and specific enthalpy are denoted by ρ_1, p_1, u_1, and h_1 in the unburnt mixture upstream of the front, and by ρ_2, p_2, u_2, and h_2 in the burnt mixture downstream of the front. Conservation of mass yields,

$$\rho_1 u_1 = \rho_2 u_2 = \dot{m} \,, \tag{1.56}$$

where $\dot{m}$ is the mass flux density through the front. Newton's second law implies that

$$p_1 + \rho_1 u_1^2 = p_2 + \rho_2 u_2^2 \,, \tag{1.57}$$

and the first law of thermodynamics, applied to this open system, yields,

$$h_1 + \tfrac{1}{2}\rho_1 u_1^2 = h_2 + \tfrac{1}{2}\rho_2 u_2^2 \,. \tag{1.58}$$

These three jump conditions can be obtained by integrating equations (1.56), (1.57) and (1.58) between the homogeneous states on the two sides of the front though a more physical approach based on a stream tube can lead to a better understanding.

The velocities, u_1 and u_2, in the momentum fluxes in Newton's second law, equation (1.57), can be eliminated in favour of the mass flux, $\dot{m}$, to yield,

$$p_2 - p_1 = -\dot{m}^2 \left(\frac{1}{\rho_2} - \frac{1}{\rho_1} \right) . \tag{1.59}$$

This equation describes a straight line in the p-v-diagram for the final, downstream, state, $(p_2, 1/\rho_2)$. This line passes through the initial, upstream, state, $(p_1, 1/\rho_1)$, and is usually refered to as the Rayleigh line.

Equations (1.56) and (1.57) can be used to eliminate u_1 and u_2 from equation (1.58), yielding,

$$h_2 - h_1 = \frac{1}{2} \left(\frac{1}{\rho_1} + \frac{1}{\rho_2} \right) (p_2 - p_1) \,. \tag{1.60}$$

When chemical reactions occur in or close to the front the jump in specific enthalpy consists of two distinct parts,

$$\begin{aligned} h_2 - h_1 = & \sum_\alpha \left[Y_{\alpha,2} \int_{T_{\text{ref}}}^{T_2} c_{p,\alpha}(T')\,\mathrm{d}T' - Y_{\alpha,1} \int_{T_{\text{ref}}}^{T_1} c_{p,\alpha}(T')\,\mathrm{d}T' \right] \\ & + \sum_\alpha (Y_{\alpha,2} - Y_{\alpha,1})\, h_{\alpha,\text{ref}} \,, \end{aligned}$$

where the first contribution is due to the rise in temperature from T_1 to T_2 and the second contribution,

$$\Delta h_{\text{chem}} = \sum_{\alpha} (Y_{\alpha,2} - Y_{\alpha,1})\, h_{\alpha,\text{ref}} \,,$$

is due to the change in composition from $\{Y_{\alpha,1}\}$ to $\{Y_{\alpha,2}\}$. If the specific heat of the reacting mixture, $c_p = \sum_\alpha Y_\alpha c_{p,\alpha}$, can be treated as being independent of composition, $\{Y_\alpha\}$, the jump in specific enthalpy can be written in the simpler form,

$$h_2 - h_1 = \int_{T_1}^{T_2} c_p(T')\, \mathrm{d}T' + \Delta h_{\text{chem}} \,.$$

If, furthermore, the reacting mixture can be modelled as a perfect gas with constant specific heats, the jump in specific enthalpy can be written in the even simpler form,

$$h_2 - h_1 = \frac{\gamma}{\gamma - 1} \left\{ \frac{p_2}{\rho_2} - \frac{p_1}{\rho_1} \right\} + \Delta h_{\text{chem}} \,.$$

Now equation (1.60) can be written in the more readily recognisable form,

$$p_2 = \frac{2\Delta h_{\text{chem}} + \left\{ \frac{1}{\rho_2} - \frac{\gamma+1}{\gamma-1}\frac{1}{\rho_1} \right\} p_1}{\frac{1}{\rho_1} - \frac{\gamma+1}{\gamma-1}\frac{1}{\rho_2}} \,. \tag{1.61}$$

This equation describes the Rankine-Hugoniot curve in the p-v-diagram for the final, downstream, state, $(p_2, 1/\rho_2)$.

Real fronts must obey the basic physical principles behind the jump conditions, equations (1.56), (1.57) and (1.58), and thus must correspond to points in figure 1.9 at which the Rayleigh line intersects the Rankine-Hugoniot curve for some value of $\dot{m}$. There are two possible branches for such intersections; the detonation branch with $p_2 > p_1$ and $1/\rho_2 < 1/\rho_1$; and the deflagration branch with $p_2 < p_1$ and $1/\rho_2 > 1/\rho_1$. When the Rayleigh line intersects the Rankine-Hugoniot curve it generally does so at two points. These two points are refered to as strong or weak depending on whether they have the larger or smaller contrasts in p and ρ.

A sonic front, $u_1 = a_1$, corresponds to a line with slope $-\gamma p_1 \rho_1$ passing through the upstream unburnt state in figure 1.9, when the unburnt gas behaves like a perfect gas. (Consider a Rayleigh line with slope $-\dot{m}^2 = -(\rho_1 a_1)^2$.) It is fairly readily seen from figure 1.9 that detonations advance into the unburnt mixture at supersonic speeds while deflagrations propagate at subsonic speeds.

Strong detonations correspond thus to strong shock waves, and weak detonations correspond to weak shocks, but observed plane detonations

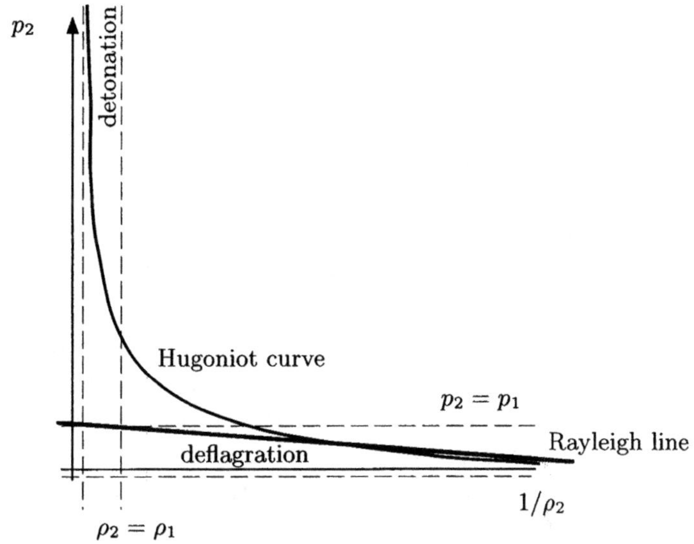

Figure 1.9. The Rankine-Hugoniot diagram.

tend to correspond to the Chapman-Jouguet point at which the Rayleigh line is a tangent to the Rankine-Hugoniot curve. Plane detonation fronts are however unstable and real detonation fronts often exhibit a remarkable cellular structure generated by transverse acoustic waves.

Observed real plane flame fronts correspond to weak deflagration fronts in the Rankine-Hugoniot diagram. From a thermodynamic point of view, flame fronts are virtually isobaric. Plane flame fronts are, however, essentially unstable. They can be realised in a laboratory by, for example, being anchored on a flat porous plate through which the fresh mixture flows.

From a practical computational point of view it is worth noting that the Rankine-Hugoniot jump conditions for a plane front, eqs (1.56), (1.57), and (1.58), do not determine the mass flux, $\dot{m}$, or the upstream flow velocity, $u_1 = \dot{m}/\rho_1$. The jump conditions determine a range of permitted values of $\dot{m}$ for a given unburnt upstream state, and, for a given value of $\dot{m}$, the jump conditions determine the burnt downstream state.

The book by Williams [11] contains thorough presentations of both detonations and the Rankine-Hugoniot analysis.

1.3.3. COMBUSTION CHEMISTRY

Combustion in gas phase often consists of the oxidisation in air of hydrocarbon molecules such as methane, CH_4, the main component of natural gas. The fuel molecules in the gas-phase reactions in Otto-cycle engines and candle flames are very much larger hydrocarbons. Typical products are carbon dioxide, CO_2, and water vapour, H_2O. In these chemical processes direct collisions between a fuel molecule and an oxygen molecule, O_2, play an insiginificant role. Instead the oxidisation takes place in a number of steps, consisting of basic reactions involving intermediary species. From a chemist's point of view, a simplified model of a flamefront can consist of 30 or so chemical species and 100 or so basic reactions! It would be daunting, and perhaps even unneccesary, to study both reaction schemes for combustion chemistry and the dynamics of the cascade in turbulence at the same time but some of the broader features of combustion chemistry are well worth considering with the view of understanding turbulent combustion.

A central feature of combustion chemistry is the so called 'pool' of radicals, the most important of which are oxygen (atoms), O, hydrogen (atoms), H, and hydroxyl, OH. Ignition, *i.e.* the initiation of self-sustaining combustion, consists to a large extent of the process of establishing the necessary pool of radicals. Once established, the pool of radicals maintains itself through reactions such as the 'chain-branching' reaction,

$$\mathrm{O_2 + H \longrightarrow OH + O}\,. \tag{1.62}$$

In this reaction radicals are produced; an H is consumed but both an OH and an O are formed. This self-production of radicals has to balance the consumption of radicals in other reactions. The radicals break down the often large hydrocarbon fuel molecules in reactions such as,

$$\mathrm{CH_4 + H \longrightarrow CH_3 + H_2}\,, \tag{1.63}$$

and oxidise carbon monoxide to carbon dioxide,

$$\mathrm{CO + OH \longrightarrow CO_2 + H}\,. \tag{1.64}$$

This is often one of the most significant exothermic single reaction steps in combustion. Radicals also play a central role in the formation of pollutants such as oxides of nitrogen.

Mathematical expressions for rates of reaction have the Arrhenius form,

$$\pm\, A\, \rho_\alpha^{\nu_\alpha} \rho_\beta^{\nu_\beta}\, T^b \exp\left(-\frac{E_{\mathrm{a}}}{RT}\right), \tag{1.65}$$

where E_{a} is called the activation energy. This slightly complicated mathematical expression is nonetheless fairly straightforward to understand.

The rate of reaction can be expected to be proportional to the frequency of molecular collisions but the vast majority of collisions between stable molecules are elastic so the collision frequency has to be multiplied by a factor equal to the fraction of collisions that have enough energy to make the collision not only inelastic but even reactive. In a reactive collision, the incoming molecules are broken up and their atoms are rearranged into different outgoing molecules, and this can only happen if they hit each other hard enough. Roughly speaking, their energies must be above a threshold, the activation energy. The exponential factor, $\exp(-E_{\mathrm{a}}/RT)$, in equation (1.65) can be seen to come from the fraction of a Maxwell distribution of molecular energies that is greater than a threshold energy given by the activation energy, E_{a}. Typically, this exponential factor varies over a great many orders of magnitude between the cold unburnt and the hot burnt gas.

The pre-exponential factor, $A\rho_\alpha^{\nu_\alpha}\rho_\beta^{\nu_\beta}T^b$, is just the collision frequency for collisions between molecules of species α and β. When the temperature dependence of the collision crossection can be ignored $b = \frac{1}{2}$ and, in general, the pre-exponential factor depends much more weakly on the temperature than the exponential factor. The powers ν_α and ν_β are the orders of the reaction, or molecular collision, and must be (positive) integers in basic reactions. For bimolecular reactions, in which just one molecule from each of a pair of species collide, $\nu_\alpha = 1$ and $\nu_\beta = 1$.

The Arrhenius expression, equation (1.65), is also used to model effective rates of overall, or compound, reactions and then the powers ν_α and ν_β need no longer be integers. Typically, for combustion of hydrocarbon fuels the effective activation energy for an overall one-step reaction is of the order of 10^6 or 10^7 Joule/kg. The corresponding temperature, $T_{\mathrm{a}} = E_{\mathrm{a}}/R$ is of the order of, or greater than, 10000 K which can be compared to adiabatic flame temperatures, $T_{\mathrm{b}} \sim 2000\,\mathrm{K}$. An increase in temperature from an ambient temperature of the order of 300 K to a flame temperature, $T_{\mathrm{b}} \sim 2000\,\mathrm{K}$, leads to an increase in $\exp(-E_{\mathrm{a}}/RT)$ by a factor of the order of 10^{12} or even greater. This strong temperature dependence is central to combustion and is one of the most significant factors leading to the bifurcation type of behaviour in ignition and extinction, or burning and non-burning.

The book by Glassman [3] gives a more detailed basic presentation of combustion chemistry. The reviews by Warnatz [10] and Seshadri & Williams [7] present the development of simplified reaction schemes to be used in numerical calculations of non-turbulent and turbulent combustion.

1.3.4. STIRRED REACTORS AND EXTINCTION

Well stirred reactors are common in a great many branches of Chemical Engineering. The various chemical species are nearly always mixed by relatively fine-scale turbulence which is maintained by some form of stirring or by the jet of the inflow. The *perfectly* stirred reactor is an idealization in which there are no spatial inhomogeneities at all, not even at the smallest length scales. This implies, for example, that the relevant time scale of the turbulence is much shorter than the time scales, *i.e.* inverse reaction rates, of all the significant reactions in the mixture.

In a well stirred reactor in a steady state, reactants have to be continuously supplied while products have to be removed. When the overall reaction is exothermic, such as in combustion, heat will also have to be removed. If this heat is removed too fast the temperature in the mixture drops, the reaction rates fall drastically, the fraction of molecular collisions that can cross the activation-energy threshold becomes tiny, the radical pool nearly disappears, and the reactions lose their combustion character. They are no longer self-sustaining and no longer strongly exothermic. We say that the combustion has been extinguished. In general, combustion is extinguished when heat is removed faster than fresh reactants are supplied. Extinction is particularly important in the understanding of turbulent combustion.

1.3.5. FLAME FRONTS

The plane, 'laminar', flame front, a weak deflagration in the Rankine-Hugoniot diagram, is a phenomenon and concept that is particularly important for the understanding of the régimes of premixed turbulent combustion. The inherent properties of a flame front which primarily determine how it interacts with turbulence are the thickness, $\delta_{\rm lam}$, of the front and the flame speed, $S_{\rm lam}$, *i.e.* the speed, $u_1 = \dot{m}/\rho_1$, with which the flame front advances into the unburnt mixture. These two quantities, $S_{\rm lam}$, and, $\delta_{\rm lam}$, are determined by the internal structure of the flame front. Mathematically, the flame speed is an eigenvalue of a full system of differential equations, eqs (1.47) & (1.13), which determines this internal structure.

In general, a flammable mixture will start to burn when it has been heated to the point at which its temperature crosses the threshold, $T_{\rm a} = E_{\rm a}/R$, where $E_{\rm a}$ is the effective activation energy of the overall reaction. Once it has started to burn it will relatively quickly burn to completion, releasing heat which is required to heat the incoming fresh mixture and keep the combustion going. In self-sustaining combustion, such as a self-propagating flame front, there is thus a 'vicious circle' in which the heat required to initiate reactions in the fresh mixture is provided by essentially the same reactions. In a spatial structure such as a flame front this 'pre-heating' has to be carried out by conduction, *i.e.* diffusion, of heat upstream against the direction of flow.

In the 'thermal' scenario of the previous paragraph, the flame front can be divided into two zones; a pre-heating zone and a reaction zone. In the pre-heating zone, the temperature in the fresh unburnt mixture is raised by diffusion of heat against the direction of flow until the reaction rate starts to become significant. Consequently, the primary balance in the temperature equation, (1.55), is between convection and diffusion,

$$\rho_{\mathrm{u}} c_p S_{\mathrm{lam}} \frac{\mathrm{d}T}{\mathrm{d}x} = \frac{\mathrm{d}}{\mathrm{d}x}\left(k \frac{\mathrm{d}T}{\mathrm{d}x}\right) + 0 \,, \tag{1.66}$$

where mass conservation has been used in the convection term and $\rho_{\mathrm{u}} = \rho_1$ is the mass density in the unburnt upstream gas.

In the reaction zone, the rate of reaction is one of the dominant terms and has to be balanced by either convection or diffusion. Both these transport terms increase when the spatial gradients increase but diffusion, based on two spatial derivatives, increases more strongly than convection, based on a single spatial derivative, so the primary balance is between upstream diffusion and reaction;

$$0 = \frac{\mathrm{d}}{\mathrm{d}x}\left(k \frac{\mathrm{d}T}{\mathrm{d}x}\right) + \Delta h \, w \,, \tag{1.67}$$

where $\Delta h \, w = -\sum_\alpha h_{\alpha,\mathrm{ref}} w_\alpha$ is the rate of heat release.

In the following $x = x_{\mathrm{i}}$ will denote the boundary between the pre-heating zone and the reaction zone, and T_{i} will denote the value of the temperature there (see figure 1.10). The temperature in the unburnt upstream gas is denoted by $T_{\mathrm{u}} = T_1$ and the temperature in the burnt downstream gas is denoted by $T_{\mathrm{b}} = T_2$.

Integration of equation (1.66) through the pre-heating zone from $x = -\infty$ to $x = x_{\mathrm{i}}$ yields,

$$S_{\mathrm{lam}} = \frac{k}{\rho_{\mathrm{u}} c_p \left(T_{\mathrm{b}} - T_{\mathrm{u}}\right)} \frac{\mathrm{d}T}{\mathrm{d}x}(x_{\mathrm{i}}) \,,$$

assuming $T_{\mathrm{i}} \approx T_{\mathrm{b}}$.

Multiplication of equation (1.67) by $\mathrm{d}T/\mathrm{d}x$ followed by integration through the reaction zone from $x = x_{\mathrm{i}}$ to $x = +\infty$ yields,

$$\frac{\mathrm{d}T}{\mathrm{d}x}(x_{\mathrm{i}}) = \left\{ \frac{2\Delta h}{k} \int_{T_{\mathrm{i}}}^{T_{\mathrm{b}}} w \,\mathrm{d}T \right\}^{1/2} .$$

Here the adiabatic relation $\Delta h = c_p(T_{\mathrm{b}} - T_{\mathrm{u}})$ can be used.

Now the flame speed is found to be given by,

$$S_{\mathrm{lam}} = \left\{ 2 \frac{k}{\rho_{\mathrm{u}} c_p} \frac{1}{\rho_{\mathrm{u}} \left(T_{\mathrm{b}} - T_{\mathrm{u}}\right)} \int_{T_{\mathrm{i}}}^{T_{\mathrm{b}}} w \,\mathrm{d}T \right\}^{1/2} = \left(2\kappa \frac{\overline{w}}{\rho_{\mathrm{u}}}\right)^{1/2} = \left(2\kappa \frac{1}{\tau_{\mathrm{r}}}\right)^{1/2} ,$$

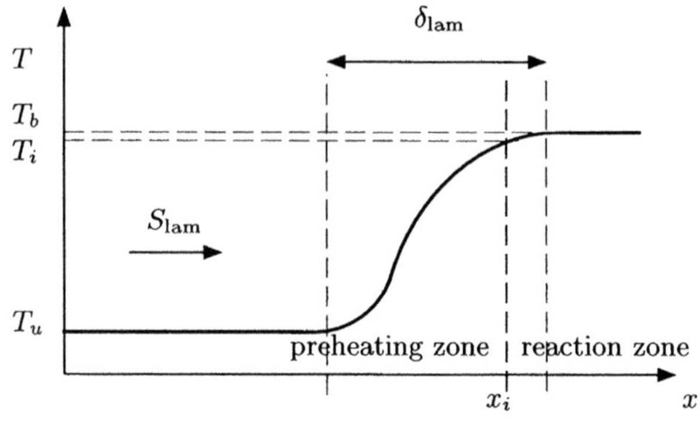

Figure 1.10. The thermal structure of a flame front.

where $\kappa = k/(\rho_u c_p)$ is a heat transport coefficient with the same physical dimensions as the diffusion coefficient, $\mathcal{D}$;

$$\bar{w} = \frac{1}{\rho_u (T_b - T_u)} \int_{T_i}^{T_b} w \, dT ,$$

is a mean reaction rate in the flame front; and $\tau_r = \rho_u/\bar{w}$ is a characteristic time scale for chemical reactions in the flame. The flame speeds of real flame fronts are influenced to some extent by the way in which they are kept plane but, all the same and roughly speaking, flame speeds in typical hydrocarbon-air mixtures can be said to be of the order of 40 cm/s for stoichiometric mixtures at atmospheric pressure.

The spatial gradients in the pre-heating zone can be estimated in terms of the full front's thickness, δ_{lam}, since burning, once initiated, takes place much faster than heating, for typical fuel-air mixtures, so that the reaction zone is much thinner than the pre-heating zone;

$$\frac{dT}{dx}(x_i) = \frac{T_i - T_u}{\delta_{lam}} .$$

Now the thickness of the flame front is given by,

$$\delta_{lam} = \frac{\kappa}{S_{lam}} = \left(\tfrac{1}{2}\kappa\tau_r\right)^{1/2} \simeq S_{lam}\tau_r .$$

Typical flame thicknesses are of the order of, though less than, 1 mm.

Flames are described in more detail in the book by Glassman [3]. The thermal analysis presented above can be placed on a firmer mathematical footing — see, *e.g.*, Williams [11] and Clavin [2].

References

1. Anderson, J. D. 1990. *Modern Compressible Flow.* 2nd ed. McGraw-Hill.
2. Clavin, P. 1994. Premixed Combustion and Gasdynamics. *Ann. Rev. of Fluid Mech.* **26**, 321–352.
3. Glassman, I. 1996. *Combustion.* Academic Press.
4. Poinsot, T., Candel, S. and Trouvé, A. 1996. Applications of Direct Numerical Simulation to Premixed Turbulent Combustion. *Prog. Energy Combust. Sci.* **12**, 531–576.
5. Prather, M. J. and Logan, J. A. 1994. Combustion's Impact on the Global Atmosphere. *25th Symp. (Int.) on Combustion*, 1513–1527. Pittsburgh: The Combustion Institute.
6. Schlichting, H. 1979. *Boundary-Layer Theory.* 7th ed. McGraw-Hill.
7. Seshadri, K. and Williams, F. A. 1994. Reduced Chemical Systems and Their Application in Turbulent Combustion. *Turbulent Reacting Flows* (eds P. A. Libby and F. A. Williams), 153–210. Academic Press.
8. Spina, E. F., Smits, A. J. and Robinson, S.K. 1994. The Physics of Supersonic Turbulent Boundary Layers. *Ann. Rev. of Fluid Mech.* **26**, 287–319.
9. Vervisch, L. and Poinsot, T. 1998. Direct Numerical Simulation of Non-premixed Turbulent Flames. *Ann. Rev. of Fluid Mech.* **30**, 655–691.
10. Warnatz, J. 1992. Resolution of Gas Phase and Surface Combustion Chemistry into Elementary Reactions. *24th Symp. (Int.) on Combustion*, 553–579. Pittsburgh: The Combustion Institute.
11. Williams, F. A. 1985. *Combustion Theory.* Benjamin/Cummings.

Chapter 2

STABILITY OF BOUNDARY LAYER FLOWS

A. HANIFI AND D.S. HENNINGSON[1]
Aeronautical Research Institute of Sweden (FFA)
Box 11021, SE-16111 Bromma, Sweden

2.1. Introduction

There exist a number of methods which are used to predict and model transition. A formulation of the complete stability and transition problem requires not only that the equations of motion are known but also that accurate description of the mean flow field, the body geometry as well as the ambient disturbance level both in terms of velocity fluctuations, sound levels, surface vibration and roughness. For the complete formulation not only the overall magnitude of these disturbances are required but also their frequency and wave number spectra. In this chapter we will develop the basic mathematical and physical ideas of stability theory and discuss how these theories can model and predict transition in various flow situations for compressible flows. Section 2 is an introduction to the stability of incompressible flow and is adapted from the lectures on stability prepared from the 1995 ERCOFTAC/IUTAM Summer School (Henningson & Alfredsson [45]) and the review article by Henningson, Gustavsson & Breuer [46]. The material is repeated here for completeness. In section 3 we proceed with an introduction to the stability of compressible flow, in section 4 with the stability of more complex flows and in section 5 with some applications.

The concept of stability originates from a laminar flow which is slightly disturbed. If the laminar flow returns to its original state one defines the flow as stable, whereas if the disturbance grows and changes the laminar state into a different state, one defines the flow as unstable. The instabil-

[1]Also at Dept. of Mechanics, Royal Inst. of Technology, SE-100 44, Stockholm, Sweden

ity often results in turbulence, but may also take the flow into a different laminar, usually more complicated state. Stability theory deals with the mathematical analysis of the evolution of disturbances on the laminar base flow. In many cases one assumes the disurbances to be small so that one can simplify the analysis by deriving linear governing equations for the disturbance evolution. As the disturbance velocities grow above a few percent of the base flow, non-linear effects usually become important and the linear equations no longer accurately predict the disturbance evolution.

Although the linear equations have a limited region of validity they are important to use to identify physical growth mechanisms and corresponding disturbance types. In fact, it can be shown that the non-linear terms of the incompressible Navier-Stokes problem are conservative and thus only redistribute energy between various spatial scales of the disturbances, and that growth of the total disturbance energy must originate from a linear growth mechanism.

We first define the concept of stability and then proceed to derive the governing equations for small amplitude disturbances on parallel shear flows in the next section.

The formal definition of stability given by Joseph [55] is based on the kinetic energy of a disturbance, $E_\mathcal{V} = \int_\mathcal{V} \frac{1}{2} u_i u_i d\mathcal{V}$, integrated over the whole domain $\mathcal{V}$.

Def. A solution U_i to the Navier-Stokes equations (2.2) and (2.3) is *stable* to perturbations if the perturbation energy satisfies

$$\lim_{t\to\infty} \frac{E_\mathcal{V}(t)}{E_\mathcal{V}(0)} \to 0 \tag{2.1}$$

2.2. Introduction to stability of incompressible parallel flows

The motion of an incompressible fluid is described by the equations of momentum and mass conservation, which in non-dimensional vector notation are given by

$$\frac{\partial \mathbf{u}}{\partial t} + (\mathbf{u} \cdot \nabla)\mathbf{u} = -\nabla p + \frac{1}{R} \nabla \cdot \left[\mu (\nabla \mathbf{u} + \nabla \mathbf{u}^{\mathrm{tr}}) \right] . \tag{2.2}$$

$$\nabla \cdot \mathbf{u} = 0, \tag{2.3}$$

Here t represents time, p pressure, μ the dynamic viscosity and $\mathbf{u}$ is the velocity vector. All flow and material quantities are nondimensionalized by the corresponding reference flow quantities, besides pressure, which is referred to twice the corresponding dynamic pressure. Reynolds number, R

is defined as

$$R = \frac{U^*_{ref} l^*_{ref}}{\nu^*_{ref}},$$

where l^*_{ref} is the reference length scale and subscript $_{ref}$ refers to the reference quantities.

2.2.1. LINEAR STABILITY EQUATIONS

In order to derive the stability equations, we decompose all flow and material quantities into a steady basic flow $\bar{\mathbf{q}}$ plus an unsteady disturbance flow component $\mathbf{q}$ according to

$$\mathbf{Q}(x, y, z, t) = \bar{\mathbf{q}}(x, y, z) + \mathbf{q}(x, y, z, t), \tag{2.4}$$

where x, y and z are the coordinates in the streamwise, normal and spanwise directions, respectively, and time is denoted by t. Specifically in the above, $\bar{\mathbf{q}}$ and $\mathbf{q}$ stand for

$$\bar{\mathbf{q}} = (U, V, W, \bar{p}), \tag{2.5}$$

$$\mathbf{q} = (u, v, w, p). \tag{2.6}$$

Here U, V, W are the basic and u, v, w the disturbance velocity components in the streamwise, normal and spanwise directions, respectively. p is the disturbance pressure.

Now we will consider the governing equations for infinitesimal disturbances in parallel flows. Let $U_i = U(y)\delta_{1i}$ be the base flow, *i.e.* a flow in the x-direction which has a variation with y. If this and the flow decomposition above are introduced into the disturbance equations (2.2) and (2.3) and the non-linear terms are dropped the resulting equations can be written,

$$\frac{\partial u}{\partial t} + U\frac{\partial u}{\partial x} + vU' = -\frac{\partial p}{\partial x} + \frac{1}{R}\nabla^2 u \tag{2.7}$$

$$\frac{\partial v}{\partial t} + U\frac{\partial v}{\partial x} = -\frac{\partial p}{\partial y} + \frac{1}{R}\nabla^2 v \tag{2.8}$$

$$\frac{\partial w}{\partial t} + U\frac{\partial w}{\partial x} = -\frac{\partial p}{\partial z} + \frac{1}{R}\nabla^2 w \tag{2.9}$$

and continuity,

$$\frac{\partial u}{\partial x} + \frac{\partial v}{\partial y} + \frac{\partial w}{\partial z} = 0. \tag{2.10}$$

Here, a prime ($'$) denotes a $y-$derivative. Taking the divergence of the linearized momentum equations, and using continuity (2.10) yields an equation for the perturbation pressure,

$$\nabla^2 p = -2U'\frac{\partial v}{\partial x}. \tag{2.11}$$

This may be used together with equation (2.8) to eliminate p, resulting in an equation for the normal velocity, v:

$$\left[(\frac{\partial}{\partial t} + U\frac{\partial}{\partial x})\nabla^2 - U''\frac{\partial}{\partial x} - \frac{1}{R}\nabla^4 \right] v = 0. \tag{2.12}$$

In order to describe the complete flow field, a second equation is also needed. This is most conveniently the equation for the normal vorticity,

$$\eta = \frac{\partial u}{\partial z} - \frac{\partial w}{\partial x} \tag{2.13}$$

where η satisfies

$$\left(\frac{\partial}{\partial t} + U\frac{\partial}{\partial x} - \frac{1}{R}\nabla^2 \right) \eta = -U'\frac{\partial v}{\partial z}. \tag{2.14}$$

This pair of equations, together with the boundary conditions:

$$v = v' = \eta = 0 \qquad \text{at a solid wall,} \tag{2.15}$$

$$v, v', \eta \qquad \text{bounded} \quad \text{in the far field} \tag{2.16}$$

and the initial conditions

$$v|_{t=0} = v_0(x, y, z) \tag{2.17}$$

$$\eta|_{t=0} = \eta_0(x, y, z) \tag{2.18}$$

form a complete description of the evolution of an arbitrary disturbance in both space and time.

Since the coordinates x and z in this problem are homogeneous and the system is linear, it is possible to work in wave number space and consider the behaviour of single Fourier-components, *e.g.*

$$v(x, y, z, t) = \hat{v}(y, t)e^{i(\alpha x + \beta z)}, \tag{2.19}$$

where α and β are the wavenumbers in the streamwise and spanwise direction, respectively. Introducing this decomposition into (2.12) and (2.14), or equivalently taking the Fourier transform in the horizontal directions, results in the following pair of equations for $\hat{v}$ and $\hat{\eta}$

$$\left[(\frac{\partial}{\partial t} + i\alpha U)(D^2 - k^2) - i\alpha U'' - \frac{1}{R}(D^2 - k^2)^2 \right] \hat{v} = 0 , \tag{2.20}$$

$$\left[(\frac{\partial}{\partial t} + i\alpha U) - \frac{1}{R}(D^2 - k^2) \right] \hat{\eta} = -i\beta U'\hat{v} \tag{2.21}$$

Here $D = {}'$ denotes a y-derivative, α and β are the components of the wavenumber vector in the streamwise and spanwise directions, respectively, and $k = \sqrt{\alpha^2 + \beta^2}$ is its modulus. Once $\hat{v}$ and $\hat{\eta}$ are known, the horizontal velocities, $\hat{u}$ and $\hat{w}$ may be obtained from

$$\hat{u} = \frac{i}{k^2}(\alpha D\hat{v} - \beta\hat{\eta}) \tag{2.22}$$

$$\hat{w} = \frac{i}{k^2}(\beta D\hat{v} + \alpha\hat{\eta}). \tag{2.23}$$

The natural measure of the size of a perturbation is its kinetic energy. Using Parseval's relation, (2.22) and (2.23), the kinetic energy can be written in terms of the Fourier coefficients of the disturbance variables as follows (Gustavsson [33]),

$$E_V = \int_\alpha \int_\beta E \, \mathrm{d}\alpha \, \mathrm{d}\beta = \int_\alpha \int_\beta \frac{1}{2k^2} \int_{-1}^{1} \left(|D\hat{v}|^2 + k^2|\hat{v}|^2 + |\hat{\eta}|^2 \right) \, dy \, \mathrm{d}\alpha \, \mathrm{d}\beta \tag{2.24}$$

where E is the energy density in Fourier space.

From the structure of equations (2.20) and (2.21), it follows that there is a fundamental difference between the normal velocity and the normal vorticity. The evolution of v is described by the homogeneous equation (2.20) with homogeneous boundary conditions, and can thus be determined if the initial data are given. In contrast, the equation for η is inhomogeneous, where the spanwise variation of v and the mean shear combine in the forcing term. Since (2.21) is the linearized form of the vorticity equation for normal vorticity, and the forcing term emanates from the linearized vortex tilting term, it may be appropriate to denote the forcing mechanism by vortex tilting. It will be given another kinematic interpretation below, in connection with the treatment of the inviscid problem.

2.2.2. INVISCID LINEAR STABILITY THEORY

Insight into the types of solutions possible for the linear problem can be found from the inviscid version of the initial value problem. We first discuss the equation for the normal velocity, the so called Rayleigh equation, and then consider solutions of the inviscid part of the equations governing the normal vorticity.

The Rayleigh equation and inflection point criterion

Assuming exponential time dependence, *i.e.* $\hat{v}(y,t) = \tilde{v}(y)e^{-i\omega t}$, where $\omega = \alpha c$ is the complex angular disturbance frequency and c the complex phase speed, the invicid part of the equation governing the normal velocity yields

$$(U - c)(D^2 - k^2)\tilde{v} - U''\tilde{v} = 0 \tag{2.25}$$

The equation above is known as the Rayleigh equation. Let us consider the discrete eigenvalues of the Rayleigh equation. For a general velocity profile the eigenvalue problem constituted by (2.25) is not tractable analytically. However, some general results for the eigenvalues can be obtained. By taking the complex conjugate of the equation one can easily prove that if c is an eigenvalue, so is its complex conjugate. Another powerful result derived by Rayleigh [80] is found by multiplying the Rayleigh equation by the complex conjugate of the solution and integrating over the interval. The imaginary part of the resulting expression becomes

$$c_i \int_{-1}^{1} \frac{U''}{|U-c|^2} |\hat{v}|^2 \mathrm{d}y = 0 \tag{2.26}$$

from which it follows that a necessary condition for instability ($c_i > 0$) is that the mean velocity has an inflection point ($U'' = 0$) somewhere in the y-domain. This is known as Rayleighs inflection point criterion, and it provides a necessary condition for inviscid exponential growth. It is one of the most important results in linear stability theory. An additional necessary condition for inviscid instability was derived by Fjørtoft [26] by considering the real part of the expression used above. He found that only inflection points that were associated with a maximum shear were unstable.

The lift-up effect and the algebraic instability

For a two-dimensional disturbance ($\beta = 0$) the solution of the Rayleigh equation for v governs the complete problem since the streamwise velocity is given by two-dimensional continuity, see equation (2.22).

When $\beta \neq 0$ the disturbance is three-dimensional, and we must also consider the normal vorticity. Eq. (2.21) can, in the inviscid case, be integrated to yield

$$\hat{\eta} = \hat{\eta}_0 e^{-i\alpha U t} - i\beta U' e^{-i\alpha U t} \int_0^t \hat{v}(y,t') e^{i\alpha U t'} \mathrm{d}t'. \tag{2.27}$$

The first term represents the advection of the initial normal vorticity field by the mean field, while the second term represents the integrated effect of the normal velocity, the so-called *lift-up* effect (Landahl [59]). This term represents the *generation* of horizontal velocity perturbations by the lifting-up of fluid elements in the presence of the mean shear. If a single Fourier component is considered this process can be illustrated as follows. Make a coordinate transformation such that one axis, x_1, is aligned with the wavenumber vector $\mathbf{k}$, see figure 2.1a. The other axis, z_1, will then be perpendicular to the wavenumber vector. The mean flow along the new coordinate axis will be denoted U_1 and W_1, respectively, and is easily found to be

$$U_1 = \frac{\alpha}{k} U \tag{2.28}$$

$$W_1 = -\frac{\beta}{k} U \tag{2.29}$$

Using equations (2.22) and (2.23) the disturbance velocities along the same axis become

$$\hat{u}_1 = \frac{i}{k} D\hat{v} \tag{2.30}$$

$$\hat{w}_1 = \frac{i}{k} \hat{\eta} \tag{2.31}$$

It is interesting to note that the velocity along the wavenumber vector is given by two-dimensional continuity in that direction, while the velocity perpendicular to the wavenumber vector is determined by the normal vorticity component. The lift-up process can now be easily visualized by considering the change in the $\hat{w}_1$ velocity during a short time Δt. Equations (2.27), (2.29) and (2.31) give

$$\Delta w_1 = -W_1' \hat{v} \Delta t \tag{2.32}$$

where terms of $\mathcal{O}(\Delta t^2)$ have been neglected and it is assumed that the observer is moving with the wave. The derived expression for Δw_1 is readily identified as the induced horizontal velocity disturbance resulting from the lift-up of a fluid particle by the normal velocity such that the horizontal momentum in the direction perpendicular to the wavenumber vector is conserved. This is illustrated in figure 2.1b. Even if the normal velocity decays this process can give rise to large amplitude perturbations in the horizontal velocity components.

For Fourier components with $\alpha = 0$ the growth may be calculated explicitly. For that case the Rayleigh equation implies that $\hat{v}$ is not a function of time, while (2.27) gives

$$\hat{\eta} = \hat{\eta}_0 - i\beta U' \hat{v}_0 t \tag{2.33}$$

We will refer to this growth as an *algebraic instability* (see also Ellingsen & Palm [22]). Landahl [60] showed that, for a localized disturbance, the algebraic instability manifests itself as a growth of the length of the disturbance, provided that the integral of the initial normal velocity in the streamwise direction does not vanish.

2.2.3. VISCOUS INSTABILITY ANALYSIS

In this section we will consider linear stability analysis of simple parallel incompressible flows, such as Poiseuille, Couette and Blasius profiles.

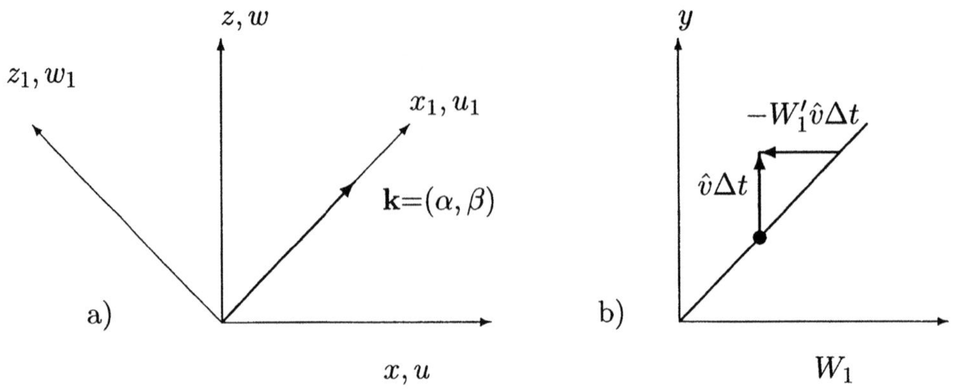

Figure 2.1. Illustration of the *lift-up* effect. a) Definition of the coordinate system aligned with the wavenumber vector. b) Creation of a horizontal velocity defect in the z_1-direction by the lift-up of a fluid element that conserves its horizontal momentum in that direction. From Henningson, Gustavsson & Breuer [46]

The Orr-Sommerfeld and Squire equations

The key ingredients in the solution to the coupled equations (2.20) and (2.21) are the eigenmodes of the respective homogeneous operators. The operator governing the normal velocity is the Orr-Sommerfeld (OS) operator and the operator governing the normal vorticity will be referred to as the Squire (SQ) operator. The respective eigenvalue problems can be found by introducing exponential time dependence, *i.e.* $\hat{v}(y,t) = \tilde{v}(y)e^{-i\omega t}$, in the homogeneous parts of the equations (2.20) and (2.21). We have

$$\left[(U-c)(D^2-k^2) - U'' - \frac{1}{i\alpha R}(D^2-k^2)^2\right]\tilde{v} = 0 \qquad (2.34)$$

where the boundary conditions are $\tilde{v} = D\tilde{v} = 0$ on solid walls and $\tilde{v} \to 0$ in the free stream. This is the so called Orr-Sommerfeld equation. In the same manner the SQ eigenvalues are found from the equations

$$\left[U - c - \frac{1}{i\alpha R}(D^2-k^2)\right]\tilde{\eta} = 0 \qquad (2.35)$$

with boundary conditions $\tilde{\eta} = 0$ on solid walls and $\tilde{\eta} \to 0$ in the free stream. From the above equations it is seen that the eigenvalues will depend on the two parameters $(k, \alpha R)$. However, one may remove the explicit k-dependence from the Squire equation by the transformation $c' = c - k^2/(i\alpha R)$.

For flows bounded in the normal direction it can be shown that the Orr-Sommerfeld and Squire eigenvalues are discrete and that their eigenfunctions form a complete set (Schensted [86], DiPrima & Habetler [17]).

Thus the evolution of an arbitrary initial condition can be obtained using an expansion in the respective eigenfunction. The eigenmodes do not, however, form an orthogonal set, which makes the expansion numerically ill-conditioned, particularly for large Reynolds numbers (Reddy, Schmid & Henningson [83]). The non-orthogonality is a result of the non-normal nature of the governing linear operator and among other things imply the possibility for transient growth of the initial energy. The instantaneous growth rates of these transients may be orders of magnitude larger than those given by the fastest growing eigenmodes of the system. This growth is related to the invisid algebraic instability and will be discussed further below. For a review of the implications of non-normal stability operators in the stability of parallel shear flows see Trefethen, Trefethen, Reddy & Driscoll [93].

For flows with an unbounded domain in the normal directions the number of discrete modes are usually finite, as shown numerically by Mack [69] for the Blasius bloundary layer. This means that the discrete modes cannot describe an arbitrary disturbance and hence the spectrum must be complemented. On basis of numerical experimentation, Mack suggested the complement to be a continuous spectrum located at $c_r = 1$ and $c_i < -\alpha/R$. Such a spectrum was derived by Grosch & Salwen [31] from the assumption that the amplitude of the eigenfunctions of the OS equation is *bounded* as $y \to \infty$ as opposed to *decaying* which is assumed when determining the discrete modes. The continuous spectrum has some noteworthy features. In the limit as $\alpha R \to 0$, no discrete modes exist and the continuous spectrum is the only component that can describe a disturbance. Hultgren & Gustavsson [53] found that large transient growth in the normal vorticity component is possible due to the forcing of the continuous SQ modes by the continuous OS modes. This is the viscous version of the algebraic instability in the boundary layer geometry.

Numerical solutions to the stability problem

The eigenvalues (and eigenfunctions) are usually obtained through numerical methods, such as local methods (*e.g.* shooting methods) giving single eigenvalues or global methods giving a large number of the eigenvalues (see *e.g.* Drazin & Reid [18], Canuto *et al.* [13]). With a shooting method, the problem is formulated as an initial value problem which is integrated from one boundary to another where the boundary conditions are applied. The procedure is repeated and an iteration scheme on the eigenvalue is applied, until the boundary conditions are fulfilled. With a global method the complete boundary value problem is discretized, usually with a spectral or finite difference metod, producing a matrix eigenvalue problem. The solution to this problem give an approximation to a large number of the

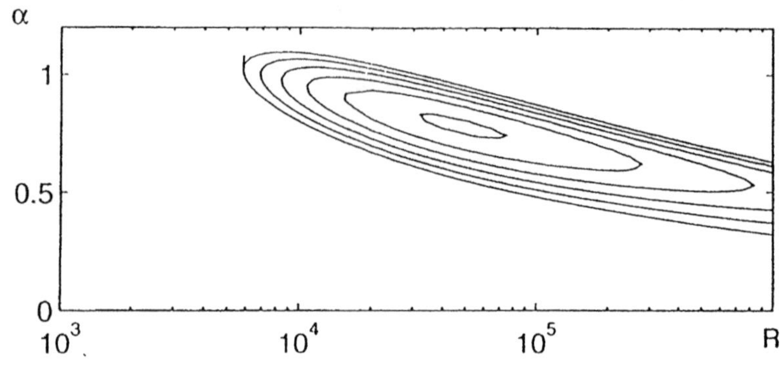

Figure 2.2. Contour lines of αc_i=0, 0.015, 0.030, 0.045, 0.060, 0.075 for plane Poiseuille flow.

eigenvalues. The discrete eigenvalues are usually straight forward to calculate, although one may often in hydrodynamic stability problems encounter large sensitivity to roundoff errors (see Reddy, Schmid & Henningson [83]). In shear flow stability this is usually related to the fact that the stability operator is not self-adjoint, but instead highly non-normal.

Example 1: plane Poiseuille flow

An example of the solution of the OS-equation will be described highlighting some of the features of such a solution. The basic flow is chosen as plane Poiseuille flow, which has a parabolic mean velocity profile

$$U = (1 - y)^2$$

The eigenvalue problem defined from the OS-equation and the parabolic velocity distribution can have two distinctly different solutions, namely solutions that are either symmetric or anti-symmetric with respect to the centreline. The boundary conditions for these two symmetries read:

symmetric: $y = -1$, $v = Dv = 0$; $y = 0$, $Dv = D^3v = 0$

anti symmetric: $y = -1$, $v = Dv = 0$; $y = 0$, $v = D^2v = 0$

It can be shown (see for instance Orszag [77]) that a symmetric mode is the most unstable in this case. It is also generally found that only one unstable mode exists as a solution to the OS-equation.

The neutral curve divides the parameter plane in the stable and unstable regions and is defined as the locus in the (α, R) plane corresponding to solutions to the OS-equation for which $c_i = 0$. This curve is shown in figure 2.2. The two parts of the curve are called branch I (lower) and II (upper) respectively. It is interesting to note that for high R the two branches seem to meet in accordance with the Rayleigh criterion which states that the

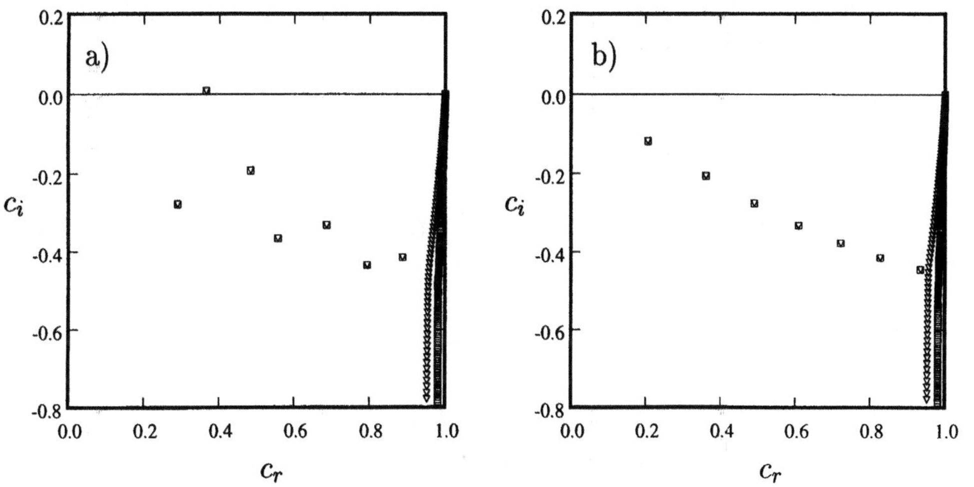

Figure 2.3. a) Orr-Sommerfeld and b) Squire eigenvalues for Blasius flow with $\alpha = 0.308, \beta = 0$ and $R = 998$. Squares: $N = 450, y_{max} = 50$; triangles: $N = 200, y_{max} = 20$. N is the number of Chebyshev points used in the expansion.

plane Poiseuille profile should be stable at infinite R, *i.e.* for zero viscosity. From this figure it is also possible to determine the critical R_{cr} for Poiseuille flow as 5772 and the critical wavenumber α_{cr} as 1.0. Also shown in figure 2.2 are the locuses of various c_i=const. It can be seen that the growth rates are fairly small for TS-waves even at high R.

Example 2: Blasius boundary layer flow

With a continuous spectrum present the numerical calculations become more difficult, although one may capture a discretized approximation if one is careful. Figure 2.3 show the least damped Orr-Sommerfeld and Squire modes for Blasius flow, obtained with an accurate spectral method. One can clearly see the finite number of discrete eigenvalues as well as an approximation of the continuous spectrum. Note the slightly unstable eigenmode of the Orr-Sommerfeld equations, this is the so called Tollmien-Schlichting wave and is the only exponentially growing mode for Blasius flow. When the value of y_{max} is increased, corresponding to a larger domain in the normal direction, the approximation of the continuous spectrum becomes better, although one is forced to use a prohibitively large number of discretization points in order to resolve the normal variation of the eigenfunctions in the boundary layer.

Squires transformation

Three-dimensional effects have to a large extent been overlooked in the early development of linear stability theory, in many instances due to what is known as Squire's theorem. Squire [89] noted that, by employing the following substitution of variables:

$$\begin{aligned} \alpha_1 R_1 &= \alpha R \\ \alpha_1 &= k \end{aligned}$$

the three-dimensional Orr-Sommerfeld equation (2.34) may be reduced to an equivalent two-dimensional equation:

$$\left[(U-c)(D^2-\alpha_1^2)-U''-\frac{1}{i\alpha_1 R_1}(D^2-\alpha_1^2)^2\right]\tilde{v}=0 \ . \tag{2.36}$$

One result which is obtained from this transformation is that the first wave to become exponentially unstable (as the Reynolds number increases from zero) is two-dimensional. On the basis of this, together with the fact that he proved that the additional eigenmodes representing purely horizontal modes, *i.e.* the Squire modes, are always damped, Squire concluded that "For the study of the stability of flow between parallel walls it is sufficient to confine attention to disturbances of two-dimensional type." This is certainly correct if only exponential growth is of interest but if one considers growth in general it puts a severe limitation on the cases worthy of study. If one is concerned with the development of three-dimensional disturbances one needs to consider other possibilities of growth which are present because of the additional mechanisms operating for such disturbances, such as the lift-up mechanism. While the growth resulting from these other mechanisms will eventually turn to decay, we emphasize that the transient effects will in many cases dominate for finite times.

Spatial stability

Previously the Orr-Sommerfeld equation was discussed in terms of a temporal problem, *i.e.* the wave disturbance was assumed to grow (or decay) with time. However for spatially developing flows like many boundary layer flows, the disturbance growth is in space rather than in time. For this case the disturbance may still be described in terms of normal modes, but with complex α and β and a real ω. The growth rate of the disturbances, σ, is given by

$$\sigma_x = -\alpha_i, \qquad \sigma_z = -\beta_i. \tag{2.37}$$

For cases where flow is homogeneous in the spanwise direction, $\beta_i = 0$. In this context one should remember that the neutral curve is still the same for both the spatial and temporal disturbances.

In addition, another complication arises for boundary layer flows as compared to channel and pipe flows, namely that the Reynolds number

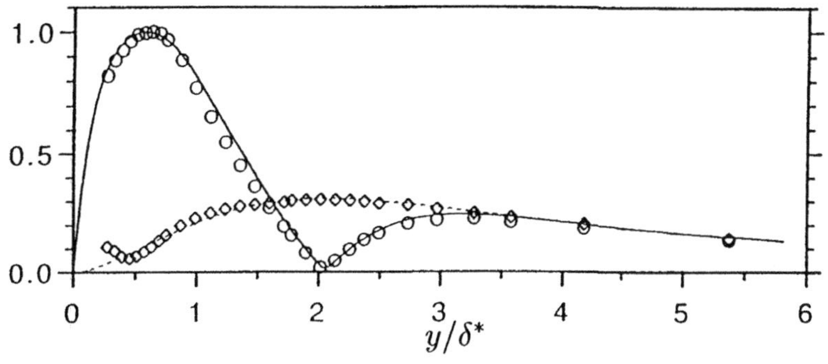

Figure 2.4. u (——) and v (- - - -) profiles for TS-waves at $F = 100$ and $R = 890$, linear parallel theory and hot-wire measurements u (○) and v (◇) (Westin, 1993, unpublished).

changes in the downstream direction and the flow is non-parallel, *i.e.* there is an x-dependence in the base flow. This complicates matters further and several attempts has been done to modify the OS-equation in order to take the streamwise variation into account. However, few of these methods have been completely successful and has not been of much practical use. The newly developed PSE method (described in a following section) can take these effects into account in a general manner, and therefore these earlier attempts to take non parallel effects into account will not be described here.

Figure 2.4 shows the u and v-distribution from both calculations based on linear spatial theory and measurements at a frequency $F = 100$ ($F = \omega\nu/{U_\infty}^2 \times 10^6$). The u-disturbance has been normalised to one. As can be seen there is a good agreement between the experiment and the parallel theory, except very close to the wall where the finite size of the probe used gives measurement errors.

Figure 2.5 shows the amplitude distributions of the streamwise velocity obtained both from linear spatial theory and measurements at three different Reynolds numbers (Klingmann *et al.* [57]). The y-coordinate is normalised with the displacement thickness. The overall shape of the distribution is similar for the three cases however it is clear that both the maximum and the minimum come closer to the wall when Re increases.

2.2.4. TRANSIENT GROWTH

Eigenvalue expansion

In addition to the dispersive effects discussed in the previous sections, the viscous initial value problem allows for the possibility for initial transient growth which, in many cases, overshadow the asymptotic behaviour predicted by the eigenmodes. The most striking example is the viscous

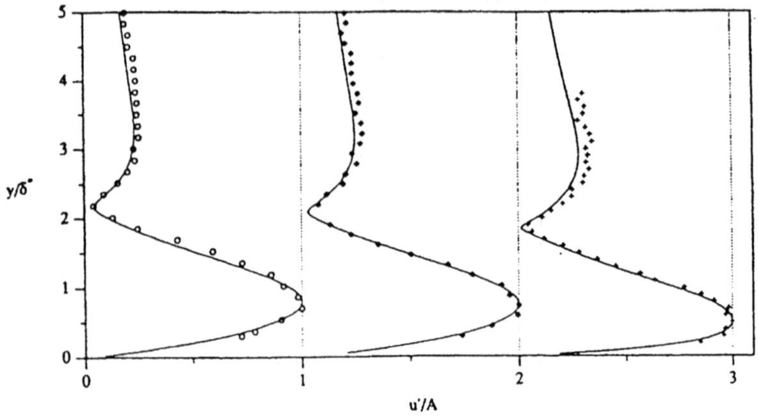

Figure 2.5. Measurements (symbols) and linear parallel theory (——) for u-distributions at $R = U_\infty \delta^*/\nu$ =343, 396, 574. F =250. From Klingmann *et al.* [57].

counterpart of the inviscid algebraic instability. For simplicity we assume that only the l:th OS-mode is excited initially. Thus we have the following solution for the normal velocity

$$\hat{v} = \tilde{v}_l e^{-i\alpha c_l t} \tag{2.38}$$

Having the solution to the normal velocity we solve the equation for the normal vorticity (2.21) using (2.38) in the right hand side. The solution consists of a homogeneous and a particular solution, which we can write in the following form

$$\hat{\eta} = \hat{\eta}_{hom} + \tilde{\eta}_l^p e^{-i\alpha c_l t} \tag{2.39}$$

where the time dependence of the particular solution has been extracted. In order to solve for $\hat{\eta}_{hom}$ and $\tilde{\eta}_l^p$ we assume that both can be expanded in the eigenmodes of the homogeneous part of the normal vorticity equation, *i.e.* in Squire modes. For the homogeneous part we have the expansion

$$\hat{\eta}_{hom} = \sum_{j=1}^{J} B_j \tilde{\eta}_j e^{-i\alpha\sigma_j t} \tag{2.40}$$

where B_j are the expansion coefficients and $\tilde{\eta}_j$ are the Squire modes corresponding to the eigenvalues of the Squire equation (2.35), σ_j. For the particular part we assume a similar expansion without the time dependence. When these expansions are substituted into equation (2.21) the following expression for the normal vorticity results

$$\hat{\eta} = \sum_{j=1}^{J} C_j \tilde{\eta}_j e^{-i\alpha\sigma_j t} + \sum_{j=1}^{J} D_{jl} \frac{e^{-i\alpha c_l t} - e^{-i\alpha\sigma t}}{\alpha c_l - \alpha\sigma_j} \tag{2.41}$$

where the expansion coefficients are

$$C_j = \int_{-1}^{1} \hat{\eta}_0 \tilde{\eta}_j dy \tag{2.42}$$

$$D_{jl} = \beta \int_{-1}^{1} U' \tilde{v}_l \tilde{\eta}_j dy \tag{2.43}$$

This expression was originally derived by Henningson [44] and a related version was used by Gustavsson [34] in his calculation of transient growth. Gustavsson also showed that this form of the solution is equivalent to the solution he derived using the Laplace transform technique.

To see the connection with the inviscid case we can consider finite times and assume that the Reynolds number approaches infinity. For $\alpha \to 0$ we find that

$$\omega_l^{OS} = \alpha c_l = -i\mu_l/R \tag{2.44}$$

$$\omega_j^{SQ} = \alpha \sigma_j = -i\nu_j/R \tag{2.45}$$

where μ_l and ν_j are both positive order one quantities, defined in terms if integrals over the eigenfunctions. This result implies that the angular frequencies for these waves are inversely proportional to the Reynolds number and that they coalesce as the Reynolds number approaches infinity. If these expressions are substituted into (2.43) and the exponents are Taylor expanded for small t/R we find

$$\hat{\eta} = \sum_{j=1}^{J} C_j \tilde{\eta}_j \left[1 - \nu_j t/R + O\left(\frac{t^2}{R^2}\right)\right] - \sum_{j=1}^{J} iD_{jl} \tilde{\eta}_j t \left[1 - (\nu_l + \mu_j)\frac{t}{2R} + O\left(\frac{t^2}{R^2}\right)\right] \tag{2.46}$$

The $\mathcal{O}(1)$ terms in the series (2.46) can be summed, we find

$$\hat{\eta} = \hat{\eta}_0 - i\beta U' \hat{v}_0 t + O(\frac{t}{R}) \tag{2.47}$$

which shows that the limit of the viscous solution as the Reynolds number approaches infinity for $\alpha = 0$ is identical to the inviscid algebraic instability.

Optimal disturbances

In the previous section we have seen that large transient growth may occur even if all of the eigenmodes of the linearized problem predict decay. We now ask the question, what is the largest growth possible? In order to answer this question we proceed to investigate the linear initial value problem (2.20)–(2.21) which we write in operator form

$$\frac{\partial \hat{\mathbf{q}}}{\partial t} = -i\mathcal{L}\hat{\mathbf{q}}, \qquad \hat{\mathbf{q}}(t=0) = \hat{\mathbf{q}}_0, \tag{2.48}$$

where $\mathcal{L}$ signifies the linear terms of the disturbance equations and $\hat{\mathbf{q}} = (\hat{v}, \hat{\eta})^T$ is a vector containing the normal velocity and the normal vorticity. An assumption of exponential time dependence of the state vector, $\hat{\mathbf{q}} = \tilde{\mathbf{q}}(y)e^{-i\omega t}$, reduces the above equation to the following eigenvalue problem

$$\mathcal{L}\tilde{\mathbf{q}} = \omega\tilde{\mathbf{q}}. \tag{2.49}$$

We can find the solution to the initial value problem in the form of an eigenfunction expansion

$$\hat{\mathbf{q}}(y,t) = \sum_{j=1}^{N} \kappa_j(t)\tilde{\mathbf{q}}_j(y), \tag{2.50}$$

where

$$\kappa_j(t) = \kappa_j(0)e^{-i\omega_j t}. \tag{2.51}$$

where $\tilde{\mathbf{q}}_j$ represents both the Orr-Sommerfeld and Squire modes, ω_j the corresponding eigenvalues and κ_j the (time-dependent) expansion coefficients. From the solution we define the growth function $G(t)$ as

$$G(t) = \max_{\mathbf{q}_0} \frac{||\mathbf{q}(t)||^2}{||\mathbf{q}_0||^2} \qquad E = \frac{1}{2}||\mathbf{q}||^2 \tag{2.52}$$

i.e. the energy growth maximized over all initial conditions at each instant of time. Here, we assume that the solution is given by the eigenfunction expansion (2.50). Thus, finding the maximum energy growth is equivalent to an optimization over the expansion coefficients κ_j. For further details on the computation of $G(t)$ see Reddy & Henningson [81].

Figure 2.6 shows contours of the maximum of $G(t)$ as a function of the wavenumbers. Note that the maximum growth occurs for $\alpha = 0$, *i.e.* structures which are independent of the streamwise direction. Gustavsson [34] calculated the growth obtained from an initial condition for the normal velocity consisting of the least stable OS-mode, normalized to unit energy. He assumed that the initial normal vorticity was zero and solved for its subsequent time dependence. He obtains a maximum which is only about 10% lower than the one shown for the optimal disturbance.

An important property of the growth function is the quadratic dependence of the energy growth on the Reynolds number. In table 2.1 this dependence, the Reynolds number dependence of the time of the maximum (t_{max}) and the wavenumber combinations for which the maximum occurs, are given for a number of wall-bounded shear flows. Note that in

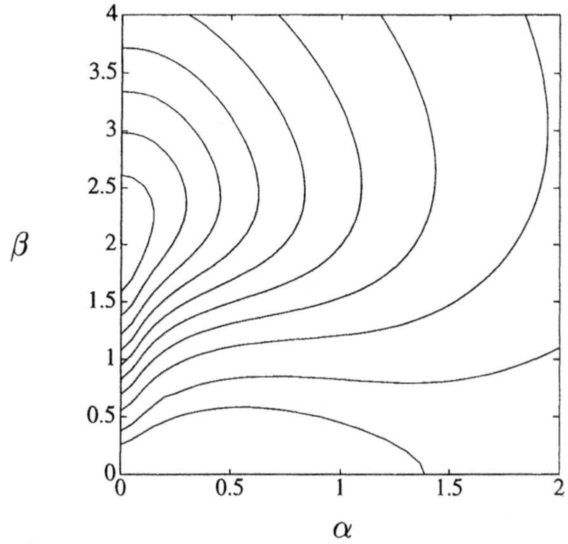

Figure 2.6. Contours of the maximum $G(t)$ in the α-β plane as a function of the wavenumbers for plane Poiseuille flow at $R = 1000$. The curves from outer to inner correspond to $\max_t G(t) = 10, 20, 40, ..., 140, 160, 180$. From Reddy & Henningson [81].

TABLE 2.1. Reynolds number dependence of $G(t)$ and t_{max} for a number of wall-bounded shear flows. Results for plane Couette and Poiseuille flow from Trefethen *et al.* (1993), for pipe flow from Schmid & Henningson (1994) and for boundary layer flow Butler & Farrell (1992). The Reynolds number in the boundary layer is based on the displacement thickness and the freestream velocity.

Flow	$\max G(t)$ (10^{-3})	t_{max}	α	β
plane Poiseuille	$0.20\ R^2$	$0.076\ R$	0	2.04
plane Couette	$1.18\ R^2$	$0.117\ R$	$35/R$	1.6
circular pipe	$0.07\ R^2$	$0.048\ R$	0	1
Blasius boundary layer	$1.50\ R^2$	$0.778\ R$	0	0.65

all flows the maximum growth occurs for structures which are essentially independent of the streamwise direction.

A number of transition scenarios utilize the lift-up effect. These include oblique transition, Schmid & Henningson [87], transition caused by the breakdown of streaks, Berlin, Wiegel & Henningson [5] and Elofsson & Alfredsson [23] and transition associated with localized disturbances, Henningson *et al.* [47], Kreiss *et al.* [58], Reddy *et al.* [82].

2.3. Stability of compressible parallel flows

In high speed flows, increase of heat transfer and skin friction, caused by laminar-turbulence transition, is greater than those in low speed flows. Existence of additional types of instability modes makes the analysis of the stability characteristics of such flows complicated. The first step to understand the transition phenomena is to identify and study different growth mechanisms of small disturbances in the flow.

Here, we will discuss different linear growth mechanism of disturbances in compressible boundary layer flows. First, the exponential instability of inviscid and viscous disturbances, in the framework of the local stability theory, are described. Then, the algebraic instability and transient growth in compressible flows is discussed. The last part of this chapter is devoted to the non-local stability theory.

The motion of a compressible fluid is described by the equations of mass, momentum and energy conservation which in non-dimensional vector notation are given by

$$\frac{\partial \rho}{\partial t} + \nabla \cdot (\rho \mathbf{u}) = 0, \tag{2.53}$$

$$\rho \left[\frac{\partial \mathbf{u}}{\partial t} + (\mathbf{u} \cdot \nabla)\mathbf{u} \right] = -\nabla p + \frac{1}{R} \nabla \left[\lambda (\nabla \cdot \mathbf{u}) \right] + \frac{1}{R} \nabla \cdot \left[\mu (\nabla \mathbf{u} + \nabla \mathbf{u}^{\mathrm{tr}}) \right], \tag{2.54}$$

$$\rho c_p \left[\frac{\partial T}{\partial t} + (\mathbf{u} \cdot \nabla) T \right] = \frac{1}{R \,\mathrm{Pr}} \nabla \cdot (\kappa \nabla T) + (\gamma - 1) M^2 \left[\frac{\partial p}{\partial t} + (\mathbf{u} \cdot \nabla) p + \frac{1}{R} \Phi \right], \tag{2.55}$$

with the viscous dissipation given as

$$\Phi = \lambda (\nabla \cdot \mathbf{u})^2 + \frac{1}{2} \mu \left[\nabla \mathbf{u} + \nabla \mathbf{u}^{\mathrm{tr}} \right]^2 . \tag{2.56}$$

The equation of state in its non-dimensional form is

$$\gamma M^2 p = \rho T. \tag{2.57}$$

Here t represents time, ρ, p, T stand for density, pressure and temperature, $\mathbf{u}$ is the velocity vector. The quantities λ, μ stand for the second and dynamic viscosity coefficient, γ is the ratio of specific heats, κ the heat conductivity, c_p the specific heat at constant pressure. All flow and material quantities are nondimensionalized by the corresponding reference flow quantities, besides pressure, which is referred to twice the corresponding dynamic pressure. The Mach number, M, Prandtl number, Pr, and Reynolds number, R are defined as

$$M = \frac{U^*_{ref}}{\sqrt{\Re \gamma T^*_{ref}}},$$

$$\mathrm{Pr} = \frac{\mu^*_{ref} c^*_{p_{ref}}}{\kappa^*_{ref}}$$

and

$$R = \frac{U^*_{ref} l^*_{ref}}{\nu^*_{ref}},$$

where $\Re$ is the specific gas constant, l^*_{ref} the reference length scale and subscript $_{ref}$ refers to the reference quantities.

In order to derive the stability equations, we decompose all flow and material quantities into a steady basic flow $\bar{\mathbf{q}}$ plus an unsteady disturbance flow component $\mathbf{q}$ according to

$$\mathbf{Q}(x, y, z, t) = \bar{\mathbf{q}}(x, y, z) + \mathbf{q}(x, y, z, t), \tag{2.58}$$

where x, y and z are the coordinates in the streamwise, normal and spanwise directions, respectively, and time is denoted by t. Specifically in the above, $\bar{\mathbf{q}}$ and $\mathbf{q}$ stand for

$$\bar{\mathbf{q}} = (U, V, W, \bar{p}, \bar{T}, \bar{\rho}), \tag{2.59}$$

$$\mathbf{q} = (u, v, w, p, T, \rho). \tag{2.60}$$

Here U, V, W are the basic and u, v, w the disturbance velocity components in the streamwise, normal and spanwise directions, respectively. We assume that c_p, μ, κ are only a function of temperature and that they can be divided into a mean and a perturbation part. The perturbation part is written as expansions in temperature according to

$$c_p = \frac{d\bar{c}_p}{d\bar{T}} T, \quad \mu = \frac{d\bar{\mu}}{d\bar{T}} T, \quad \kappa = \frac{d\bar{\kappa}}{d\bar{T}} T. \tag{2.61}$$

For the ratio of the coefficients of second and dynamic viscosity

$$\frac{\bar{\lambda}}{\bar{\mu}} = \frac{\lambda}{\mu} = \frac{\mu_v}{\mu} - \frac{2}{3}. \tag{2.62}$$

is assumed. Here, μ_v is the bulk viscosity. Ususally, under Stokes' hypothesis, μ_v is chosen to be zero. For a discussion on choices of μ_v, readers are referred to *e.g.* Bertolotti [8].

2.3.1. DERIVATION OF STABILITY EQUATIONS

To obtain the linearized disturbance equations, the decomposition (2.58) is introduced into (2.53)–(2.55), the equations for the basic flow are subtracted and products of disturbance quantities are neglected. Thus we obtain the linear disturbance equations

$$\frac{\mathrm{D}\rho}{\mathrm{D}t} + \rho\nabla\cdot\bar{\mathbf{u}} + \bar{\rho}\nabla\cdot\mathbf{u} = 0, \tag{2.63}$$

$$\bar{\rho}\left[\frac{\mathrm{D}\mathbf{u}}{\mathrm{D}t} + \mathbf{u}\cdot\nabla\bar{\mathbf{u}}\right] + \rho\bar{\mathbf{u}}\cdot\nabla\bar{\mathbf{u}} = -\nabla p + \frac{1}{R}\nabla\left[\lambda(\nabla\cdot\bar{\mathbf{u}}) + \bar{\lambda}(\nabla\cdot\mathbf{u})\right]$$
$$+\frac{1}{R}\nabla\cdot\left[\bar{\mu}(\nabla\mathbf{u} + \nabla\mathbf{u}^{\mathrm{tr}}) + \mu(\nabla\bar{\mathbf{u}} + \nabla\bar{\mathbf{u}}^{\mathrm{tr}})\right] \tag{2.64}$$

$$\bar{\rho}\bar{c}_p\left[\frac{\mathrm{D}T}{\mathrm{D}t} + \mathbf{u}\cdot\nabla\bar{T}\right] + (\rho\bar{c}_p + \bar{\rho}c_p)\bar{\mathbf{u}}\cdot\nabla\bar{T} = \tag{2.65}$$
$$\frac{1}{R\,\mathrm{Pr}}\left[\nabla\cdot(\kappa\nabla\bar{T}) + \nabla\cdot(\bar{\kappa}\nabla T)\right] + (\gamma-1)M^2\left[\frac{\mathrm{D}p}{\mathrm{D}t} + \mathbf{u}\cdot\nabla\bar{p} + \frac{1}{R}\Phi\right]$$

with

$$\frac{\mathrm{D}}{\mathrm{D}t} = \frac{\partial}{\partial t} + \bar{\mathbf{u}}\cdot\nabla, \tag{2.66}$$

and

$$\begin{aligned}\Phi &= 2\bar{\lambda}\left[(\nabla\cdot\bar{\mathbf{u}})(\nabla\cdot\mathbf{u})\right] + \lambda(\nabla\cdot\bar{\mathbf{u}})^2 + \bar{\mu}\left[(\nabla\bar{\mathbf{u}} + \nabla\bar{\mathbf{u}}^{tr}) : (\nabla\mathbf{u} + \nabla\mathbf{u}^{tr})\right]\\ &+ \frac{1}{2}\mu\left[(\nabla\bar{\mathbf{u}} + \nabla\bar{\mathbf{u}}^{tr}) : (\nabla\bar{\mathbf{u}} + \nabla\bar{\mathbf{u}}^{tr})\right]\end{aligned} \tag{2.67}$$

where the operator : is defined such that $A : B = A_{ij}B_{ij}$. The linearized equation of state for the disturbance quantities is given as

$$\rho = \gamma M^2\frac{p}{\bar{T}} - \bar{\rho}\frac{T}{\bar{T}}. \tag{2.68}$$

These equations are subjected to the following boundary conditions

$$u = v = w = T = 0; \qquad y = 0, \tag{2.69}$$

$$u,\ v,\ w,\ T \to 0; \qquad y \to \infty. \tag{2.70}$$

2.3.2. EXPONENTIAL INSTABILITIES

Here, we restrict us to a *quasi-parallel flow* which yields

$$\bar{\mathbf{q}} = \bar{\mathbf{q}}(y), \qquad V = 0, \tag{2.71}$$

The disturbances are taken to be harmonic waves as

$$\mathbf{q} = \tilde{\mathbf{q}}(y)e^{i(\alpha x + \beta z - \omega t)} \tag{2.72}$$

where α and β are the wavenumbers in the streamwise and spanwise direction, respectively. ω represents the angular disturbance frequency. The stability equations with above disturbance ansatz, constitute an eigenvalue problem for ω or α which can be solved in different ways.

Before we go further, some additional concepts will be introduced. Let us define the *relative Mach number* as in Mack [70]

$$\hat{M} = \frac{\alpha U + \beta W - \omega}{\sqrt{\alpha^2 + \beta^2}} \frac{M}{\sqrt{T}}. \tag{2.73}$$

For the neutral disturbances the physical meaning of $\hat{M}$ can be understood if it is rewritten as

$$\hat{M} = M_\psi - \frac{c^*}{a^*}, \tag{2.74}$$

where M_ψ is the local Mach number of the mean flow in the direction of wavenumber vector $\mathbf{k}$, c^* the dimensional phase velocity and a^* the local velocity of sound. Here, the nondimensional phase velocity, c, is defined as

$$c = \frac{\omega}{k}, \tag{2.75}$$

where $k = |\mathbf{k}|$. Note that this is now the phase velocity in the direction of the wavenumber vector and not in the x-direction. The angle between the direction of $\mathbf{k}$, and the x-axis, ψ, is found from the relation

$$\tan\psi = \frac{\beta_r}{\alpha_r}, \tag{2.76}$$

where subscript $_r$ refers to the real part of quantities.

Then, $\hat{M}$ denotes the local Mach number of the mean flow in the direction of $\mathbf{k}$ relative to the phase velocity. Now, we can classify the disturbances based on the value of the relative Mach number at the edge of the boundary layer, $\hat{M}_e$. A disturbance is called *subsonic, sonic* or *supersonic* if $\hat{M}_e^2$ is less than, equal or greater than unity, respectively.

Compressible Rayleigh equation

The stability equations become simplified in the inviscid limit ($R \to \infty$) which makes it possible to analytically investigate the behavior of the disturbance quantities. The results for large Reynolds numbers should approach those from inviscid theory. The results discussed below are taken from Lees & Lin [61] and Mack [70].

In the inviscid limit it is possible to derive a second order equation for the normal velocity,

$$\left[\frac{(U_\psi - c)\tilde{v}' - U_\psi'\tilde{v}}{\bar{T} - M^2(U_\psi - c)^2}\right]' = \frac{k^2}{\bar{T}}(U_\psi - c)\tilde{v}, \tag{2.77}$$

with boundary conditions $\tilde{v}(0) = \tilde{v}(\infty) = 0$. Here, $U_\psi = (\alpha U + \beta W)/k$ is the velocity component in the direction of the wavenumber vector. Introducing the relative Mach number to the equation above yields

$$\left[\frac{(U_\psi - c)\tilde{v}' - U_\psi'\tilde{v}}{\bar{T}(1 - \hat{M}^2)}\right]' = \frac{k^2}{\bar{T}}(U_\psi - c)\tilde{v}, \tag{2.78}$$

This equation is the three-dimensional compressible counterpart of the incompressible Rayleigh equation. It also emphasis that the inviscid instability is governed by the mean flow in the direction of **k**.

Critical layer singularity

The compressible version of the Rayleigh equation also has a singularity at the critical layer $y_c : U_\psi(y_c) = c$ of the same type as for incompressible flow. This can be seen by rewriting equation (2.78) in the self-adjoint form given by Lees & Lin. We have,

$$(F\tilde{v}')' - (H + \frac{k^2}{\bar{T}})\tilde{v} = 0, \tag{2.79}$$

where

$$F(y) = \frac{1}{1 - \hat{M}^2}\frac{1}{\bar{T}} \quad \text{and} \quad H(y) = \frac{1}{U_\psi - c}(FU_\psi')'.$$

Note that unless the quantity

$$\left[(FU_\psi')'\right]_{U_\psi = c} = \left[(U_\psi'/\bar{T})'\right]_{U_\psi = c}$$

vanishes at the critical layer, the point $y = y_c$ in the complex plane is a regular singular point of equation (2.79).

Lees & Lin's [61] investigations of 2D disturbances, $\beta = 0$, in a parallel two dimensional basic flow showed that the quantity $(U'/\bar{T})'$, usually written as $(\bar{\rho}U')'$, plays the same role as U'' in incompressible flows. The position where $(\bar{\rho}U')' = 0$ is called *generalized inflection point.* They showed that the existence of a generalized inflection point in the boundary layer, *i.e.* ,

$$\frac{\mathrm{d}}{\mathrm{d}y}\left(\bar{\rho}\frac{\mathrm{d}U}{\mathrm{d}y}\right)\bigg|_{y=y_s} = 0, \qquad U(y_s) > U_e - a_e, \tag{2.80}$$

is a necessary condition for existence of a *neutral subsonic* wave. Here, subscript $_e$ refers to the values at the edge of the boundary layer. This wave has a phase velocity, c_s, equal to the mean flow at y_s. If $\hat{M}^2 < 1$ through the entire boundary layer the above condition is also a sufficient condition for existence of a such wave with a unique wavenumber α_s. It was also shown that the above conditions are sufficient for existence of an *unstable subsonic* disturbance. Furthermore, they found that there is a *neutral sonic* wave with $\alpha = 0$ and the phase velocity $c_0 = U_e - a_e$.

As a result of the generalized inflection point criteria, it can be shown that the compressible boundary layer over an adiabatic flat plate is inviscidly unstable, in contrast to what is found in the incompressible case.

The inflection point criterion is also valid for three-dimensional disturbances. For these waves the phase velocity c_s is $U_s \cos\psi$, and the phase velocity c_0 is $U_e \cos\psi - a_e$.

Inward and outward propagating waves

The v-disturbance equation (2.78) in the free-stream takes the form

$$\tilde{v}'' = (1 - \hat{M}_e^2)k^2\tilde{v} \tag{2.81}$$

and its solutions $\tilde{v} \sim e^{\pm k\eta y}$, where $\eta^2 = (1 - \hat{M}_e^2)$. The complete solution for the normal velocity thus becomes

$$v = \tilde{v}(y)e^{k(c_i t+\eta_r y)}e^{ik(-c_r t+\eta_i y)}e^{i(\alpha x+\beta z)} \tag{2.82}$$

It is easily seen that the boundary conditions imply that $\eta_r < 0$, and that if $\eta_i \neq 0$ the waves in the free-stream propagate toward or away from the boundary layer, depending on the sign of η_i.

For $\hat{M}_e < 1$ we have,

$$\eta = \pm\sqrt{1 - M^2(U_{e_\psi} - c_r)^2} \pm i\frac{c_i M^2(U_{e_\psi} - c_r)}{\sqrt{1 - M^2(U_{e_\psi} - c_r)^2}} + \mathcal{O}(c_i^2) \tag{2.83}$$

and for $\hat{M}_e > 1$ we have

$$\eta = \pm\frac{c_i M^2(U_{e_\psi} - c_r)}{\sqrt{M^2(U_{e_\psi} - c_r)^2 - 1}} \pm i\sqrt{M^2(U_{e_\psi} - c_r)^2 - 1} + \mathcal{O}(c_i^2) \tag{2.84}$$

This implies that damped subsonic and supersonic waves propagate out of the boundary layer and that amplified subsonic and supersonic waves propagate towards the boundary layer. In contrast neutral subsonic waves propagate parallel to the x-axis, whereas neutral supersonic waves propagate both in and out of the boundary layer. Although the prediction by Lees & Lin [61] only applies to subsonic waves , numerical calculations have shown that both types of neutral waves are inflectional, *i.e.* have their critical layer located at the generalized inflection point (see Mack [70]).

Higher modes

As it was mentioned above, the uniqueness of α_s were based on the condition that $\hat{M}^2 < 1$ through the entire boundary layer. Lees & Reshotko [62] conjectured that α_s may not be unique for $\hat{M}^2 > 1$. These multiple solutions were first found by extensive numerical work of Mack [65],[66], [67]. About the same time, Gill [30] independently found similar multiple solutions for 'Top-Hat' jets and wakes. Alternatively, one can find evidence for the existence of such multiple modes by studying the inviscid equation for a pressure disturbance (see Mack [70])

$$\tilde{p}'' - \left[\ln(\hat{M}^2)\right]' \tilde{p}' - (\alpha^2 + \beta^2)(1 - \hat{M}^2)\tilde{p} = 0. \tag{2.85}$$

One can easily see that equation (2.85) changes its behavior when $(1 - \hat{M}^2)$ changes sign. If the second term on the LHS is neglected (which can be justified from numerical calculations), the remaining equation is a wave equation for $\hat{M}^2 > 1$. Consequently, multiple solutions may be found. The first of these additional modes, usually called the second mode, turns out to be particularly interesting. The numerical studies of Mack showed that this mode is the most amplified mode for supersonic boundary layers at high Mach numbers, see figure 2.7. He also found that the most unstable first mode, which is an extension of the incompressible Tollmien-Schlichting (TS) wave, is oblique and not two-dimensional as in incompressible boundary layer flows, see figure 2.8.

In addition to the above mentioned multiple modes, Mack found another group of neutral waves with phase velocities in the range $U_e \leq c \leq U_e + a_e$. For each of these phase velocities there is an infinite sequence of wavenumbers. The existence of these modes is not related to the inflection point, therefore these are called *noninflectional waves.* According to Mack "the importance of the $c = U_e$ neutral waves is that in the absence of an interior generalized inflection point they are accompanied by a neighboring family of unstable waves with $c < U_e$. Consequently, a compressible boundary layer is unstable to inviscid waves whenever $\hat{M}^2 > 1$, regardless of any other feature of the velocity and temperature profiles". This family of modes does not have any counterpart in incompressible theory.

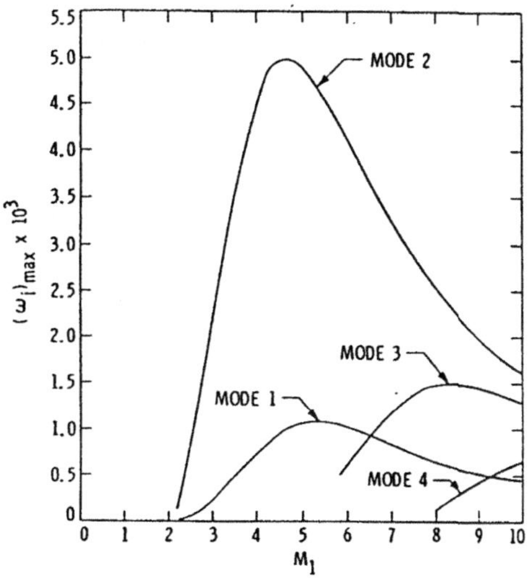

Figure 2.7. Effect of Mach number on maximum temporal amplification rate of 2D waves for first four modes. Insulted wall, wind-tunnel temperatures. From Mack [70]

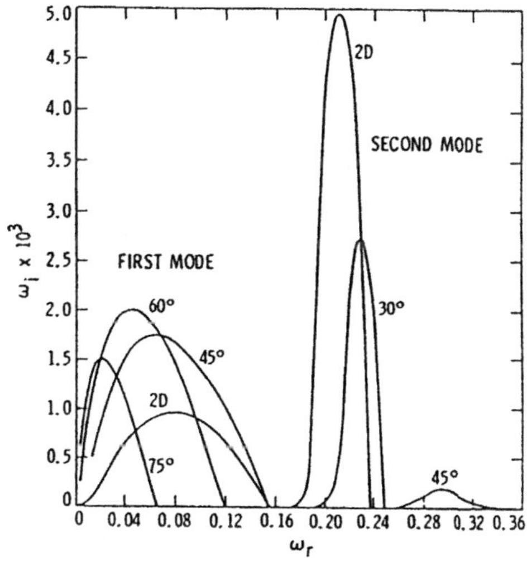

Figure 2.8. Temporal amplification rate of first and second modes vs. frequency for several wave angles at $M_\infty = 4.5$. Insulated wall, $\bar{T}_\infty = 311$ K. From Mack [70].

Viscous instability

The early results of viscous compressible stability investigations were based on the asymptotic theories of Lees & Lin [61], Dunn & Lin [19] and

Lees & Reshotko [62]. None of these theories deals with the complete set of the stability equation. Brown [11],[12] solved the stability equations numerically, first according to the theory of Dunn & Lin [19] and later the complete set of equations. Mack [67] presented detailed numerical results for Mach numbers up to 10. These results include both first and second viscous modes which are unstable. Similar to the inviscid results, the most unstable of the first-mode disturbances were found to be the oblique ones. For the supersonic boundary layers over a flat plate, the waveangle ψ of these disturbances is about 50-60 degree. Among the second-mode disturbances the 2D waves were shown to have the largest growth rate, see figure 2.9. The lowest Mach number at which the second-mode instability at finite Reynolds numbers exists is about 3. This occurs for a Reynolds number about 13,900. Then, as the Mach number increases the unstable region moves rapidly to lower Reynolds numbers. However, the *critical Reynolds number* (the lowest Reynolds number for which an unstable disturbance exists) has a minimum for a Mach number about 4.5. As shown in figure 2.10, the first- and second-mode unstable regions are separated at low Mach numbers whereas at Mach numbers above 4.6 they have merged.

The stability of the compressible boundary layer is known to be a viscous instability at lower Mach numbers. Whereas, at Mach numbers above 3 the inviscid instability is dominant. By viscous instability, as in Mack [70], we mean that the maximum amplification rate increases as the Reynolds number decreases. An indication of viscous instability is a neutral stability curve with an upper-branch wave number which increases as Reynolds number decreases. In figure 2.11 the neutral stability curves of 2D waves for different Mach numbers are given. At $M_\infty = 1.6$, the neutral curve is similar to that of incompressible flow. The viscous instability weakens as the Mach number increases and at $M_\infty = 3.8$ it has completely vanished. The same behavior have been observed for 3D disturbances.

Effects of heat transfer on stability characteristics

We start to discuss the effects of heat transfer on the flow instability by examine its effects on the boundary layer profiles. As mentioned before, a compressible flow is inviscidly unstable if it has a generalized inflection point at $y_s > y_0$ where $U(y_0) = U_e - a_e$. A compressible boundary layer on an insulated surface contains only one generalized inflection point. As the surface temperature is reduced, another generalized inflection point appears inside the boundary layer at $y = y_{s2}$, see figure 2.12. Since this point is located in the portion of the boundary layer where $U(y) < U_e - a_e$, it does not alter an inviscid instability. The distance between y_s and y_{s2} decreases as the surface cooling increases. For a given Mach number, there is

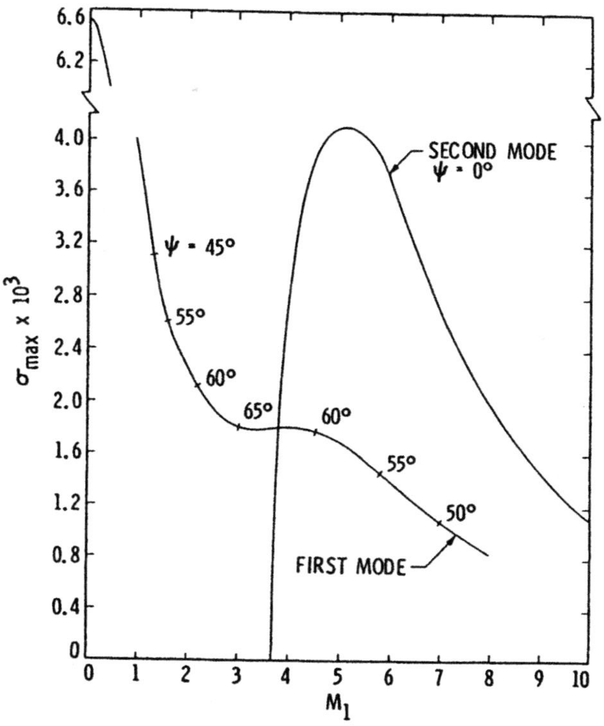

Figure 2.9. Maximum spatial growth rate, $R = 1500$. From Mack [70].

a level of cooling at which these two generalized inflection points cancel each other (see figure 2.13) and consequently no unstable first-mode solutions will exist. However, existence of unstable higher-modes depends only on a region of relative supersonic flow and can not be eliminated by wall cooling.

These conclusions have been confirmed by numerical investigations, (*e.g.* Mack [68], Malik [71],Masad, Nayfeh & Al-Maaitah [76]), see figure 2.14. These investigations showed that the maximum growth rate of the second-mode disturbances increases with increasing surface cooling, see figure 2.15. It was also found that the frequency of the most unstable first-mode disturbances decreases with increasing wall cooling. The effect on the frequency of the second-mode was shown to be opposite to this.

Effects of suction

The effects of suction on the mean profiles is close to those of surface cooling. Consequently, the first-mode instability decays as the suction rate increases. Suction has also a dumping effect on the instability of the second-mode disturbances. However, the damping effect of suction on the first-

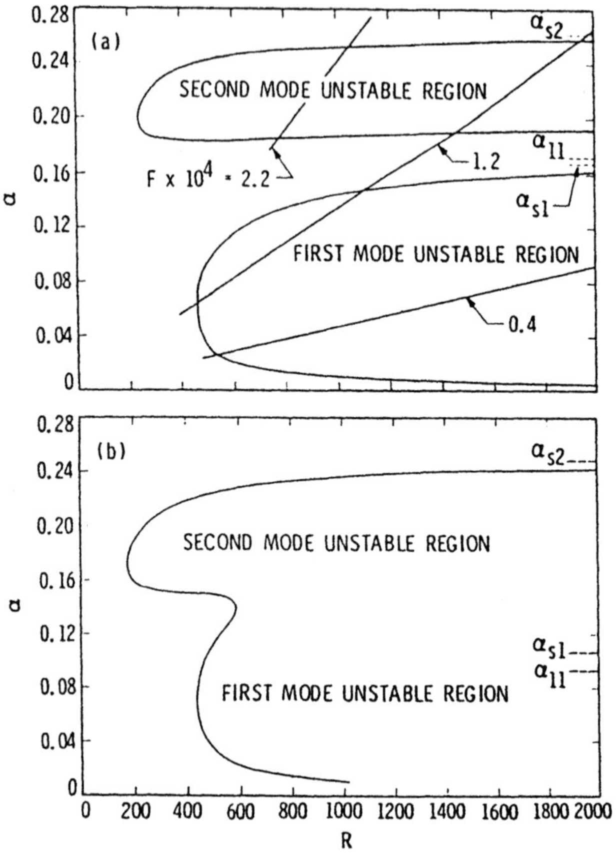

Figure 2.10. Neutral stability curve. a)$M_e = 4.5$, b) $M_e = 4.8$. From Mack [70].

and second-mode disturbances decays as the Mach number increases, see figures 2.16 and 2.17. The numerical investigations have shown that the maximum growth rate of these disturbances varies almost linearly with the level of suction.

Detailed numerical results can be found in *e.g.* Malik [71], Masad, Nayfeh & Al-Maaitah [75], and Al-Maaitah, Nayfeh & Masad [2].

Effects of pressure gradient

A favorable pressure gradient is known to stabilizes the first- and second-mode disturbances by reducing the peak growth rate and decreasing the band of unstable frequencies (see *e.g.* Malik [71] and Zurigat, Nayfeh & Masad [95]). Moreover, the instability of the second-mode waves is shifted

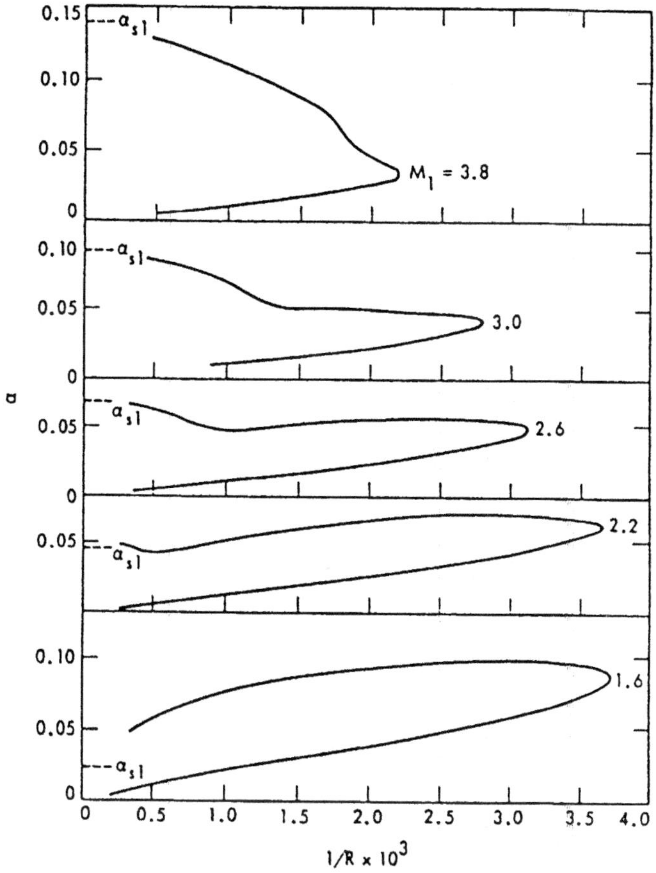

Figure 2.11. Effect of Mach number on 2D neutral stability curves. Insulated wall, wind tunnel temperature. From Mack [70].

to higher frequency. The oblique first-mode waves, which are the most unstable first-mode waves, are much more affected by pressure gradients. The efficiency of a favorable pressure gradient in stabilizing the disturbances decreases at high Mach numbers.

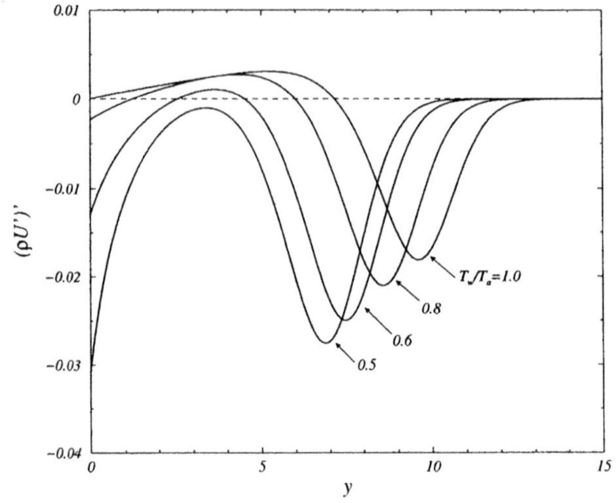

Figure 2.12. Effects of cooling on distribution of $(\rho U')'$ through a flat-plate boundary layer; $M_\infty = 4$ for $\bar{T}_\infty = 300$ K and Pr = 0.72.

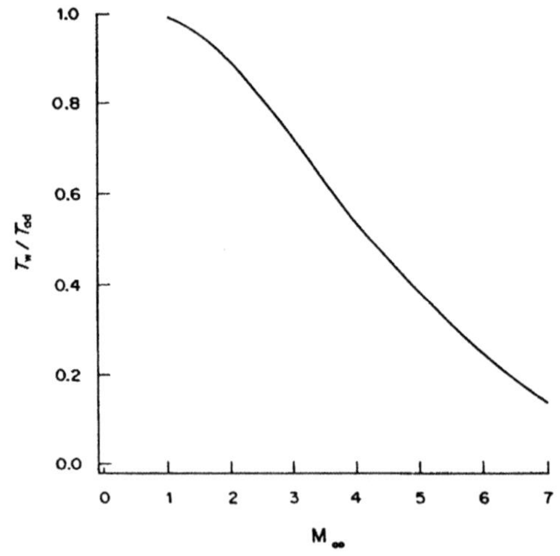

Figure 2.13. The cooling level needed to eliminate the generalized inflection points as a function of M_∞ for $\bar{T}_\infty = 300$ K and Pr = 0.72. From Masad *et al.* [76].

Effects of approximation to thermodynamical properties

The results of the stability calculations for supersonic flows are known to be sensitive to the choice of the thermodynamic properties of the flow. These effects are discussed in works by *e.g.* Bertolotti [6], Malik & Anderson [72] and Stuckert & Reed [91], Chang, Vinh & Malik [15], Hudson,

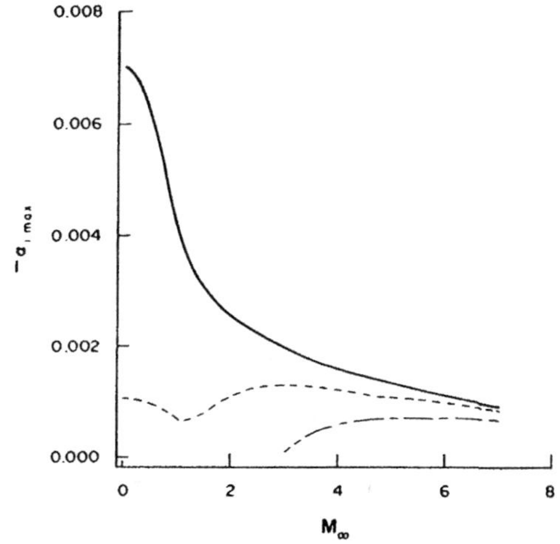

Figure 2.14. Variation of the maximum growth rates of first-mode waves with wall temperature at $R = 1500$, $\bar{T}_\infty = 300$ K and Pr = 0.72, with (—),$\bar{T}_w/\bar{T}_{ad} = 1.0$; (- -), $\bar{T}_w/\bar{T}_{ad} = 0.8$ and (— - —), $\bar{T}_w/\bar{T}_{ad} = 0.6$. From Masad *et al.* [76].

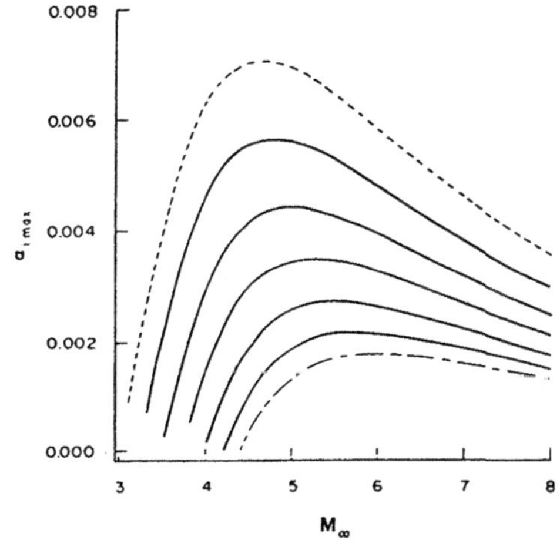

Figure 2.15. Variation of the maximum growth rates of second-mode waves with wall temperature at $R = 1500$, $\bar{T}_\infty = 300$ K and Pr = 0.72, with (- -),$\bar{T}_w/\bar{T}_{ad} = 0.4$; (— - —), $\bar{T}_w/\bar{T}_{ad} = 1.6$. The values of $\bar{T}_w/\bar{T}_{ad}$ proceeding downwards are 0.4, 0.6, 0.8, 1.0, 1.2, 1.4 and 1.6. From Masad *et al.* [76].

Chokani & Candler [52] and Bertolotti [9]. In Malik & Anderson's work ($M = 10$ and 15) the effects of vibrational excitation and dissociation of the gas molecules were accounted through the mean flow and gas property variations. They found that the high temperature effect on the second-mode

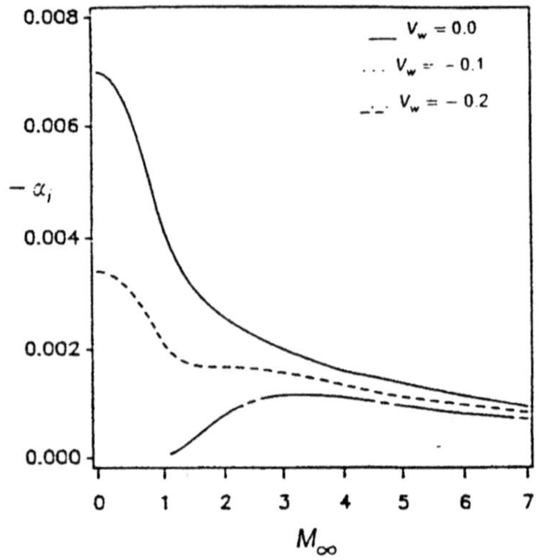

Figure 2.16. Variation of the maximum growth rates of First-mode instability waves with suction level at $R = 1500$, $\mathrm{Pr} = 0.72$ and $\bar{T}_\infty = 150$ K. From Masad *et al.* [75].

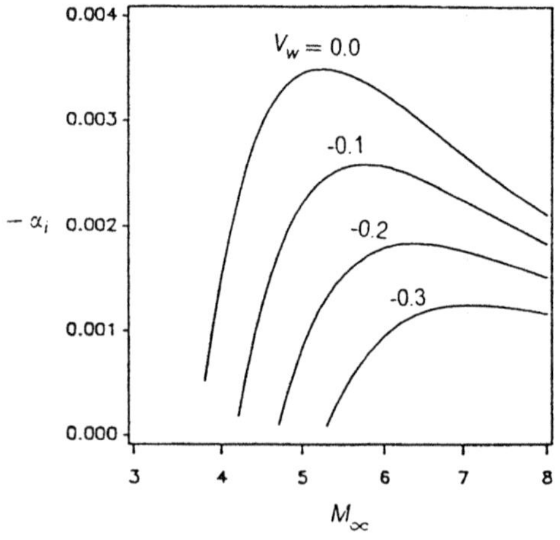

Figure 2.17. Variation of the maximum growth rates of Second-mode instability waves with suction level at $R = 1500$, $\mathrm{Pr} = 0.72$ and $\bar{T}_\infty = 125$ K. From Al-Maaitah *et al.* [2].

disturbances was destabilizing while it was stabilizing for the first-mode stability waves. Calculations of Stuckert & Reed [91] were performed for a 10-degree half-angle sharp cone at $M = 25$ and effects of chemical non-equilibrium on the stability of the shock-layer was studied. They found that in both the equilibrium and non-equilibrium air the second-mode instability is shifted to lower frequencies. In their equilibrium air calculations supersonic modes with oscillating magnitude in the inviscid region of the shock layer was found. They argued that "this oscillatory behavior was possible only because the shock standoff distance is finite".

Bertolotti [9] investigated the influence of rotational and vibrational energy relaxation on the stability of the supersonic boundary layers flows in thermal non-equilibrium. The rotational relaxation, which was modelled by approximating temperature dependence of the bulk viscosity with a formula, had a damping effect on the high-frequency instabilities. He also found that the vibrational relaxation, which acts by changing the characteristics of the laminar mean flow, increased the instability. This effct is strongest when the flow field contains a region at, or near, stagnation conditions, followed by a rapid expansion, such as inside wind tunnels and around bodies with a blunt leading edge.

2.3.3. NON-MODAL INSTABILITIES

Inviscid algebraic instabilities

The non-modal instability of incompressible flows was discussed in section 2. Here, we will discuss the non-eigenmode behavior of the disturbances in a compressible flow by directly looking at the initial boundary value problem describing the evolution of infinitesimal disturbances in compressible boundary layers.

Let us consider an inviscid compressible parallel shear flow independent of the stream- and spanwise directions. Furthermore, let the perturbations be independent of the streamwise coordinate. The linearized equations of motion can then be written

$$\frac{\partial \rho}{\partial t} + v\frac{\partial \bar{\rho}}{\partial y} + \bar{\rho}\frac{\partial w}{\partial z} + \rho\frac{\partial v}{\partial y} = 0,$$

$$\frac{\partial u}{\partial t} + v\frac{\partial U}{\partial y} = 0,$$

$$\frac{\partial w}{\partial t} = -\frac{1}{\bar{\rho}}\frac{\partial p}{\partial z},$$

$$\frac{\partial v}{\partial t} = -\frac{1}{\bar{\rho}}\frac{\partial p}{\partial y},$$

$$\frac{\partial T}{\partial t} + v\frac{\partial \bar{T}}{\partial y} = \frac{(\gamma - 1)M^2}{\bar{\rho}\bar{c}_p}\frac{\partial p}{\partial t}.$$

with boundary conditions

$$u = w = v = 0 \qquad \text{on} \qquad \Omega$$

and initial conditions

$$u = 0, \quad w = w_0(y,z), \quad v = v_0(y,z); \qquad t = 0.$$

Here, Ω denotes the boundary of the domain of interest. It is easily verified that the incompressible solution of Ellingsen & Palm [22] also satisfies these equations, with the addition that ρ and T also grow algebraic in time, we have

$$u = -v_0 \frac{\partial U}{\partial y} t, \tag{2.86}$$

$$v = v_0, \tag{2.87}$$

$$w = w_0, \tag{2.88}$$

$$p = 0, \tag{2.89}$$

$$\rho = \rho_0 - v_0 \frac{\partial \bar{\rho}}{\partial y} t, \tag{2.90}$$

$$T = T_0 - v_0 \frac{\partial \bar{T}}{\partial y} t. \tag{2.91}$$

Thus, not only is the streamwise momentum conserved as a fluid particle is lifted up by the normal velocity but also the perturbation density and temperature. Note that it is still necessary for the initial condition to be incompressible in order to satisfy the equation of state. Here, we have used that $\bar{\rho} = 1/\bar{T}$ in the boundary layer approximation.

Transient growth

Above, it was mentioned that inviscid compressible infinitesimal perturbations may experience unbounded algebraic growth. However, this growth is bounded when viscosity is present. If we define a measure of the size of a compressible disturbance, we can investigate transient growth in that case. In the case of compressible flows, there is no obvious choice of how a disturbance measure is defined. Hanifi, Schmid & Henningson [41] suggested to use the following quantity

$$E = \int_{\mathcal{V}} \left\{ \bar{\rho}(|\hat{u}|^2 + |\hat{w}|^2 + |\hat{v}|^2) + \frac{\bar{T}}{\bar{\rho}\gamma M^2}|\hat{\rho}|^2 + \frac{\bar{\rho}}{\gamma(\gamma-1)\bar{T}M^2}|\hat{T}|^2 \right\} \mathrm{d}\mathcal{V}. \tag{2.92}$$

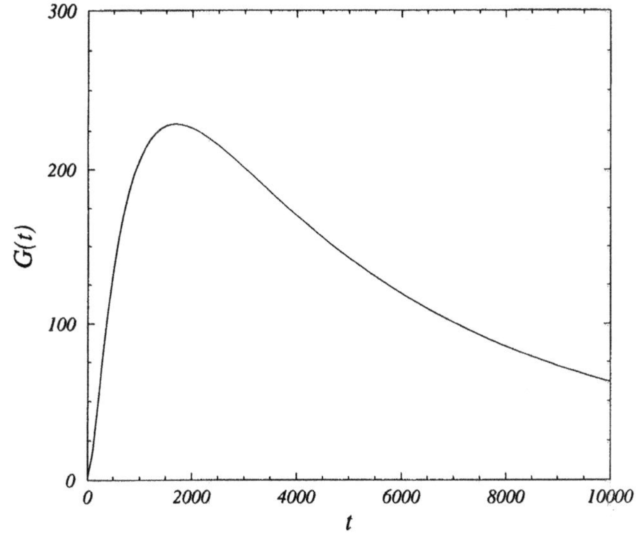

Figure 2.18. Transient growth as a function of time. $R = 300$, $M = 2.5$, $\alpha = 0$ and $\beta = 0.1$. For these parameters all eigenmodes are stable.

Using the above measure and the method described in section 2.2.4, Hanifi *et al.* [41] showed that the largest transient growth occurs for $\alpha = 0$. They also showed that the maximum transient growth scales with R^2 while t^{max} varies with R. Here, t^{max} is the time the maximum growth is reached. The optimal disturbances, those which give the largest growth, are initially streamwise vortices with small streamwise velocity. All of these characteristics agree well with those found for incompressible shear flows. Figure 2.19 shows contours of the maximum transient growth. The region of the unstable first- and second-mode disturbances are also shown in this figure. As can be seen, the transient growth is concentrated to low streamwise wavenumbers.

The relation between the transient growth and the algebraic instability of compressible streamwise independent disturbances were investigated by Hanifi & Henningson [40]. They showed that for small time, when the viscosity effects are small, the optimal transient growth is close to the optimal growth of inviscid disturbances given by solutions (2.86)–(2.91). As shown in figure 2.20, the higher the Reynolds number the longer the viscous curves are close to the inviscid one. For large times, however, the viscous results deviate from that of inviscid theory and viscosity start to damp the growth.

The results regarding transient growth for compressible boundary layer indicate that transition starting with two oblique waves may be of par-

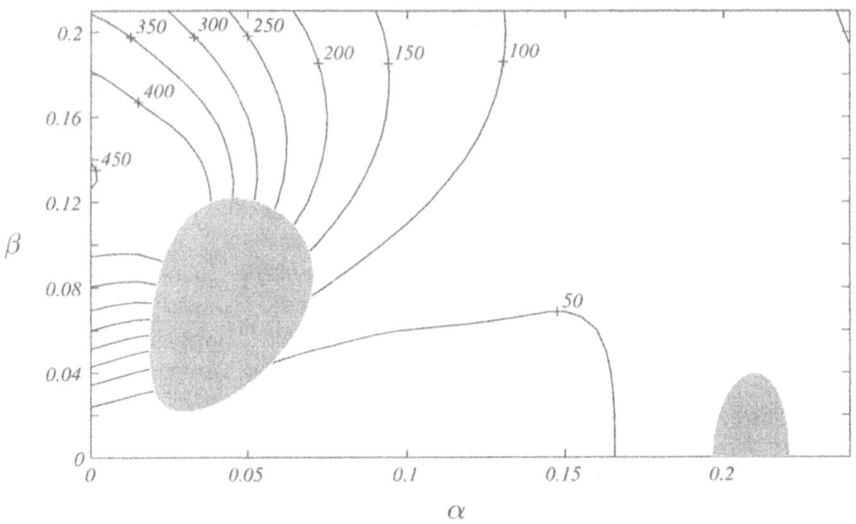

Figure 2.19. Contour plots of maximum transient growth at $R = 300$ and $M = 4.5$. The shaded region corresponds to the parameter space where the flow is linearly unstable. The shaded region at the left corresponds to the first- and at the right to the second-mode unstable disturbances. From Hanifi *et al.* [41].

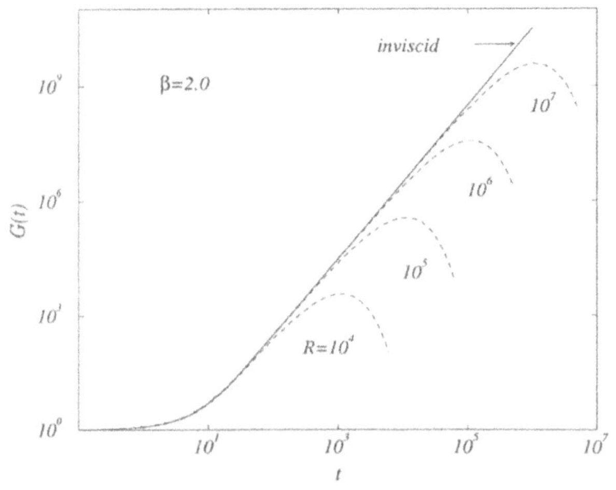

Figure 2.20. Comparison of the growth of the viscous and inviscid disturbances. $M_e = 2.5$, $\alpha = 0$, $\beta = 2.0$. From Hanifi & Henningson [40].

ticular importance. Since the most unstable first-mode disturbances are oblique, small amplitude naturally growing three-dimensional disturbances can create streamwise vortices through a non-linear interaction. These vortices may then trigger rapidly growing streaks. In figure 2.19, it can be seen that the most unstable first mode disturbance has a spanwise wavenumber approximately half of the one for the optimally growing streaks. Thus, a pair of the most unstable oblique first mode waves naturally transfers energy into the Fourier component where optimal transient growth occurs.

Recent numerical simulations of transition in compressible shear flows (*e.g.* Fasel *et al.* [25], Gathmann *et al.* [29],Chang & Malik [14], Sandham *et al.* [84]) have shown that oblique transition in many cases is a more rapid transition mechanism than traditional secondary instability of two-dimensional waves. In particular, for the case of the compressible confined shear layer excited by weak broadband noise, Gathmann *et al.* [29] found that the oblique transition scenario appeared naturally even though the oblique waves were not always the most amplified modes.

2.4. Stability of non-parallel compressible flows

2.4.1. NON-LOCAL STABILITY THEORY

The traditional stability theory, based on the quasi-parallel flow assumption, does not account for the growth of the boundary layer. The *local* character of this theory exclude any effects associated with the varying properties of the basic flow or consistent way of taking curvature into account. The effect of a growing boundary layer on the characteristics of the disturbances has among other been studied by Gaster [28], Saric & Nayfeh [85], Gaponov [27] and El-Hady [21]. Gaster used an iterative method to generate an asymptotic series solution in powers of $R^{-1} = \sqrt{\nu/Ux}$. In the other investigations the non-parallel effects were introduced into the stability equations using a multiple-scales method.

The idea of solving parabolic evolution equations for disturbances in the boundary layer was first introduced by Hall [36] for steady Görtler vortices. Itoh [54] derived a parabolic equation for small-amplitude Tollmien-Schlichting waves. Herbert & Bertolotti [51], Herbert [48], Bertolotti [7] and Bertolotti, Herbert & Spalart [10] developed this method further and derived the non-linear parabolized stability equations (PSE). Simen [88] independently developed a similar theory. His contribution was to consistently and in a general way model convectively amplified waves with divergent and/or curved wave-rays and wave-fronts propagating in non-uniform flow. Due to the fact that the disturbance characteristics predicted by such a method are influenced by local and upstream flow conditions, this theory is called *non-local.*

Within the last years the non-local method has been applied to various types of problems such as receptivity, instability and transition analysis, see *e.g.* Crouch & Bertolotti [16], Malik & Li [73], Wang & Herbert [94], Chang & Malik [14], Airiau [1].

Below, we discuss the basics of the non-local method, a detailed discussion on this subject can be found in Bertolotti [7] and Herbert [49, 50].

2.4.2. DERIVATION OF STABILITY EQUATIONS

Assumptions

Here, we restrict ourselves to quasi-three dimensional flows (basic flows which are independent of the z coordinate). We start the derivation of the non-local stability equations by introducing the following assumptions:

1. The first assumption is of WKB type, where the dependent variables are divided into an amplitude function and an oscillating part, *i.e.*

$$\mathbf{q} = \tilde{\mathbf{q}}(x, y)e^{i\Theta}, \tag{2.93}$$

with

$$\Theta = \int_{x_0}^{x} \alpha(x')\mathrm{d}x' + \beta z - \omega t. \tag{2.94}$$

Here, as in the spatial local theory, α is a complex number. Note that in non-local theory, in contrast to the local theory, both the amplitude function, $\tilde{q}$, and the phase function, Θ, are allowed to vary in streamwise direction.

2. As a second assumption, a scale separation $1/R_0$ is introduced between the weak variation in the x-direction and the strong variation in the y-direction analogous to the multiple-scales method. Here, R_0 is the local Reynolds number defined as $\sqrt{Ux_0/\nu}$. Important implications of this assumption are that the x-dependence of scale factors associated with coordinate curvature, h_i, the basic and disturbance flow, and the normal basic flow velocity component v are also assumed to scale with the small parameter $1/R_0$. Thus, we introduce the slow scale (subscripted with "s")

$$x = x_s R_0 \tag{2.95}$$

and the new dependent variables:

$$\bar{\mathbf{q}} = \bar{\mathbf{q}}(x_s, y), \qquad V = V_s(x_s, y)/R_0$$

$$\alpha = \alpha(x_s), \qquad \tilde{\mathbf{q}} = \tilde{\mathbf{q}}(x_s, y). \tag{2.96}$$

With the above scaling introduced into the governing equations, considering only terms which scale with powers of Reynolds number up to $1/R_0$, we obtain the non-local, linear stability equations. They describe the kinematics, nonuniform propagation and amplification of wave-type disturbances with divergent or curved wave rays in a nonuniform basic flow. If (2.93) and the scaling assumption described above are introduced into the linearized Navier-Stokes equations we obtain the following *nearly* parabolic equations for the amplitude functions

$$A\tilde{\mathbf{q}} + B\frac{\partial \tilde{\mathbf{q}}}{\partial y} + C\frac{\partial^2 \tilde{\mathbf{q}}}{\partial y^2} + D\frac{\partial \tilde{\mathbf{q}}}{\partial x} = 0. \tag{2.97}$$

Here, A, B, C and D are functions of α, β, ω, mean flow and metric quantities. The entries of these matrices can be found in Hanifi *et al.* [39]. However, the major part of the ellipticity of the total disturbances $\tilde{\mathbf{q}}$ is taken into account by the α^2 terms appearing in equation (2.97).

Normalization condition

As both the amplitude and phase function depend on the x-direction, we follow [7], [10] and [49] and remove this ambiguity by specifying various forms of 'normalization' conditions. The normalization condition can be based either on the amplitude of one of the disturbance quantities $\tilde{q}_k$ of equation (2.97) at some fixed y, *i.e.*

$$\left(\frac{\partial \tilde{q}_k}{\partial x}\right)_{y=y_m} = 0, \tag{2.98}$$

or on the integral quantity

$$\int_0^\infty \left(\tilde{u}^\dagger \frac{\partial \tilde{u}}{\partial x} + \tilde{v}^\dagger \frac{\partial \tilde{v}}{\partial x} + \tilde{w}^\dagger \frac{\partial \tilde{w}}{\partial x} + \tilde{T}^\dagger \frac{\partial \tilde{T}}{\partial x} + \tilde{\rho}^\dagger \frac{\partial \tilde{\rho}}{\partial x}\right) \mathrm{d}y = 0, \tag{2.99}$$

where the superscript † refers to the complex conjugate. The normalization condition enforce the variation of the amplitude functions to remain small enough to justify the $(1/R_0)$ scaling of $\partial\tilde{q}/\partial x$.

Definition of growth

In the frame of the non-local theory, the physical growth rate, σ, of an arbitrary disturbance quantity ξ can be defined as

$$\sigma = -\alpha_i + Real\left\{\frac{1}{\xi}\frac{\partial \xi}{\partial x}\right\}, \tag{2.100}$$

where the first RHS term is the contribution of the exponential part of the disturbance. The second term is the correction due to the changes of the amplitude function. Usually, ξ is taken to be $\tilde{u}, \tilde{v}, \tilde{w}, \tilde{T}$ or $\tilde{\rho}\bar{u} + \bar{\rho}\tilde{u}$, either at some fixed y or at the location where it reaches its maximum value. Moreover, the disturbance kinetic energy E, here defined as

$$E = \int_0^\infty \bar{\rho}\left(|\tilde{u}|^2 + |\tilde{w}|^2 + |\tilde{v}|^2\right) dy,$$

can be used as a measure for the disturbance growth. Then, equation (2.100) becomes

$$\sigma = -\alpha_i + \frac{\partial}{\partial x}\ln(\sqrt{E}).$$

It should be noticed that the growth rate based on equation (2.100) is a function of the normal coordinate, y.

The corresponding physical streamwise wavenumber is given by

$$\bar{\alpha} = \alpha_r + Imag\left\{\frac{1}{\xi}\frac{\partial \xi}{\partial x}\right\}, \tag{2.101}$$

The direction of wave propagation is given by the physical waveangle

$$\psi = \tan^{-1}\left(\frac{\beta}{\bar{\alpha}}\right). \tag{2.102}$$

The waveangle ψ also describes wave-front distortion or wave-ray curvature as ψ can vary both in the x- and y-directions.

Comparison with Direct Numerical Simulations (DNS)

The most accurate stability results are those from the DNS calculations. A drawback of these methods is the long CPU time needed in such calculations, and the limitation to simple geometries. The comparisons of the DNS and non-local results over a wide range of Mach numbers have shown the great potential of the non-local method for prediction of the stability characteristics as well as transition location.

Pruett & Chang [78] compared the DNS results for flat plate boundary layer with $M_e = 4.5$ with those of PSE. The number of points in the streamwise direction used for DNS was more than twice of that for PSE. The PSE calculations gave almost identical results consuming about 100 seconds of CPU time on a Cray Y-MP compared to some 20 hours needed for the DNS. Hanifi *et al.* [39] compared their non-local results with those from the DNS calculations of Fasel & Konzelmann [24] for an incompressible boundary layer, Thumm [92] for $M_e = 1.6$, Guo, Kleiser & Adams [32] for $M_e = 4.5$, and Eissler & Besteck [20] for $M_e = 4.8$ compressible flat plate boundary layer, see fig. 2.21. In all these calculations the growth rates of the

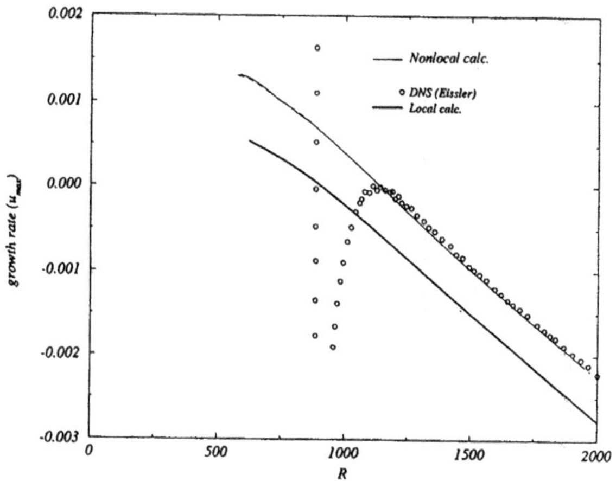

Figure 2.21. Growth rate of highly oblique wave in compressible flat plate boundary layer flow from DNS, local and non-local stability calculations. $M_e = 4.8$, $T_e = 55.41$ K, $\Pr = 0.71$, $\omega/R = 5 \times 10^{-5}$ and $\beta/R = 2 \times 10^{-4}$.

non-local method were in good agreement with those of DNS. Berlin *et al.* [4] calculated the neutral stability curves for two-dimensional disturbances in an incompressible flat plate boundary layer. Their results are given in figure 2.22. As it can be seen there, a close agreement between the results from the DNS and non-local calculations is found. These results are also in good agreement with experimental data of Klingmann *et al.* [57]. The CPU time of the DNS calculations were about hundred times larger than that of the non-local calculations.

The results of non-linear calculations of PSE and DNS for high speed boundary-layer flows were compared by Pruett et al. [79]. They reported a good agreement for all harmonics in regions of the flow which are characterized by weakly or moderately strong non-linear interactions. Only results for the mean-flow distortion from these two methods diverged from each other as the flow approached transition.

2.4.3. MATHEMATICAL CHARACTER OF THE NON-LOCAL STABILITY EQUATIONS

Although no second x-derivatives of $\tilde{\mathbf{q}}$ are present in (2.97), still some ellipticity is left in the stability equations. This is demonstrated by the rapid oscillations of the solutions when the marching step size becomes too small, see figure 2.23. The residual ellipticity is due to the upstream propagation either through the pressure terms or viscous diffusion. The ill-posedness of the non-local stability equations has been studied by *e.g.* Haj-Hariri [35],

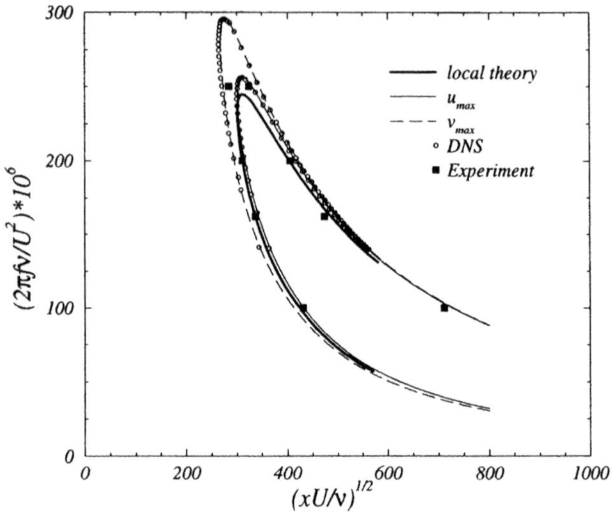

Figure 2.22. Neutral stability curves based on the maximum of $\tilde{u}$ and $\tilde{v}$. The non-local results are compared to the DNS calculations (Berlin *et al.* [4]) and experimental data (Klingmann *et al.* [57]).

Airiau [1], Herbert [49] and Li & Malik [63, 64]. Haj-Hariri concluded that the residual ellipticity can be removed by multiplying the streamwise derivatives of $\tilde{p}$ by

$$\min\left\{1, \frac{\gamma M^2 \bar{\rho} U^2}{(\gamma - 1) + \bar{\rho}\bar{T}}\right\},$$

whereas Li & Malik concluded that this is not sufficient to eliminate the step size restriction. However, it was shown that the dropping of $\partial\tilde{p}/\partial x$ relaxes the step size restriction. They also gave the limit for the smallest stable step size to be $1/|\alpha_r|$. Recently, Andersson, Hennigson & Hanifi [3] suggested a modification of the PSEs which make the equations well-posed and eliminate the step size problem. This is done by approximating the streamwise derivative by a first-order implicit scheme and including a term propotional to the leading truncation error of the scheme. In discrete form, the equation (2.97) is written as

$$\frac{\tilde{\mathbf{q}}^{n+1} - \tilde{\mathbf{q}}^n}{\Delta x} + \tau = \bar{\mathcal{L}}\tilde{\mathbf{q}}^{n+1}, \tag{2.103}$$

where

$$\tau = \frac{\Delta x}{2}\tilde{\mathbf{q}}_{xx} + \cdots$$

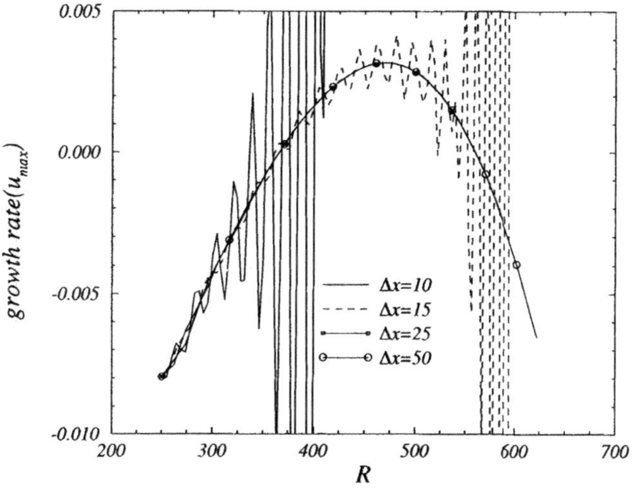

Figure 2.23. Effect of step size on the growth rates of 2D waves in compressible flat plate boundary layer flow, $M = 0.1$, $T_0 = 300K$, $F = 1.4 \times 10^{-4}$, $\Pr = 0.7$ and constant c_p (from Hanifi *et al.* [39]).

and

$$\bar{\mathcal{L}} = -D^{-1}\left(A + B\frac{\partial}{\partial y} + C\frac{\partial^2}{\partial y^2}\right).$$

Here, the superscript ^{n} refers to n-th point in streamwise direction. Since we have assumed that the streamwise variation of the amplitude function is of order $\mathcal{O}(R^{-1})$, the terms signified by $\cdots$ in the expression above will be of order $\mathcal{O}(R^{-3}\Delta x^2)$. Dropping all but the leading terms of the truncation error, we get

$$\tau = \frac{\Delta x}{2}\tilde{\mathbf{q}}_{xx} = \frac{\Delta x}{2}(\bar{\mathcal{L}}_x\tilde{\mathbf{q}} + \bar{\mathcal{L}}\tilde{\mathbf{q}}_x) \approx \frac{\Delta x}{2}\bar{\mathcal{L}}\tilde{\mathbf{q}}_x, \tag{2.104}$$

Here, the term $\bar{\mathcal{L}}_x\tilde{\mathbf{q}}$ has been neglected for simplicity. However, equality holds if $\bar{\mathcal{L}}$ is independent of x. Again, the assumption of small x-derivative terms implies that the added truncation error is of the order $\mathcal{O}(R^{-2})$. Since terms of this order were neglected in the original approximation, the addition of τ does not introduce any extra error at this order of approximation, and we can introduce the new set of equations

$$(I - \Delta x\bar{\mathcal{L}} - s\bar{\mathcal{L}})\tilde{\mathbf{q}}^{n+1} = (I - s\bar{\mathcal{L}})\tilde{\mathbf{q}}^n, \tag{2.105}$$

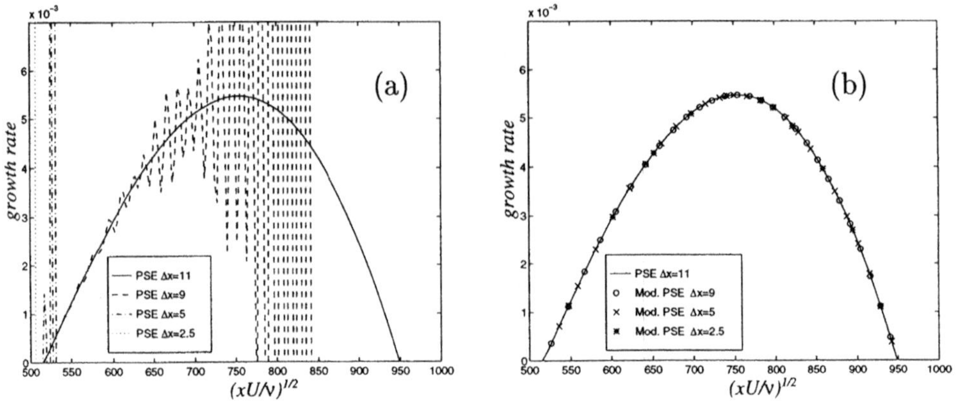

Figure 2.24. Growth rate *vs* streamwise position, for incompressible boundary-layer flow, obtained from the PSE method without (a) with (b) stabilizing terms. From Andersson *et al.* [3].

where s is a positive real number. Based on the discussion given above, the differences between the solution of equations (2.97) and discrete form of (2.105) are of order $\mathcal{O}(R^{-2})$. Note that, although s take the place of Δx in the added truncation error term, this term is small, even if $s = \mathcal{O}(1)$, since $\bar{\mathcal{L}}\tilde{\mathbf{q}}_x$ was shown to be of order $\mathcal{O}(R^{-2})$. Anderson *et al.* [3] found that the critical step size for a stable numerical scheme is

$$\Delta x > \frac{1}{|\alpha_r|} - 2s. \qquad (2.106)$$

Some numerical examples of this method are given in figure 2.24.

2.5. Applications

In this section, we give some examples of the stability analysis. The aim is to direct your attention toward some of important issues.

Example 1: RADUGA D2 rocket

Here, we will demonstrate the effects of mean flow resolution on the stability characteristics of the boundary layer. A detailed report on the material given here can be found in Hanifi & Wallin [42].

The geometry is a simplified model of RADUGA D2 vehicle consisting of a blunt conical forebody and a cylindrical part with total length of 8.74m. The mean flow was calculated using a Navier-Stokes solver on a grid of 305 × 161 points. About 60 to 70 points were placed inside the boundary layer to resolve the normal derivative of the mean profiles. The mean flow

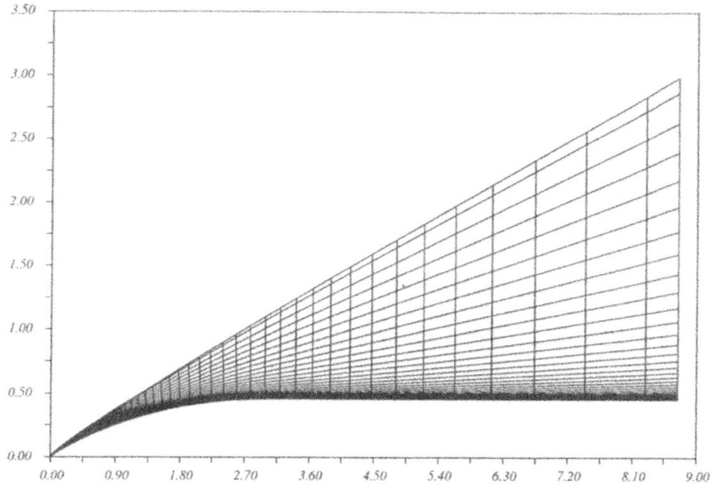

Figure 2.25. Computational grid over the RADUGA D2 geometry. Every fourth grid line is plotted, all measures in meter. From Hanifi & Wallin [42].

calculations has been repeated for a medium (153×81) and a coarse (77×41) resolution. The velocity and G_i profiles from these calculations, obtained at $x = 3.05\ m$, are shown in figure 2.26. Here, G_i is defined as

$$G_i = (\bar{\rho}U')' - m_{23}\bar{\rho}U', \tag{2.107}$$

with

$$m_{23} = \frac{1}{h_2}\frac{\partial h_2}{\partial y}.$$

For compressible flows on axisymmetric bodies, the quantity G_i plays the same role as the generalized inflection point for flat plate boundary layer (see Malik & Spall [74]). The generalized inflection point, $(\bar{\rho}U')' = 0$, is a necessary condition for inviscid instability. From the behavior of these solutions, it can be concluded that the results of the finest grid are fairly well converged. It should be noticed that even if the velocity profile for the medium size grid deviates by a few percent from that of the finest grid, the G_i profile, which is the key quantity for stability calculations (see *e.g.* Mack [70]), differs by 50%.

The N-factor for disturbances with a frequency of 30 and 40 kHz, calculated using local theory, are given in figure 2.27. The profiles from the coarse mesh found to be stable to these disturbances. For the medium size

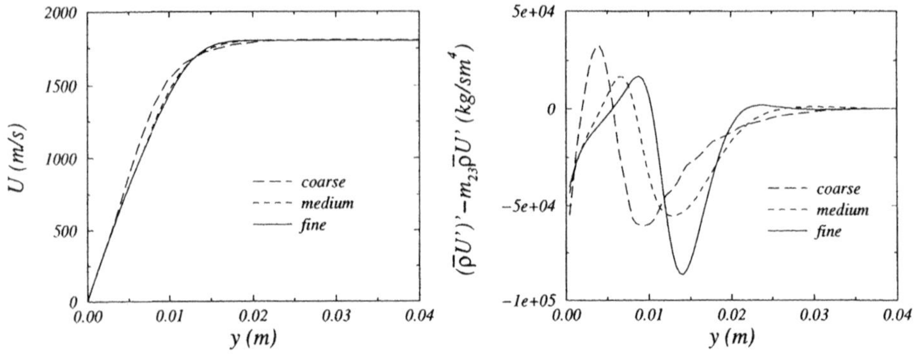

Figure 2.26. Effects of grid resolution on the boundary layer profile. From Hanifi & Wallin [42].

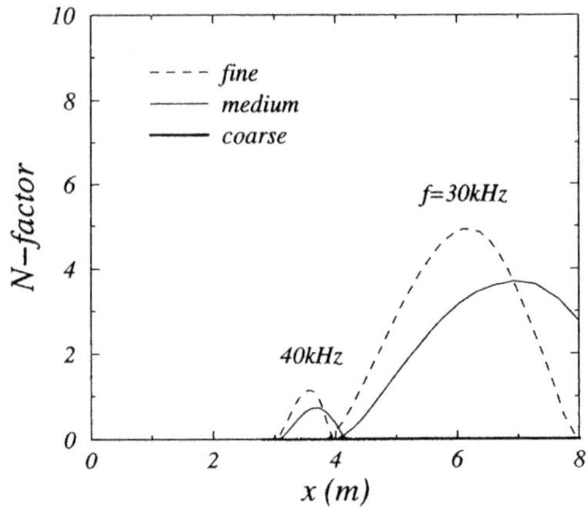

Figure 2.27. Disturbance amplifications for the boundary-layer profiles calculated using different grid resolutions. From Hanifi & Wallin [42].

mesh, the maximum amplifications were about 30% lower than those for the fine mesh. The peaks of the amplification were also shifted upstream when the fine mesh was used.

Example 2: Infinite swept wing

Stolte *et al.* [90] studied the convective cross-flow and Tollmin-Schlichting instability of compressible, three-dimensional flow past the laminar glove of the ATTAS/VFW 614 wing. The calculations were based on the infinite swept wing approximation and performed within the framework of the nonlocal theory. Stolte *et al.* studied the effects of different degree of approximation about the geometry and the mean flow by including or excluding the curvature or/and terms related to the nonparallel boundary layer.

In figure 2.28a results for a typical stationary disturbance with frequency $f = 0HZ$ and transverse wavenumber $\beta^* = 4839m^{-1}$ are given. There, the N-factors calculated by different stability theory are compared. As the curvature effects are included in the theory, the disturbance become totally stable. The nonparallel terms have a strong destabilizing effect on the cross-flow disturbances.

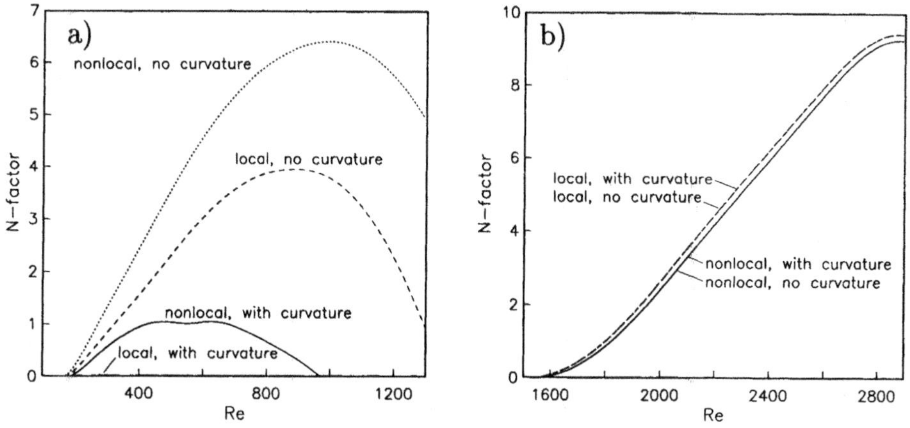

Figure 2.28. N-factors of approximately maximum amplified, a) cross-flow disturbances ($f = 0Hz$, $\beta* = 4839m^{-1}$), b) 'Tollmien-Schlichting' disturbances ($f = 2500Hz$, $\beta* = 53.1m^{-1}$) in transonic ATTAS laminar glove flow. From Stolte *et al.* [90].

Stolte *et al.* reported a totally different behavior of a representative 'Tollmien-Schlichting' disturbance. As shown in figure 2.28b, the amplification of this disturbance is only slightly affected by including curvature or nonlocal terms.

Example 3: Sharp cone at angles of attack

The different effects, like curvature and growth of the boundary layer, are treated differently in theories discussed in the previous sections. Here, we will demonstrate the effects of applications of these theories on prediction of the transition location. We choose a rather complicated generic case namely a sharp cone at angle of attack. The complexity is due to the flow three-dimensionality. Note that the nonlocal stability theory discussed in the previous sections does not account for flow variations in the spanwise direction. A modification to this theory was introduced by Herbert [50] which includes effects of small spanwise variations of the mean flow. This

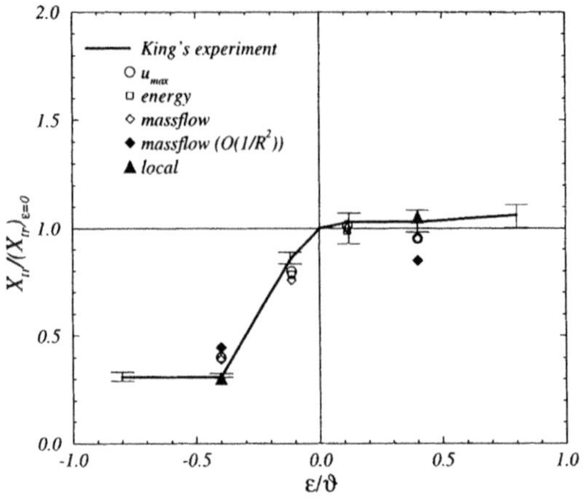

Figure 2.29. Comparison of the predicted position of transition onset (Hanifi [37]) with the data from King's experiments [56] (negative ε/ϑ: leeward meridian).

results to a set of partial differential equations which are parabolic in span- and streamwise directions. Here we will use the approximate method.

The results given here are from Hanifi [37]. The geometry is a sharp cone with a half-angel of $\vartheta = 5$ degrees which is placed in a supersonic flow at Mach 3.5. The mean flow for around the cone at a small angle of attack, ϵ, is calculated using a perturbation method.

Hanifi [37] showed that the spanwise derivatives of the amplitude functions can be removed from the set of stability equations by assuming that these derivatives scales as $\mathcal{O}(1/R^2)$. This assumption is very restrictive and may not be valid for slender cones at relative large incidence. For the symmetric disturbances at the leeward and windward rays, a less restrict scaling assumption, $\mathcal{O}(1/R)$, gives the same type of the equations as (2.97). However, extra terms related to the mean flow variation in the spanwise direction will appear in the equations.

In figure 2.29, the relative movement of the transition point on the yawed cone is plotted as a function of ε/ϑ. The location of transition is predicted using the e^N method (see lecture notes by Arnal). At the windward meridian, the theoretical results for $\varepsilon = 0.57^o$ predict rather small changes in the transition Reynolds number, R_{tr}, compared to that for $\varepsilon = 0^o$ (about 1% only). King's data also shows small effects of yaw angle in this case. The nonlocal results for $\varepsilon = 2^o$ show an upstream shift of the transition onset relative to that for zero incident case. This contradicts the general finding in the experiments and previous theoretical investigations (see Hanifi & Dahlkild [38]). The experimental data for this yaw angle show a relative small downstream displacement of the transition front (the relative movement of the transition onset is less than 5%).

At the leeward meridian, changes in the R_{tr} is more significant. The experimentally obtained values of $X_{tr}/(X_{tr})_{\varepsilon=0}$ are about 85% and 30% for $\varepsilon/\vartheta = 0.12$ and 0.4, respectively. Here, the theoretically predicted X_{tr} agrees fairly well with the experimental findings.

References

1. Airiau, C. 1994. *Linear and Weakly Nonlinear Stability of an Incompressible Laminar Boundary Layer by the Parabolized Stability Equations (PSE).* PhD thesis, L'École Nationale Supérieure de l'Aéronautique et de l'Espace.
2. Al-Maaitah, A. A., Nayfeh, A. H. and Masad, J. A. 1991. Effect of suction on the stability of supersonic boundary layers. part i second-mode waves. *J. Fluids Engng.* **113**, 591–597.
3. Andersson, P., Henningson, D. S. and Hanifi, A. 1998. On a stabilization procedure for the parabolic stability equations. *J. Eng. Math.* **33**, 311–332.
4. Berlin, S., Hanifi, A. and Henningson, D. S. The neutral stability curve for non-parallel boundary layer flow. in Oblique Waves in Boundary layer Transition, Ph.D. Thesis, Royal Institute of Technology, Stockholm, 1998.
5. Berlin, S., Wiegel, M. and Henningson, D. S. 1998. Numerical and experimental investigation of oblique boundary layer transition. Submitted to jfm.
6. Bertolotti, F. 1989. The effects of approximations to the thermodynamic properties on the stability of compressible boundary layer flow. In *Instability and Transition* (Hussaini, M. Y. and Voigt, R. G., editors), volume II. Springer-Verlag.
7. Bertolotti, F. P. 1991. *Linear and Nonlinear Stability of Boundary Layers with Streamwise Varying Properties.* PhD thesis, The Ohio State University, Department of Mechanical Engineering, Columbus, Ohio.
8. Bertolotti, F. P. 1997. The influence of rotational and vibrational energy relaxation on the instability of boundary layers in supersonic flows. Technical Report DLR-FB-97-18, Deutsches Zentrum für Luft und Raumfahrt, D-37073 Göttingen, Germany.
9. Bertolotti, F. P. 1998. The influence of rotational and vibrational energy relaxation on boundary-layer stability. *J. Fluid Mech.* **372**, 93–118.
10. Bertolotti, F. P., Herbert, T. and Spalart, S. P. 1992. Lnear and nonlinear stability of the blasius boundary layer. *J. Fluid Mech.* **242**, 441–474.
11. Brown, W. B. 1961. A stability criterion for three-dimensional laminar boundary layers. In *Boundary Layer and Flow Control* (Lachmann, G., editor), volume 2, pp. 913–329. Pergamon Press, London.
12. Brown, W. B. 1962. Exact numerical solution of the complete linearized equations for the stability of compressible boundary layers. *Norair Report.* **NOR-62-15**.
13. Canuto, C., Hussaini, M. Y., Quarteroni, A. and Zang, T. A. 1988. *Spectral Methods in Fluids Dynamics.* Springer.
14. Chang, C.-L. and Malik, M. R. 1994. Oblique-mode breakdown and secondary instability in supersonic boundary layers. *J. Fluid Mech.* **273**, 323–360.
15. Chang, C. L., Vinh, H. and Malik, M. R., 1997. Hypersonic boundary-layer stability with chemical reactions using PSE. AIAA Paper 97-2012.
16. Crouch, J. D. and Bertolotti, F. P. 1992. Nonlocalized receptivity of boundary layers to three-dimensional disturbances. *AIAA Paper.* **92-0740**.
17. DiPrima, R. C. and Habetler, G. J. 1969. A completeness theorem for non-selfadjoint eigenvalue problems in hydrodynamic stability. *Arch. Rat. Mech. Anal.* **32**, 218–227.
18. Drazin, P. G. and Reid, W. H. 1981. *Hydrodynamic Stability.* Cambridge University

Press.
19. Dunn, D. W. and Lin, C. C. 1955. On the stability of the laminar boundary layer in a compressible fluid. *J. Aero. Sci.* **22**, 455–477.
20. Eissler, W. and Bestek, H. 1993. Spatial numerical simulation of nonlinear transition phenomena in supersonic boundary layers. In *Symposium on Transitional and Turbulent Compressible Flows, 1993 ASME Fluids Engineering Conference, Washington, D.C., June 21-24.*
21. El-Hady, N. M. 1991. Nonparallel instability of supersonic and hypersonic boundary layers. *Phys. Fluids A.* **3**, 2164–2178.
22. Ellingsen, T. and Palm, E. 1975. Stability of linear flow. *Phys. Fluids.* **18**, 487–488.
23. Elofsson, P. A. and Alfredsson, P. H. 1998. An experimental study of oblique transition in plane poiseuille flow. *J. Fluid Mech.* **358**, 177–202.
24. Fasel, H. and Konzelmann, U. 1990. Non-parallel stability of a flat-plate boundary layer using the complete navier-stokes equations. *J. Fluid Mech.* **221**, 311–347.
25. Fasel, H., Thumm, A. and Bestek, H. 1993. Direct numerical simulation of transition in supersonic boundary layers oblique breakdown. *FED.* **151**, 77–92.
26. Fjørtoft, R. 1950. Application of integral theorems in deriving criteria for instability for laminar flows and for the baroclinic circular vortex. *Geofys. Publ., Oslo.* **17** (6), 1–52.
27. Gaponov, S. A. 1981. The influence of flow non-parallelism on disturbance development in the supersonic boundary layers. In *Proceedings of the 8-th Canadian Congress of Applied Mechanics.*
28. Gaster, M. 1974. On the effects of boundary-layer growth on flow stability. *J. Fluid Mech.* **66**, 465–480.
29. Gathmann, R. J., Si-Ameur, M. and Mathey, F. 1993. Numerical simulations of three-dimensional natural transition in the compressible confined shear layer. *Phys. Fluids A.* **5**, 2946–2968.
30. Gill, A. E. 1965. Instabilities of "top-hat" jets and wakes in compressible fluids. *Phys. Fluids.* **8**, 1428–1430.
31. Grosch, C. E. and Salwen, H. 1978. The continuous spectrum of the Orr-Sommerfeld equation. Part 1. The spectrum and the eigenfunctions. *J. Fluid Mech.* **87**, 33–54.
32. Guo, Y., Kleiser, L. and Adams, N. A. 1994. A comparison study of an improved temporal DNS and spatial DNS of compressible boundary layer transition. *AIAA Paper.* **94-2371**.
33. Gustavsson, L. H. 1986. Excitation of direct resonances in plane Poiseuille flow. *Stud. Appl. Math.* **75**, 227–248.
34. Gustavsson, L. H. 1991. Energy growth of three-dimensional disturbances in plane Poiseuille flow. *J. Fluid Mech.* **224**, 241–260.
35. Haj-Hariri, H. 1994. Characteristics analysis of the parabolized stability equations. *Stu. in App. Math.* **92**, 41–53.
36. Hall, P. 1983. The linear development of görtler vortices in growing boundary layers. *J. Fluid Mech.* **130**, 41–58.
37. Hanifi, A. 1995. *Local and Non-local Stability Analysis and Transition Prediction of Compressible Boundary Layer Flows.* PhD thesis, Royal Institute of Thechnology, KTH.
38. Hanifi, A. and Dahlkild, A. A., 1993. Some stability characteristics of the boundary layer on a yawed cone. AIAA Paper 93-3048.
39. Hanifi, A., Henningson, D., Hein, S., Bertolotti, F. P. and Simen, M. 1994. Linear nonlocal instability analysis - the linear nolot code. Technical Report FFA TN 1994-54, FFA. See also Hein *et al.* [43].
40. Hanifi, A. and Henningson, D. S. 1998. The compressible inviscid algebraic instability for streamwise independent disturbances. *Phys. Fluids.* **10**, 1784–1786.
41. Hanifi, A., Schmid, P. J. and Henningson, D. S. 1995. Transient growth in compressible boundary layer flow. *Phys. Fluids.* **8**, 826–837.
42. Hanifi, A. and Wallin, S., 1996. Stability analysis of the boundary layer on

RADUGA D2. FFA TN 1996-28, Technical Report from the Aeronautical Research Institute of Sweden (FFA), Bromma.

43. Hein, S., Bertolotti, F. P., Simen, M., Hanifi, A. and Henningson, D. 1994. Linear nonlocal instability analysis - the linear nolot code-. Technical Report DLR-IB 223-94 A 43, Deutsche Forschunganstalt für Luft- und Raumfahrt.
44. Henningson, D. S. 1991. An eigenfunction expansion of localized disturbances. In *Advances in turbulence 3* (Johansson, A. V. and Alfredsson, P. H., editors), pp. 162–169, Stockholm. Springer-Verlag.
45. Henningson, D. S. and Alfredsson, P. H. 1996. Stability and transition of boundary layer flows. In *Turbulence and transition modelling* (Hallbčk, M., Henningson, D. S., Johansson, A. V. and Alfredsson, P. H., editors), pp. 13–80. Kluwer Academic Publishers.
46. Henningson, D. S., Gustavsson, L. H. and Breuer, K. S. 1994. Localized disturbances in parallel shear flows. *Appl. Sci. Res.* **53**, 51–97.
47. Henningson, D. S., Lundbladh, A. and Johansson, A. V. 1993. A mechanism for bypass transition from localized disturbances in wall bounded shear flows. *J. Fluid Mech.* **250**, 169–207.
48. Herbert, T., 1991. Boundary-layer transition - analysis and prediction revisited. AIAA Paper 91-0737.
49. Herbert, T., 1994. Parabolized stability equations. AGARD Report 793.
50. Herbert, T. 1997. Parabolized stability equations. *Ann. Rev. Fluid Mech.* **29**, 245–283.
51. Herbert, T. and Bertolotti, F. P. 1987. Stability analysis of nonparallel boundary layers. *Bull. Am. Phys. Soc.* **32**, 2079.
52. Hudson, M. L., Chokani, N. and Candler, G. V. 1997. Linear stability of hypersonic flow in thermochemical non-equilibrium. *AIAA J.* **35**, 958–964.
53. Hultgren, L. S. and Gustavsson, L. H. 1981. Algebraic growth of disturbances in a laminar boundary layer. *Phys. Fluids.* **24**, 1000–1004.
54. Itoh, N. 1986. The origin and subsequent development in space of tollmien-schlichting waves in a boundary layer. *Fluid Dyn. Res.* **1**, 119–130.
55. Joseph, D. D. 1976. *Stability of Fluid Motions I.* Springer.
56. King, R., 1991. Mach 3.5 boundary-layer transition on a cone at angle of attack. AIAA Paper 91-1804.
57. Klingmann, B. G. B., Boiko, A., Westin, K., Kozlov, V. and Alfredsson, P. 1993. Experiments on the stability of Tollmien-Schlichting waves. *Eur. J. Mech., B/Fluids.* **12**, 493–514.
58. Kreiss, G., Lundbladh, A. and Henningson, D. S. 1994. Bounds for threshold amplitudes in subcritical shear flows. *J. Fluid Mech.* **270**, 175–198.
59. Landahl, M. T. 1975. Wave breakdown and turbulence. *SIAM J. Appl. Math.* **28**, 735–756.
60. Landahl, M. T. 1980. A note on an algebraic instability of inviscid parallel shear flows. *J. Fluid Mech.* **98**, 243–251.
61. Lees, L. and Lin, C. 1946. Investigation of the stability of the laminar boundary layer in a compressible fluid. Technical Report NACA Tech. Note 1115.
62. Lees, L. and Reshotko, E. 1962. Stability of the compressible laminar boundary layer. *J. Fluid Mech.* **12**, 555–590.
63. Li, F. and Malik, M. R. 1995. Mathematical nature of parabolized stability equations. In *Laminar-Turbulent Transition* (Kobayashi, R., editor), pp. 205–212. Springer-Verlag.
64. Li, F. and Malik, M. R. 1996. On the nature of PSE approximation. *Theor. Comp. Fluid Dyn.* **8**, 253–273.
65. Mack, L. M. 1963. The inviscid stability of the compressible laminar boundary layer. In *Space Programs Summary*, number 37-23, p. 297. Jet Propulsion Laboratory, Pasanda, CA.
66. Mack, L. M. 1964. The inviscid stability of the compressible laminar boundary layer

part ii. In *Space Programs Summary*, volume IV, p. 165. Jet Propulsion Laboratory, Pasanda, CA.

67. Mack, L. M. 1965. Stability of the laminar compressible boundary layer according to a direct numerical solution. *AGARDograph.* **I** (97), 329–362.
68. Mack, L. M. 1975. Linear stability theory and the problem of supersonic boundary-layer transition. *AIAA J.* **13**, 278–289.
69. Mack, L. M. 1976. A numerical study of the temporal eigenvalue spectrum of the Blasius boundary layer. *J. Fluid Mech.* **73**, 497–520.
70. Mack, L. M. 1984. Boundary-layer stability theory. In *AGARD Report*, number 709, pp. 3–1–3–8.
71. Malik, M. R. 1989. Prediction and control of transition in supersonic and hypersonic boundary layers. *AIAA J.* **27**, 1487–1493.
72. Malik, M. R. and Anderson, E. C. 1991. Real gas effects on hypersonic boundary-layer stability. *Phys. Fluids A.* **3**, 803–821.
73. Malik, M. R. and Li, F., 1993. Transition studies for swept wing flows using PSE. AIAA Paper 93-0077.
74. Malik, M. R. and Spall, R. E. 1991. On the stability of compressible flow past axisymmetric bodies. *J. Fluid Mech.* **228**, 443–463.
75. Masad, J. A., Nayfeh, A. H. and Al-Maaitah, A. A. 1991. Effect of suction on the stability of supersonic boundary layers. part ii first-mode waves. *J. Fluids Engng.* **113**, 598–601.
76. Masad, J. A., Nayfeh, A. H. and Al-Maaitah, A. A. 1992. Effect of heat transfer on the stability of compressible boundary layers. *Computers Fluids.* **21**, 43–61.
77. Orszag, S. A. 1971. Accurate solution of the Orr-Sommerfeld stability equation. *J. Fluid Mech.* **50**, 689–703.
78. Pruett, C. D. and Chang, C.-L. 1993. A comparison of PSE and DNS for high-speed boundary-layer flows. *FED.* **151**, 57–67.
79. Pruett, C. D. and Chang, C.-L. 1995. Spatial direct numerical simulation of high-speed boundary-layer flows, part i algorithmic considerations and validation. *Theoret. Comput. Fluid Dynamics.* **7**, 49–76.
80. Rayleigh, J. W. S. 1880. On the stability, or instability, of certain fluid motions. *Proc. Math. Soc. Lond.* **11**, 57–70.
81. Reddy, S. C. and Henningson, D. S. 1993. Energy growth in viscous channel flows. *J. Fluid Mech.* **252**, 209–238.
82. Reddy, S. C., Schmid, P. J., Bagget, P. and Henningson, D. S. 1998. On stability of streamwise streaks and transition thresholds in plane channel flows. *J. Fluid Mech.* **365**, 269–303.
83. Reddy, S. C., Schmid, P. J. and Henningson, D. S. 1993. Pseudospectra of the Orr-Sommerfeld operator. *SIAM J. Appl. Math.* **53**, 15–47.
84. Sandham, N. D., Adams, N. A. and Kleiser, L. 1995. Direct simulation of breakdown to turbulence following oblique instability waves in a supersonic boundary layer. *Appl. Sci. Res.* **54**, 223–234.
85. Saric, W. S. and Nayfeh, A. H. 1975. Non-parallel stability of boundary-layer flows. *Phys. Fluids.* **18**, 945–950.
86. Schensted, I. V. 1961. *Contribution to the theory of hydrodynamic stability.* PhD thesis, University of Michigan, Ann Arbor.
87. Schmid, P. J. and Henningson, D. S. 1992. A new mechanism for rapid transition involving a pair of oblique waves. *Phys. Fluids A.* **4**, 1986–1989.
88. Simen, M. 1992. Local and nonlocal stability theory of spatially varying flows. In *Instability, Transition and Turbulence* (Hussaini, M., Kumar, A. and Streett, C., editors), pp. 181–201. Springer-Verlag.
89. Squire, H. B. 1933. On the stability for three-dimensional disturbances of viscous fluid flow between parallel walls. *Proc. Roy. Soc. Lond. Ser. A.* **142**, 621–8.
90. Stolte, A., Bertolotti, F. P., Hein, S. and Simen, S., 1995. Nonlocal instability analysis of transonic flow past the ATTAS laminar glove. DLR-IB 223-95 A07.

91. Stuckert, G. and Reed, H. 1994. Linear disturbances in hypersonic, chemically reacting shock layers. *AIAA J.* **32**, 1384–1393.
92. Thumm, A. 1991. *Numerische Untersuchungen zum laminar-turbulenten Strömungsumschlag in transsonischen Grenzschichtströmungen.* Ph.D. dissertation, Stuttgart University.
93. Trefethen, L. N., Trefethen, A. E., Reddy, S. C. and Driscoll, T. A. 1993. Hydrodynamic stability without eigenvalues. *Science.* **261**, 578–584.
94. Wang, M. and Herbert, T., 1994. PSE analysis of receptivity and stability in swept wing flows. AIAA Paper 94-0180.
95. Zurigat, Z. H., Nayfeh, A. H. and A., M. J., 1990. Effect of pressure gradient on the stability of compressible boundary layers. AIAA Paper 90-1451.

Chapter 3

TRANSITION PREDICTION IN INDUSTRIAL APPLICATIONS

D. ARNAL
ONERA CERT DMAE
2 av. Ed. Belin, BP 4025, 31055, Toulouse Cedex, France

3.1. Introduction

Since the classical experiments performed by Osborne Reynolds (1883), the instability of laminar flow and the transition to turbulence have maintained a constant interest in fluid mechanics problems. This interest results from the fact that transition controls important aerodynamic quantities such as drag or heat transfer. For example, the heating rates generated by a turbulent boundary layer may be several times higher than those for a laminar boundary layer, so that the prediction of transition is of great importance for hypersonic reentry spacecraft, because the thickness of the thermal protection is strongly dependent upon the altitude where transition occurs. For high-subsonic speed, commercial transport aircraft, the achievement of laminar flow can significantly reduce the drag on the wings and hence the specific consumption of the aircraft. The potential benefits are important, because transition separates the laminar flow region with low drag from the turbulent region where skin friction dramatically increases, see figure 3.1. A good knowledge of the transition mechanisms is also required to reduce hydrodynamically generated noise beneath transitional flows.

These examples explain why for many years a large amount of theoretical, numerical and experimental work has been devoted to the laminar-turbulent transition problems. It is now established that transition is the result of a sequence of complex phenomena which depend on many parameters, such as flow quality, vibrations, surface polishing, etc. Therefore a rigorous modelling of the transition process appears to be a very difficult

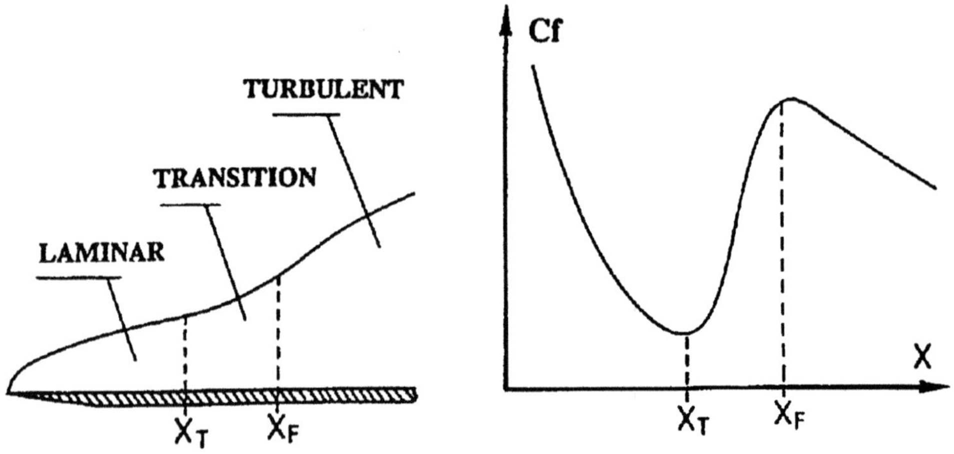

Figure 3.1. Typical evolution of the skin friction coefficient in laminar, transitional and turbulent flows

task. A pessimistic conclusion could be that a "good" prediction of transition is impossible. In spite of this negative situation, transition predictions must be made for many industrial problems.

The objective of this paper is to give an overview of the practical tools which could be useful for a designer. More precisely, the paper is mainly devoted to the analysis of the advantages and shortcomings of the popular "e^N method", that is currently used by people who are assigned the job of making transition prediction.

Paragraph 3.2 gives a short description of the different mechanisms leading to turbulence. When transition results from the amplification of unstable waves (Tollmien-Schlichting waves), the linear stability theory can be used to determine the characteristics of these disturbances and also to estimate the transition location. This theory (as well as some other theoretical elements dealing with receptivity and nonlinear phenomena) and the problem of transition prediction by the e^N method are presented in paragraphs 3.3 and 3.4, respectively. Paragraphs 3.5 and 3.6 present applications of the practical prediction tools for two series of industrial problems, namely transonic aircraft (paragraph 3.5) and high speed vehicles (paragraph 3.6). The important problem of leading edge contamination is also addressed in the first case.

3.2. Qualitative description of some transition mechanisms

For the sake of simplicity, this paragraph will be restricted to the case of low speed flows. The transition phenomena described below are qualitatively similar for high speed flows.

There are several routes to turbulence, which can be summarized as follows. Let us consider a laminar flow developing along a given body. It is strongly affected by the various types of forced disturbances generated by the model itself (roughness, vibrations...) or existing in the free stream. The first stage of the transition process is the boundary layer *receptivity* (Morkovin [62]). Receptivity describes the means by which the forced disturbances enter the laminar boundary layer, as well as their signature in the disturbed flow. This signature constitutes the initial conditions for the development of more or less complex mechanisms which ultimately lead to turbulence.

After the forced disturbances have been internalized by the boundary layer, two kinds of transition processes can be distinguished :

- If the amplitude of the forced disturbances is small, one can observe the appearance of more or less regular oscillations which start to develop downstream of a certain critical point. These waves constitute the eigenmodes of the laminar boundary layer and the first stage of their evolution can be described by a linear theory. They are amplified up to the point where transition occurs (breakdown). This process is called *natural transition.* It will be described in more detail in paragraph 3.2.1.
- If the amplitude of the forced disturbances is not weak , nonlinear phenomena may immediatly be observed and transition can occur rapidly. Two typical examples of this type of transition will be described in paragraph 3.2.2.

3.2.1. "NATURAL" TRANSITION

Let us consider first the simple case of *two-dimensional* flows. As stated before, the instability leading to transition starts with the development of wave-like disturbances, the existence of which was first demonstrated by the now classic experiments of Schubauer and Skramstad (1948 [82]).

Figure 3.2 shows some records delivered by a hot wire placed at six streamwise positions on a flat plate, at a constant distance above the wall. When the hot wire is moved downstream, regular oscillations become more and more visible, with increasing amplitude as the distance increases. The amplitude of these so-called Tollmien-Schlichting (TS) waves exhibit at first an exponential growth, which can be computed by the linear stability theory, see paragraph 3.3.1. Further downstream, the disturbances reach a finite amplitude, so that their development begins to deviate from that predicted by the linear theory. Low speed experiments have shown that the initially two-dimensional TS waves are distorted into a series of "peaks"

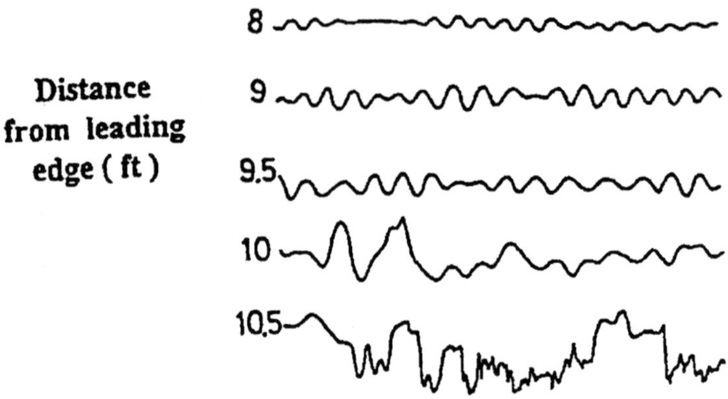

Figure 3.2. Records showing laminar boundary layer oscillations. Distance from surface : 0.023 in. $U_e = 53$ ft/s. From Schubauer and Skramstad [82].

and "valleys". Further downstream, three-dimensional and nonlinear effects become more and more important. The nonlinear development of the disturbances terminates with the "breakdown" phenomenon : experiments and Direct Numerical Simulations (DNS) indicate that the peak-valley structures are stretched and form horseshoe vortices which break down into smaller vortices, which again break down into smaller vortices. The fluctuations finally take a random character and form a turbulent "spot". The streamwise location where the first spots appear can be defined as the onset of transition. At this point, the shape factor of the mean velocity profile begins to decrease, while the skin friction (and also the wall heat transfer for high speed flows) starts to exceed its laminar value. The last stage is the development of the turbulent spots up to the fully turbulent regime.

In fact, the distance between the end of the linear region and the breakdown to turbulence is rather short : for flat plate conditions, the streamwise extent of the linear amplification covers about 75 to 85 percent of the distance between the leading edge and the beginning of transition. This explains why most of the practical transition prediction methods are based on linear stability only.

When the mean flow is *three-dimensional*, for instance on a swept wing, the basic phenomena are qualitatively the same in the linear regime. The major difference is that the unstable waves now develop in a very wide range of propagation directions. The linear stability theory shows that the most unstable propagation direction strongly depends if the flow is accelerated or decelerated. A peculiarity of three-dimensional flow instability is that zero frequency disturbances become highly unstable as soon as the crossflow velocity component is large enough. In the experiments, these stationary disturbances can be visualized as more or less regularly spaced streaks practically aligned in the streamwise direction.

A large amount of experimental work has been devoted during the last years to the analysis of nonlinear mechanisms occurring in three-dimensional flows, see for instance [17], [18], [47], [50], [63]. A detailed discussion of these problems is out of the scope of the present paper. However, it can be noticed that the extent of the region where nonlinearities play a dominant role is much larger than for two-dimensional flows- at least when transition is triggered by a pure crossflow instability.

3.2.2. TRANSITION CAUSED BY LARGE AMPLITUDE DISTURBANCES

When the amplitude of the forced disturbances is too large, periodic waves of the TS type no longer appear. In this paragraph, attention is focused on two particular types of transition mechanisms which need to be taken into account for practical problems, namely leading edge contamination and boundary layer tripping by roughness elements.

Leading edge contamination

Leading edge contamination is a particular phenomenon which can occur along the attachment line of a swept wing. Roughly speaking, the attachment line is a particular streamline which divides the flow into one branch following the upper surface and another branch following the lower surface, see figure 3.3.

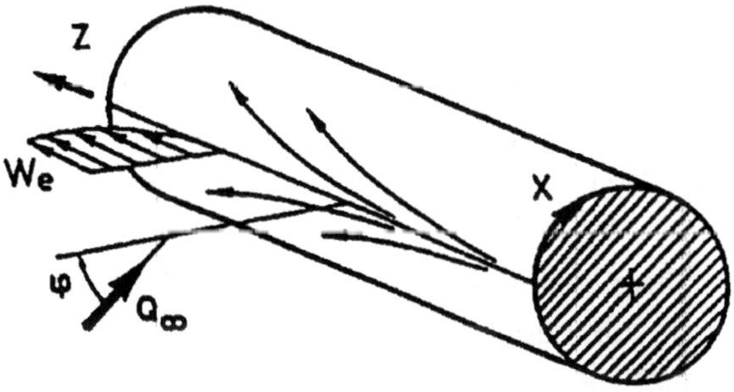

Figure 3.3. Attachment line of a swept cylinder

When a swept wing is in contact with a solid wall (fuselage, wind tunnel wall...), it has been observed that the large turbulent structures coming from the wall at which the wing is fixed may develop along the attachment line : this is the so-called *leading edge contamination.* Many wind tunnel or flight experiments have demonstrated that the main parameter is a leading edge Reynolds number usually denoted as $\bar{R}$. $\bar{R}$ increases when the leading edge radius and/or the unit Reynolds number and/or the sweep angle φ increases. Experimental results (Pfenninger [65], Poll [66] among others) indicate that leading edge contamination occurs for $\bar{R} > 250$; in this case, the turbulent structures coming from the fuselage or from the wind tunnel wall become self-sustaining, develop in the spanwise direction and make the leading edge (and possibly the whole wing) fully turbulent. These turbulent structures are damped and disappear for $\bar{R} < 250$. It is interesting to notice that this very simple criterion was validated by DNS performed by Spalart [86]. For two-dimensional wings, $\bar{R} = 0$ (because $\varphi = 0°$), so that leading edge contamination is unlikely to occur.

Boundary layer tripping

Although modern manufacturing techniques can provide smooth surfaces which are compatible with natural laminar flow, many surface imperfections are unavoidable. These imperfections include waviness and bulges, steps and gaps at structural junctions, and three-dimensional roughness elements such as screws or rivets. Other discontinuities arise from leading edge panels or access panels... In addition, environmental factors such as ice crystals, rain, insects, dirt... can create localized surface irregularities. Because some of these imperfections cannot be avoided, it is necessary to study their effects on transition and to develop calculations methods or criteria in order to estimate their effects.

There are in fact two mechanisms for boundary layer tripping :

- For *two-dimensional* irregularities such as bumps, gaps or steps of "infinite" span located normal to the mean flow, the surface imperfections do not generate any "new" disturbances, but they amplify existing or potentially existing TS waves according to the linear theory. Their main effect is to locally modify the pressure field in such a way that the boundary layer instability is enhanced. Here we still observe a *natural transition.*
- By contrast, *three-dimensional roughness elements* such as rivets, insects or ice crystals generate streamwise vortices which trigger transition. The admissible size of such protuberances can be estimated from empirical correlations only. A summary of the boundary layer tripping criteria can be found in [5].

3.3. Some theoretical elements for "natural" transition

This paragraph presents a summary of the most important theoretical results which are aimed at explaining the mechanisms of *natural transition* (there is no general theory for *bypass*-induced transition).

3.3.1. LINEAR STABILITY THEORY : LOCAL APPROACH

A comprehensive description of this theory can be found in [54], see also the lecture notes by Hanifi and Henningson.

Assumptions. Eigenvalue problem

The principle of linear stability theory is to introduce small sinusoidal disturbances into the Navier-Stokes equations in order to compute the range of unstable frequencies. It is assumed that any fluctuating quantity r' (velocity, pressure, density or temperature) is expressed by :

$$r' = r(y)\ \exp[i(\alpha x + \beta z - \omega t)] \tag{3.1}$$

x, y, z is an orthogonal coordinate system, which can be either cartesian or curvilinear, y being normal to the surface. The complex amplitude function r depends on y only. In the general case, α, β and ω are complex numbers.

The fluctuating quantities are very small, so that the quadratic terms of the disturbances can be neglected in the Navier-Stokes equations. It is also assumed that the mean flow quantities do not vary significantly over a wavelength of the disturbances ; therefore U and W (mean flow components in the the x and z directions) as well as the mean temperature T are functions of y alone, and the normal velocity V is equal to zero.

The implication of this *parallel flow approximation* is that the stability of the flow at a particular station (x, z) is determined by the **local** conditions at that station independently of all others.

This leads to a system of homogeneous, ordinary differential equations for the amplitude functions $r(y)$. There are 4 equations for an incompressible flow (continuity + the x, y, z momentum equations) and 6 equations for a compressible flow (the four previous equations + energy equation + equation of state). For the simplest case of a two-dimensional, low speed flow with $\beta = 0$, the stability equations can be combined to obtain a single equation for the vertical component v of the velocity fluctuation. This is the well known *Orr-Sommerfeld equation* :

$$v'''' - 2\alpha^2 v'' + \alpha^4 v - iR\left[(\alpha U - \omega)(v'' - \alpha^2 v) - \alpha U'' v\right] = 0 \tag{3.2}$$

with boundary conditions :

$$v(0) = 0 \qquad v'(0) = 0 \tag{3.3}$$

$$v(y) \to 0 \qquad v'(y) \to 0 \qquad as \quad y \to \infty \tag{3.4}$$

R is the Reynolds number and the ' denotes differentiation with respect to y. Due to the homogeneous boundary conditions (the disturbances must vanish at the wall and in the free stream, except the pressure fluctuations which have a non-zero amplitude at the wall), the problem is an eigenvalue one : when the mean flow is specified, non trivial solutions exist only for certain combinations of the parameters α, β, ω and R. This constitutes the *dispersion relation.*

In this paper, the discussion will be restricted to the so-called *spatial theory,* which is more relevant than the temporal theory for boundary layer flows. ω is real, α and β are complex : $\alpha = \alpha_r + i\alpha_i$ and $\beta = \beta_r + i\beta_i$. r' is expressed by :

$$r' = r(y)\ \exp(-\alpha_i x - \beta_i z)\ \exp[i(\alpha_r x + \beta_r z - \omega t)] \tag{3.5}$$

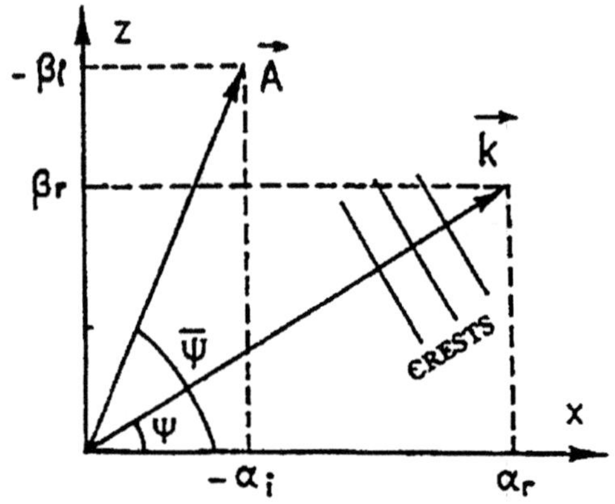

Figure 3.4. Notations

As can be seen in figure 3.4, we define a wavenumber vector $\vec{k} = (\alpha_r, \beta_r)$ and an amplification vector $\vec{A} = (-\alpha_i, -\beta_i)$, with angles ψ and $\bar{\psi}$ with respect to the x direction :

$$\psi = \tan^{-1}(\beta_r/\alpha_r) \tag{3.6}$$

$$\bar{\psi} = \tan^{-1}(\beta_i/\alpha_i) \tag{3.7}$$

If β_i is set equal to zero, the waves can be amplified ($\alpha_i < 0$), neutral ($\alpha_i = 0$) or damped ($\alpha_i > 0$) in the x direction.

From these definitions, it is now clear that any eigenvalue problem involves six real parameters (α_r, α_i, β_r, β_i, ω, R). This set of parameters is often replaced by (α_r, α_i, ψ, $\bar{\psi}$, ω, R). Numerically, the input data are the mean velocity and mean temperature profiles, the free stream Mach number, the stagnation temperature, the fluid properties and all but two of the previous real parameters. The computation gives the values of the two remaining parameters (eigenvalues) as well as the disturbance amplitude profiles (eigenfunctions). Some examples of the numerical procedures are described in [54].

Inviscid theory

In this theory, it is assumed that the viscosity acts only on the mean profiles. Furthermore, the terms of the order $1/R$ are neglected in the stability equations. As a consequence of this simplification, a number of mathematical results can be established. This analysis demonstrates up the importance of two parameters :

a) The first one is the so-called "generalized inflection point", which corresponds to the altitude y_s where :

$$\frac{d}{dy}\left[\rho\frac{dU}{dy}\right]_{y_s} = 0 \tag{3.8}$$

where ρ and U are the mean density and the mean velocity, respectively. It was demonstrated by Lees and Lin [49] that the presence of such a point is a sufficient condition for the appearance of unstable disturbances. It is also a necessary and sufficient condition for the existence of a neutral subsonic (the definition of a subsonic disturbance will be given below). The phase velocity of this wave is the mean velocity at y_s. These results are the generalization of Rayleigh's condition for incompressible flows. However, they are valid only if :

$$\frac{U}{U_e}(y_s) > 1 - \frac{1}{M_e} \tag{3.9}$$

Let us consider the case of a flat plate. When the Mach number increases, the generalized inflection point goes from the wall ($M_e = 0$) towards the outer edge of the boundary layer, and the numerical results indicate that the range of unstable frequencies is enlarged at high Reynolds numbers. It is clear that this "inflectional instability" plays a crucial role for instability and transition in hypersonics.

b) The second parameter is the "relative Mach number", $\hat{M}$, which is defined as :

$$\hat{M} = M - \frac{c}{a} \tag{3.10}$$

where M is the local Mach number. For classical boundary layers, it increases from 0 to M_e between the wall and the free stream. c is the phase velocity of the waves ; it does not depend on y. a represents the local

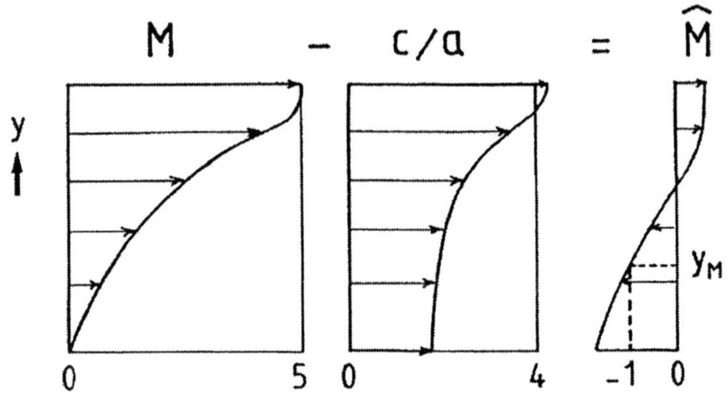

Figure 3.5. Typical evolution of the relative Mach number

speed of sound, which depends on the mean temperature distribution. Obviously, it takes a non-zero value at the wall. The disturbances are subsonic if $\hat{M}^2 < 1$, sonic if $\hat{M}^2 = 1$ and supersonic if $\hat{M}^2 > 1$.

If a wave is locally supersonic, say between $y = 0$ and $y = y_M$ (figure 3.5), the mathematical nature of the stability equations changes, and any eigenvalue problem admits an infinite sequence of neutral solutions. For instance, in the case of two-dimensional disturbances, there is an infinity of neutral waves having the same phase velocity but different wavenumbers α. These multiple solutions (the higher modes) were first discovered by Mack for boundary layer flows [51]. In the case of adiabatic wall conditions, they appear when the free stream Mach number exceeds 2.2.

Typical results for two-dimensional flows

Figure 3.6 shows stability diagrams computed for flat plate flow on adiabatic wall for $M_e = 0$ and $M_e = 1.1$, where M_e is the Mach number at the boundary layer edge. These results were obtained for two-dimensional waves, *i.e.* for $\psi = 0°$. For the sake of clarity, only some curves of equal amplification rates ($\alpha_i < 0$) are plotted in the (Reynolds number, wavenumber) plane. The Reynolds number $R\delta_1$ is computed from the displacement thickness δ_1, the free stream velocity U_e and the kinematic viscosity ν_e of the outer flow. α_r and α_i are made dimensionless with δ_1. For both Mach numbers, there is a critical Reynolds number $R\delta_{1cr}$ below which all disturbances are damped.

It can be seen that the shape of the stability diagrams does not change very much from incompressible to transonic flow. The generalized inflection point, which is at the wall for $M_e = 0$, remains very close to it for $M_e = 1.1$, so that inflectional instability has practically no effect. In fact,

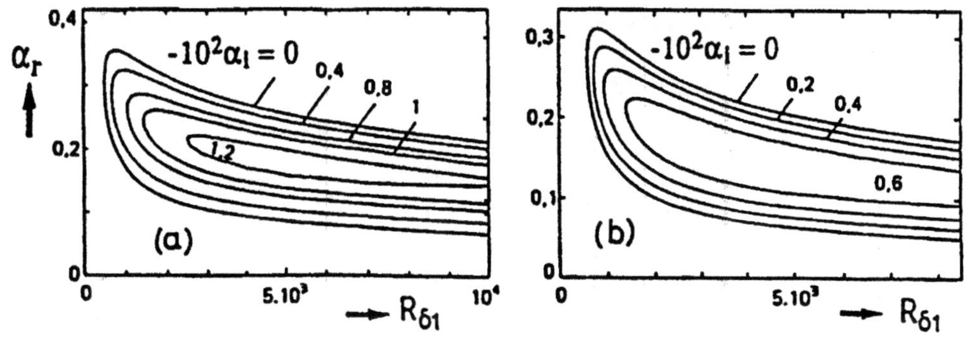

Figure 3.6. Stability diagrams : flat plate, adiabatic wall, two-dimensional waves. a) $M_e = 0$, b) $M_e = 1.1$. From Arnal [4].

the boundary layer is unstable essentially through the action of viscosity (viscous instability). Another interesting feature is that the growth rates are smaller in the transonic range than in incompressible flow : compressibility has a stabilizing effect.

An important aspect of instability for transonic Mach numbers is the effect of the wavenumber direction ψ on the amplification rates. Up to Mach numbers of the order of 0.6-0.8, the maximum value of $-\alpha_i$ corresponds to ψ $=0°$. At higher Mach numbers, the largest amplification rates are obtained for non zero values of ψ (oblique waves). Typically, the most unstable direction is around 40 or 50° for M_e close to 1. In supersonic and hypersonic conditions, the situation is particularly complex, as it can be seen in figure 3.7, where two stability diagrams are represented for $M_e = 4.5$ and two values of ψ : 0 and 60°. For $\psi = 0°$, first and second mode disturbances are unstable at low and high values of ω, respectively. Changing the wave orientation stabilizes the second mode, but increases the instability of the first one, so that it becomes difficult to define "the most unstable diection".

The linear stability theory makes it possible to study the effects of some parameters acting on transition. As it will be shown in paragraph 3.5, negative pressure gradients (accelerated flows), suction or cooling in air have a stabilizing influence. Positive pressure gradients (decelerated flows), blowing or heating in air have a destabilizing effect.

Typical results for three-dimensional flows

Let us consider the case of a laminar boundary layer developing on a swept wing. As it is illustrated in figure 3.8, the mean velocity profile is now decomposed into a *streamwise velocity profile* u (in the direction x of the external streamline) and a *crossflow velocity profile* w (in the direction z normal to this streamline).

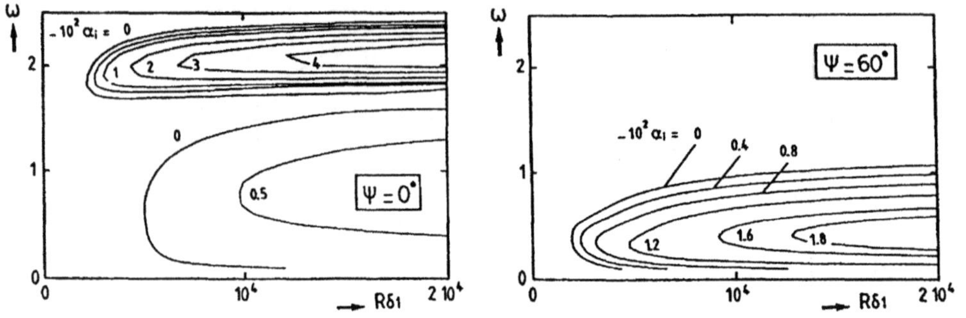

Figure 3.7. Effect of the wave orientation on the growth rate ($M_e = 4.5$). From Arnal [4].

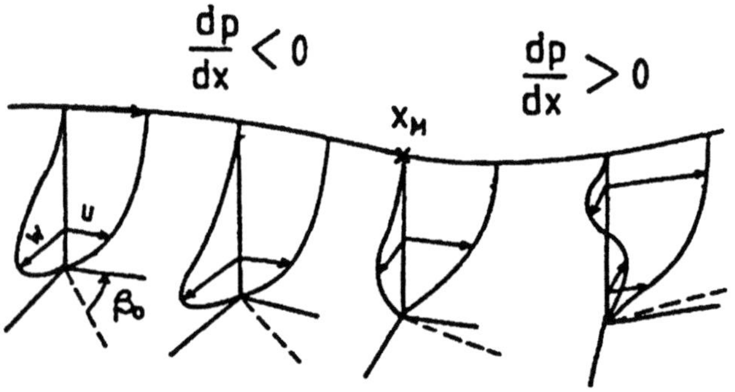

Figure 3.8. Laminar boundary layer development on a swept wing. x_M is the location of the inviscid streamline inflection point. β_0 is the angle between wall and potential streamlines

From the leading edge to the chordwise location x_M where the free stream velocity is maximum, the crossflow is directed towards the concave part of the external streamline. Its magnitude is zero at the attachment line, then it increases more or less rapidly due to the flow acceleration. In the region of negative pressure gradient, the maximun value of the crossflow velocity component remains rather weak, about 5 to 10 percent of the free stream velocity ; it will be shown, however, that this is sufficient to create a strong crossflow instability.

As x_M is approached, the intensity of the pressure gradient decreases, leading to a decrease in the crossflow amplitude. At $x = x_M$, the pressure gradient becomes positive, the curvature of the external streamline changes and the velocity close to the wall reverses (S-shaped profiles). If the positive pressure gradient is strong enough, the crossflow velocity profile can be completely reversed. In the same region, an inflection point appears on the streamwise profile u.

The mechanisms of three-dimensional transition are now relatively well understood, see [67], [2], [78], [73]. As a first approximation, it can be assumed that transition is triggered either by *streamwise instability* (or *TS instability*) or by *crossflow instability (CF instability)* :

- Streamwise instability : as the streamwise mean velocity profiles u look like classical two-dimensional profiles, they are essentially unstable in decelerated flows (positive pressure gradients), *i.e.* downstream of point x_M in figure 3.8. The wavelengths of the streamwise disturbances are about 10 times the boundary layer thickness.
- Crossflow instability : as an inflection point is always present in the crossflow mean velocity profile w, a powerful inflectional instability is expected to occur in regions where the crossflow velocity develops rapidly. This phenomenon is observed, for instance, in the vicinity of the leading edge of a swept wing, *i.e.* upstream of point x_M in figure 3.8. The most unstable disturbance propagation direction ψ_M (relative to the external streamline) is never exactly equal to 90°. It lies in a narrow range close to the crossflow direction z, say between 85 and 89°. The instability is dominated by the properties of the inflection point (altitude, local value of the velocity derivative dw/dy...) which are not very much affected by compressibility effects. The wavelengths of the crossflow disturbances are 3 to 4 times the boundary layer thickness.

Let us notice that the most unstable frequencies are usually lower for crossflow instability than for streamwise instability. In particular, linear stability theory shows that crossflow instability can amplify zero frequency disturbances. As explained in paragraph 3.2.1, this leads to the formation of stationary vortices, the axes of which are close to the streamwise direction.

Immediately downstream of the point of maximum free stream velocity, CF disturbances are decaying while TS disturbances start to be amplified. Complex wave interactions are likely to occur in this region.

3.3.2. LINEAR STABILITY : NONLOCAL APPROACH

A new formulation for the stability analysis was independently proposed by Herbert [40] and by Dallmann at DLR Göttingen [28], [84], see also the lecture notes by Hanifi and Henningson. In this approach, the general expression of the disturbances is :

$$r' = r(x,y)\ \exp[i(\theta(x) + \beta z - \omega t)] \qquad \text{with} \quad \frac{d\theta}{dx} = \alpha(x) \tag{3.11}$$

In contrast to the local approach expressed by relation (3.1), the amplitude functions depend on y and x, and α depends on x. Substituting the previous expression into the stability equations, neglecting $\partial^2 r \ / \ \partial x^2$ and linearizing in r yield a partial differential equation of the form :

$$Lr + M\frac{\partial r}{\partial x} + \frac{d\alpha}{dx}Nr = 0 \tag{3.12}$$

where L, M and N are operators in y with coefficients that depend on x and y through the appearance of the basic flow profiles. When $d\alpha/dx$ is computed from a so-called normalization condition, the previous equation can be solved using a marching procedure in x with prescribed initial conditions : this constitutes the PSE (Parabolized Stability Equations) approach. The interest of this procedure is that the nonparallel effects are taken into account. It is also possible to introduce the *non linear* terms, see paragraph 3.3.4. As the results of the PSE approach at a given x station depend on the upstream history of the disturbances, this approach is called *nonlocal.*

3.3.3. RECEPTIVITY

How are the unstable waves excited by the available disturbance environment? This question is a part of the problem which is usually addressed under the term "receptivity", introduced by Morkovin [62]. The receptivity describes the means by which the forced disturbances (free stream turbulence, sound field...) enter the laminar boundary layer. It also describes their signature in the disturbed flow. A part of this signature is the development of unstable waves, which constitute the eigenmodes of the boundary layer.

For *two-dimensional*, low speed flows, theoretical studies by Goldstein [33] and Kerschen [45] have demonstrated that the receptivity process occurs in regions of the boundary layer where the mean flow exhibits rapid changes in the streamwise direction. This happens near the body leading edge and/or in any region farther downstream where some local feature forces the boundary layer to adjust on a short streamwise lengthscale (sudden change in the wall slope or in the wall curvature, suction strip for instance). Two-dimensional roughness elements (normal to the main flow direction) are widely used to experimentally study the localized receptivity mechanisms in low speed wind tunnels. The height of these devices is very small (adhesive Scotch tapes are often used), so that the wall can be considered as hydraulically smooth. However, when such a roughness element is placed on the surface, it generates regions with rapid local changes of the boundary layer flow, which in turn generate unstable TS waves.

In *three-dimensional* flows, experiments performed by Radetsky *et al.* [71] on a swept wing demonstrated that three-dimensional, micron-sized roughness elements around the leading edge provide the initial amplitude of the stationary vortices which can be mainly responsible for transition.

This constitutes another receptivity mechanism which is typical of crossflow instability.

3.3.4. NON LINEAR EFFECTS

Today, several approaches are available to analyse the behavior of the disturbances when they begin to deviate from the linear amplification regime. These numerical tools include, for instance, the method of multiple scales or the secondary instability theories [39]. DNS are of course particularly helpful for improving the understanding of the nonlinear mechanisms. For practical applications, nonlinear PSE are particularly attractive ; although they only describe *weakly nonlinear phenomena*, they are less time-consuming than DNS and they can be used for rather complex geometries.

The basic ideas of the nonlinear PSE have been explained in many papers, see for instance [40]. The disturbances are written as double Fourier expansions containing two- and three-dimensional discrete normal modes denoted as (n, m) modes :

$$r' = \sum_{n=-\infty}^{\infty} \sum_{m=-\infty}^{\infty} r_{n,m}(x,y) \exp\left[i\left[\int_{x_0}^{x} \alpha_{n,m}(\xi)d\xi + m\beta z - n\omega t\right]\right] \quad (3.13)$$

$\alpha_{n,m}$ is complex, β and ω are real and constant. Introducing (3.13) into the Navier-Stokes equations leads to a system of coupled partial differential equations which are solved by a marching procedure. This method makes it possible to treat nonlinear mode interaction with a numerical effort which is much less than that which is required by DNS. However, it cannot been yet used as a "black box" for practical problems.

3.4. The e^N method

The so-called e^N method is widely used by people who are assigned the job of making transition prediction. It was first developed by Smith and Gamberoni [85] and by van Ingen [89] for low speed flows and then extended to compressible and/or three-dimensional flows.

The e^N method is based on *linear theory* only, so that many fundamental aspects of the transition process are not accounted for. However, one has to keep in mind *that there is no other practical method presently available for industrial applications.*

The e^N method can be used either with the local (classical) stability approach or with the nonlocal (PSE) stability approach. Both aspects are examined successively.

3.4.1. LOCAL APPROACH

Two-dimensional, incompressible flows

In this case, only two-dimensional waves ($\beta = 0$) are considered. The disturbances are amplified or damped according to the sign of the spatial growth rate $-\alpha_i$.

For a given mean flow, it is possible to compute a stability diagram (upper part of figure 3.9) showing the range of unstable frequencies f as a function of the streamwise distance x. Let us now consider a wave which propagates downstream with a fixed frequency f_1. This wave passes at first through the stable region. It is damped up to x_0, then amplified up to x_1, and then it is damped again downstream of x_1. At a given station x, the total amplification rate of a spatially growing wave can be defined as :

$$\ln(A/A_0) = \int_{x_0}^{x} -\alpha_i \, dx \tag{3.14}$$

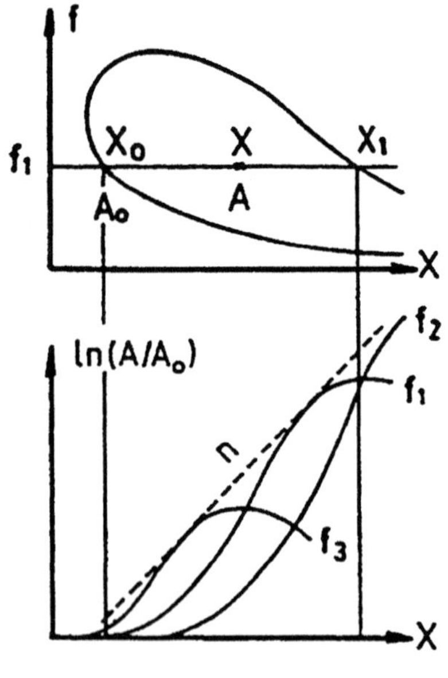

Figure 3.9. Typical stability diagram in physical coordinates. Definition of the total growth rate and of the envelope curve

In the previous expression, A is the wave amplitude, and the index 0 refers to the spanwise position where the wave becomes unstable. The envelope of the total amplification curves is :

$$n = \underset{f}{\text{Max}} \; [\ln(A/A_0)] \tag{3.15}$$

In incompressible flow, with a low disturbance environment, it is assumed that transition occurs as soon as the N factor reaches a critical value in the range $7 - 10$, *i.e.* when the locally most unstable frequency is amplified by a factor e^7 to e^{10}.

Two-dimensional, compressible flows

When compressibility begins to play a role, the problem becomes more complex, because the most unstable waves are often oblique waves. As a consequence, a new parameter enters the dispersion relation : the angle ψ between the streamwise direction and the wavenumber vector. It is still assumed that the amplification takes place in the x-direction only, *i.e.* $\beta_i = 0$ or $\bar{\psi} = 0$, but β_r (or ψ_r) needs to be specified. To compute the local growth rate of a given frequency, the following procedure is often used : at each streamwise location, one seeks for the direction ψ_M of the wave which gives the maximum value $-\alpha_{iM}$ of α_i, and the N factor is defined by :

$$\begin{aligned} n &= \underset{f}{\text{Max}} \int_{x_0}^{x} -\alpha_{iM}\, dx \qquad (3.16) \\ &= \underset{f}{\text{Max}} \int_{x_0}^{x} \underset{\psi}{\text{Max}}\; (-\alpha_i)\, dx \end{aligned}$$

This is the so-called *envelope method.*

Three-dimensional flows

The extension of the e^N method to three-dimensional flows is not straight forward. The first reason is that the assumption $\beta_i = 0$ is not necessarily correct. Hence β_i must be assigned or computed. Several solutions have been proposed to solve this problem, see review in [6]. For instance, it is possible to use the wave packet theory and to impose the ratio $\partial\alpha/\partial\beta$ to be real. A simpler solution is to assume that the growth direction is the group velocity direction or the potential flow direction. In the case of infinite swept wings, it is often assumed that there is no amplification in the spanwise direction.

After one of the previous assumptions for β_i has been adopted, one has to integrate the local growth rates in order to compute the N factor. Several strategies are available :

- Envelope method : this strategy was previously described for two-dimensional, compressible flows.
- Fixed frequency/fixed spanwise wavenumber method (fixed ω/fixed β method) : the integrations are performed by following waves having a given dimensional frequency f and a constant value of the dimensional component β_r of the wavenumber vector in the spanwise direction. The N factor is computed by maximizing the total amplification with respect to both f and β_r. In other words, the N factor represents the envelope of several envelope curves. In the following, this method will be referred to as the CSW (Constant Spanwise Wavenumber) strategy.
- Fixed frequency/fixed wavelength method (fixed ω/fixed λ method) and fixed frequency/ fixed direction method (fixed ω/ fixed ψ method) : as a wave of fixed frequency moves downstream, the wavelength or the propagation direction of the disturbances are kept constant. These strategies resemble the previous one in this sense that the N factor represents the envelope of several envelope curves.
- Streamwise N factors/crossflow N factors (N_{TS}/N_{CF} method) : the principle is to compute a N factor for streamwise disturbances (the so-called N_{TS}) and another N factor for crossflow disturbances (the so-called N_{CF}). Transition is assumed to occur for particular combinations of these parameters.

As it could be expected, each strategy gives a different value of the N factor at the onset of transition.

Shortcomings of the e^N method

As the e^N method is based on linear stability only, receptivity and non-linear mechanisms are not taken into account. In addition the nonparallel effects are neglected in the local procedure, and it has been shown that they could be important for oblique waves.

Several particular problems arise for three-dimensional flows. The first one is to choose the "best" strategy of integration of the N factor. The envelope method is widely used, but its physical meaning is not clear in many cases, especially for swept wing flows in the vicinity of the point of minimum pressure : rapid variations in ψ_M can be observed around this point when the dominant instability suddenly changes from the crossflow to the streamwise type. The other strategies are often used in order to avoid these discontinuous (and probably unphysical) evolutions.

From a practical point of view, the most important issue is the value of the N factor at the onset of transition. Concerning two-dimensional flows, it is now admitted that the e^N method with $N \simeq 10$ can be applied to predict transition for a wide range of flows if the background disturbance level is small enough. In three-dimensional flows, the results depend, of course, on the strategy which is chosen to compute the N factor, but, even with the same strategy, the results are not very clear. In fact, all the strategies produce a large scatter in the values of N at the onset of transition. A possible reason is that crossflow and streamwise instabilities are not initiated by the same type of forced disturbances. For instance, the crossflow disturbances are very sensitive to micron-sized roughness elements which have no effect on streamwise disturbances. Another reason is that there exists a multiplicity of nonlinear processes before the breakdown to turbulence, depending on the relative part of stationary and travelling unstable modes.

However, the e^N method remains a very practical and efficient tool, especially for parametric studies. For a given test model and for a given disturbance environment, it is often able to predict the variation of the transition location when changing a parameter which governs the stability properties of the mean flow (pressure gradient, suction rate...).

3.4.2. NONLOCAL APPROACH

By contrast with the local approach, it is no longer necessary to impose additional conditions for computing nonlocal N factors in three-dimensional flows. This is due to the fact that β is real and constant ; because β is real, the amplification vector has only one component in the x direction ; because β is constant, the N factor will be computed in a way which is similar to the CSW strategy described previously for the local theory. However, this implies that the PSE equations should be solved in a coordinate system which is coherent with the assumptions made on the mathematical nature of β. For the case of an infinite swept wing, the obvious choice is to consider x as the direction normal to the leading edge and z as the spanwise direction.

Another difficulty arises from the treatment of curvature terms in the stability equations. In in coordinate system adapted to the wing, these terms take into account the surface curvature of the body. Because the curvature effects are of the same order of magnitude as the nonparallel effects, it is not coherent to include them into the local stability equations, but they have to be accounted for in nonlocal stability computations.

To summarize, PSE results with curvature can be compared only with local results obtained without curvature by using the CSW strategy.

3.5. Application to transonic flows : laminar flow control

The drag of transport aircraft originates from various sources, such as induced drag, skin friction drag, interference drag, wave drag or parasitic drag. Over the last few decades and particularly since the 70s, civil aircraft manufacturers have made great efforts to reduce aircraft drag. The objective of these efforts is to decrease the specific consumption, because the potential reduction of over 10% would represent savings of several million dollars for the airlines [76]. Although the importance of the different drag sources vary according to the type of aircraft and the type of flight mission, skin friction drag usually plays a dominant role. It represents about 50% of the total drag for a commercial transport aircraft of the Airbus type. This percentage is somewhat lower for a supersonic aircraft (about 40%), it is larger for small transport aircraft (about 60%).

There are essentially two methods that can be used to reduce skin friction drag (a complete account of this problem can be found in [1]). The first method consists in modifying the structure of the turbulent boundary layer in such a way that the skin friction coefficient at the wall is reduced. This topic will not be covered in this paper, see [26] or [27] for details. The principle of the second technique is to maintain laminar flow on the surface, *i.e.* to control the laminar-turbulent transition mechanisms in order to increase the extent of the region with low skin friction boundary layer (see figure 3.1).

Laminar flow control is essentially a war against the inflection points. As indicated by the inviscid theory, the inflectional mean velocity profiles are highly unstable at infinite Reynolds numbers, but even including viscous effects they are unstable at finite Reynolds numbers of practical interest. The main objective is to eliminate the inflection points or, at least, to reduce their destabilizing effect, for instance by modifying their position into the boundary layer.

Clearly, the problem is not easy to solve for crossflow instability, because the crossflow profile always exhibits an inflection point, generally towards the outer edge of the boundary layer. For the streamwise instability, it can be assumed that the possibility of an inflectional instability will be less if the second derivative of the mean velocity near the wall becomes more negative. Thus the *stability modifiers* can be identified by examining the two-dimensional momentum equation close to the surface (Reshotko [74]) :

$$\left(\rho V_w - \frac{d\mu}{dT}\frac{\partial T}{\partial y}\right)\frac{\partial U}{\partial y} + \frac{dP}{dx} = \mu_w \frac{\partial^2 U}{\partial y^2} \tag{3.17}$$

This equation shows that negative pressure gradients, suction ($V_w < 0$), cooling in air ($\frac{\partial T}{\partial y} > 0$) or heating in water ($\frac{\partial T}{\partial y} > 0$) tend to make the second derivative of U more negative. The effects of pressure gradient and of suction are discussed below (see [5] for an overview of cooling/heating effects).

3.5.1. EFFECT OF STREAMWISE PRESSURE GRADIENTS

Because this stabilization technique does not require any external energy source, such as suction, cooling or heating, it is often referred to as *"Natural Laminar Flow " (NLF)* control. The aim is to delay the onset of transition by optimizing the pressure distribution along the body, *i.e.* by "shaping" the surface.

Two-dimensional flows

According to the results of the inviscid stability theory, a generalized inflection point generates strongly amplified TS waves. For low speed flow, the inflection point exists as soon as the free stream velocity is decreasing, and the laminar boundary layer becomes highly unstable at low Reynolds numbers. On the contrary, there is no inflection point if the main flow is accelerated, and the laminar boundary layer remains stable up to very large Reynolds numbers. In the latter case, TS waves, if they exist, are generated by a viscous instability. To give a more precise idea, the critical Reynolds number $R\delta_{1cr}$ is less than 70 for the Falkner-Skan profile at separation, while it is close to 12.000 for the Falkner-Skan profile associated to a planar stagnation point.

It follows that stabilization of the laminar boundary layer is a rather simple problem for two-dimensional flows : as TS waves are damped (or only slightly amplified) in negative pressure gradient, the distance between the stagnation point and the point of minimum pressure must be as long as possible. On shaped airfoils or bodies, the flow usually remains laminar up to the streamwise location where the free stream velocity is maximum, then it undergoes transition as it enters the positive pressure gradient region where separation can occur.

The exploitation of this simple principle dates back to the 1930's and led to the development of the NACA6-series airfoils. Wind tunnel experiments performed on these airfoils showed transition Reynolds numbers of 14 to 16 million- four or five times the value which is often measured on flat plates without pressure gradient. It is also possible to apply this method for shaping turbofan nacelles, which can be considered, as a first approximation, as two-dimensional, axisymmetric bodies.

Three-dimensional flows

Shaping a swept wing constitutes a much more difficult task than shaping a two-dimensional airfoil, because one has to find a compromise between two opposite effects. Close to the leading edge, a negative pressure gradient damps out TS disturbances, but it creates a crossflow velocity component which amplifies CF disturbances : negative pressure gradients are "favourable" for two-dimensional flows only! Downstream of the location where the free stream velocity reaches its maximum, a positive pressure gradient reduces crossflow instability, but streamwise disturbances start to grow rapidly.

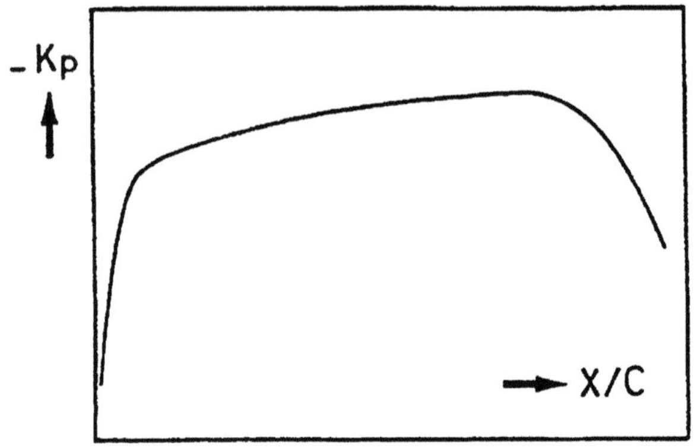

Figure 3.10. Typical pressure distribution for Natural Laminar Flow

Figure 3.10 shows an example of upper surface pressure distribution that might be used for NLF applications to swept transonic wings. Near the leading edge, the pressure falls very rapidly in order to minimize the development of the crossflow velocity component and consequently of the crossflow disturbances. This implies that the leading edge radius must be as small as possible. In addition, this strong acceleration helps to eliminate leading edge contamination by reducing the attachment line Reynolds number, see paragraph 3.5.3. Downstream of the "knee" of the pressure distribution, the flow is slowly accelerated in order to avoid the development of streamwise disturbances.

3.5.2. SUCTION

Stabilization by suction is a method which is referred to as *"Laminar Flow Control" (LFC)*.

Fundamental aspects

Suction acts in two ways. First, it reduces the boundary layer thickness and the associated Reynolds numbers ; however, as the suction rates are usually very small, this effect is weak. The second, and more important effect, is to modify the shape of the mean velocity profile in such a way that its stability is improved. For example, let us consider the so-called *asymptotic suction profile* in two-dimensional, incompressible flow. This profile is obtained when a continuous suction is applied at the wall of a flat plate. Some distance downstream of the leading edge, the boundary layer characteristics become independent of the streamwise coordinate, and it is easy to demonstrate that :

$$V(y) = -V_w \quad (V_w \text{ is the absolute value of the suction velocity}) \tag{3.18}$$

$$U(y) = U_e \left[1 - \exp\left(\frac{V_w\, y}{\nu}\right)\right] \tag{3.19}$$

The displacement thickness and the shape factor are equal to $-\nu/V_w$ and 2, respectively. Linear stability computations indicate that the critical Reynolds number $R\delta_{1cr}$ is about 42.000. Other numerical results give higher values, of the order of 50.000. Let us recall that the value of $R\delta_{1cr}$ for the Blasius flow without suction is 520! The values of $R\delta_{1cr}$ for the asymptotic suction profile are two orders of magnitude larger than that obtained without suction. The strong stabilizing influence of suction is explained not only by the increase of the critical Reynolds number, but also by the decrease in the range of unstable frequencies and by the decrease in the disturbance growth rates. Nonlinear studies, see [20] for instance, demonstrate that suction is efficient if it is applied in the linear growth rate regime, *i.e.* before the appearance of the peak-valley system. Fundamental wind tunnel experiments performed by Reynolds and Saric [75] indicated that suction is more effective when applied at Reynolds numbers close to the lower branch of the neutral curve, in qualitative agreement with the theoretical results.

In three-dimensional flows, however, the inflectional nature of the crossflow mean velocity profile makes stabilization by suction more difficult. Suction is able to suppress the inflection point of a streamwise profile in a mild positive pressure gradient, but it is obviously unable to eliminate the inflection point of the crossflow velocity profile. In fact, the main effect of suction is to reduce the crossflow Reynolds number.

Slots and perforated surfaces

The best way of removing a small portion of the boundary layer flow is to develop a *continuous* porous surface. This solution is not always easy to apply for both structural and aerodynamic difficulties. At the present time, the best methods to approach a continuous suction are the use of spanwise slots or strips of perforated material.

When *slots* are used, their spacing (in the main flow direction) is usually much larger than their width, so that the suction distribution is spatially discontinuous. On the other side, it is obvious that a well-designed slot must avoid to generate disturbances which could reduce the extent of laminar flow. Extensive research performed by Northrop before and during the so-called X-21 program yielded useful design criteria in which many parameters are accounted for : the slot width, the height of the slot lip, the sucked height, the pressure drop, the slot Reynolds number and the spacing between slots.

Perforated strips represent a closer approach to ideal continuous suction, because the spacing between the strips is generally smaller than the width of each strip. In addition, the tolerance to off-design conditions is better. However, two problems need to be solved :

- Close to the leading edge of an aircraft wing, the streamwise pressure gradient may be strong enough to cause inflow in the forward region of the strip (as expected) and outflow in the aft region. As blowing is destabilizing, the streamwise extent of the strip must be reduced and/or the pressure drop across the perforated surface must be increased.
- For "large" suction rates, the suction holes generate three-dimensional disturbances which can fix the transition on the perforated surface. To avoid this "negative roughness" effect, both the hole diameter and the hole spacing must be small compared to the boundary layer thickness. Early investigations in the 1950's showed this tripping effect with holes of relatively large diameter. At this time, the diameter of the smallest holes that could be manufactured practically was of the order of 0.5 mm. Today, electron and laser beams drill closely spaced holes as small as 0.025 mm diameter. As it will be shown in paragraph 3.5.4, such suction panels have demonstrated their practicability for aircraft applications.

A more theoretical problem is the enhancement of receptivity to free stream disturbances at any junction between a perforated strip and the impermeable wall. This is a typical example of localized receptivity which involves abrupt adjustement of the boundary layer to the sudden change of the transverse velocity at the wall.

The e^N method with suction

The estimation of suction quantities for LFC applications can be calculated by imposing that the integrated amplification rates (N factor) of the unstable waves remain below a prescribed level. The main result is that very small suction rates are often sufficient to considerably delay the onset of transition. Usual values of the ratio $-V_w/Q_\infty$ range between 10^{-4} to 10^{-3}. It appears that the suction flows required for LFC are weak and have a rather limited effect on the energy balance. The computations also demonstrate that the best efficiency is obtained when suction is applied as soon as unstable disturbances start to develop.

It is not obvious, however, that the N factor remains constant without and with suction. To investigate this problem, wind tunnel experiments were performed at ONERA, see [31] and [9] for details. The main results are summarized below.

Experimental setup The experiments have been carried out in the F2 wind tunnel at Le Fauga-Mauzac ONERA Centre. The working section size is 1,8 $\times$ 1,4 m^2 with a length of about 4 m. The wind tunnel speed can be varied from 0 to 100 ms^{-1}, and the free stream turbulence level is about 0,07%.

The model is a swept wing which was especially designed at ONERA for this study ; it is sketched in figure 3.11. The chord normal to the leading edge is 0,7 m, and the span is about 2,5 m. The pressure side of the wing is equipped with 7 independent suction chambers (in the chordwise direction) from 5 to 25 percent chord. A constant suction velocity was used to obtain the results reported below. The pressure distribution on the pressure side (*i.e.* on the side equipped with the suction panel) was measured by three rows of pressure taps. These rows are placed at three different spanwise locations in order to ensure that the inviscid flow is uniform in the spanwise direction.

Two techniques were used to detect transition location on the pressure side. Firstly, the wing surface downstream of the suction panel was black painted in order to perform wall visualizations by infrared thermography. Secondly, 12 hot film probes were stuck on the model between 30 and 85 percent chord ; in the following results, the transition abscissa x_T is taken as the first appearance of turbulent spots.

Examples of experimental results A systematic variation of suction velocity V_w and of angle of attack α has been performed for wind tunnel speeds of 75 and 95 ms^{-1} anf for sweep angles φ of 40 and 50°. As a typical example of result, figure 3.12 shows the pressure distributions measured without suction for $Q_\infty = 70$ ms^{-1}, $\varphi = 50°$ and several angles of attack.

Figure 3.13 shows the variation of the transition location as a function of V_w and α for a series of experiments corresponding to $Q_\infty = 95$ ms^{-1}, φ = 40°. The symbols represent the experimental data, whilst the lines correspond to numerical results obtained by using simple prediction methods ; for TS-type disturbances, a two-dimensional simplified method [3] was applied along the external streamline (full line) ; for crossflow disturbances, the analytical criterion described in [10] was used (dotted lines). As it could be expected from figure 3.12, transition is triggered by streamwise insta-

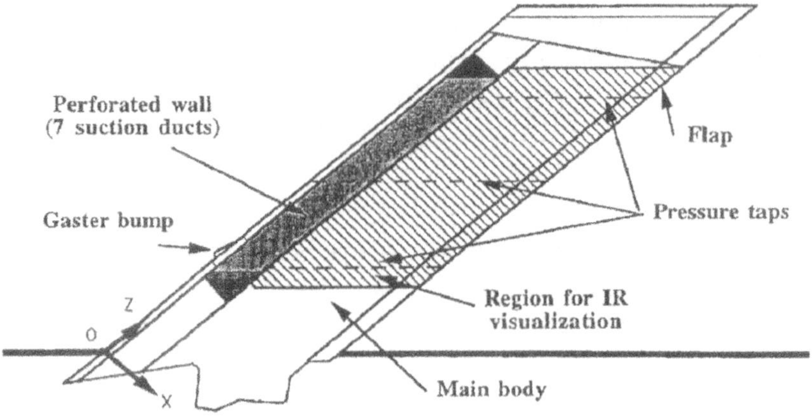

Figure 3.11. Experimental setup. From [31, 9].

bilities at low values of α, while crossflow instability dominates at higher angles of attack. It can be observed that for α around 4°, suction modifies the nature of the dominant instability.

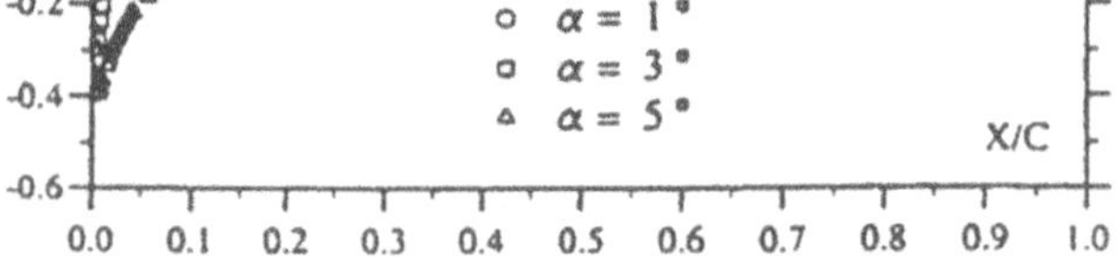

Figure 3.12. Example of pressure distributions. From [31, 9].

Linear, local analysis Two integration strategies were used : the envelope method and the constant spanwise wavenumber (CSW) strategy. The numerical results demonstrated that, in many cases, the use of the envelope method led to a strong decrease of the N factor at transition when suction was applied. Typical values were 12-13 without suction and 7-8 with suction. On the other side, the N factor at transition was more or less the same (around 7-8) with and without suction by employing the CSW strategy. In

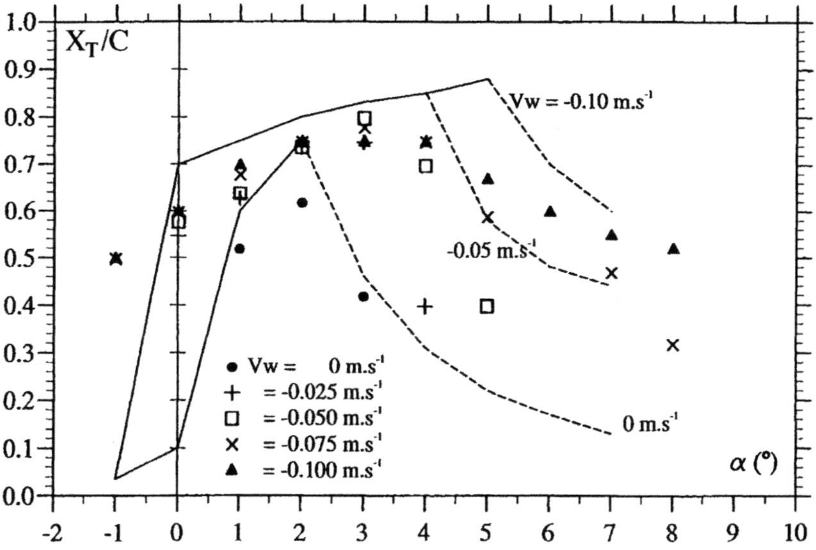

Figure 3.13. Transition location. From [31, 9].

order to explain this behaviour, the evolution of the wavenumber direction ψ was systematically computed for the most amplified frequency leading to transition. It turned out that the ψ direction computed with the CSW strategy was nearly constant from the neutral curve to the measured onset of transition. When the envelope method is considered, the most unstable ψ direction is seeked at each chordwise location ; it follows that ψ is likely to exhibit very large variations. Without suction, ψ typically decreases from about 85° close to the leading edge (CF disturbances) to about 0° at the transition location ; in other words, the N factor in this case represents the cumulative effects of CF and TS disturbances. When suction is applied, the CF contribution is practically eliminated, and the instability process is governed by quasi-TS waves.

From these observations the idea arose to consider the mean value of ψ, denoted as $\bar{\psi}$, for the waves leading to transition. $\bar{\psi}$ is simply defined as :

$$\bar{\psi} = \frac{1}{x_T - x_0} \int_{x_0}^{x_T} \psi \, \mathrm{dx} \tag{3.20}$$

x_0 and x_T represent the critical and the transition abscissas, respectively.

The values of the N factor at transition are plotted as function of $\bar{\psi}$ in figure 3.14 (envelope method) and 3.15 (CSW strategy). The results obtained with the envelope strategy exhibit large values of N for $\bar{\psi}$ between 40 and 80°, *i.e.* for cases where the N factor is the sum of a CF contribution and of a TS contribution. When there is only one dominant instability

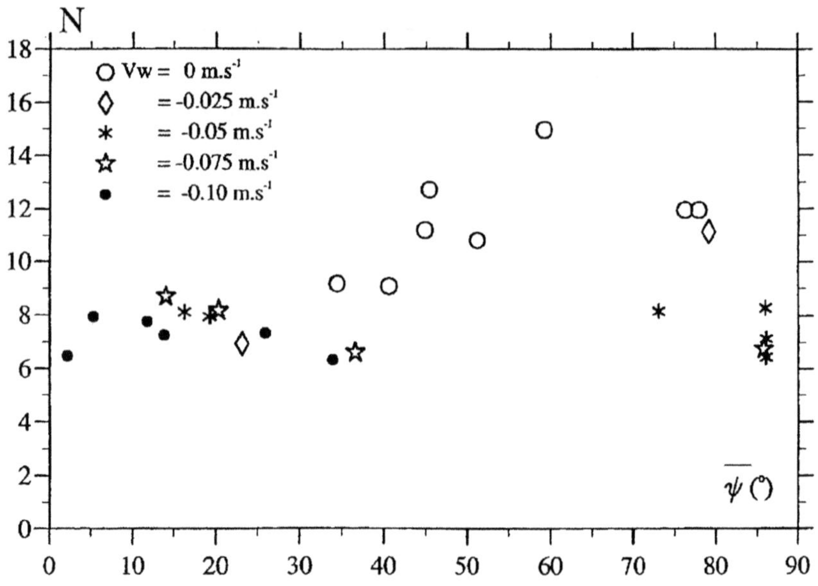

Figure 3.14. Local N factors at transition (envelope method). From [31, 9].

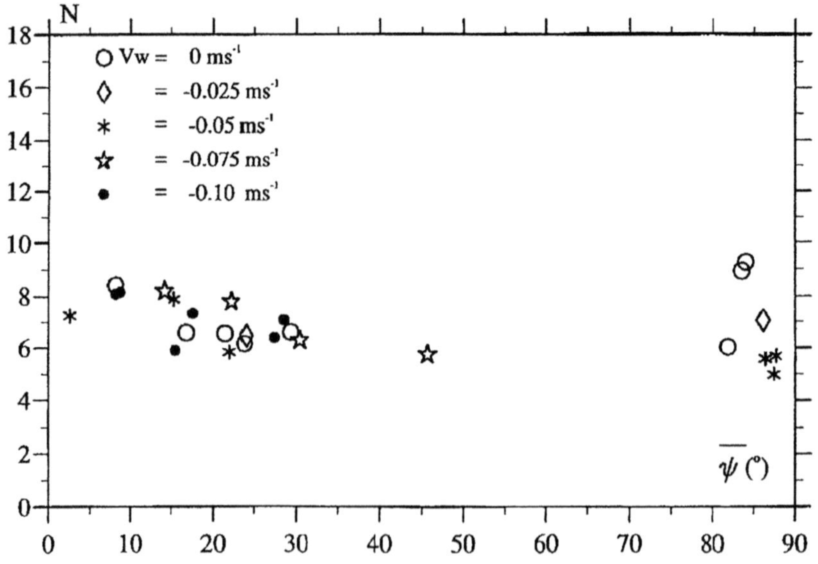

Figure 3.15. Local N factors at transition (CSW strategy). From [31, 9].

mechanism ($\bar{\psi}$ smaller than 40° and $\bar{\psi}$ close to 90°), the N factor at transition is smaller and around 8. On the other side, the N factors given by the CSW strategy take into account a single instability mechanism, which is either TS or CF (there is only one point at an intermediate value of $\bar{\psi}$), and the scatter in the values of N is smaller. In the present investigation, the values of N for "pure" TS and "pure" CF transition processes are not very different, but it cannot claimed that this result is universal : a different wind tunnel with a different model could provide smaller N factors in the TS range and larger N factors in the CF range, or the contrary.

The main conclusion is that the use of the envelope method often leads to a change in the instability mechanisms, from a "cumulative" type without suction to a "pure" TS or CF type with suction. The decrease is the N factor is mainly a result of this change, and not an effect of suction, at least at the first order. In the range $80° < \bar{\psi} < 90°$, however, the general trend of the CSW strategy results is a decrease of N when the suction velocity increases. This could be due to disturbances generated by the suction holes, which could enhance the amplitude of the stationary vortices generated by the crossflow instability. This seems to be confirmed by nonlinear computations (see [77]) and by recent experiments performed at ONERA (to be published).

Linear, nonlocal analysis Some experimental configurations have been analysed by using a linear, nonlocal stability code developed at ONERA. The results demonstrated that nonparallel effects are strongly destabilizing in regions where CF disturbances are dominant, *i.e.* close to the leading edge. However, introducing curvature terms in the PSE equations reduces the difference between local and nonlocal N factors. Figure 3.16 shows the nonlocal N factors at transition given by the nonlocal approach with curvature terms included. If these results are compared with those given by the local CSW strategy without curvature (figure 3.15), the difference is negligeable fot TS dominated cases, whilst nonlocal effects lead to an increase $\Delta N \simeq 2$ for CF dominated cases. But the local and nonlocal values of $\bar{\psi}$ are always close together, and the unstable frequency ranges are quite similar.

3.5.3. HOW TO PREVENT LEADING EDGE CONTAMINATION?

It has been shown in paragraph 3.2.2 that leading edge contamination occurs on swept wing as soon as a certain Reynolds number $\bar{R}$ exceeds a critical value close to 250. Unfortunately, $\bar{R}$ is larger than this value for most of the commercial aircraft (at least near the root). Typical values are of the order of 400-600. As a consequence, it becomes necessary to develop specific tools to delay the onset of leading edge contamination. This is in fact the first problem to solve for maintaining laminar flow on a wing : if the attachment line boundary layer is turbulent, turbulence will spread over

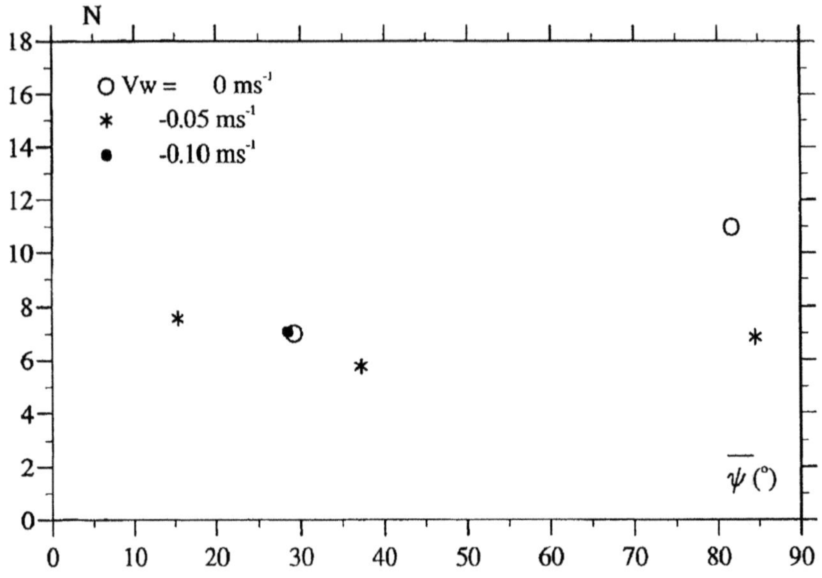

Figure 3.16. Nonlocal N factors at transition. From [31, 9].

the whole wing and the benefits of the systems described in the previous paragraph will be lost.

As a first approximation, it can be demonstrated that $\bar{R}$ is proportional to $(r \sin\varphi \tan\varphi)^{0.5}$, where r is the leading edge radius and φ is the sweep angle. Therefore the first idea is to reduce the leading edge radius and/or the sweep angle near the root. Technological problems can make this solution difficult to apply.

A successful device to prevent leading edge contamination is the Gaster bump, [32]. It consists of a small fairing which is placed on the leading edge close to the wing root. It is shaped in such a way that the contaminated turbulent boundary layer is brought to rest at a stagnation point on the upstream side, whilst a "clean" boundary layer is generated on the downstream side.

Seyfang [83] tested several passive devices aimed at restoring laminar flow on a contaminated attachment line (step-up, step-down, square trips, grooves...). In all cases, the objective is to create a stagnation point from which a laminar boundary layer starts to develop. The main problem is to optimize the dimensions of these devices in order to avoid boundary layer tripping effects.

Another solution is to relaminarize the turbulent boundary layer developing along the leading edge by applying suction along the attachment line. The efficiency of this process was first demonstrated by Spalart's direct numerical simulations [86]. These computations showed that contamination could be delayed up to $\bar{R} \simeq$ 350-400 for K = -1. K is a dimensionless suction parameter :

$$K = \frac{V_w}{W_e}\bar{R} \tag{3.21}$$

where V_w is the vertical mean velocity at the wall (it is negative for suction) and W_e is the freestream velocity component parallel to the leading edge ($W_e = Q_\infty \sin\varphi$). Due to the limited number of available experimental data, ONERA decided to perform tests in the F2 wind tunnel at Le Fauga-Mauzac in order to study this phenomenon at large values of $\bar{R}$ [11], [31].

The chosen experimental support was a constant chord swept wing model generated from a symmetrical airfoil with a radius of 0.2 m near the leading edge. The phenomenon of leading edge contamination was studied at sweep angles of 40 and 50° by fixing the model to the wind tunnel wall.

The objective of the tests was to delay leading edge contamination either by the use of a Gaster bump or by applying suction along the leading edge or a combination of both. Two leading edges which have been tested; the first one consists of six independent suction chambers fitted along the leading edge and the second one combines a Gaster bump with three leading edge suction chambers downstream of the bump. The chordwise width of the suction panel was about 70 mm, i.e 35 mm on each side of the attachment line. The titanium perforated panel was laser drilled by AS&T company and the mean diameter of the holes was about 50 μm. The model instrumentation consisted of 3 rows of surface pressure taps aligned normal to the leading edge. Leading edge contamination was detected by flush-mounted surface hot films. Figure 3.17 shows the evolution of $\bar{R}$ corresponding to the onset of leading edge contamination (first spots) as a function of the suction parameter K. The results obtained without Gaster bump at $\varphi = 50°$ are compared with the DNS results by Spalart [86] and with the experimental data currently available [68], [44]. Without suction, leading edge contamination occurs for $\bar{R} \simeq 250$, as expected. Application of suction causes the onset of contamination to be delayed to $\bar{R} \simeq 550$ for the maximum suction rate attainable in the experiments.

For the leading edge fitted with a Gaster bump at $\varphi = 50°$, leading edge contamination in the absence of suction occurs at $\bar{R} \simeq 320$, a value which is lower than that obtained in other previous experiments. As soon as the flow over the bump is fully turbulent, the data with and without bump become close together (within the experimental uncertainty). The porosity of the porous leading edge fitted with the bump was larger than that of the leading edge without bump, so that the dimensionless suction parameter could be increased up to K = -3.07. This allowed to delay the onset of

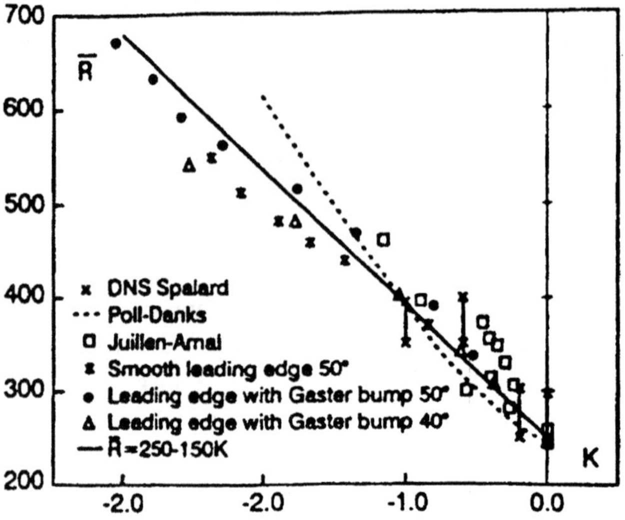

Figure 3.17. Leading edge contamination Reynolds numbers : summary of the results. From [31, 11].

leading edge contamination up to $\bar{R} = 670$. The results can be correlated with the following expression :

$$\bar{R} = 250 - 150\,K \tag{3.22}$$

The conclusion of these studies is that with rather modest suction rates, boundary layers that are contaminated by turbulence at the wing root can be relaminarized and kept in the laminar state up to very large values of $\bar{R}$. This technique still remains to be validated under flight conditions.

3.5.4. EXAMPLES OF FLIGHT EXPERIMENTS WITH TRANSITION CONTROL

The final objective of the theoretical, numerical and experimental studies described in the previous sections is to develop the tools required for practical applications on airplanes. The best way to check the validity of these methods is to perform free flight tests. This paragraph gives a short survey of recent flight experiments for which the NLF, LFC and HLFC concepts have been applied. More details can be found in [25] for some of these experiments.

NLF experiments

Falcon 50 experiments NLF flight experiments were conducted in France (1985-1987) by Dassault Aviation with state aid and in cooperation with ONERA [19], [7]. A wing section was installed on the fin of a Falcon 50 aircraft and tested in transonic conditions for two sweep angles. Due to the rather low values of the chord Reynolds number, streamwise disturbances dominated for $\varphi = 25°$, while transition was induced by crossflow disturbances for $\varphi = 35°$. The results agreed well with theoretical predictions.

ATTAS experiments In 1986, a national research program investigating NLF at transonic speed and high Reynolds number was funded by the German Ministry of Research and Technology [38], [41]. In collaboration between DLR and Deutsche Airbus, a special glove was designed and installed on the right wing of the DLR research aircraft VFW 614/ATTAS (ATTAS : Advanced Technologies Testing Aircraft System). A large number of flight test data were obtained during two test periods in summer 1987.

The ATTAS aircraft is a swept wing aircraft with a leading edge sweep $\varphi = 18°$. The maximum Mach number for these tests was 0.7, leading to a Reynolds number based on wing chord close to 30 million. The glove covered a spanwise region of 3 m. Transition was detected by infrared image technique and by hot films. Changing the yaw angle provided sweep angle variations from $\varphi = 13°$ to $\varphi = 23°$.

The linear stability analysis of many flight tests was carried out by means of an *incompressible* stability code. The principle of this analysis was described in paragraph 3.4 : the N factor for TS waves (N_{TS}) and the N factor for CF disturbances (N_{CF}) are computed separately. Further investigations using a *compressible* stability code showed that the N_{CF}/N_{TS} strategy was difficult to apply in transonic flow, because the propagation directions for the TS waves and for the CF disturbances sometimes are close together (Schrauf [79]). Evaluation of ATTAS flight test data using nonlinear PSE analysis is reported in [80].

The ELFIN and EUROTRANS projects A European effort has been done to study the laminar flow technology within the so-called ELFIN program (European Laminar Flow INvestigation). This program constituted a collaborative venture, bringing together the majority of European airframe manufacturers, research institutes and universities. The essential goals were to prove the basic concepts of laminar flow technology and to prepare the tools, methods and systems required for its application.

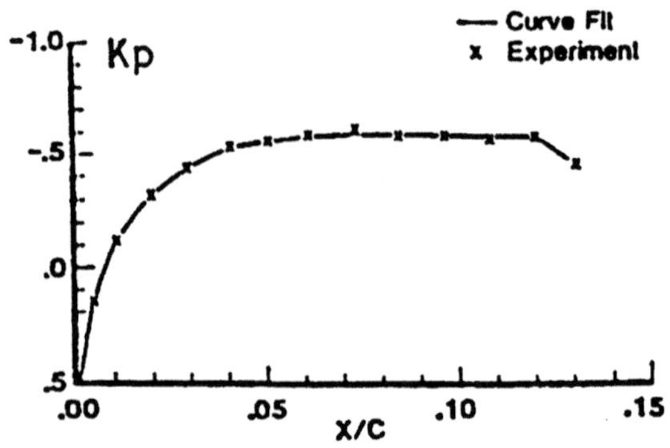

Figure 3.18. Pressure coefficient distribution, from Maddalon *et al.* [55].

In the framework of the first phase of this project (ELFIN I), NLF experiments were conducted on a glove bonded to the original wing surface of a Fokker 100 aircraft. The aircraft was instrumented with 2 infrared cameras for transition detection, one above and one below the wing. A series of flight tests (1991) provided a large amount of data which have been analyzed within the second phase of the project (ELFIN II) by using local stability methods. One of the objective of the theoretical work was to apply different strategies of integration of the N factor. The results are summarized in [81]. HLFC wind tunnel experiments (on a wing and on a nacelle) have been also performed within ELFIN II.

The main objective of the EUROTRANS research project (1997-1999), chaired by ONERA CERT, is to apply nonlocal methods to selected experiments carried out within ELFIN I and ELFIN II, see [8]. Cooperation between participating research institutes and universities allows all to compare and validate their nonlocal codes (linear and nonlinear) for realistic configurations.

LFC and HLFC experiments

Jet Star experiments In 1980, NASA initiated the Leading Edge Flight Test (LEFT) program as a flight validation of the LFC systems [55]. This program was the first attempt to use suction control on an aircraft since the X-21 program ended in 1965. In these experiments, suction was applied in the leading edge region through perforated strips. The leading edge suction panel which was installed in the right wing of a Jet Star aircraft was a titanium sheet perforated with over 1 million holes of 0.0025 in. diameter spaced about 0.035 in. between centers ; it extended from just below the attachment line to 13 percent chord. Fifteen independent perforated strips of 0.62 in. chord were separated by impermeable strips of 0.38 in. chord.

A Gaster bump was installed on the leading edge in order to avoid leading edge contamination. This bump was combined with a leading edge "notch".

The sweep angle of the test article was 30°. Transition location was detected by near surface Pitot tubes on the suction panel only. Figure 3.18 shows the measured pressure distribution for $M_\infty = 0.775$ at an altitude of 29.000 ft. The flow is accelerated up to $X/C = 0.05$, then it is practically uniform.

High initial suction levels were required to control crossflow disturbances at the leading edge. Late suction yielded less laminar flow than did earlier suction for the same amount of sucked air. Figure 3.19 presents three typical distributions of the suction coefficient C_q together with the corresponding boundary layer state on the test model (white = laminar, black = turbulent). Fully laminar flow over the entire perforated area is achieved with the third distribution.

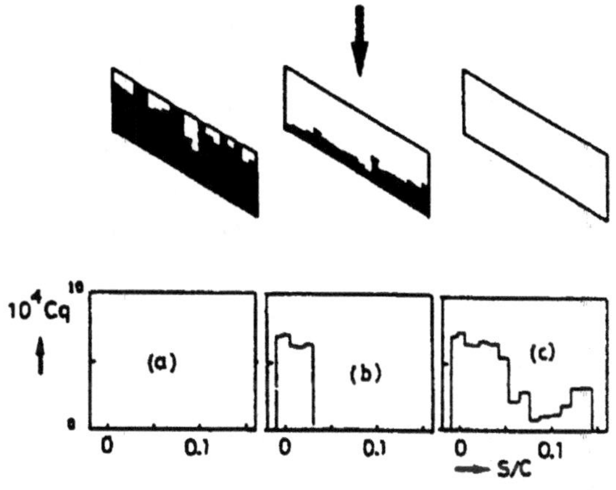

Figure 3.19. Transition location without and with suction, $M_\infty = 0.77$, altitude = 29.000 ft. a) No suction, b) Suction on flutes 1-5, c) Suction on flutes 1-15. From Maddalon *et al.* [55].

Linear stability computations were carried out using the envelope method to analyze the flight data. N factors larger than 12 were computed at the onset of transition with curvature terms neglected in the stability equations.

Boeing 757 experiments Some years ago, high Reynolds numbers flight experiments have been performed on a Boeing 757 aircraft, after more than 300 hr of wind tunnel testing on a simulated section of the wing. A 22-ft span section of the left wing was modified for the flight tests. Suction was applied in the leading edge region, *i.e.* to the front spar, with a structural concept similar to that used for the Jet Star experiments : the suction surface was a microperforated titanium skin with over $10\ 10^6$ laser-drilled, closely spaced holes of 0.0016 in. diameter. Over the wing box, *i.e.* downstream of the front spar, the pressure distribution was "tailored" to maintain laminar flow to the wing shock. This combination of suction (LFC) with wing shaping (NLF) is referred to as *"Hybrid Laminar Flow Control" (HLFC)*.

Flight tests were conducted over a 5-month period to altitudes around 40.000 ft, Mach numbers to 0.82 and chord Reynolds numbers up to $30\ 10^6$. In June 1990, the Boeing 757 aircraft achieved laminar boundary layer flow up to 65 percent wing chord.

Falcon 50 and Falcon 900 experiments As described in paragraph 3.5.4, a first phase of the flight tests on a Falcon 50 aircraft was aimed at demonstrating the feasibility of the NLF concept on a wing section. The second phase (1987-1990) was much more ambitious since its objective was to use HLFC on the inboard right wing of the same airplane. To perform these tests, Dassault Aviation designed a new wing shape and developed a suction system as well as a leading edge cleaning and anti-icing system. Transition was detected with 36 hot films flush mounted on the wing up to 30 percent chord, downstream of the suction panel, and a Gaster bump was installed close to the wing-fuselage junction to prevent leading edge contamination. For sweep angles around 30° and weak suction rates, laminar flow was maintained over nearly the whole test surface.

New HLFC experiments were then performed by Dassault Aviation on a Falcon 900 aircraft. Both wings were equipped with a suction system in the leading edge region. A first series of flight tests (1994) provided promising results, which have been confirmed during a second series of flight experiments performed in 1996. Details can be found in [19], [7].

A320 fin experiments A program consisting of theoretical analysis, wind tunnel experiments and flight tests was initiated by Airbus Industrie in close collaboration with ONERA and DLR in 1987. The vertical fin of the A320 aircraft was chosen as the candidate for evaluation of the feasibility of HFLC. Wind tunnel tests with leading edge suction were carried out in the ONERA S1MA facility. Flight experiments conducted in the second half of 1998 showed a large extent of laminar flow in cruise conditions, in agreement with numerical predictions based on the e^N method.

3.6. Application to high speed flows

The importance of transition and its effects on skin friction in subsonic vehicle drag has been investigated for many years and is relatively well known. The prominence of transition in hypersonic vehicle drag is more uncertain, because of the small amount ot hypersonic flight experience we have. In this paragraph, recent efforts in modeling hypersonic transition physics are described, with the objective of outlining some of the most important challenges to the transition community. Emphasis will be given on transition prediction using the e^N method. All the results presented below were obtained with the local approach, except in paragraph 3.6.2. Real gas effects will not be discussed.

3.6.1. FACTORS ACTING ON THE STABILITY PROPERTIES

As it was already the case for subsonic and transonic flows, linear stability theory makes it possible to study the influence of some parameters which can modify the stability properties of hypersonic flows.

Streamwise pressure gradient

The effects of streamwise pressure gradients are well known for low speed flows. In a negative pressure gradient (accelerated flow), the mean velocity profile has no inflection point and the instability -if it exists- is very weak. In a positive pressure gradient (decelerated flow), the appearance of an inflection point gives rise to a strong inflectional instability.

The stability results showing the influence of streamwise pressure gradients at high speeds are not numerous. A major difficulty is that self-similar boundary layer profiles no longer exist, and it becomes necessary to deduce the basic profiles from non similar boundary layer equations. Vignau [90], for instance, analyzed the effects of positive pressure gradients for M_e decreasing linearly with x. For a local free stream Mach number equal to 5.8, it was found that the first mode was stabilized in comparison with the flat plate case, but that the instability of the second mode was increased. On the other hand, Malik [57] demonstrated that the second mode could be stabilized by favorable, *i.e.* negative, pressure gradients.

Wall temperature

To illustrate the influence of the wall temperature, figure 3.20 shows experimental results collected by Potter [70]. The ratio Rx_T/Rx_{T0} is plotted as a function of T_w/T_{aw} for Mach numbers between 1.4 and 11. Rx_T is the transition Reynolds number computed with the free stream conditions and the transition location x_T. Rx_{T0} is the value for Rx_T measured under adiabatic conditions. T_w and T_{aw} denote the wall temperature and the adiabatic wall temperature, respectively. Although there is some scatter in the data, it appears that cooling the wall delays the onset of transition. This

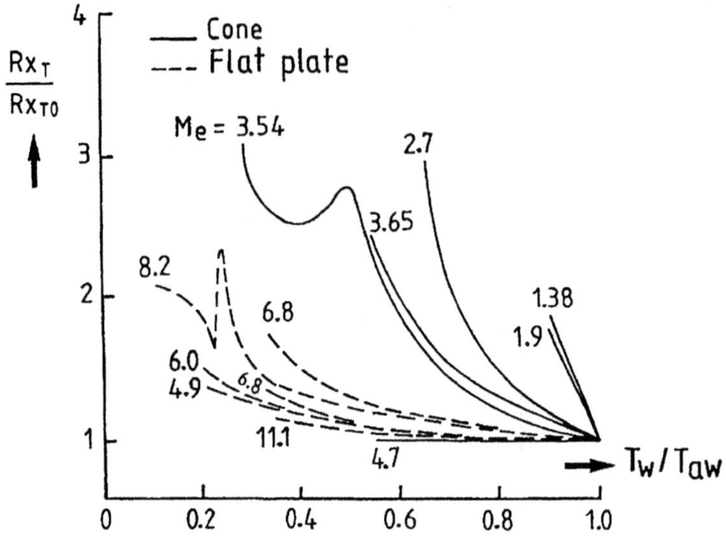

Figure 3.20. Effect of wall cooling on the transition Reynolds number. From Potter [70].

effect is rather strong in the transonic range, but it is reduced when the Mach number increases. Another interesting feature which can be observed in figure 3.20 is the appearance of "transition reversals" and "transition re-reversals" ($M_e = 3.54$ and 8.2). The origin of this behavior has not yet been clearly established.

The stability of laminar boundary layers on cooled walls was studied by Mack [52], Wazzan *et al.* [91], Malik [57] and Vignau [90] for supersonic flows. The stability computations show a qualitative agreement with the experiments, in this sense that the growth rates are reduced when the wall temperature decreases. This effect is linked with the evolution of the height y_s of the generalized inflection point. When y_s increases, the instability is enhanced for a fixed value of M_e. At high Mach numbers, y_s remains close to the outer edge of the boundary layer, and the stability properties become less sensitive to the wall temperature.

3.6.2. TRANSITION PREDICTION

First application of the e^N method : flat plate with adiabatic wall

Figure 3.21 shows an application of the e^N method for adiabatic flat plates. The stability results were used to compute the theoretical streamwise Reynolds numbers Rx which correspond to different values of the N factor, as a function of the free stream Mach number. The dotted lines

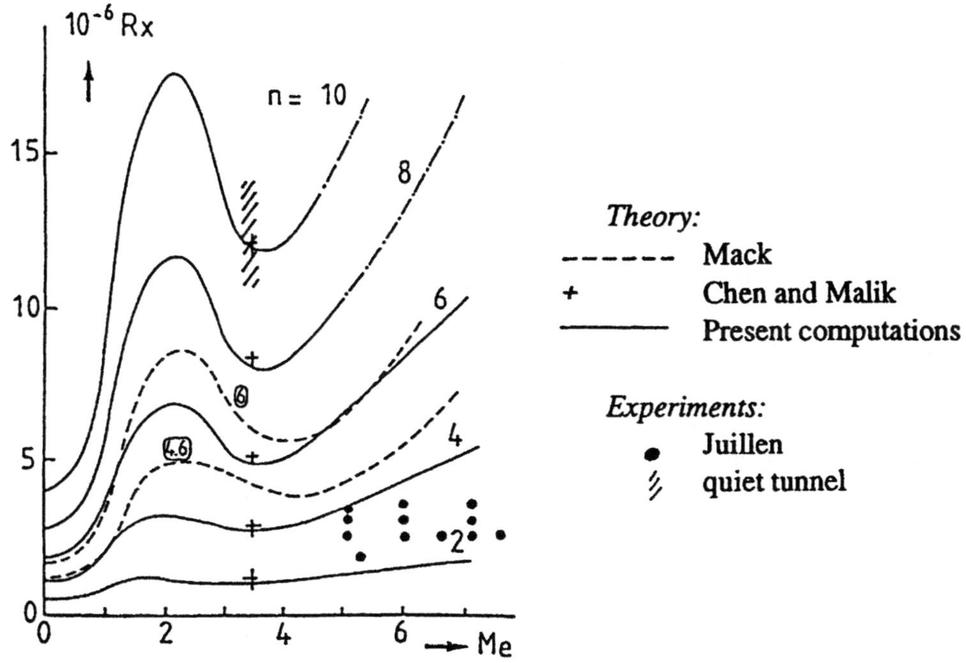

Figure 3.21. Application of the e^N method (flat plate, adiabatic wall). From [4, 53].

represent theoretical results given by Mack [53] for $N = 4.6$ and 6, and the crosses are computations performed by Chen and Malik [24] for $M_e = 3.5$ and $N = 2, 4, 6, 8$ and 10. If it is assumed that transition occurs for a fixed value of the N factor, each curve represents the evolution of the transition Reynolds number with increasing M_e. It can be observed that compressibility has a strong stabilizing effect for transonic flows. It becomes destabilizing for Mach numbers between 2 and 3.5, and stabilizing again at hypersonic conditions.

It has been previously observed that the value of N at the onset of transition is usually between 7 and 10 in "clean" subsonic wind tunnels. The problem is obviously to know if similar values of N are observed at high speeds. The solid symbols in figure 3.21 represent experimental data obtained at ONERA [43]. They correspond to low values of the N factor, between 2 and 4. Numerous examples could be given to illustrate the fact that, in conventional hypersonic wind tunnels, the measured transition Reynolds numbers are rather low, say between 2 and 3 million. Possible origins of this behavior are considered below.

Wind tunnel simulation. Empirical criteria

It is well known that transition on a smooth surface can be triggered by disturbances which are present in the free stream: velocity fluctuations u' (free stream turbulence), pressure fluctuations p' (acoustic disturbances), temperature fluctuations T'. The problem is very complex, because the effects of these various disturbances depend on the Mach number range (see Pate [64]).

At low speeds, the transition Reynolds number is very sensitive to the free stream turbulence level, which is denoted as Tu. However, if Tu becomes very low, pressure fluctuations (fan noise in a wind tunnel for instance) can be of major importance for inducing transition.

In transonic flows, acoustical phenomena linked with slotted or perforated walls give rise to strong pressure disturbances, which cause early transition. Better results are obtained in transonic wind tunnels with solid walls.

In supersonic and hypersonic flows, the main factor affecting transition is also the noise, the origin of which lies in the pressure disturbances radiated by the turbulent boundary layers developing along the wind tunnel walls. The effects of u' and T' have not yet been firmly established for $M_e > 3.5$.

The strong effect of the wind tunnel noise on the transition Reynolds number is illustrated in figure 3.22 (Harvey, [37]), where values of Rx_T measured on cones and flat plates are given as a function of $\tilde{p'}_\infty/q_\infty$ for $4 < M_e < 23$; q_∞ is the mean dynamic pressure in the free stream and the $\tilde{}$ denotes a root mean square value. There are two separate mean curves for cones and flat plates, but the effect of Mach number disappears completely in this representation: transition is governed by $\tilde{p'}_\infty$ rather than by M_∞.

Pate also analysed available wind tunnel data and developed an empirical criterion for "natural" transitions measured in supersonic and hypersonic facilities. This criterion is a correlation between the transition Reynolds number Rx_T and the parameters acting on the noise intensity, *i.e.* the tunnel test section circumference P, and two characteristic parameters of the turbulent boundary layer on the nozzle walls : the mean skin friction coefficient C_F and the displacement thickness $\delta 1$ [64]. If the free stream Mach number and the wall temperature ratio T_w/T_{aw} are constant, this criterion reduces to :

$$Rx_T = (\frac{U_e}{\nu_e})^{0.5} g(P) \tag{3.23}$$

where g is a function of P which increases as the test section circumference increases. This shows that the transition Reynolds number increases with increasing unit Reynolds number U_e/ν_e and increasing wind tunnel size.

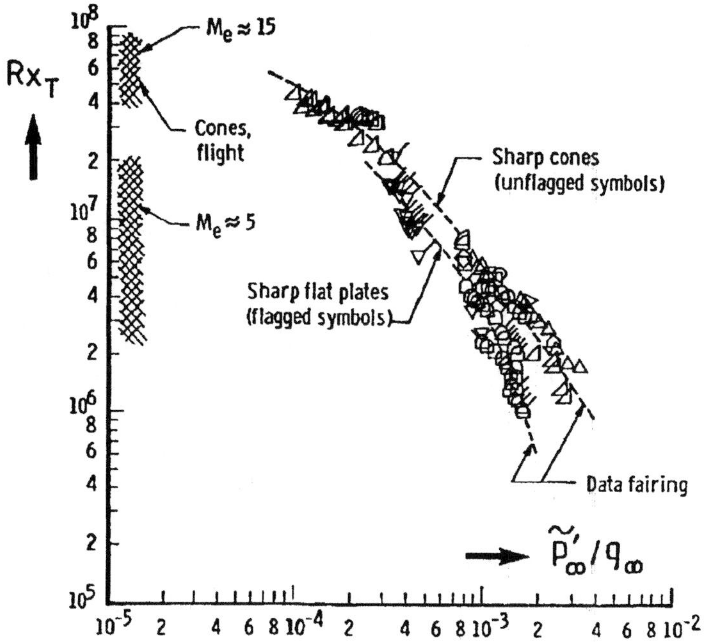

Figure 3.22. Effect of wind tunnel noise on the transition Reynolds number. From Harvey [37].

Another correlation, which was used during wind tunnel tests of a smooth model of Columbia space shuttle, is expressed by the simple relationship:

$$\left[\frac{R_\theta}{M_e}\right]_T = constant \tag{3.24}$$

R_θ is the momentum thickness Reynolds number of the laminar boundary layer which develops on the model. The constant is of the order of 200. The values of $R_{\theta T}$ deduced from this expression are too low when compared with free flight data on smooth models.

Since the radiated noise is inherent in the presence of walls around the model, there is little doubt concerning the incapacity of wind tunnels to properly simulate free flight conditions. In order to reduce this noise level, it is necessary to delay transition on the nozzle walls, since a laminar boundary layer is less noisy than a turbulent one. This was done in the "quiet tunnel" built at NASA Langley with a free stream Mach number $M_\infty = 3.5$. A detailed description of the wind tunnel was given by Beckwith *et al.* [15]. Notable features are the use of boundary layer bleed slots upstream of the throat, a careful polishing and a careful design of the nozzle walls contour in order to minimize the amplification of Görtler vortices. With a laminar boundary layer on the nozzle walls, the measured pressure fluctuations can be one or two orders of magnitude below those measured in conventional facilities. Design and operational details of new low-disturbance wind tunnels are described in [14].

Further applications of the e^N method

Several transition experiments were carried out in the "quiet tunnel". On a flat plate, transition Reynolds numbers as high as 12 10^6 were measured. This value is nearly an order of magnitude larger than those obtained in conventional, noisy facilities [24]. It is interesting to notice that this result corresponds to the theoretical value of Rx_T given by the e^N method for $M_e = 3.5$ and $N = 10$ (figure 3.21).

The flow on supersonic sharp cones constitutes a second case where the free stream Mach number is constant in the streamwise direction. Measurements performed in the "quiet tunnel" on a 5 deg. half angle sharp cone indicated values of Rx_T close to 7 or 8 10^6, two or three times greater than the values obtained in conventional wind tunnels. The predicted transition Reynolds number computed for $N = 10$ is 8 10^6, in good agreement with experimental data [24].

The problem is to know whether or not low disturbance level wind tunnels are representative of free flight conditions. As direct comparisons of the disturbance environment are difficult to perform, indirect comparisons are made by looking at the value of the N factor at the onset of transition. In this respect, the flight experiments on the so-called AEDC cone (Fisher and Dougherty [30]) provided some interesting information. This cone was mounted on the nose of an F-15 aircraft and flown at Mach numbers from 0.5 to 2 and at altitudes from 1500 to 15000 m. Malik [56] computed the N factor for four supersonic Mach numbers and found that the transition Reynolds numbers at the onset of transition were correlated with N factors between 9 and 11. In the experiments, surface pressure fluctuations indicated the formation of TS waves , the frequency band of which was in agreement with linear theory.

Unfortunately, reliable and accurate free flight data are not numerous at higher, hypersonic Mach numbers. Figure 3.23 shows flight transition results which were collected for sharp cones by Beckwith [13]. The transition Reynolds numbers are plotted as a function of the free stream Mach number. The figure also contains a correlation for wind tunnel transition data, which lies much below the flight experiments. The range of results obtained in the "quiet tunnel" are shown for comparison. It is clear that there is a very important scatter in the free flight results, essentially for Mach numbers between 2 and 4. This is mainly due to the fact that these data have been obtained for various wall temperatures, the distribution

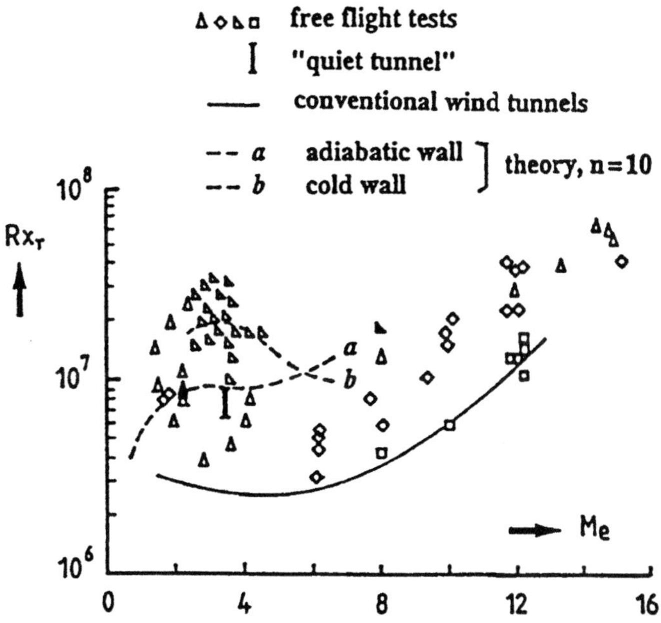

Figure 3.23. Comparison between measured and predicted Reynolds numbers on sharp cones. From Malik [57].

of which is not known in many cases. Malik [57] calculated the transition Reynolds numbers corresponding to $N = 10$ for a 5 deg. half cone and various Mach numbers up to 7. He made two series of computations: one by assuming that the wall was adiabatic and the other by assuming that the wall temperature depended on the free stream Mach number according to a purely empirical relationship:

$$\frac{T_w}{T_{aw}} = 1 - 0.05M_\infty - 0.0025{M^2}_\infty \tag{3.25}$$

The numerical results are reported in figure 3.23. Relation (3.25) makes it possible to reproduce the trends exhibited by the flight results.

The e^N method was used by Malik *et al.* [58] for the rather complex reentry-F experiments. The reentry-F vehicle [42]) consisted of a 5 deg. semi-vertex cone with an initial nose of 2.54 mm. Computations were carried out for an altitude of 30.48 km, where the free stream Mach number was close to 20. The base flow was calculated by equilibrium gas Navier-stokes and PNS codes. At the measured transition location, the N factor was approximately 7.5. Roughness effects probably affected the transition mechanisms at the junction between the nose and the cone, and so that the value of N would be somewhat larger for a perfectly smooth surface.

These computations extend the e^N method into the hypersonic, reacting gas regime.

Examples of parametric studies : bluntness effects

Even if the exact value of the N factor is not known for practical applications, the e^N method remains a very efficient tool for parametric studies : for a given test model and for a given disturbance environment, it is often able to predict the variation of the transition location as a function of the variation of a key parameter which governs the stability properties of the mean flow.

In hypersonic flows, the nose bluntness of a cone at zero angle of attack strongly affects the transition location. For instance, Malik et al [59], [61] performed a linear stability analysis for the experimental conditions studied by Stetson et al [87]. In these experiments ($M_\infty = 8$), the transition mechanisms were investigated on a cone which could be equipped with interchangeable spherically blunted noses of various radii. By using the e^N method, Malik *et al.* found that the predicted transition Reynolds number increased due to small nose bluntness, in qualitative agreement with experimental results. They also demonstrated that nose bluntness could explain the unit Reynolds number effect observed in the aeroballistic range data of Potter [69].

Quite recently, ESA (European Space Agency) launched a TRP (Technological Research Programme) involving several research centers under the responsability of Aerospatiale. The general objective was to analyze laminar-turbulent transition problems for hypersonic flow on slender lifting configurations. One of the experimental studies was devoted to the effect of nose bluntness for a cone placed at zero angle of attack in a wind tunnel at Mach 7 (unit Reynolds number = $25\ 10^6$/m). It was found that the e^N method was able to reproduce the rearward movement of the transition point when the nose radius increased from (nearly) 0 to 0.5 mm ; the chosen value of the N factor was that corresponding to the sharp cone case [88] [12].

Nonlocal effects

PSE results were published by Bertolotti [16] and by Chang et al [23] for supersonic Mach numbers. A good agreement was achieved with the multiple scale analysis of El Hady [34] at $M_e = 1.6$. Figure 3.24 shows the integrated growth rates for a two-dimensional wave (the dimensionless frequency $F = 2\pi f \nu_e / U_e^2$ is $5\ 10^{-5}$) at several Mach numbers. The streamwise Reynolds number R_s is defined as the square root of $\rho_e U_e x/(\mu_e M_e)$. The curves obtained under the parallel flow assumption are compared with

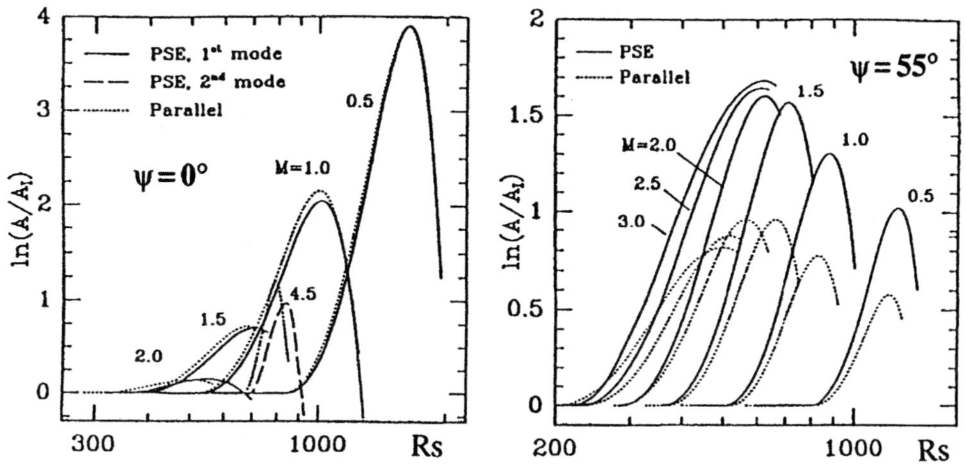

Figure 3.24. Integrated amplification rates for two-dimensional ($\psi = 0°$) and for oblique waves ($\psi = 55°$). From Bertolotti [16].

the PSE results of Bertolotti [16] with the growth rate computed at the maximum of the mass flux. The nonparallel effects become appreciable at $M_e = 4.5$ only.

This conclusion, however, is no longer valid for oblique waves, as it can be seen in figure 3.24 (Bertolotti [16]) which displays the evolution of the total gtrowth rate for the same frequency, but with $\psi = 55°$. The mean flow nonparallelism exerts a strong destabilizing effect, and this effect increases with increasing Mach number. The importance of this difference on the N factor could be significant for supersonic flows since the oblique waves are the most unstable ones and for hypersonic Mach numbers when first mode disturbances are dominant.

A detailed study of the nonparallel effects using PSE was also performed by Chang and Malik [22]. Trends qualitatively similar with those described before were reported for Mach 1.6 and 4.5 flat plate flows. In general, nonparallel effects appear to be less significant for oblique waves near the lower branch of the neutral curve but become more important at higher Reynolds numbers near the upper branch.

The e^N method in three-dimensional flows

Up to now there are only a few results dealing with the application of the e^N method for three-dimensional supersonic and hypersonic flows. The geometries which have been studied in these investigations are cones at angle of attack, swept (or delta) wings or rotating bodies without angle of attack.

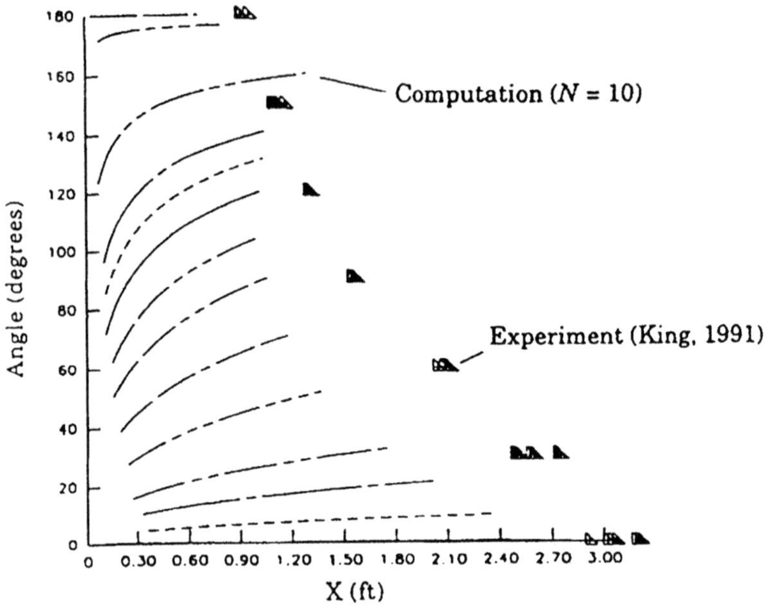

Figure 3.25. Comparison of experimental transition data with the computed ($N = 10$) location of transition. From Malik and Balakumar [60].

Transition on a cone at incidence usually occurs earlier on the leeward line of symmetry than on the windward line. As there is no azimuthal mean velocity component along these lines, their stability properties are those of two-dimensional flows (at least in the framework of the classical stability theory). Away from the windward and leeward rays, crossflow instability can dominate and cause transition.

Malik and Balakumar [60] studied the linear stability of the three-dimensional flow field on a 5° half-angle cone at 2° incidence, for freestream conditions corresponding to those of King's experiments [46] ($M_\infty = 3.5$, unit Reynolds number = 2.5 10^6/ft). As an example of result, figure 3.25 shows a comparison between measured and predicted transition fronts. The N factor trajectories are plotted in the $x - \theta$ plane (θ is the azimuthal angle, which is 0° for the windward ray), with each line ending at $N = 10$. Along each line, the frequency is held constant, and the N factor integration is carried out along the inviscid streamlines. The agreement is satisfactory, even if the predicted transition location is underestimated in the windward ray region. Early transition on the leeward ray is caused by the fact that the mean flow profiles are highly inflectional.

The problem becomes more complicated at higher Mach numbers due to the appearance of second mode disturbances. In the framework of the ESA TRP mentioned before, experiments and computations (local theory) have been performed for a cone at Mach 7 and 2° angle of attack [88] [12]. Figure 3.26 shows the integrated growth rates of first and second mode disturbances at the measured transition locations on the leeward ray (denoted as L in the figure) and on the windward ray (denoted as W). Results at zero angle of attack are given for comparison. The most striking feature is that N increases from windward to leeward ray for first mode disturbances, whereas it decreases for second mode disturbances. As a consequence, predicted transition occurs earlier on the leeward ray than on the windward ray (in agreement with the experiments) if one assumes that it is triggered by first mode disturbances. But transition would appear earlier on the windward ray if it is induced by second mode disturbances. This could be an indication that second mode, high frequency disturbances do not play any major role in these experiments, even if their N factors are much larger than those of first mode disturbances. More details on these computations (including results along the equatorial ray) can be found in [29].

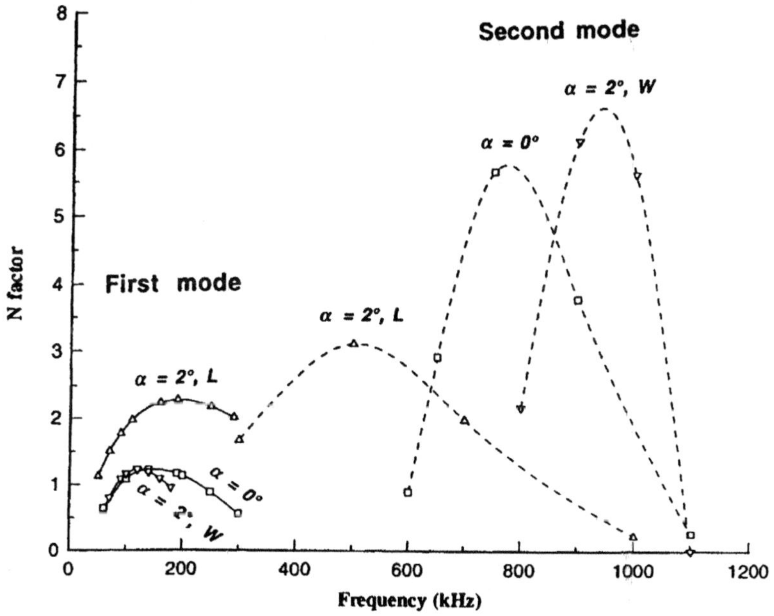

Figure 3.26. N factors at transition on a cone at angle of attack (windward and leeward rays). From [88, 12].

The stability characteristics along the leeward and windward rays of a cone at incidence were investigated by Hanifi [35] using PSE approach (NOLOT code, developed at FFA/KTH in Sweden and DLR in Germany [36]). The computations were performed for conditions corresponding to King's experiments [46] and to Krogmann's experiments [48]. Due to the rather low values of M_∞ (3.5 and 5), only first mode disturbances were present. The movement of transition as a function of angle of attack was fairly well predicted by the e^N method. It was also observed that the non-local effects were larger on the windward meridian than on the leeward meridian.

Transition on a swept wing leading edge model at Mach 3.5 was investigated by Cattafesta et al [21]. Numerical results obtained with the e^N method (local theory, envelope strategy) were compared to experimental transition location measured in the NASA Langley "quiet tunnel". It was found that traveling disturbances with $N \simeq 13$ provided a good correlation with experiments over a range of unit Reynolds numbers and angles of attack.

The transition process on a delta wing was investigated at Imperial College gun tunnel within ESA TRP [88]. A large number of parameters (angle of attack, angle of sweep, unit Reynolds number, leading edge bluntness) was investigated. Linear, local stability computations have then been performed for a few experimental configurations [12]. With sharp leading edges, the boundary layer development was nearly two-dimensional, and experimental transition locations were correlated with N factors between 1 and 2. As soon as the the leading edge radius was increased, the boundary layer flow became highly three-dimensional, and transition was induced by crossflow instability at much larger values of the N factor. From this (rather limited) series of computations, it appears that the N factors at transition are likely to be very different depending on the type of dominant instability.

Reed and Haines [72] investigated the local instability properties of a supersonic rotating cone at zero angle of attack, which was used as a model of a swept wing. The results were then used to develop a simple criterion which was applied to several experiments on cones at angle of attack.

3.7. Conclusion

The e^N method was first developed in 1955 with regard to Görtler vortices and in 1956 for Tollmien-Schlichting waves. About 40 years later, it remains the most widely used method to estimate the transition location, although many experimental and numerical studies have clearly pointed out its deficiencies : the receptivity mechanisms are not accounted for, and nonlinear phenomena as well as higher order instabilities are replaced by a continuous linear amplification up to the onset of transition.

It seems, however, that the e^N method can be applied with some confidence for two-dimensional flows. In the presence of a "good" flow quality, N factors from 9 to 11 correlate experimental data for subsonic, transonic and supersonic flows. This can be explained by the relatively short extent of the nonlinear region (at least for low speed flows) and by the fact that the results were obtained in rather similar disturbance environment.

The situation is not so clear for three-dimensional flows. In the framework of the local approach, the first problem is to know if there exists an "optimum" strategy for the computation of the N factor. The envelope method is widely used (it can be used as a black box), but we think that it is physically meaningless when the wavenumber angle changes rapidly from the crossflow to the streamwise direction. Other procedures separate more or less implicitely TS and CF disturbances and need to be more systematically tested. The linear, nonlocal approach provides results which are qualitatively similar to those given by the constant spanwise wavenumber strategy. The second problem is that the transition mechanisms in a three-dimensional boundary layer are much more intricate than in two-dimensional flows. For instance, micron-sized roughness elements play a major role in the formation of stationary vortices, and the transition process depends on the ratio between the amplitude of the stationary and travelling modes. In addition, nonlinearities strongly affect the development of disturbances. Nonlinear PSE constitute a very efficient tool to identify the possible scenarios leading to transition.

How is it possible to improve the transition prediction methods? Whatever the case might be, a key problem is to know the initial amplitude of the disturbances in the laminar boundary layer. Clearly, receptivity is a long term issue. Beside the development of new numerical tools, more and more detailed experiments need to be performed in order to understand the rich physics of the instability mechanisms. In the long run, the final objective is to replace the actual "amplification methods" by "amplitude methods" which are more relevant from a physical point of view.

References

1. Special Course on Skin Friction Drag Reduction. AGARD Report 786, 1992.
2. Arnal, D. 1986. Three-dimensional boundary layers: laminar-turbulent transition. In *Computation of Three-Dimensional Boundary Layers Including Separation.* AGARD Report No 741.
3. Arnal, D. 1988. Transition prediction in transonic flow. In *Transsonicum III, IUTAM Symp., Göttingen* (Zierep, . and Oertel, ., editors). Springer-Verlag.
4. Arnal, D. June 1989. Laminar-turbulent transition problems in supersonic and hypersonic flow s. In *Special Course on Aerothermodynamics of Hypersonic Vehicles.* AGARD Report No 761.
5. Arnal, D. 1992. Boundary layer transition : prediction, application to drag reduction. In *Skin Friction Drag Reduction.* AGARD Report No 786.
6. Arnal, D. 1993. Prediction based on linear theory. In *Progress in Transition Modelling.* AGARD Report No 793.
7. Arnal, D. and Bulgubure, C. May 1996. Drag reduction by boundary layer laminarization. *La Recherche Aérospatiale.* .

8. Arnal, D., Casalis, G. and Schrauf, G., June 1996. The EUROTRANS project. 2nd European Forum on Laminar Flow Technology, Bordeaux.
9. Arnal, D., Gasparian, G. and Salinas, H., 1998. Recent advances in theoretical methods for laminar-turbulent transition prediction. AIAA Paper 98-0223.
10. Arnal, D., Habiballah, M. and Coustols, E., 1984. Laminar instability theory and transition criteria in two- and three-dimensional flows. La Recherche Aérospatiale 1984-2.
11. Arnal, D., Juillen, J., Reneaux, J. and Gasparian, G. 1997. Effect of wall suction on leading edge contamination. *Aerospace Science and Technology.* **1** (8), 505–517.
12. Arnal, D., Kufner, E., Oye, I. and Tran, P., March 1996. PROGRAMME TRP TRANSITION : Computational results for transition prediction. Study Note 7.
13. Beckwith, I. 1975. Development of a high Reynolds number quiet tunnel for transition re search. *AIAA Journal.* **13** (3).
14. Beckwith, I., Chen, F., Wilkinson, S., Malik, M. and Tuttle, D., 1990. Design and operational features of low-disturbance wind tunnels at NAS A Langley for Mach numbers from 3.5 to 18. AIAA Paper 90–1391.
15. Beckwith, I., Creel Jr, T., Chen, F. and Kendall, J. 1983. Freestream noise and transition measurements on a cone in a Mach 3.5 pilot low-disturbance tunnel. TP 2180, NASA.
16. Bertolotti, F., 1991. Compressible boundary layer analyzed with the PSE equations. AIAA Paper 91–1637.
17. Bippes, H. 1990. Instability features appearing on swept wing configurations. In *Laminar-Turbulent Transition, IUTAM Symp., Toulouse* (Arnal, D. and Michel, R., editors). Springer-Verlag.
18. Bippes, H., 1991. Experiments on transition in three-dimensional accelerated boundary layer flows. Proc. R.A.S. Boundary Layer Transition and Control, Cambridge.
19. Bulgubure, C. and Arnal, D. March 1992. Dassault Falcon 50 laminar flow flight demonstrator. In *First European Forum on Laminar Flow Technology,* Hamburg.
20. Casalis, G., Copie, M., Airiau, C. and Arnal, D., July 1995. Nonlinear analysis with PSE approach. IUTAM Symposium on Nonlinear Instability and Transition in Three-dimensional Boundary Layers, Manchester.
21. Cattafesta III, L., Iyer, V., Masad, J., King, R. and Dagenhart, J. November 1995. Three-dimensional boundary-layer transition on a swept wing at mach 3.5. *AIAA Journal.* **33** (11).
22. Chang, C. and Malik, M., 1993. Non-parallel stability of compressible boundary layers. AIAA Paper 93–2912.
23. Chang, C., Malik, M., Erlebacher, G. and Hussaini, M., 1991. Compressible stability of growing boundary layers using parabolized stability equations. AIAA Paper 91–1636.
24. Chen, F. and Malik, M., 1988. Comparison of boundary layer transition on a cone and flat plate at M ach 3.5. AIAA Paper 88–0411.
25. Collier Jr, F., 1993. An overview of recent subsonic laminar flow control flight experiments. AIAA Paper 93-2987.
26. Coustols, E., 1995. Control of turbulent flows for skin friction drag reduction. CISM Course on Control of Flow Instabilities and Unsteady Flows.
27. Coustols, E. and Savill, A., 1992. Turbulent skin friction drag reduction by active and passive means. AGARD Report 786.
28. Dallmann, U., Hein, S., Koch, W., Bertolotti, F., Simen, M., Stolte, A., Gordner, A. and Nies, J., June 1996. Status of the theoretical work within DLR's non empirical transition prediction project. 2nd European Forum on Laminar Flow Technology, Bordeaux.
29. Dussillols, L. 1999. Calculs de stabilité et transition sur des configurations aérodynamiques complexes. Master's thesis, SUPAERO, France.
30. Fisher, D. and Dougherty, N. 1982. In-flight transition measurements on a 10 deg. cone at Mach numbers f rom 0.5 to 2. TP 1971, NASA.

31. Gasparian, G. 1998. Etude expérimentale et modélisation des mécanismes de transition sur aile en flèche, avec application au maintien de la laminarité. Master's thesis, SUPAERO, France.
32. Gaster, M. May 1967. On the flow along leading edges. *The Aeron. Quarterly.* **XVIII, Part2**.
33. Goldstein, M. 1983. The evolution of TS waves near a leading edge. *Journal of Fluid Mechanics.* **127**, 59–81.
34. Hady, N. E., 1991. Non-parallel instability of supersonic and hypersonic boundary layers. AIAA Paper 91–0324.
35. Hanifi, A. 1995. *Non-local stability analysis of the compressible boundary layer on a yawed cone.* PhD thesis, KTH, Stockholm.
36. Hanifi, A., Henningson, D., Hein, S., Bertolotti, F. and Simen, M. 1994. Linear nonlocal instability analysis – the linear NOLOT code. Report 1994–54, FFA.
37. Harvey, W. 1978. Influence of free stream disturbances on boundary layer transition. TM 78635, NASA.
38. Henke, R., Münch, F. and Quast, A., 1990. Natural laminar flow: a wind tunnel test campaign and comparison with flight test data. AIAA Paper 90-3045.
39. Herbert, T. 1988. Secondary instability of boundary layers. *Ann. Rev. Fluid Mech.* **20**, 487–526.
40. Herbert, T. 1993. Parabolized Stability Equations. In *Progress in Transition Modelling.* AGARD–FDP–VKI Special Course.
41. Horstmann, K., Redeker, G., Quast, A., Dressler, U. and Bieler, H., 1990. Flight tests with a natural laminar flow glove on a transport aircraft. AIAA Paper 90-3044.
42. Johnson, C., Stainback, P., Wiker, K. and Bony, L. 1972. Boundary layer edge conditions and transition Reynolds number data fo r a flight test at Mach 20 (Reentry F). TM X–2584, NASA.
43. Juillen, J. 1969. Détermination expérimentale de la région de transition sur une pl aque plane à M = 5, 6 et 7. Technical Report 10/2334 AN, ONERA.
44. Juillen, J. and Arnal, D. 1995. Experimental study of boundary layer suction effects on leading edge contamination along the attachment line of a swept wing. In *Laminar-Turbulent Transition, IUTAM Symp., Sendai* (Kobayashi, R., editor). Springer-Verlag.
45. Kerschen, E., 1989. Boundary layer receptivity. AIAA Paper 89–1109.
46. King, R., 1991. Mach 3.5 boundary layer transition on a cone at angle of attack. AIAA Paper 91-1804.
47. Kohama, Y., Saric, W. and Hoos, J., 1991. A high frequency, secondary instability of crossflow vortices that leads to transition. Proc. R.A.S. Boundary Layer Transition and Control, Cambridge.
48. Krogmann, P., 1977. An experimental study of boundary layer transition on a slender cone at mach 5. AGARD CPP 224.
49. Lees, L. and Lin, C. 1946. Investigation of the stability of the laminar boundary layer in a compressible fluid. TN 1115, NACA.
50. Lerche, T. and Bippes, H. 1996. Experimental investigation of cross-flow instability under the influence of controlled disturbance excitation. In *Transitional Boundary Layers in Aeronautics* (Henkes, R. and van Ingen, J., editors). Royal Netherlands Academy of Arts and Sciences.
51. Mack, L. 1965. Stability of the compressible laminar boundary layer according to a direct numerical solution. In *AGARDograph 97, Part 1*, pp. 329–362.
52. Mack, L., November 1969. Boundary layer stability theory (2 volumes). Jet Propulsion Laboratory, Calif. Inst. of Techn., Pasadena, California.
53. Mack, L. 1975. Linear stability and the problem of supersonic boundary layer transition. *AIAA Journal.* **13** (3), 278–289.
54. Mack, L. 1984. Boundary layer linear stability theory. In *AGARD Report No 709.*
55. Maddalon, D., Collier Jr, F., Montoya, L. and Putnam, R. 1990. Transition flight experiments on a swept wing with suction. In *Laminar-Turbulent Transition,*

IUTAM Symp., Toulouse (Arnal, D. and Michel, R., editors). Springer-Verlag.
56. Malik, M. February 1984. Instability and transition in supersonic boundary layers. In *Laminar Turbulent Boundary Layers.* Energy Sources Technology Conf., New Orleans, Louisiana.
57. Malik, M., 1987. Prediction and control of transition in hypersonic boundary layers. AIAA Paper 87–1414.
58. Malik, M. 1990. Stability theory for chemically reacting flows. In *Laminar-Turbulent Transition, IUTAM Symp., Toulouse* (Arnal, D. and Michel, R., editors). Springer-Verlag.
59. Malik, M. 1990. Transition in hypersonic boundary layers. In *Numerical and Physical Aspects of Aerodynamic Flows IV* (T.Cebeci, ., editor). Springer-Verlag.
60. Malik, M. and Balakumar, P., 1992. Instability and transition in three-dimensional supersonic boundary layers. AIAA Paper 92–5049.
61. Malik, M., Spall, R. and Chang, C., 1990. Effect of nose bluntness on boundary layer stability and transition. AIAA Paper 90–0112.
62. Morkovin, M. 1968. Critical evaluation of transition from laminar to turbulent shear layer s with emphasis on hypersonically travelling bodies. Report AFFDL–TR–68–149, Wright-Patterson Air Force Base, Ohio.
63. Müller, B. and Bippes, H., 1988. Experimental study of instability modes in a three-dimensional boundary layer. AGARD CP No 438.
64. Pate, S., 1980. Effects of wind tunnel disturbances on boundary layer transition with e mphasis on radiated noise: a review. AIAA Paper 80–0431, Colorado Springs.
65. Pfenninger, W., 1965. Flow phenomena at the leading edge of swept wings. AGARDograph 97, Part 4.
66. Poll, D. August 1978. Some aspects of the flow near a swept attachment line with particular reference to boundary layer transition. Technical Report 7805./K, Cranfield, College of Aeronautics.
67. Poll, D., 1984. Transition description and prediction in three-dimensional flows. AGARDo Report 709.
68. Poll, D. and Danks, M. 1995. Relaminarisation of the swept wing attachment-line by surface suction. In *Laminar-Turbulent Transition, IUTAM Symp., Sendai* (Kobayashi, R., editor). Springer-Verlag.
69. Potter, J. 1974. The unit Reynolds number effect on boundary layer transition. Master's thesis, Vanderbildt University.
70. Potter, J. 1980. Review of the influence of cooled walls on boundary layer transition. *AIAA Journal.* **18** (8).
71. Radetsky, R., Reibert, M., Saric, W. and Takagi, S., 1993. Effect of micron-sized roughness on transition in swept wing flows. AIAA Paper 93-0076.
72. Reed, H. and Haynes, T., 1993. Transition correlation in three-dimensional boundary layers. AIAA Paper 93–3054.
73. Reed, H., Saric, W. and Arnal, D. 1996. Linear stability theory applied to boundary layers. *Annu. Rev. Fluid. Mech.* **28**, 389–428.
74. Reshotko, E., 1984. Laminar flow control– viscous simulation. Special Course on Stability and Transition of Laminar Flow, AGARD–R–709.
75. Reynolds, W. and Saric, W., 1982. Experiments on the stability of the flat plate boundary layer with suction. AIAA Paper 82-1026.
76. Robert, J., 1992. Drag reduction : an industrial challenge. AGARD Report 786.
77. Salinas, H., 1998. Stabilité linéaire et faiblement non linéaire d'une couche limite pour un fluide compressible, avec l'approche PSE. Thèse en préparation.
78. Saric, W. and Reed, H., 1987. Three-dimensional stability of boundary layers. Perspectives in Turbulence Studies, Springer-Verlag.
79. Schrauf, G., 1994. Transition prediction using different linear stability analysis strategies. AIAA Paper 94-1848.
80. Schrauf, G., Herbert, T. and Stuckert, G., 1995. Evaluation of transition in flight tests using nonlinear PSE analysis. AIAA Paper 95-1801.

81. Schrauf, G., Perraud, J., Vitiello, D., Lam, F., Stock, H. and Abbas, A., June 1996. Transition prediction with linear stability theory. Lessons learned from the ELFIN F100 flight demonstrator. 2nd European Forum on Laminar Flow Technology, Bordeaux.
82. Schubauer, G. and Skramstad, H. 1948. Laminar boundary layer oscillations and transition on a flat plate. Report 909, NACA.
83. Seyfang, G. September 1987. Turbulence reduction on swept leading edges. In *Turbulent Drag Reduction by Passive Means*, London.
84. Simen, M. 1992. *Local and nonlocal stability theory of spatially varying flows*, volume Instability, Transition and Turbulence. Springer-Verlag.
85. Smith, A. and Gamberoni, N. 1956. Transition, pressure gradient and stability theory. Rept. ES 26388, Douglas Aircraft Co., El Segundo, California.
86. Spalart, P., 1988. Direct numerical study of leading edge contamination. AGARD CP No 438.
87. Stetson, K., Thompson, E., Donaldson, J. and Siler, L., 1984. Laminar boundary layer stability experiments on a cone at Mach 8– Part 2: blunt cone. AIAA Paper 84–0006.
88. Tran, P., Seraudie, A., Wendt, V. and Poll, D., 1995. PROGRAMME TRP TRANSITION : Experimental results for transition prediction. Study Note 6.
89. van Ingen, J. 1956. A suggested semi-empirical method for the calculation of boundary layer transition region. Rept. UTH–74, Univ. of Techn., Dept. of Aero. Eng., Delft.
90. Vignau, F. 1989. *Etude théorique et expérimentale de la transition en écoulement bidimensionnel compressible.* PhD thesis, ENSAE Toulouse.
91. Wazzan, A. and Taghavi, H. 1982. The effect of heat transfer on three–dimensional spatial stability and transition on flat plate boundary layer at Mach 3. *Int. J. Heat Mass Transfer.* **25** (9), 1321–1331.

Chapter 4

AN INTRODUCTION TO TURBULENCE MODELLING

A.V. JOHANSSON AND A.D. BURDEN
Department of Mechanics, Royal Institute of Technology,
SE-100 44 Stockholm, Sweden

4.1. Introduction

The main aim of the present chapter is to introduce engineering turbulence modelling emphasizing the underlying physics and methodology of development of this type of turbulence models. A complete state-of-the-art review is not attempted, but rather to give a basic understanding of turbulence closures and a tutorial into the foundations of basic enginering turbulence modelling. We will will focus on models for incompressible flows, but also introduce some basic concepts and equations for compressible turbulence.

In section 4.2 some typical features of turbulent flows will be exemplified and discussed in relation to their relevance for various aspects of turbulence modelling. For compressible flows, the mass-weighted, or Favre, averaging is introduced to obtain the basic equations in a form as similar as possible to the incompressible counterparts. This is especially motivated since the modelling approach for compressible flows usually can be seen as modifications of incompressible closures. Also the mean flow equations for both incompressible and compressible flow are discussed.

The exact transport equations for the statistical moments of fluctuating quantities used in single-point turbulence modelling are presented and discussed in section 4.3. The hierarchy of single-point turbulence closures and the associated concepts are introduced in section 4.4. Fundamental features of the exact equations that should be conserved in the model formulation are discussed in section 4.5. So called invariant modelling is also introduced to illustrate, e.g., the consequencies of requiring the solutions to be physically realizable. Purely algebraic models are briefly discussed in section

4.6 and two-equation models based on the eddy-viscosity concept are presented in section 4.7. Also, the modelling of the individual terms and the calibration of model constants are discussed in some detail. The basics of differential Reynolds stress models are discussed in section 4.8 and explicit algebraic Reynolds stress models are presented in section 4.9.

More detailed aspects of modelling of turbulent transport or diffusion terms in the above mentioned equations, and for passive scalar transport, were discussed in chapter 3, section 10 of the book from the 1995 Summer School on Turbulence and Transition Modelling.

We here also wish to acknowledge that the present text contains a large portion of revised text from the previous Summer School in 1995 where Dr Magnus Hallbäck gave significant contributions to the turbulene modelling chapter.

4.2. Basic properties of turbulence and the mean flow equation

Turbulence is characterized by a multitude of scales in time and space and associated mixing and diffusion of momentum, heat etc that are orders of magnitude stronger than in laminar flows. Turbulence is present in most flows in technical applications, but occurs also in a variety of other situations, such as geophysical flows and even in astrophysics. In physical three-dimensional turbulence the dynamics of the eddies is also characterized by an energy cascade from large to small eddies and finally, through the action of viscosity, into heat. One may note that this energy cascade is coupled to the action of vortex stretching and thereby not present in strictly twodimensional (two-component) turbulence.

The span of lengthscales grows as the macroscale Reynolds number (Re_l) to a power of 3/4. This means that one may think of an effective number of 'degrees of freedom' as increasing with Reynolds number as fast as $Re_l^{9/4}$ in three dimensions. A direct computation of the turbulent flow without approximations of the equations of motion hence requires a number of grid points that increases 'prohibitively' fast with increasing Reynolds number in the range typical for technical applications. Taking the requirements also for the time-step into account one finds that the computational 'work' increases as fast as the cube of the Reynolds number. The need for turbulence modelling is, hence, illustrated by the fact that an increase of the Reynolds number by one order of magnitude results in a thousandfold increase of the computational effort in DNS.

Nonetheless, direct numerical simulation (DNS) of turbulent flows at low to moderate Reynolds numbers plays an important role, together with physical experiments, in the development and validation of turbulence models at different levels of closure. A major advantage of DNS is that it allows direct term-by-term comparison between the exact and modelled transport equations for various turbulence quantities. For compressible turbulent flows experiments are extremely dificult to carry out in a controlled manner. DNS here plays a primary role in model testing, despite the rather severe limitations in Reynolds number.

Compressibility effects on turbulence are complex with many different facets (see *e.g.* the review of Lele[64]). The increased complexity is readily understood from the fact that in this case both thermodynamic and thermophysical fluid properties are fluctuating quantities. Among the former we have the density and specific entropy, and among the latter we have *e.g.* the viscosity. The compressible turbulence field involves motions associated with (Kovasznay[52]) vorticity, acoustic and entropy modes. The dissipation of turbulent kinetic energy has a dilatational part, and the pressure-dilatation correlation, which can have either sign, is involved in the transfer of energy between the turbulent kinetic energy and the mean internal energy. Explicit modelling of such terms is referred to the chapter of Friedrich in the present book.

The relative intensity of the density fluctuations can be used as a measure of the effects of compressibility. In turbulence closures, and for collapsing data from compressible turbulent flows, various compressibility parameters are used. A common parameter is the turbulence Mach number, which is the ratio of the characteristic turbulence velocity scale (for the energetic eddies) and the speed of sound. The compressibility effects typically become strong when this parameter excedes a value of 0.2-0.3. As pointed out by Lele[64] an increase in turbulence Mach number naturally leads to an increase in the relative density fluctuations, but is independent from the density fluctuations associated with temperature fluctuations. Large imposed temperature differences between a solid wall and ambient fluid will, of course, lead to significant density fluctuations. This source is dominating also in turbulent flames.

In wall-bounded compressible flows the compressibility effects show up at considerably larger convective (or free-stream) Mach numbers than in free shear flows. For a compressible mixing layer the growth rate significantly decreases with increasing convective (based on velocity difference between the twwo streams) Mach number. The scatter in available data is considerable but the growth rate is roughly decreased to half its incompressible value at a convective Mach number of 0.5 (see Lele[64]). This is associated with a decrease in the relative turbulence levels. For turbulent boundary layers the compressibility effects are rather small up to Mach numbers of about 5 (Wilcox[137]).

An approach that is becoming increasingly interesting with growing computer capacities, is the so called large eddy simulation (LES) technique in which only the small scales are modelled. These are normally less anisotropic and less influenced by the geometry of the flow than the large scales, which encourages optimism about the possibility of constructing simpler models for these small scales than those that can be thought of as adequate for the complete turbulence field. The LES technique is undergoing rapid development and is becoming an alternative to traditional turbulence modelling, in particular for non-stationary flows.

Otherwise in turbulence modelling the underlying ideas and inspiration are taken both from traditional continuum mechanics and statistical mechanics/physics. In contrast to LES other closures attempt to model the complete turbulence field. The most ambitious approaches (for incompressible flows) involve closure of the evolution equations for the two-point velocity correlations (even two-time has been analyzed), or equivalently moments of Fourier coefficients in spectral space. Such approaches involve the full range of lengthscale dynamics and may accurately describe the cascade process in three-dimensional turbulence, and are also capable of accounting for the inverse cascade that occurs in the mathematically idealized case of two-dimensional turbulence. This type of closure has in some ambitious attempts been used as a subgrid model in LES, but can otherwise be regarded more as a tool for improving the understanding of turbulence dynamics, and for obtaining information that in turn can be used for the development of engineering turbulence models based on Reynolds decomposition and transport equations for single-point statistics.

4.2.1. DECOMPOSITION AND MEAN FLOW EQUATION FOR INCOMPRESSIBLE FLOW

For incompressible flows the only natural decomposition into a mean field and a fluctuating part is through the use of ensemble averaging

$$u_i = \overline{u_i} + u_i' = U_i + u_i' \quad \text{and} \quad p = \overline{p} + p' = P + p'.$$

We will use dash to denote ensemble averages in general and prime to denote fluctuations around this value. Capital letters are also used for mean values of velocity and pressure. The above split is normally referred to as Reynolds decomposition.

The average mass-conservation equation reduces to the incompressibility condition

$$U_{i,i} = 0$$

where the short-hand notation with subscripts after a comma denote corresponding derivatives.

The ensemble-averaged incompressible Navier-Stokes equation, often called the Reynolds equation, takes the form

$$\rho\left(\frac{\partial U_i}{\partial t}+U_j\frac{\partial U_i}{\partial x_j}\right)=-\frac{\partial P}{\partial x_i}+\frac{\partial}{\partial x_j}\left(2\mu S_{ij}-\rho\overline{u_i'u_j'}\right)+\rho F_i \qquad (4.1)$$

where μ is the dynamic viscosity, F_i is the mean value of the body force and the unknown correlation

$$-\rho\overline{u_i'u_j'}$$

can be regarded as the turbulence contribution to the stress tensor, and is ususally referred to as the Reynolds stress tensor. It is seen that the viscous stress only depends on the mean strain rate tensor ($S_{ij}\equiv\frac{1}{2}(\frac{\partial U_i}{\partial x_j}+\frac{\partial U_j}{\partial x_i})$), which is the symmetric, *i.e.* the deformational part of the mean velocity gradient tensor.

For incompressible flows the density, ρ, is of no qualitative importance and one often refers to the second order, single-point, velocity correlation

$$-\overline{u_i'u_j'}\equiv -R_{ij}$$

as the Reynolds stress tensor (or sometimes the kinematic Reynolds stress tensor). Correspondingly, we define the kinetic energy (per unit mass as)

$$K=\frac{1}{2}\overline{u_k'u_k'}$$

In contrast to the viscous stress tensor R_{ij} depends in a complex manner on the dynamics of the turbulent flow. In a turbulent flow the dynamics involves, for instance, energy transfer between the mean flow and the fluctuations, intercomponent energy transfer, spatial redistribution of energy (and other quantities) and the dynamics of energy transfer between scales of different sizes. Although the Reynolds stress is a single-point measure the dynamics governing this quantity does in fact depends also on features like the mentioned energy transfer between different scales.

In traditional engineering models of turbulence only single-point turbulence statistics are used to describe the effects of the turbulence field. This is in essence equivalent to an assumption of self-similar turbulence spectra for which the dynamics are governed by one single lengthscale.

A transport equation for R_{ij} may readily be derived, but involves third-order statistics etc, revealing the well-known closure problem caused by the non-linearity of the Navier-Stokes equations.

4.2.2. DECOMPOSITION AND MASS-WEIGHTED, FAVRE, AVERAGING FOR COMPRESSIBLE FLOW

When the compressibility effects are felt, for instance in terms of a significant relative density fluctuation level the method for decompostion into mean and fluctuating parts is no longer obvious. The conventional decomposition based on un-weighted ensemble averaging, as used in the incompressible case, is here not the only alternative.

For an arbitrary quantity, ϕ we write this decomposition as

$$\phi = \overline{\phi} + \phi' \tag{4.2}$$

where the first part represents the ensemble mean and prime denotes fluctuation around this value. For the density itself this is the only (natural) choice

$$\rho = \overline{\rho} + \rho' \tag{4.3}$$

With u_i as the total instantaneous velocity we may write the ensemble averaged mass conservation equation (simply by adding dashes) as

$$\frac{\partial \overline{\rho}}{\partial t} + \frac{\partial \overline{\rho u_k}}{\partial x_k} = 0$$

from which the Favre, or mass-weighted, decomposition of the velocity field is naturally suggested

$$u_i = \tilde{u}_i + u_i'' \tag{4.4}$$

where the first part, the Favre average, is defined as

$$\tilde{u}_i = \frac{\overline{\rho u_i}}{\overline{\rho}} \tag{4.5}$$

and u_i'' is the flucuation around this mass-weighted average. This type of notation will be used also for decomposition of other mass-weighted quantities. The Favre decomposition allows the mean mass-conservation equation to be witten in the form

$$\frac{\partial \overline{\rho}}{\partial t} + \frac{\partial}{\partial x_k}(\overline{\rho}\tilde{u}_k) = 0 \tag{4.6}$$

The corresponding equation written with the aid of a Renolds (ensemble) decomposition of the velocity field would involve the density-velocity correlation $\overline{\rho' u_k'}$.

We define the Favre average of an arbitrary scalar (or vector or tensor) quantity as

$$\tilde{\phi} = \frac{\overline{\rho\phi}}{\overline{\rho}} \tag{4.7}$$

The Favre decomposition is then written as

$$\phi = \tilde{\phi} + \phi'' \tag{4.8}$$

which also means that $\overline{\rho\phi''} = 0$ and $\widetilde{\phi''} = 0$ (also true for the velocity).

4.2.3. THE MEAN FLOW EQUATION FOR COMPRESSIBLE FLOWS

The ensemble average of the momentum balance equation (see chapter 1) reads

$$\frac{\partial \overline{\rho u_i}}{\partial t} + \frac{\partial \overline{\rho u_j u_i}}{\partial x_j} = -\frac{\partial \overline{p}}{\partial x_i} + \frac{\partial \overline{t_{ji}}}{\partial x_j} \tag{4.9}$$

where t_{ij} is the viscous stress tensor

$$t_{ij} = 2\mu\left(s_{ij} - \frac{1}{3}s_{kk}\delta_{ij}\right) + \mu_v s_{kk}\delta_{ij}.$$

$s_{ij} = \frac{1}{2}\left(u_{i,j} + u_{j,i}\right)$ and μ_v is the bulk viscosity which we here will take to be zero (true for monatomic gases) so that In the averaging process we should here keep in mind that also the viscosity in general is a fluctuating quantity.

In equation (4.9) we introduce ensemble average based decomposition for the density ($\rho = \overline{\rho} + \rho'$) and pressure ($p = P + p'$) and Favre decomposition for the velocity ($u_i = \tilde{u}_i + u_i''$), yielding

$$\frac{\partial \overline{\rho}\tilde{u}_i}{\partial t} + \frac{\partial \overline{\rho}\tilde{u}_j\tilde{u}_i}{\partial x_j} = -\frac{\partial P}{\partial x_i} + \frac{\partial}{\partial x_j}\left(\overline{t_{ji}} - \overline{\rho u_j'' u_i''}\right) \tag{4.10}$$

where the identity $\overline{\rho u_i''} = 0$ has been used and we define the Reynolds stress tensor as

$$\tau_{ij} = -\overline{\rho u_j'' u_i''}$$

Equation (4.10) is the compressible counterpart to the incompressible form (4.1) of the Reynolds equation. The similarity becomes even more obvious if we (by use of the continuity equation) rewrite the left hand side of (4.10) as

$$\overline{\rho}\left(\frac{\partial \tilde{u}_i}{\partial t} + \tilde{u}_j \frac{\partial \tilde{u}_i}{\partial x_j}\right).$$

In analogy with the incompressible case we here define the turbulent kinetic energy (per unit volume) as

$$\overline{\rho}K = \frac{1}{2}\,\overline{\rho u_k'' u_k''}.$$

4.2.4. AVERAGED CONSERVATION EQUATIONS FOR e, h, Y_α

The conservation equations (see chapter 1) for the specific internal energy, the specific enthalpy and the mass fraction of a species, Y_α, all have the general form,

$$\frac{\partial}{\partial t}(\rho\phi) + \frac{\partial}{\partial x_k}(\rho\phi u_k) = -\frac{\partial j_k}{\partial x_k} + s_\phi \tag{4.11}$$

where $j_k(\vec{x}, t)$ is the (molecular) flux of ϕ and s_ϕ is a possible source term. The form of this equation is essentially equivalent to the momentum balance, and suggests that a Favre decomposition is a natural choice for ϕ. This yields an ensemble average of equation (4.11)

$$\frac{\partial}{\partial t}\left(\bar{\rho}\tilde{\phi}\right) + \frac{\partial}{\partial x_k}\left(\bar{\rho}\tilde{\phi}\tilde{u}_k\right) = -\frac{\partial}{\partial x_k}\left(\overline{\rho\phi'' u_k''} - \overline{j_k}\right) + \overline{s_\phi} \tag{4.12}$$

where the form of the equation is the same as for the original equation. An ensemble ('Reynolds') decomposition would would also yield an equation that conserves the form of (4.11), but with two extra terms that involve correlations between the fluctuations of density and ϕ or u_k.

We see that equation (4.12) includes a mean turbulent flux term

$$\overline{\rho\phi'' u_k''}$$

which has to be modelled just as the Reynolds stress in the Reynolds equation.

From the modelling viewpoint a significant aspect is that (4.12) is virtually identical to the incompressible counterpart for a scalar variable

$$\frac{\partial}{\partial t}\left(\rho\overline{\phi}\right) + \frac{\partial}{\partial x_k}\left(\rho\overline{\phi}\overline{u}_k\right) = -\frac{\partial}{\partial x_k}\left(\rho\overline{\phi' u_k'} - \overline{j_k}\right) + \overline{s_\phi}. \tag{4.13}$$

In the development of computational models for compressible flow the first iteration, hence, often consists of the incompressible forms applied to the Favre averages.

Some insight into the usefulness of mass-averaged modelling can perhaps be gained by noting that the variables e and h are specific, *i.e.* defined per unit mass rather than per unit volume. Hence, when they appear in conservation equations they are normally multiplied by the mass density. The same is true for the mass fraction, Y_α, of a chemical species.

We will return to the specific forms of the mean equations for e, h and Y_α.

4.2.5. METHODOLOGY OF SINGLE-POINT MODEL DEVELOPMENT

The methodology in the development of single-point turbulence closures is to a large extent inspired by the approaches used in continuum mechanics for development of constitutive relations for (non-fluid) materials. One may there distinguish three basic principles. Firstly, the formulation should be independent of the specific coordinate system, which leads to proper tensor formulation of the equations. Much of the general theory of tensor relations and invariant theory has also been used for turbulence modelling purposes. Secondly, the material response should be independent of solid body rotations, and thirdly, the behaviour of a material element should only depend on the history of that element and not on its neighbouring elements. The last two principles are, however, not valid for the modelling of the Reynolds stress tensor.

Turbulence modelling bears many similarities with rheology and modelling of non-Newtonian behaviour. The principle of fading memory often used there is fundamental also for turbulence modelling. One often requires the rheological models to comply with basic physical constraints, for instance to ensure positive energy dissipation. This is equally relevant here, where one nowadays often requires the models to satisfy 'realizability'. For instance, in models based on transport equations for the individual Reynolds stresses one often requires that the models for the terms involved ensure negative values of the diagonal components of the stress tensor.

In the cascading flow of energy to smaller and smaller scales the viscosity of the fluid sets a lower limit to the size of scales that can exist in the flow. The viscosity, *i.e.* the internal friction in the fluid will convert kinetic enrgy into heat. The total energy dissipation is dominated by direct contributions from the small scales, and plays an important role in the dynamics of the turbulence. It enters, either directly or indirectly in all single-point turbulence models. Before analyzing the transport equations associated with the models it may therefore be worthwhile to take a closer look at this quantity and the interpretation that we should give to it in the context of single-point turbulence models.

For thc purpose of studying the dissipation of energy at small scales and the way in which this energy is supplied by the larger scales we may analyze the situation of homogeneous incompressible turbulence in a spectral formulation. We will restrict our attention to isotropic turbulence.

4.2.6. BASIC PROPERTIES OF NEAR-WALL INCOMPRESSIBLE TURBULENCE

Many of the modelling difficulties are associated with the regions near solid walls. It is in this region that due to the structure of the turbulence is highly anisotropic and influenced by the presence of the wall. The mixing of the turbulent fluctuations smears out the velocity gradients in the region far from the walls, which due to the viscous no-slip condition causes steep gradients and associated strong turbulence production near the walls. The strong inhomogeneity, the damping effects of the wall etc. form extremely difficult problems for the modeller.

From a global view we have two distinctly different regions. Close to the wall the viscosity is of significant influence, wheras far from the wall it is only of secondary importance. Correspondingly there is a disparity of governing scales in these two regions.

In order to study near-wall turbulence it is instructive to choose the case of a fully developed plane turbulent channel flow. In that case the mean velocity field is parallel and independent of the downstream coordinate (x) and the Reynolds averaged momentum equations read

$$0 = -\frac{\partial P}{\partial x} + \frac{d}{dy}\left(\mu\frac{dU}{dy} - \rho\overline{u'v'}\right) \tag{4.14}$$

$$0 = -\frac{\partial P}{\partial y} - \rho\frac{d}{dy}\overline{v'v'} \tag{4.15}$$

where U and $\overline{u_i'u_j'}$ are functions of y only and $P = P(x, y)$. The continuity equation is trivially satisfied. Let the channel height be $2h$ and $y = 0$ at the lower wall. Integration of the second equation gives the pressure

$$P(x, y) = -\rho\overline{v'v'} + P_{\mathrm{w}}(x) \tag{4.16}$$

where $P_{\mathrm{w}}(x)$ is the pressure at the wall ($\overline{v'v'}_{\mathrm{w}} = 0$). Integrating the first equation one gets

$$\mu\frac{dU}{dy} - \rho\overline{u'v'} = \frac{dP_{\mathrm{w}}}{dx}y + \tau_{\mathrm{w}} \tag{4.17}$$

where

$$\tau_{\mathrm{w}} \equiv \mu\frac{dU}{dy}\bigg|_{y=0} \tag{4.18}$$

is the wall shear stress. Evaluating the equation at the centerline ($y = h$) one finds that $\frac{dP_{\mathrm{w}}}{dx} = -\frac{1}{h}\tau_{\mathrm{w}}$ and the equation may be rewritten

$$\mu\frac{dU}{dy} - \rho\overline{u'v'} = \left(1 - \frac{y}{h}\right)\tau_{\mathrm{w}} \tag{4.19}$$

The total shear stress, *i.e.* the left hand side of (4.19), varies linearly across the channel.

The characteristic length-scale, l_*, in the near wall region should be independent of the outer geometric restrictions of the flow and therefore only determined by the wall shear stress and the viscosity. Introducing the so called friction velocity

$$u_\tau \equiv \sqrt{\frac{\tau_{\mathrm{w}}}{\rho}} \tag{4.20}$$

we obtain $l_* \equiv \nu/u_\tau$. Introducing the inner (viscous) scaling of variables

$$y^+ = y/l_* \tag{4.21}$$
$$U^+ = U/u_\tau \tag{4.22}$$

equation (4.19) reads

$$\frac{dU^+}{dy^+} - \overline{u'v'}^+ = 1 - y^+/\frac{u_\tau h}{\nu} \tag{4.23}$$

We see that there may be an appreciable range of y^+ units near the wall in which the total shear stress is nearly constant if the Reynolds number, $u_\tau h/\nu$, is high enough. One may argue that in this range the parametric dependence on the Reynolds number is weak and that in this region there exist universal functions

$$U^+ = f_1(y^+) \tag{4.24}$$
$$-\overline{u'v'}/u_\tau^2 = g_1(y^+) \tag{4.25}$$

This is usually referred to as the "law of the wall", and is obviously consistent with the above assumption that the governing length-scale near the wall should be independent of the details of the outer flow. There is numerous experimental evidence that near-wall velocity distributions are compatible with the "law of the wall".

Well away from the wall, it is reasonable to assume that the channel height is the relevant length scale. The viscous stress is here negligible in comparison with the turbulent one, which, as we can see from equation (4.19) varies linearly in this region. The velocity distribution, or rather the distribution of the normalized deviation from the centreline velocity (U_{CL}) and the distribution of the normalized turbulent stress should, hence, be uniqe functions of $Y = y/h$ in the outer region.

$$(U_{CL} - U)/u_\tau = f_2(Y) \tag{4.26}$$
$$-\overline{u'v'}/u_\tau^2 = g_2(Y) \tag{4.27}$$

The first of these two relations is usually referred to as the "velocity defect law".

For high enough Reynolds numbers there should be a region for which y^+ becomes very large (in principle $\rightarrow \infty$) simultaneously as Y becomes very small (in principle $\rightarrow 0$). In this region both sets of functions (f_1, g_1) and (f_2, g_2) should be valid (Millikan 1939). We may approach this matching problem by studying the nondimensional quantity $\frac{y}{u_\tau}\frac{dU}{dy}$ in the overlap region

$$\frac{y}{u_\tau}\frac{dU}{dy} = y^+\frac{df_1}{dy^+} = -Y\frac{df_2}{dY} = \text{const} \tag{4.28}$$

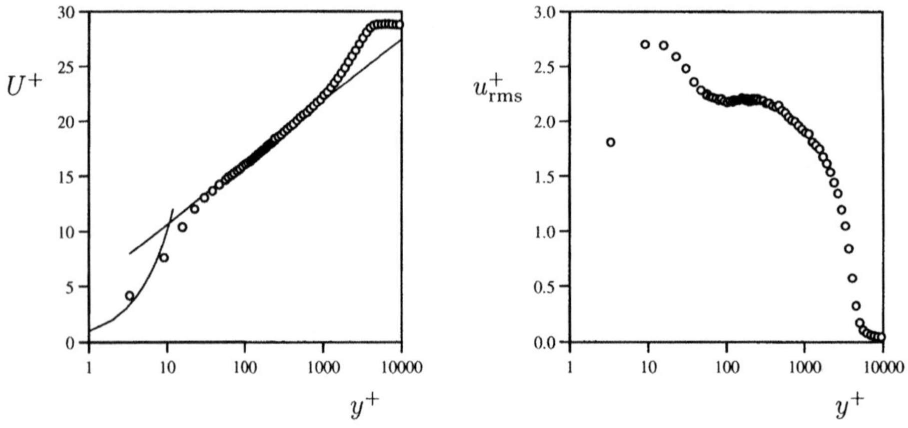

Figure 4.1. The mean velocity and streamwise turbulence intensity profiles in a zero-pressure gradient turbulent boundary layer. Momentum loss Reynolds number is 14000 (Österlund 1995, private communication).

Consistency in the above relation requires the quantity to be a constant, usually written as $1/\kappa$, where κ is known as the von Kármán constant. This yields

$$f_1(y^+) = \frac{1}{\kappa}\ln y^+ + B \tag{4.29}$$

$$f_2(Y) = -\frac{1}{\kappa}\ln Y + C \tag{4.30}$$

where $\kappa \approx 0.41$ and $B \approx 5.0$ for smooth surfaces (and high enough Reynolds numbers). These values hold well also for boundary layer flows. The constant C has a value of about 1 in channel flow, but is considerably larger in boundray layers. The region in which (4.29,b) holds is known as the "inertial sublayer" due to the independence of both viscous effects and the details of the outer flow. This independence of lengthscale is analogous with the inertial range in the wave-number spectrum as discussed above. The log-law was derived also by von Kármán (1930) although in a somewhat more heuristic manner.

In this case the matching also requires that $g_1 = g_2 =$ const. in the overlap region. Hence, the normalized Reynolds stress should be constant in the overlap region (and approach unity with increasing Reynolds number).

The characteristic log-region in the mean velocity distribution in a zero pressure-gradient boundary layer flow is seen in figure 4.1. Also the variation of the streamwise turbulence intensity is shown, and exhibits a plateau in the log-region. This is actually a characteristic feature for all Reynolds stresses in this region.

A further consequence of the existence of the overlap region may be derived by adding the relations (4.29,b). We thereby obtain an implicit relation between U_{CL}/u_T and the Reynolds number, $Re = U_{CL}h/\nu$

$$\frac{U_{CL}}{u_T} = -\frac{1}{\kappa}\ln\frac{U_{CL}}{u_T} + \frac{1}{\kappa}\ln Re + B + C \tag{4.31}$$

Hence, it follows directly from the two-layer hypothesis that the friction coefficient is uniquely determined as a function of the Reynolds number.

From (4.23) it is clear that there is a region closest to the wall, denoted the 'viscous sub-layer', in which $\overline{u'v'} \to 0$ and the viscous stress term, $\frac{dU^+}{dy^+}$, dominates, hence, giving a linear variation of the velocity, $U^+ = y^+$.

We may estimate the turbulence production term in the near-wall region by multiplying (4.23) by $\frac{dU^+}{dy^+}$. For reasonable Reynolds numbers the y^+/h^+ term is here negligible, and we obtain the normalized production rate as

$$\mathcal{P}^+ = \frac{dU^+}{dy^+} - \left(\frac{dU^+}{dy^+}\right)^2 \tag{4.32}$$

The maximum production rate is found where

$$\frac{d\mathcal{P}^+}{dy^+} = \frac{d^2U^+}{dy^{+2}}\left(1 - 2\frac{dU^+}{dy^+}\right) = 0$$

This obviously occurs where $\frac{dU^+}{dy^+} = \frac{1}{2}$, whence

$$\mathcal{P}^+_{\max} = \frac{1}{4} \tag{4.33}$$

The location of maximum production is, hence, found where the viscous and turbulent stresses are equal. This location is found at $y^+ \approx 10$, inbetween the viscous sublayer and the log-layer. This region is usually referred to as the buffer region and extends from $y^+ \approx 5$ to $y^+ \approx 30$.

In the inertial sub-range (or the log-layer), which extends out to $y/\delta \approx 0.15$ the distribution of energy among the components of the Reynolds stress tensor is approximately (table 1, Launder *et al.* 1975)

$$\frac{\overline{u'u'}}{2K} = 0.59, \qquad \frac{\overline{u'v'}}{2K} = -0.12, \qquad \frac{\overline{v'v'}}{2K} = 0.12, \qquad \frac{\overline{w'w'}}{2K} = 0.29$$

The main discrepancy between these data and those of homogeneous shear flow (section 4.3) is particularly that the $\overline{v'v'}$ component is substantially damped by the presence of the wall. The randomizing effect of the pressure field, for instance quantified by the pressure-strain correlation, is thus inhibited by the presence of the wall in this region. To further illustrate, quantitatively, the difference between a free shear layer and a shear layer along a wall boundary one may compare the two-dimensional plane free jet and wall jet flows. The spreading rate of the 2D plane wall jet is about 30% lower than that of the free jet. Again, the reason is that the lateral motions, and thereby also the shear stress $-\overline{u'v'}$, are damped by the wall which influences also the outer parts of the wall jet. One may note that although the shear stress is significantly lower in the wall shear layer the correlation coefficient $-\overline{u'v'}/u'_{\rm rms}v'_{\rm rms}$ is just slightly lower, 0.44, compared to 0.47 in the homogeneous (free) shear flow.

The near wall region exhibits a streaky structure for $y^+ < 20$, with long streamwise structures of alternating high and low velocity. These have a mean relative spanwise spacing of about 100 viscous units. Knowledge about this is important in, for instance, LES of the near wall layer where the filter width has to be small enough in the spanwise direction to resolve the dominating structures.

Another feature of the immediate vicinity of the wall is that the turbulence Reynolds number is low. One consequence of this in modelling contexts is that the standard length scale estimate ($l \sim K^{3/2}/\varepsilon$) is not appropriate here. The remedy is usually to introduce Reynolds number corrections. In the limit of very low Reynolds numbers the dissipation rate should adhere to a scaling like $\varepsilon \sim \nu K/l^2$.

4.2.7. THE COMPRESSIBLE TURBULENT BOUNDARY LAYER

A recent review on supersonic turbulent boundary layers is given by Spina *et al.* [122] and is discussed in Wilcox[137]. A typical feature of this flow is the presence of a density gradient caused by the dissipative heating near the solid wall. For adiabatic walls (Lele[64]) the ratio of wall to free-stream temperature rises from $T_{\rm W}/T_\infty = 1.9$ at $M_\infty = 2.2$ to $T_{\rm W}/T_\infty = 4.7$ at $M_\infty = 4.5$.

The density variation also contributes to a reduced turbulence intensity near the wall, as compared to the incompressible case, and therby to a reduction in skin friction coefficient.

For adiabatic walls the mean velocity profiles can be made to collapse with the incompressible counterparts reasonably well by use of the van Driest[130] rescaling (see *e.g.* discussion in Wilcox[137])

$$\frac{u*}{U_\infty} = \frac{1}{A}\sin^{-1}\left(A\frac{\tilde{u}}{U_\infty}\right), \quad A^2 = \frac{\gamma - 1}{2}M_\infty^2\left(\tilde{T}_\infty/\tilde{T}_{\rm W}\right).$$

The friction velocity is here defined as

$$u_\tau = \sqrt{\tau_{\rm W}/\overline{\rho}_{\rm W}}$$

where subscript w signifies wall values.

4.2.8. THE ENERGY CASCADE IN TURBULENCE

At high Reynolds numbers, Re_{T}, of the turbulence itself, turbulence possesses a very wide range of significant length, time, and velocity scales and this allows us to discuss these scales in a fairly loose, order-of-magnitude, manner. The kinetic energy, K, of the turbulence is more or less concentrated to 'eddies' with a size, ℓ, corresponding to the formal integral length scales of the fluctuating part of the velocity field. These energy-bearing eddies have a velocity scale, $u \sim K^{1/2}$, where '$\sim$' is used here to denote 'of the order of' and even 'varies like as Re_{T} varies (but is still large)'.

Bearing in mind the Navier-Stokes equation, eq. (1.8) in Chapter 1, it is clear that two mechanisms act on the eddies; inertial effects such as vortex-stretching and instability, and viscous effects which damp the motion. At high values of $Re_{\mathrm{T}} \simeq u\ell/\nu$, viscous damping has a negligible effect on the energy-bearing eddies but their energy is spread by the inertial effects over a wide range of scales. The upper bound to the spread of turbulent energy is the restriction provided by the boundaries of the flow, though these extremely large eddies are much less intensive than the energy-bearing eddies. The lower bound to the spread of turbulent energy is effectively set by the viscous damping. Roughly speaking, the effects of viscosity are as significant as the inertial effects when the particular Reynolds number of the relevant small eddies is of the order of one, $u'\ell'/\nu \sim 1$. Assuming now that these small eddies have a time scale, ℓ'/u', which is much smaller than the time scale, ℓ/u, of the energy-bearing eddies, their mean length, time, and velocity scales will be determined by ε_f, the mean rate at which energy is transfered to the small dissipating eddies from the large energy-bearing eddies. This dominant process of energy transfer is refered to as the cascade. The cascade concept, ε_f, is important in the modelling of rates of turbulent mixing and, consequently, is sometimes applicable in the modelling of mean rates of chemical reactions in turbulent combustion.

At high Reynolds numbers, Re_{T}, the direct influence of viscous dissipation on the eddy-bearing eddies is negligible so the total mean rate of viscous dissipation must be equal to the rate at which energy is transfered out of the energy-bearing range, $\varepsilon \simeq \varepsilon_f$. Furthermore, the quasi-equilibrium assumption made above, based on $\ell'/u' \ll \ell/u$, implies that the rate of transfer out of the energy-bearing eddies can be estimated in terms of the scales of these eddies, $\varepsilon_f \sim u^2/(\ell/u)$. Putting these estimates together, we find,

$$\varepsilon \simeq \varepsilon_f \sim \frac{u^2}{\ell/u} \simeq \frac{K^{3/2}}{\ell}, \tag{4.34}$$

which is a central relationship in two-equation models of turbulent flow.

The quasi-equilibrium assumption also implies that the length, time and velocity scales of the small dissipating eddies can be determined in terms of ε and ν and this leads to the definitions,

$$\eta = \ell_{\mathrm{K}} = \left(\frac{\nu^3}{\varepsilon}\right)^{1/4}, \qquad \tau_{\mathrm{K}} = \left(\frac{\nu}{\varepsilon}\right)^{1/2} \quad \text{and} \quad v_{\mathrm{K}} = (\nu\varepsilon)^{1/4}. \tag{4.35}$$

These are refered to as the Kolmogorov microscales and the hypothesis that the small scales are in quasi-equilibrium with the large scales is also attributed to Kolmogorov. The estimate for ε in eq. (4.34) leads to the estimates,

$$\frac{\eta}{\ell} \sim Re_{\mathrm{T}}^{-3/4}, \qquad \frac{\tau_{\mathrm{K}}}{\ell/u} \sim Re_{\mathrm{T}}^{-1/2} \quad \text{and} \quad \frac{v_{\mathrm{K}}}{u} \sim Re_{\mathrm{T}}^{-1/4}. \tag{4.36}$$

Note that the discussion above is not a rigorous deriviation. It is logically self-consistent but it is primarily descriptive. The qualitative hypotheses are confirmed by experiments but more quantitative comparison with experiment reveals the need to take temporal and spatial intermittency of the small scales into account.

Kolmogorov eddies play an even more significant role in turbulent mixing and combustion than they do in turbulent flow. They are, of course, responsible for the finest-scale mixing that the turbulence can achieve but they also dominate the rate-of-strain field,

$$\overline{s_{ij}s_{ij}} \sim \overline{s'_{ij}s'_{ij}} \sim \frac{1}{\tau_{\mathrm{K}}^2}.$$

This is why they provide the largest contribution to the viscous dissipation and it also means that they play a decisive role in determining which local instantaneous burning structures exist in turbulent combustion.

The quasi-equilibrium hypothesis, together with the assumption of high Re_{T}, implies that scales, u'' and ℓ'', which lie in the inertial range, inbetween the energy-bearing eddies and the dissipating eddies, can be determined in terms of $\varepsilon_f \simeq \varepsilon$ alone. Dimensional analysis yields,

$$u'' \sim \left(\varepsilon\,\ell''\right)^{1/3}. \tag{4.37}$$

(Possible intermittency corrections in the power 1/3 are relatively small.)

4.3. Transport equations for single-point moments

In chapter 1 of this book we discussed the basic equations necessary to describe the motion in a compressible medium, *i.e.* conservation of mass, momentum and energy, the latter involving the internal energy (and enthalpy). A relation between the state variables is also needed.

Single-point turbulence closures for incompressible flows are based on the Reynolds equation and the continuity condition together with a set of transport equations for turbulence quantities that forms a closed set of equations. This is the context to which most of the discussion in the following sections will be devoted.

For the more general case when compressibility effects are non-negligible we need to complement these with equations for further quantities, such as the heat flux vector. The degree of compressibility can be measured in numerous ways, but perhaps a reasonably general measure is the relative density fluctuation level. Most of the turbulence model development for compressible flows has been based on the experience, and equations, for the incompressible case, complemented with corrections for the extra explicit compressibility terms that arise. The hypersonic regime is left outside the scope of this presentation.

The most common starting point is to formulate the equations in so called Favre, or mass-weighted, averages (see *e.g.* Wilcox[137]). For instance, this means that one replaces the velocity as a primary variable by the momentum, which is natural since the net effect of the imposed forces is a rate of change of the momentum rather than of the velocity.

We will see that the Favre averaging leads to basic transport equations that take on a form quite similar to that for the incompressible case. It is, however, not a totally undisputed starting point. For instance, Yoshizawa[138] discusses the possibility of, instead using the 'incompressible' approach of ensemble averaging, which would lift out more clearly the effects of density fluctuations. The price to pay is, however, rather high in terms of increased complexity of the equations. We will not analyze this alternative here.

We may refer to the quantities for which we solve transport equations as primary quantities. In so called two-equation models a transport equation for a velocity scale (or rather the kinetic energy) is complemented by an equation for the lengthscale of the energetic eddies or some other quantity such as the total dissipation rate. If instead one chooses to solve a transport equation for the Reynolds stress tensor, this still needs to be complemented by an ε-equation or some equivalent quantity. For incompressible flow the exact forms of the kinetic energy Reynolds stress tensor and dissipation rate equations will be presented below and the grouping of the terms and their physical interpretation will be discussed there. In all these equations we will use the notation

$$\frac{\mathrm{D}}{\mathrm{D}t} \equiv \frac{\partial}{\partial t} + U_m \frac{\partial}{\partial x_m} \tag{4.38}$$

to denote rate of change plus advection by the mean flow. Also the kinetic energy equation for compressible flow will be discussed.

4.3.1. THE EXACT K-EQUATION FOR INCOMPRESSIBLE FLOW

In turbulence modelling contexts the standard way of grouping the terms in the kinetic energy equation is

$$\frac{\mathrm{D}K}{\mathrm{D}t} = \mathcal{P} - \varepsilon - \frac{\partial}{\partial x_m}\left(J_m - \nu\frac{\partial K}{\partial x_m}\right) + \overline{u_i' f_i'} \tag{4.39}$$

where

$$\mathcal{P} \equiv -\overline{u_i' u_m'} S_{im} \tag{4.40}$$

$$\varepsilon \equiv \nu \overline{u_{i,m}' u_{i,m}'} \tag{4.41}$$

$$J_m \equiv \frac{1}{2}\overline{u_i' u_i' u_m'} + \frac{1}{\rho}\overline{u_m' p'} \tag{4.42}$$

The first term on the right hand side of (4.39) represents production of turbulent kinetic energy, *i.e.* transfer of energy from the mean flow to the turbulent fluctuations. Thus, in the mean flow kinetic energy equation (see chapter 1) we have a corresponding term with the opposite sign. The production term is practically always positive, but can under some conditions temporarily and locally become negative.

As we should expect it is only the deformational, symmetric part of the mean velocity gradient tensor that contributes to the energy production. We also see that for isotropic turbulence, where we have $\overline{u_i' u_m'} = \frac{2}{3}K\delta_{ij}$, the turbulence production is identically zero (since $S_{ii} = 0$). On the other hand, a non-zero mean strain will drive the turbulence away from isotropy.

An advantage of the above grouping of terms is that one gets an explicit term representing molecular diffusion of the kinetic energy. In this formulation the dissipation rate term ε is positive semi-definite. The difference between this dissipation term and the true dissipation, $\varepsilon^{\mathrm{true}} \equiv 2\nu\overline{s_{im}' s_{im}'}$, is

$$\varepsilon^{\mathrm{true}} - \varepsilon = \nu\frac{\partial^2 \overline{u_i' u_m'}}{\partial x_i \partial x_m} \tag{4.43}$$

By estimating orders of magnitude this discrepancy can be shown to be small in realistic situations. The ratio ε/ν scales as K/λ^2, where λ is the Taylor micro scale (cf. Batchelor 1953), whereas the right hand side of (4.43) scales as K/L^2, where L is a length scale over which R_{ij} exhibits $\mathcal{O}(1)$ changes, *i.e.*

$$\varepsilon^{\mathrm{true}} = \varepsilon\left(1 + \mathcal{O}\left(\tfrac{\lambda^2}{L^2}\right)\right)$$

In all realistic situations L, which is typically determined by the geometry of the flow problem, is larger or of the same order of magnitude as the turbulence integral length scale l. At worst, the length scale ratio scales as $l/\lambda \sim Re_T^{1/2}$ at high Reynolds numbers. In homogeneous turbulence $L \to \infty$ and ε becomes the true dissipation rate. Hence, "homogeneous dissipation rate" is an appropriate designation of ε. One can easily show that, in fact, $\varepsilon = \varepsilon^{\text{true}}$ also at a solid wall. In homogeneous turbulence the following identities hold

$$\frac{\varepsilon}{\nu} \equiv \overline{u'_{i,m}u'_{i,m}} = 2\overline{s'_{im}s'_{im}} = 2\overline{\omega'_{im}\omega'_{im}} = \overline{\omega'_i\omega'_i} \tag{4.44}$$

where the last quantity is referred to as the enstrophy.

The turbulent flux of kinetic energy, J_m, is driven both by turbulent velocity and pressure fluctuations.

4.3.2. THE EXACT REYNOLDS STRESS TRANSPORT EQUATION FOR INCOMPRESSIBLE FLOW

In a similar manner the Reynolds stress transport equation may be written

$$\frac{\mathrm{D}\,R_{ij}}{\mathrm{D}t} = \mathcal{P}_{ij} - \varepsilon_{ij} + \Pi_{ij} - \frac{\partial}{\partial x_m}\left(J_{ijm} - \nu R_{ij,m}\right) + \overline{u'_i f'_j} + \overline{u'_j f'_i} \tag{4.45}$$

where

$$\mathcal{P}_{ij} \equiv -\overline{u'_i u'_m}\frac{\partial U_j}{\partial x_m} - \overline{u'_j u'_m}\frac{\partial U_i}{\partial x_m} \tag{4.46}$$

$$\varepsilon_{ij} \equiv 2\nu\overline{u'_{i,m}u'_{j,m}} \tag{4.47}$$

$$\Pi_{ij} \equiv \frac{2}{\rho}\overline{p' s'_{ij}} \tag{4.48}$$

$$J_{ijm} \equiv \overline{u'_i u'_j u'_m} + \frac{1}{\rho}\left(\overline{u'_j p'}\delta_{im} + \overline{u'_i p'}\delta_{jm}\right) \tag{4.49}$$

Here again, one obtains a molecular diffusion term that is explicit in the Reynolds stress tensor. All other viscous effects are lumped into one single term, denoted by ε_{ij}, normally referred to as the 'dissipation rate tensor'.

The production tensor represents the direct interaction between the turbulence and the mean flow, the trace being equal to $2\mathcal{P}$. In contrast to the energy production term, the individual $\mathcal{P}_{ij}$-terms are influenced both by the mean strain and mean rotation rate tensors.

In the RST-equations we have one term that has no counterpart in the energy equation. This pressure-strain rate correlation tensor, which hence has zero trace, is associated with intercomponent energy redistribution.

Since this correlation involves the pressure fluctuations, it represents a term associated with non-local interactions.

The spatial redistribution is represented by J_{ijm} and is driven both by turbulent velocity and pressure fluctuations.

Example: homogeneous shear flow

In a two-dimensional, plane, steady, homogeneous shear-flow, the mean flow is given by

$$\frac{\partial U}{\partial y} = S, \qquad V = 0, \qquad W = 0$$

where S here denotes the constant mean shear rate. The K-equation reads

$$\frac{\partial K}{\partial t} = -\overline{u'v'}S - \varepsilon \tag{4.50}$$

Ideally, in a homogeneous shear flow there are no outer constraints limiting the turbulence length or velocity scales and observations from experiments show that turbulence tends to reach a quasi-equilibrium state where the kinetic energy, and the dissipation, grow at a rate such that $SK/\varepsilon \approx 6$ (Tavoularis & Corrsin 1981). The transport equation for the $\overline{u_i'u_j'}$ tensor here reads

$$\frac{\partial \overline{u'u'}}{\partial t} = \Pi_{11} - \varepsilon_{11} - 2\overline{u'v'}S \tag{4.51}$$

$$\frac{\partial \overline{u'v'}}{\partial t} = \Pi_{12} - \varepsilon_{12} - \overline{v'v'}S \tag{4.52}$$

$$\frac{\partial \overline{v'v'}}{\partial t} = \Pi_{22} - \varepsilon_{22} \tag{4.53}$$

$$\frac{\partial \overline{w'w'}}{\partial t} = \Pi_{33} - \varepsilon_{33} \tag{4.54}$$

These equations show that all transfer of energy from the mean flow to the turbulence goes through the $\overline{u'u'}$ component. Although there is no direct transfer to the $\overline{v'v'}$ and $\overline{w'w'}$ components (zero production: $\mathcal{P}_{22} = \mathcal{P}_{33} = 0$) the typical distribution of energy among the velocity components in a homogeneous shear flow is (Tavoularis & Corrsin 1981)

$$\frac{\overline{u'u'}}{2K} = 0.53, \qquad \frac{\overline{u'v'}}{2K} = -0.15, \qquad \frac{\overline{v'v'}}{2K} = 0.19, \qquad \frac{\overline{w'w'}}{2K} = 0.28$$

This is due to the action of the pressure field, manifested by the pressure-strain correlation terms, that tends to randomize the turbulence field and redistribute energy from the rich to the poor velocity components.

The kinetic energy grows indefinitely in this case, and a closer look at the numbers above yields that the production to dissipation ratio is about two,

$$\frac{\mathcal{P}}{\varepsilon} = \frac{-\overline{u'v'}S}{\varepsilon} = -2\frac{\overline{u'v'}}{2K}\frac{SK}{\varepsilon} \approx 2 \times 0.15 \times 6 = 1.8$$

The homogeneous turbulent shear flow is one of the corner-stone flow cases that are often used in model calibration.

4.3.3. THE DISSIPATION RATE EQUATION FOR INCOMPRESSIBLE FLOW

The transport equation for the dissipation rate, ε, can be written

$$\frac{\mathrm{D}\varepsilon}{\mathrm{D}t} = \mathcal{P}^{\varepsilon(1)} + \mathcal{P}^{\varepsilon(2)} + \mathcal{P}^{\varepsilon(3)} + T^{\varepsilon} - D^{\varepsilon} - \frac{\partial}{\partial x_m}\left(J_m^{\varepsilon} - \nu\frac{\partial \varepsilon}{\partial x_m}\right) + 2\nu\overline{u'_{i,m}f'_{i,m}} \quad (4.55)$$

where on the right-hand side we have grouped the various terms in the 'standard' way

$$\mathcal{P}^{\varepsilon(1)} = -2\nu S_{ij}\overline{u'_{i,k}u'_{j,k}} \quad (4.56)$$

$$\mathcal{P}^{\varepsilon(2)} = -2\nu S_{ij}\overline{u'_{k,i}u'_{k,j}} \quad (4.57)$$

$$\mathcal{P}^{\varepsilon(3)} = -2\nu U_{i,jk}\overline{u'_j u'_{i,k}} \quad (4.58)$$

$$T^{\varepsilon} = -2\nu\overline{u'_{i,j}u'_{i,k}u'_{j,k}} \quad (4.59)$$

$$D^{\varepsilon} = 2\nu^2\overline{u'_{i,jk}u'_{i,jk}} \quad (4.60)$$

$$J_m^{\varepsilon} = \nu\left\{\overline{u'_{i,j}u'_{i,j}u'_m} + \tfrac{2}{\rho}\overline{p'_{,i}u'_{m,i}}\right\} \quad (4.61)$$

The physical interpretation of the terms is mean flow related production ($\mathcal{P}^{\varepsilon(1)}$, $\mathcal{P}^{\varepsilon(2)}$ and $\mathcal{P}^{\varepsilon(3)}$), turbulence related production through vortex stretching (T^{ε}), viscous destruction (D^{ε}), turbulent transport flux (J_m^{ε}) and viscous diffusive flux ($-\nu\frac{\partial\varepsilon}{\partial x_m}$). All terms except the viscous diffusive flux term need to be modelled.

4.3.4. THE K-EQUATION FOR COMPRESSIBLE FLOW

In order to derive the $K(\equiv \overline{\rho u_i'' u_i''}/2\overline{\rho})$-equation for compressible flow we may start by multipling the momentum balance (u_i-) equation by u_i'' and take the ensemble average of this equation. After some algebra (see *e.g.* Wilcox[137]) one arrives at

$$\overline{\rho}\frac{\partial K}{\partial t} + \overline{\rho}\tilde{u}_j\frac{\partial K}{\partial x_j} = -\overline{\rho u_i'' u_j''}\frac{\partial \tilde{u}_i}{\partial x_j} - \overline{t_{ji}\frac{\partial u_i''}{\partial x_j}}$$

$$- \frac{\partial}{\partial x_j}\left(-\overline{t_{ji}u_i''} + \frac{1}{2}\overline{\rho u_j'' u_i'' u_i''} + \overline{p' u_j''} + P\overline{u_i''}\right)$$

$$+ \quad P\overline{\frac{\partial u_i''}{\partial x_i}} + \overline{p'\frac{\partial u_i''}{\partial x_i}} \tag{4.62}$$

where we keep in mind from section 4.2 that t_{ij} denotes the viscous stress tensor and P is the mean pressure.

We recognize the first term on the right hand side as the production of turbulent kinetic energy. The second term defines the Favre averaged dissipation rate

$$\varepsilon = \frac{1}{\bar{\rho}}\overline{t_{ji}\frac{\partial u_i''}{\partial x_j}}$$

The second line represents transport and the two terms on the third line represent pressure work and pressure dilatation, respectively.

4.4. The hierarchy and history of single-point closures

The history of turbulence model development has followed a route towards an increased complexity accompanied by an increased generality of the models. In the simplest approaches the Reynolds stresses are described explicitly in terms of the mean flow quantities. Various types of such models have been proposed and are still used in practical applications, not least in aeronautics. Most of them are constructed to deal with boundary-layer type shear flows. All early models are based on the concept of a turbulent viscosity or eddy-viscosity. This class of models still dominates in CFD-codes for technical applications.

4.4.1. THE EDDY VISCOSITY HYPOTHESIS

The idea of a turbulent viscosity is based on the notion of the momentum transfer in a turbulent flow being dominated by the mixing caused by the large energetic turbulent eddies. The first ideas along these lines seem to have been proposed by Boussinesq (1877) who in analogy with the Newtonian fluid hypothesis introduced a turbulent (kinematic) viscosity, ν_T, such that the turbulent shear stress in an incompressible boundary-layer type flow can be described as a product of ν_T and the cross stream mean velocity gradient,

$$\overline{u'v'} = -\nu_T\frac{\partial U}{\partial y} \tag{4.63}$$

One should keep in mind that the Boussinesq hypothesis was put forward in the early days of turbulence research when Reynolds pipe experiments still were ongoing (see Reynolds 1883).

In early work the turbulent viscosity was taken to be a constant. This turns out to yield useful results only in a very limited class of free shear flows. In wall-bounded flows and other general flow situations more advanced approaches are needed to predict at least some of the essential effects of the turbulence. Almost fifty years after the classical paper of Boussinesq the, for wall-bounded flows, quite successful idea of a mixing length to describe the turbulent mixing and therewith associated diffusion coefficient, ν_T, was put forward by Prandtl (1925).

A coordinate invariant formulation of the incompressible Boussinesq hypothesis may be written

$$R_{ij} = \frac{2}{3} K \delta_{ij} - 2\nu_T S_{ij} \tag{4.64}$$

where $K = R_{ii}/2$ is the kinetic energy (per unit mass). The tensor relation (4.64) is equivalent to five scalar relations for the relative energy distribution among the components of the Reynolds stress tensor. It is illustrative to introduce the stress anisotropy tensor

$$a_{ij} \equiv \frac{R_{ij}}{K} - \frac{2}{3}\delta_{ij} \tag{4.65}$$

with the use of which we may rewrite the generalized Boussinesq hypothesis (4.64) for incompressible flows as

$$a_{ij} = -2\frac{\nu_T}{K} S_{ij} \tag{4.66}$$

For the compressible case we write the Boussinesq hypthesis with the aid of the earlier introduced Favre averages (and $K \equiv \overline{\rho u_i'' u_i''}/2\overline{\rho}$)

$$\tau_{ij} \equiv -\overline{\rho u_i'' u_j''} = \mu_T \left(\tilde{u}_{i,j} + \tilde{u}_{j,i} - \frac{1}{3}\tilde{u}_{k,k}\delta_{ij} \right) - \frac{2}{3}\overline{\rho} K \delta_{ij} \tag{4.67}$$

or

$$a_{ij} = -\frac{\mu_T}{\overline{\rho} K} \left(\tilde{u}_{i,j} + \tilde{u}_{j,i} - \frac{1}{3}\tilde{u}_{k,k}\delta_{ij} \right) \tag{4.68}$$

where we keep in mind that S_{ij} is the Reynolds average of the strain tensor, and the anisotropy tensor here is defined as

$$a_{ij} = \frac{R_{ij}}{K} - \frac{2}{3}\delta_{ij} \equiv \frac{-\tau_{ij}/\overline{\rho}}{K} - \frac{2}{3}\delta_{ij}$$

For the compressible case a corresponding eddy-diffusivity hypothesis for the turbulent heat-flux vector can be formulated.

It is obvious from the relations (4.66,4.68) that an anisotropy of the Reynolds stress tensor can only be sustained by a local mean strain at this level of closure. If, for instance, the turbulence is subjected to a mean strain field for an interval of time after which the strain field is suddenly removed, the turbulence is predicted to abruptly change to an isotropic state. Hence, the features of the relaxation phase, where the turbulence may be said not to be in 'equilibrium' with the mean strain field, cannot be captured by this type of local modelling of the stress anisotropy. In many engineering applications though, the turbulence state is reasonably close to such an equilibrium and it has proven sufficient to use this type of local modelling concept.

It is interesting to note the analogy between (4.66) and the Newtonian hypothesis for the viscous stress. In both cases we have an insensitivity to solid body rotations following from the linear dependence on the mean strain rate. It is important to keep in mind though that the relation (4.66) is not a constitutive relation for a material, but rather a description of features of the flow. The independence of rotation is therefore an artifact of this level of modelling, and the linearity of the relation must be seen only as a first approximation. Generalizations in this respect will be discussed in section 4.9.

We may think of the turbulent (eddy) viscosity as a diffusivity determined by the macroscopic velocity and length scales of the large energetic eddies of the turbulence, in contrast to molecular viscosity which is determined by the scales of the Brownian motion on the molecular level. At the above discussed level of closure these scales are modelled as functions of the local mean flow variables. Such models are therefore referred to as zero-equation or algebraic models, and are completely local in their description of the turbulence. Well known, and much used, forms of zero-equation models are the Cebeci-Smith and the Baldwin-Lomax models (see e.g. Wilcox[137]). These were developed as late as in the mid 60's and mid 70's, respectively. Any non-locality in the description within the eddy-viscosity framework of models enters through transport equations of the velocity and length scales. Other equivalent pairs of quantities may alternatively be used that together form the eddy-viscosity.

The natural velocity scale, used in practically all existing eddy-viscosity based models, is the square root of the kinetic energy, and the transport equation used is consequently some modelled form of the turbulent kinetic energy equation. We may express the eddy viscosity as

$$\nu_T = C'_\mu K^{1/2} L = C''_\mu K^n Z^m \tag{4.69}$$

where L is a characteristic measure of the lengthscale of the energetic eddies, say, the macroscale. Alternatively we may use some other related quantity, Z. The exponents m and n are then chosen so as to ensure correct dimension. Common choices of Z are the total dissipation rate ε, τ $(= K/\varepsilon)$ and ω $(= \varepsilon/K)$.

4.4.2. ONE-EQUATION MODELS

Models in which only one of the two quantities, from which the turbulent viscosity is formed, is determined from a transport (or evolution) equation, are referred to as one-equation models. They suffer from one serious problem, in that (e.g.) the velocity scale has to be taken from some ad-hoc empirical argument. This shortcoming, which is shared with the zero-equation models, has empirically proven to seriously limit the usefulness and generality of this type of model. The one-equation modelling idea goes back to Prandtl[81] who proposed the use of a modelled K-equation to determine the velocity scale.

An interesting idea at the one-equation closure level is that of constructing a transport equation for the eddy-viscosity itself. A model of this kind that has attracted considerable attention is that of Spalart & Allmaras[113]. It contains a number of ad-hoc relations containing empirical constants and the square root of the second invariant of the mean rotation rate tensor. The model is discussed also in Wilcox[137]. It has been calibrated against a number of boundary layer type situations (also compressible cases) and has shown some success as compared to lower level models and comparable predictions with two-equation models for this class of flows.

4.4.3. TWO-EQUATION MODELS

The lowest level of closure where history effects can be accounted for in some reasonable manner, and that has proven to give a reasonable generality for engineering predictions in a large variety of flows, is the level of two-equation models. Here, transport equations are formulated both for the velocity and the lengthscale, or some alternative pair of quantities that make up the eddy viscosity. For practically all models the K-equation is used for the velocity scale. The first two-equation model was formulated 1942 by Kolmogorov[51] who used a quantity ω with the dimension of an inverse time scale. According to Wilcox[137] Kolmogorov referred to this quantity as "the rate of dissipation of energy per unit volume and time".

A number of other alternative complementing quantities have also been tried, such as the lengthscale itself (L), a timescale (τ) and the dissipation rate (ε), of which the latter so far has been the most widely used. When some other quantity is used the transport equation for this quantity can often be derived from the corresponding ε-equation. Two-equation models will be discussed further in section 4.7.

4.4.4. REYNOLDS STRESS TRANSPORT MODELS

The main shortcomings at the standard two-equation model level is the locality of the stress anisotropy description, as follows from the Boussinesq hypothesis (4.66) (or 4.68), and the poor description of effects of system rotation or rotational mean flows. The latter is partly caused by the linearity of the Boussinesq hypothesis, which results in an independence of the rotational, antisymmetric part of the mean velocity gradient tensor in this relation. A natural remedy for these shortcomings is to remove the eddy-viscosity hypothesis altogether, and formulate modelled transport equations for the Reynolds stress tensor. The six scalar equations for the stress components must still be complemented by (at least) an equation for the length scale, or *e.g.* , as is most commonly used, an equation for the dissipation rate of kinetic energy. One may see the basis of this approach with six equations for the energy distribution as a strong emphasis on the description of the dynamics of the large scales. This may be motivated by the fact that the large scales normally are more anisotropic than the small scales, and more directly affected by imposed strain, rotation and geometrical constraints.

In the Reynolds stress transport equations the rotational part of the mean flow enters naturally in the production term, in contrast to eddy-viscosity based models. Alternatively, an imposed system rotation enters explicitly through the body force term via the Coriolis term. The trace of this is, of course zero, and does not affect the kinetic energy equation. Also, the production term is explicit in the Reynolds stresses and consequently does not need to be modelled at this level in the hierarchy of closures.

The first model based on the Reynolds stress transport equation was devised 1951 by Rotta[90], although much of the development of Reynolds stress models occurred during the 1970's (see, *e.g.* , Launder, Reece & Rodi[55]). Continued development of more advanced forms of this type of model is still going on (see, *e.g.* , Johansson & Hallbäck[44]).

Even at this level of closure the description of rotational effects form the major challenge and difficulty, and generalizations have been analyzed where further transport equations are incorporated into the model (see *e.g.* Cambon, Jacquin & Lubrano[8]). These approaches offer improvements of the description of effects of rotation, but will not be further discussed in this chapter.

4.4.5. ALGEBRAIC REYNOLDS STRESS MODELS

On the one hand, the considerable success of two-equation models for engineering flow predictions, and their relative ease of use and numerical robustness have motivated a strong interest in continued use of this level of closure in computational fluid dynamics. It is clear from what has been discussed above, on the other hand, that standard two-equation models give a rather poor description of, *e.g.* , rotation-associated effects on the turbulence. Algebraic Reynolds stress models are based on two transport equations, for K and some auxiliary quantity, but do not make use of the standard Boussinesq hypothesis nor the concept of an eddy viscosity. Instead a generalization of the constitutive-like relation between the stress anisotropy and the mean flow quantities (S_{kl}, Ω_{kl}) is sought (see section 4.9).

The first approach to this problem was devised by Rodi[84],[85] who took as a starting point the Reynolds stress transport equations and made use of the existing modelling of the terms involved there. His approach lead to an implicit set of algebraic equations for the stress anisotropies. Recent work has been directed towards finding adequate explicit forms of this relation. Based on the Rodi approach Pope[78] 1975 studied an explicit form for two-dimensional mean flows. This pioneering work was practically forgotten until the last few years when also more general forms for three-dimensional flows have been derived. A number of open issues remain and it may still be regarded as an open field for research, although promising results have already been obtained.

The basic idea of this level of closure is to remove the requirement of an explicit linear relation for the stress anisotropy, and through nonlinear terms incorporate effects also of the rotational, antisymmetric part of the mean velocity gradient tensor. As compared to models based on the standard Boussinesq hypothesis one can thereby hope to substantially improve the description of rotation-associated effects.

4.5. What should a closure fulfill?

4.5.1. COORDINATE INVARIANCE

A basic principle of continuum mechanics, and all physical theories, is that all mathematically formulated relations must be independent of the coordinate system used for their description. This can be ensured by a proper tensorial formulation of the relations. Cartesian tensor formalism will be used throughout the present text. A vector $\vec{v}$ will be expressed in terms of its Cartesian tensor components v_i or v'_i,

$$\vec{v} = \sum_{i=1}^{3} v_i \vec{e}_i = \sum_{j=1}^{3} v'_j \vec{e}'_j$$

where $\vec{e}_i$ and $\vec{e}'_j$ are orthonormal base vectors of two different cordinate systems. In the following the summation over two repeated indices is implied. The relations between the tensor components in the two different systems are given by

$$v'_j = v_i \vec{e}_i \cdot \vec{e}'_j \qquad \text{and} \qquad v_i = v'_j \vec{e}'_j \cdot \vec{e}_i$$

where the transformation matrix $q_{ij} \equiv \vec{e}_i \cdot \vec{e}'_j$ satisfies $q_{ik}q_{jk} = q_{ki}q_{kj} = \delta_{ij}$.

Applied to turbulence modelling this means that any exact equation and the modelled version of it should be expressed in tensorial form. Also any term or group of terms in the modelled equation should satisfy the same tensor index symmetry and contraction properties as the corresponding term of the exact equation.

For instance the coordinate invariant form of the original Boussinesq hypothesis was given in the previous section.

4.5.2. MATERIAL FRAME INDIFFERENCE

A more complicated issue than coordinate invariance is the question of frame invariance of the exact and modelled equations, *i.e.* whether or not superimposed time dependent rigid body motions affect the physical phenomena under consideration and how to take this into account in the construction of models. In rheology frame invariance is normally assumed and required of the modelled constitutive relations. In turbulence modelling one seeks constitutive relations describing features of the flow field rather than of the material, which makes the situation more complicated. The Navier-Stokes, the Reynolds, and the RST equations are all Galilean (G) invariant which means that model expressions for the various terms of the RST equations must also be G-invariant. This, for instance, precludes inclusion of the mean velocity U_i (which is not G-invariant) but allows $\frac{\partial U_i}{\partial x_j}$ and R_{ij} in any model expression.

From the governing equation for the turbulent velocity component $\vec{u}'$ in a non-inertial frame undergoing time-dependent motions one sees that it is actually only through the Coriolis force that the turbulent field is affected

$$\begin{aligned} \frac{\partial u'_i}{\partial t} + U_j \frac{\partial u'_i}{\partial x_j} &= -u_j \frac{\partial U_i}{\partial x_j} - \frac{1}{\rho}\frac{\partial p'}{\partial x_i} - \frac{\partial}{\partial x_j}\left(u'_i u'_j - R_{ij}\right) \\ &\quad + \nu \frac{\partial^2 u'_i}{\partial x_j \partial x_j} - 2\epsilon_{ijk}\Omega^{\mathrm{s}}_j u'_k \end{aligned} \tag{4.70}$$

$$\frac{\partial u_i'}{\partial x_i} = 0 \tag{4.71}$$

where $\vec{\Omega}^S(t)$ is the system rotation rate vector. The centrifugal force and the inertial force due to translational acceleration can be included in the mean pressure field and the angular acceleration rate affects only the mean velocity field. Any Reynolds stress model should hence be invariant under translational acceleration (or *extended Galilean transformation*). However, in contrast to constitutive relationships in continuum mechanics, invariance under rotations is generally not applicable to turbulence modelling.

In the limit of two-dimensional turbulence the Reynolds stress tensor is, in fact, unaffected by rotation around the axis perpendicular to the plane of turbulent motion (Speziale[115]) and thus satisfies the principle of material frame indifference. Any closure relation for the Reynolds stress tensor in two-dimensional turbulence should hence be form invariant under superimposed rigid body motions. In three-dimensional turbulence this is not the case and there are many examples in which superimposed rotation indeed significantly affects the turbulence field, for instance, the decay of homogeneous isotropic turbulence which exhibits a lower decay rate when subjected to a system rotation. This is caused by an inhibition of the energy cascade. Also the relaxation process of anisotropic homogeneous turbulence (Jacquin *et al.* 1990) is affected by rotation.

It should be pointed out that on physical grounds an analogy between a molecular viscosity, a microscopic feature of the fluid, and a turbulent viscosity, a flow relaed feature of the turbulence field, is in principle not justified since the turbulent eddies are not individual rigid objects retaining their identity. Furthermore, the separation of time and length scales between the turbulent part and mean part of the flow field is not as distinct as for molecular motions in a continuum. The "mean free path" of a turbulent eddy is of the same order as the size of the large energy-containing eddies, and hence not small compared to the flow domain. During the travel along such a "mean free path", which takes on the order of an eddy-turnover time, an eddy will most certainly be redirected due to Coriolis forces unless the system rotation is extremely slow in contrast to the situation for a molecule which will not be appreciably redirected during a travel along a mean free path unless the system rotation is extremely high.

4.5.3. INVARIANT MODELLING

Invariant modelling is a concept that is used extensively in development of single-point turbulence models, especially in RST-closures. The Reynolds stress transport equations may equivalently be expressed as an equation for the kinetic energy (half the trace of R_{ij}) and a transport equation for the

Reynolds stress anisotropy tensor $a_{ij} \equiv \frac{R_{ij}}{K} - \frac{2}{3}\delta_{ij}$. This tensor is obviously traceless and symmetric, and its diagonal components are limited to a finite interval. A diagonal component with vanishing energy has an anisotropy value of $-2/3$, and when all the energy is contained in one component the anisotropy attains a value of $4/3$.

The a_{ij}-tensor has two independent scalar measures, referred to as 'invariants', *i.e.* invariant with respect to the choice of coordinate system.

$$II_a \equiv a_{ij}a_{ji}, \qquad III_a \equiv a_{ij}a_{jk}a_{ki} \tag{4.72}$$

These are often used in model expressions. All possible anisotropy states are bound to a finite region in the II_a, III_a-plane, the so called anisotropy invariant map (Lumley 1978). Two of the boundaries of this region, as shown in figure 4.2, represent axisymmetric turbulence, where $II_a^{1/2} = 6^{1/6} \mid III_a \mid^{1/3}$, whereas the third represents the two-component limit, where $II_a = 8/9 + III_a$.

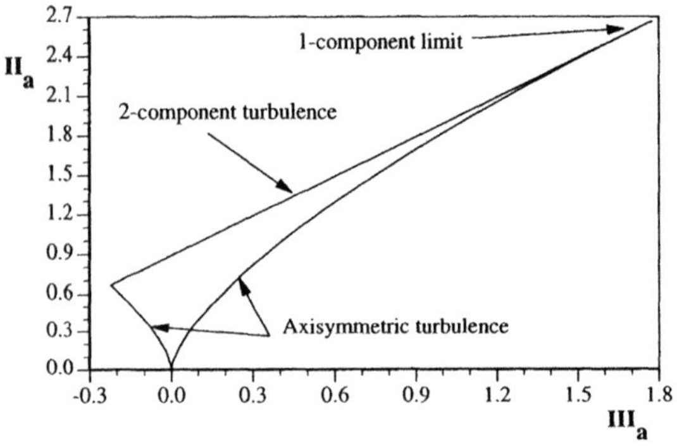

Figure 4.2. The anisotropy invariant map.

For modelling purposes it is also convenient to introduce

$$A = 6^{1/6} \mid III_a \mid^{1/3} / II_a^{1/2}, \qquad F = 1 - \frac{9}{8}(II_a - III_a) \tag{4.73}$$

where A is the "degree of axisymmetry" and F is the "degree of two-componentality" being zero at the two-component limit.

A typical modelling problem would be to find a suitable model for a symmetric second rank tensor, A_{ij} say, that is assumed to depend on only one other second rank tensor, B_{ij} say. The general complete tensorial form may be written as

$$A_{ij} = f_1\delta_{ij} + f_2 B_{ij} + f_3 B_{ik}B_{kj} \tag{4.74}$$

where the f_i:s are functions of the three invariants of the B-tensor, $I_B = B_{ii}, II_B = B_{ij}B_{ji}, III_B = B_{ij}B_{jk}B_{ki}$. The completeness of the above form rests on the validity of the Cayley-Hamilton theorem that states that a matrix satisfies its own characteristic equation. In this case we may express the theorem as

$$B_{ik}B_{kl}B_{lj} = \frac{1}{6}\left(2III_B - 3I_B II_B + I_B^3\right)\delta_{ij} + \frac{1}{2}\left(II_B - I_B^2\right)B_{ij} + I_B B_{ik}B_{kj} \tag{4.75}$$

By use of the relation (4.75) third and higher order tensor products may be reduced to expressions containing $\delta_{ij}, B_{ij}, B_{ik}B_{kj}$ and the invariants. Hence, the expression (4.74) is tensorially complete.

Other scalar measures such as

$$Re_T \equiv \frac{4K^2}{\nu\varepsilon}, \quad S \equiv \sqrt{2S_{ij}S_{ji}}, \quad S^* \equiv S\frac{K}{\varepsilon}, \quad \Omega \equiv \sqrt{\Omega_{ij}\Omega ij/2}, \quad \Omega^* \equiv \Omega\frac{K}{\varepsilon}$$

are also often used in model expressions. In the above expression we could, hence, have $f_i = f_i(II_B, III_B, Re_T, S^*, \Omega^*)$.

4.5.4. REALIZABILITY

A general requirement in constructing models of various terms in the exact dynamic equations of different statistical moments, such as the Reynolds stresses, is that the modelled equations should not give rise to physically unrealizable solutions, such as negative energies. When dealing with closures based on the K and ε equations or the R_{ij} and ε equations one should thus at least demand that K and ε are non-negative, and that any normal component $R_{\alpha\alpha} \geq 0$ regardless of the choice of unit vector $e_i^{(\alpha)}$. This is what is normally referred to as realizability of turbulence models.

In terms of the anisotropy tensor the diagonal components must, hence, attain values in the interval from $-2/3$ to $4/3$. Also the off-diagonal components are bounded. With $\alpha \neq \beta$ we have

$$| a_{\alpha\beta} | = \frac{| R_{\alpha\beta} |}{K} \leq \frac{2 | R_{\alpha\beta} |}{\left(\overline{u_\alpha^2} + \overline{u_\beta^2}\right)} \leq \frac{2\left(\overline{u_\alpha^2}\,\overline{u_\beta^2}\right)^{1/2}}{\left(\overline{u_\alpha^2} + \overline{u_\beta^2}\right)} \leq 1$$

The realizability condition on the anisotropy tensor guarantees that the path in the anisotropy invariant map (AIM) does not cross the two-component line (see figure 4.2).

The need for models to satisfy realizability becomes accentuated when the model equations are integrated all the way to solid walls. This can be illustrated by the AIM-path for tubulent channel flow. Figure 4.3 shows this path for the DNS data of [50]. It can be observed that the points from the log-region essentially are clustered to a single point in the AIM, and that points from the viscous sublayer come very close to the two-component line.

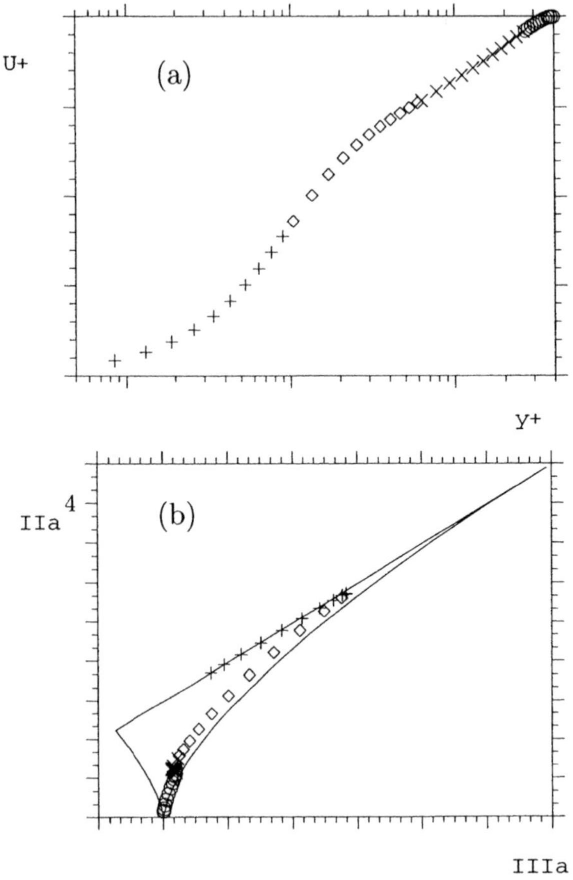

Figure 4.3. DNS data of [50] for turbulent channel flow a) The mean velocity profile b) The anisotropy invariant map.

When an extreme state, a minimum possible value say, of a physical quantity governed by a transport equation is reached, its first-order time derivative must be zero and the lowest order non-zero time derivative has to be positive to ensure that no states beyond the extreme state may be reached (figure 4.4). This is the strong realizability principle according to Pope (1985) who also discusses the principle of weak realizability, which states that in an extreme state the first-order time derivative must be non-negative. This also ensures that the limit of realizable states is not exceeded, but may lead to situations in which the extreme states are not accessible.

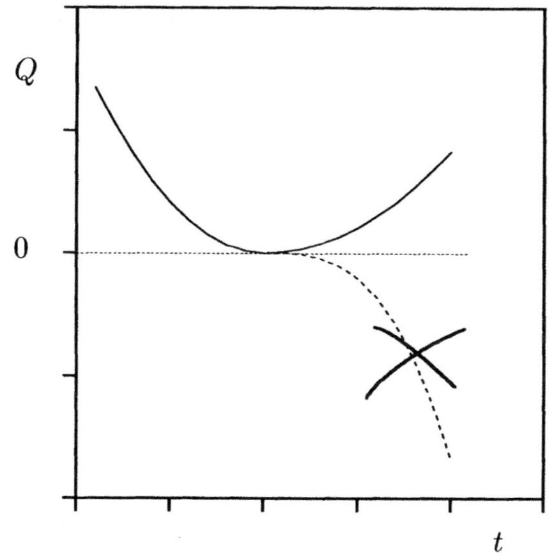

Figure 4.4. The time evolution of a quantity ($Q \geq 0$) near an extreme state.

The Boussinesq hypothesis at large strain rates The linear, local modelling of the Reynolds stress in terms of the mean strain rate in the form of the Boussinesq hypothesis does not satisfy the realizability constraint. The coordinate invariant tensor formulation of this hypothesis may be written,

$$a_{ij} = -2\frac{\nu_t}{K}S_{ij}$$

It is obvious that for strong strains suddenly imposed on the turbulence this model may produce unrealizable values of the anisotropy tensor before the turbulence adjusts to the change in mean strain.

For instance, with the $K - \varepsilon$ model description of the eddy viscosity, $\nu_t = C_\mu \frac{K^2}{\varepsilon}$, the anisotropy tensor can be written as

$$a_{ij} = -2C_\mu \frac{K}{\varepsilon} S_{ij} \equiv -2C_\mu S^*_{ij}$$

where S^*_{ij} denotes the mean strain rate normalized by the turbulence time scale.

A quantity of central interest in model predictions is the production to dissipation ratio, $\mathcal{P}/\varepsilon \equiv -a_{ij}S^*_{ij}$. Since the anisotropy components in the real physical situation always are bounded to the interval $-2/3$ to $4/3$ the true production to dissipation ratio for large strain rate values will be of the order of the nondimensional mean strain rate. In a $K-\varepsilon$ model, on the other hand, this quantity can be written as

$$\mathcal{P}/\varepsilon = -2C_\mu S^*_{ij}S^*_{ij}$$

from which we see that the value predicted by the eddy viscosity model will be of the order of the square of the strain rate.

The prediction of the anisotropy components can be illustrated by the following examples. For a parallel shear flow with a nondimensional mean strain, $\sigma = \frac{1}{2}\frac{K}{\varepsilon}\frac{dU}{dy}$, the eddy viscosity model will predict a shear stress anisotropy component

$$a_{12} = -2C_\mu \sigma$$

Hence, the Boussinesq hypothesis will here generate unphysical results for $\sigma > 1/2C_\mu \approx 5.6$. This is obvously a primary reason for the strong need for near-wall damping functions in this type of models. A simple remedy for this problem would be to require, by use of a limiter, that

$$| a_{\alpha\beta} | \leq 1 \quad \alpha \neq \beta$$

We can find a similar approach in the SST model of Menter[71], which adopts the Bradshaw assumption, $a_{12} = -0.3$ for $\mathcal{P}/\varepsilon$ ratios greater than unity. This model has also been motivated by the observation that in flows with an adverse pressure gradient, the production to dissipation ratio is greater than one and eddy viscosity models with constant C_μ overestimates the turbulent viscosity or the a_{12} anisotropy.

For flows with large irrotational strain, flows in nozzles etc, the inability of the Boussinesq hypothesis to satisfy realizability may cause severe problems. We may, for instance, consider axisymmetric strain with $S^*_{11} = \sigma$ $(S^*_{22} = S^*_{33} = -\sigma/2)$. The Boussinesq hypothesis here yields (with $\nu_t = C_\mu \frac{K^2}{\varepsilon}$)

$$a_{11} = -2C_\mu \sigma$$

which shows that unphysical results are produced for $\sigma > 1/3C_\mu \approx 3.7$.

Realizability of the exact RST-equations Since the eigenvalues of R_{ij} have

the physical meaning of kinetic energies in three perpendicular directions they must always remain positive or, in an extreme state, be equal to zero. From an analysis of the exact Reynolds stress transport equations (see chapter 3.6.4 in the book of the 1995 Summer School) one finds that the realizability condition is fulfilled individually for the advection term, the production term, the triple correlation term $(\overline{u'_i u'_j u'_k}_{,k})$, the total pressure related term (*i.e.* sum of pressure strain and pressure diffusion) and the total viscous term (*i.e.* sum of dissipation and viscous diffusion).

In constructing turbulence models for the various terms, although rigorously not correct, it is convenient to impose the realizability condition directly on the pressure-strain-rate tensor or the dissipation-rate tensor;

This means that sufficient but somewhat unnecessarily hard constraints are imposed on the turbulence model.

This approach can be justified by reasoning in the following way. The extreme states of the stress tensor where the realizability condition becomes important are characterized by the velocity field being plane, *i.e.* , $n_i u_i = 0$ for some vector n_i being constant over the complete ensemble of velocity fields at the specific spatial location at the specific time in question. We may ask ourselves in what types of situations such plane velocity fields may occur. In homogeneous turbulence a strong positive mean strain in the direction of n_i may cause a plane velocity field perpendicular to n_i. This is approximately the case when turbulence is strained by the passage through a wind tunnel contraction. In homogeneous turbulence the spatially redistributive terms are zero and we may therefore impose the realizability on the 'homogeneous' parts of the source terms, *i.e.* , the pressure strain correlation and the dissipation rate.

The requirement that the first non-zero time-derivative of the normal stress must be positive is normally not accounted for and is of course much more difficult to guarantee in an RST closure scheme.

In the extreme state at a solid wall all velocity components are zero and what is of interestest here is the limiting behaviour of the energy distribution among the components, for instance quantified in terms of the anisotropy tensor a_{ij}. Even in a state with zero turbulent kinetic energy the anisotropy state must not be forced out of the allowed region of the anisotropy invariant map. At the solid wall the two-component limit must be reached but not exceeded, *i.e.* $a_{nn} \geq -2/3$, which is the relevant extreme state at the wall (in addition to $K \geq 0$).

From a detailed scrutiny of the limiting behaviour of exact transport equation of a_{nn} in the near-wall vicinity it turns out that ensuring $\frac{\partial a_{nn}}{\partial t} = 0$ is equivalent to requiring that $\frac{1}{K}(-\frac{1}{\rho}\langle v' \frac{\partial p'}{\partial y}\rangle + \nu \langle v' \frac{\partial^2 v'}{\partial y^2}\rangle) = 0$. Thus, according to the properties of the exact equations, for the near wall extreme state one should impose the realizability condition on the pressure and viscosity related terms lumped together and normalized by K.

We will return to the issue of realizability of models of the terms in the RST-equations in section 4.8. For further reading on realizability we refer to, for instance, Schumann[97] and Durbin & Speziale[19].

4.5.5. NEAR-WALL ASYMPTOTICS

When devising turbulence models to be used all the way down to a solid boundary one may impose the requirement that the modelled and exact terms should have the same limiting behaviour as the wall is approached.

By use of the expansion the near wall asympotics of the different terms in, for instance, the RST-equations may be established. In the immediate vicinity of a solid wall the fluctuating velocity and pressure may be written as $(a_1, 0, b_1)y$ and p_0 to leading order. Hence, the following Reynolds stresses are found;

$$\begin{aligned} \langle u_1' u_1' \rangle &= \langle {a_1'}^2 \rangle y^2 + \dots \\ \langle u_2' u_2' \rangle &= \langle {b_1'}^2 \rangle y^4 + \dots \\ \langle u_3' u_3' \rangle &= \langle {c_1'}^2 \rangle y^2 + \dots \\ \langle u_1' u_2' \rangle &= \langle a_1' b_1' \rangle y^3 + \dots \end{aligned} \tag{4.76}$$

The turbulent kinetic energy and the dissipation rate are given by

$$\begin{aligned} K &= \frac{1}{2}\left(\langle {a_1'}^2 + {c_1'}^2 \rangle y^2 + 2 \langle a_1' a_2' + c_1' c_2' \rangle y^3 + \dots \right) \\ \varepsilon/\nu &= \langle {a_1'}^2 + {c_1'}^2 \rangle + 4 \langle a_1' a_2' + c_1' c_2' \rangle y + \dots \end{aligned} \tag{4.77}$$

Similarly the asymptotic behaviour of all terms of the exact RST-equations may be derived. The components of the dissipation rate tensor satisfies, at increasing orders of y:

$$\begin{aligned} y^0 &: \quad \frac{\varepsilon_{11}}{\varepsilon} = \frac{R_{11}}{K}, \quad \frac{\varepsilon_{33}}{\varepsilon} = \frac{R_{33}}{K}, \quad \frac{\varepsilon_{13}}{\varepsilon} = \frac{R_{13}}{K} \\ y^1 &: \quad \frac{\varepsilon_{12}}{\varepsilon} = 2\frac{R_{12}}{K}, \quad \frac{\varepsilon_{23}}{\varepsilon} = 2\frac{R_{23}}{K} \\ y^2 &: \quad \frac{\varepsilon_{22}}{\varepsilon} = 4\frac{R_{22}}{K} \end{aligned} \tag{4.78}$$

A reasonable condition on low Reynolds number models that are to be integrated all the way down to the wall, is that they should be asymptotically consistent with the above type of expansions. The limiting behaviour in the immediate wall vicinity may also be used to prescribe adequate boundary conditions for closure schemes.

For twodimensional wall bounded parallel shear flow both DNS data (see Antonia & Kim 1994) and experiments (see Alfredsson *et al.* 1988) show that (u_τ is the friction velocity)

$$\frac{a_{1\mathrm{rms}}}{u_\tau} = \left.\frac{u'^{+}_{\mathrm{rms}}}{U^+}\right|_{\mathrm{wall}} \approx 0.40, \qquad \frac{c_{1\mathrm{rms}}}{u_\tau} = \left.\frac{w'^{+}_{\mathrm{rms}}}{U^+}\right|_{\mathrm{wall}} \approx 0.25 \tag{4.79}$$

If one may assume that the rms-values of the expansion coefficients are universal in the same sense as the von Kármán constant then one may use such empiricism for calibration of model parameters and for imposing wall boundary conditions.

4.6. Purely algebraic models

The general trend towards more generality in the model formulations have penetrated into the general use of commercial CFD-codes. Aeronautics is a field that has been dominated by boundary layer type problems, often with effects of pressure gradients and compressibility. A number of lower level models, one-equation or purely algebraic, have been developed with the main focus on boundary layer applications in aeronautics. These are still in active use, although two-equation models are gaining ground also in this field, partially because of increased ambitions to obtain accurate predictions for the complex parts of the flow field, such as wing-body junction flows, shock-induced separation zones etc.

The basic assumption involved in the use of algebraic, or equivalently, zero-equation models is that the Reynolds stress tensor is determined by the Boussinesq hypothesis with an eddy viscosity, μ_T , that is allowed to depend only in the mean flow variables and some geometrical parameters of the problem. It is supposed too that its dependence can be established through an algebraic relation. Most of the algebraic models for wall-bounded flow are based on the ideas of the Prandtl mixing length model, usually combined with damping functions. As an illustration of the basic concepts of this type of modelling we may therefore give the following simple example.

4.6.1. THE MIXING LENGTH MODEL WITH A VAN DRIEST DAMPING FUNCTION

It may be worthwhile to first look at the basic mixing length model in its incompressible form applied to a simple (two-dimensional) wall-bounded flow. The turbulent viscosity in the Boussinesq hypothesis for the turbulent shear stress is then given by

$$-\overline{u'v'} = \nu_t \frac{\partial U}{\partial y}, \qquad \nu_t = l_{\mathrm{m}}^2 \left| \frac{\partial U}{\partial y} \right| \tag{4.80}$$

and the total shear stress thus becomes

$$\tau = \mu \frac{\partial U}{\partial y} + \rho l_{\mathrm{m}}^2 \left(\frac{\partial U}{\partial y} \right)^2 \tag{4.81}$$

The basic arguments behind this model can be found in numerous textbooks on turbulence. The standard choice for the mixing length is as follows. Far from the wall the mixing length l_{m} is taken as proportional to the boundary layer thickness (in a free shear flow it is usually be taken as proportional to the width of the jet or wake). Closer to the wall the mixing length is reduced due to the fact that the presence of the wall hinders motions normal to the wall. To account for this inviscid, kinematic, near-wall effect the mixing length is normally taken as proportional to the distance to the wall, $l_{\mathrm{m}} = \kappa y$. Introducing this model into (4.81) we obtain after scaling by the friction velocity, u_τ, and the viscous length scale, $l_* \equiv \nu/u_\tau$

$$\frac{\tau}{\tau_{\mathrm{w}}} = \frac{\partial U^+}{\partial y^+} + \kappa^2 y^{+2} \left(\frac{\partial U^+}{\partial y^+}\right)^2 \tag{4.82}$$

In the nearwall region of a boundary layer without pressure gradient the normalized shear stress is approximately constant, which together with (4.82) yields

$$\frac{\partial U^+}{\partial y^+} = \frac{2}{1 + \sqrt{1 + 4\kappa^2 y^{+2}}} \quad \rightarrow \frac{1}{\kappa y^+}, \quad y^+ \rightarrow \infty \tag{4.83}$$

That is, in the near-wall region, at sufficiently large y^+ the log-layer profile is reproduced upon integration of (4.83). The constant κ is consequently chosen as the von Kármán constant.

The predicted value of the additive constant in the log-profile with this simple approach is, however, far too low. The reason for this is easy to understand. With the mixing length proportional to the wall distance, it is obvious from (4.80) that the turbulent shear stress will be proportional to y^2 as the wall is approached, whereas we know from the near-wall asymptotics (see section 4.5) that it should vary as y^3, *i.e.* be more rapidly damped near the wall. Hence, we recognize the need for so called wall damping.

Following van Driest (1956) the range of influence of the visous damping may be estimated by reasoning in analogy with the problem of a oscillating wall in a fluid at rest. In that case it is well known the fluid motion due to the oscillating wall decreases with the distance, y, to the wall as $\exp(-y/A)$, where the constant A depends on the fluid viscosity and the frequency of oscillation ($A = \sqrt{2\nu/\omega}$). If, instead, the wall is at rest and the fluid is oscillating the fluid motion will be damped by a factor of $[1 - \exp(-y/A)]$ near the wall. Assuming that this damping factor will influence each of the velocity components in the turbulent shear stress the Prandtl mixing length model should be changed to

$$-\overline{u'v'} = \kappa^2 y^2 [1 - \exp(-y/A)]^2 \left(\frac{\partial U}{\partial y}\right)^2 \tag{4.84}$$

We may interprete this as the mixing length being damped by viscous forces near the wall. It is, however, not equivalent to saying that the integral length scales are damped to the same extent. It rather reflects that mixing in the wall normal direction is strongly inhibited by viscous forces. Expressed in terms of scaled quantities we have

$$l_{\mathrm{m}}^+ = \kappa y^+ [1 - \exp(-y^+/A^+)] \tag{4.85}$$

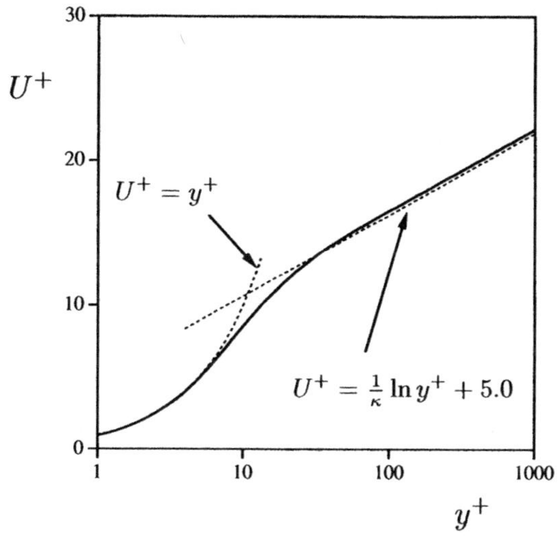

Figure 4.5. The van Driest velocity profile (4.86) with $A^+ = 26$, solid curve, and the viscous sub-layer linear profile and the log-profile.

By comparison between the predicted mean velocity profile, obtained upon insertion of the mixing length model (4.84) into the momentum equation (4.81) with $\tau \approx \tau_w$,

$$U^+(y^+) = \int_0^{y^+} \frac{2dy'}{1 + \sqrt{1 + 4\kappa^2 y'^2 [1 - \exp(-y'/A^+)]^2}} \tag{4.86}$$

and standard forms of the viscous sublayer and log-layer profiles, the value of the remaining constant was determined to $A^+ \approx 26$ (see figure 4.5). van Driest type of damping functions

$$1 - \exp(-y^+/A^+)$$

are frequently used in low Reynolds number near wall modelling.

4.7. Eddy-viscosity based two-equation models

Standard two-equation models based on the Boussinesq hypothesis

$$R_{ij} = \frac{2}{3}K\delta_{ij} - 2\nu_T S_{ij} \qquad \text{or} \qquad a_{ij} = -2\frac{\nu_T}{K}S_{ij} \tag{4.87}$$

and transport equations for the kinetic energy and an auxiliary quantity still dominate in commercial CFD codes. We will here restrict the discussion to the $K-\varepsilon$ and $K-\omega$ models. More exhaustive descriptions of two-equation models can be found in Launder & Spalding[56] or Wilcox[137].

The use of the Boussinesq hypothesis (4.87) results in a mean flow equation of the form

$$\frac{\mathrm{D}U_i}{\mathrm{D}t} = \frac{\partial}{\partial x_j}\left[-\left(\frac{1}{\rho}P + \frac{2}{3}K\right)\delta_{ij} + 2(\nu + \nu_T)S_{ij}\right] \tag{4.88}$$

We note that the role of the trace of the normal stresses (or K) is equivalent to that of a pressure-contribution. The analogy between the this form of the modelled turbulent stresses and the Newtonian fluid hypothesis is apparent in (4.88) although one should keep in mind that ν_T depends on flow related features whereas the molecular viscosity is a material property. The modelled form of the K-equation can for two-equation models of this type be written

$$\frac{\mathrm{D}K}{\mathrm{D}t} = 2\nu_T S_{ij}S_{ij} - \varepsilon + \frac{\partial}{\partial x_i}\left[\left(\nu + \frac{\nu_T}{\sigma_k}\right)\frac{\partial K}{\partial x_i}\right] \tag{4.89}$$

The modelling of the production term simply follows from the form of the Reynolds stress given by the Boussinesq hypothesis. The flux term is modelled by a gradient diffusion expression, where all spatial variation of the turbulent diffusivity coefficient is assumed to be given by the eddy viscosity. The diffusivity coefficient is assumed to differ from ν_T by a constant factor given by the Schmidt number σ_k. Since K is simply (minus) twice the trace of the Reynolds stress tensor one normally assumes σ_k to be close to unity.

In the $K-\varepsilon$-model, equation (4.89) needs no further modelling, whereas for the $K-\omega$ and other two-equation models the dissipation term needs to be modelled in terms K and the auxilliary quantity.

If we denote the auxilliary quantity by Z, the standard structure of the transport equation for Z (at high Re) is written

$$\frac{\mathrm{D}Z}{\mathrm{D}t} = C_{Z1}\frac{Z}{K}\mathcal{P} - C_{Z2}\frac{Z}{K}\varepsilon + \frac{\partial}{\partial x_i}\left[\frac{\nu_T}{\sigma_Z}\frac{\partial Z}{\partial x_i}\right] + \text{Source} \tag{4.90}$$

where the form of the possible source term depends on the specific choice of Z. There are essentially two different schools of approaches. The Z-equation is either constructed in an *ad-hoc* manner with the aim of mimicking some of the physics that are believed to be essential, or one may try to model the exact transport equation for Z as closely as possible. The two often lead to similar results due to the restraint that the models of the specific terms involved in the exact transport equation by necessity have to be described in terms of the primary transported quantities. A drawback of the *ad-hoc* approach is that it offers no natural method of generalization of the equation if, for instance, further transported quantities are included. The latter occurs, *e.g.* , when we go from a two-equation model to a Reynolds stress closure. The modelling of the Z-equation is one of the major weaknesses in both two-equation and Reynolds stress models.

The most common choice of the auxilliary quantity is the turbulence dissipation rate ε from which a length scale may be extracted through use of the high Reynolds number relation $l \sim K^{3/2}/\varepsilon$. The advantage of this choice is of course that ε is anyway needed in the closure (also true in RST-closures) and that it is a well defined scalar measure. It is not at all obvious, however, that it is the optimal choice in all applications, and many of the differences in the behaviour of different two-equation models may be ascribed to differences in the near-wall treatment and boundary conditions.

The two different choices for the Z-quantity discussed below have quite different behaviour near solid walls and as a free-stream is approached. The limiting value of ε near a solid wall is finite (and non-zero), whereas ω becomes infinite at the wall.

4.7.1. THE $K - \varepsilon$ MODEL

In the history of the $K - \varepsilon$ model much of the essential work was done in the seventies (e.g Jones & Launder[47]) although the origin may be traced back to Chou[14]. The model is defined by the modelled mean flow equation (4.88) (and continuity), the kinetic energy equation (4.89), the ε-equation

$$\frac{\mathrm{D}\varepsilon}{\mathrm{D}t} = C_{\varepsilon 1}\frac{\mathcal{P}}{K}\varepsilon - C_{\varepsilon 2}\frac{\varepsilon^2}{K} - \frac{\partial}{\partial x_m}\left(J_m^{\varepsilon} - \nu\frac{\partial \varepsilon}{\partial x_m}\right) \tag{4.91}$$

and the eddy-viscosity relation

$$\nu_T = C_\mu \frac{K^2}{\varepsilon} \tag{4.92}$$

In the eddy-viscosity relation ($\nu_T \sim K^{1/2} l$) the length-scale is taken as a constant times $K^{3/2}/\varepsilon$. This approximation for the macro-scale is valid for high Reynolds numbers. For low Reynolds numbers this would in principle need modification.

The kinetic energy production term is defined through the Boussinesq hypothesis as

$$\mathcal{P} = 2\nu_T S_{ij} S_{ij} \tag{4.93}$$

and the turbulent flux of ε is given by the gradient diffusion hypothesis as

$$J_m^{\varepsilon} = -\frac{\nu_T}{\sigma_\varepsilon}\frac{\partial \varepsilon}{\partial x_m} \tag{4.94}$$

A set of standard values for the model parameters may be said to be:

$$C_\mu = 0.09, \quad \sigma_k = 1.0, \quad \sigma_\varepsilon = 1.3, \quad C_{\varepsilon 1} = 1.44, \quad C_{\varepsilon 2} = 1.92. \qquad (4.95)$$

The above set of parameter is arrived at by considering a few basic arguments and the model predictions for a set of 'corner-stone' flow cases, namely decaying isotropic turbulence, homogeneous shear flow and the log-layer.

A traditional way of arriving at the above value of C_μ is to consider a thin shear flow with approximate balance between production and dissipation. With y as the cross-stream coordinate we then get

$$C_\mu = \nu_T \frac{\varepsilon}{K^2} = \nu_T \frac{\nu_T \left(\frac{\partial U}{\partial y}\right)^2}{K^2} = \left(\frac{-\overline{uv}}{K}\right)^2 \qquad (4.96)$$

where use has been made of (4.92) and the Boussinesq hypothesis. The shear stress to kinetic energy ratio for thin shear flows is typically ≈ 0.3 (the so-called Bradshaw hypothesis) from which we get the above value of C_μ. This is probably a good compromise value for a range of shear flows. A closer look at the situation in the inner part of the log-layer where often the 'wall boundary condition' is applied yields that a suitable value of C_μ there would be close to 0.07.

One should keep in mind that this calibration of C_μ which is used in an equivalent mannner in other two-equation models is a rather serious restriction of the range of validity of this type of models. It obviously suggests that the primary target for this type of models should be turbulent shear flows without strong effects of rotation or other external body forces that may remove the turbulent state from this equilibrium value.

The modelling of the other terms and the calibration of the corresponding parameter values are of relevance also when the ε equation is used as an auxilliary equation in Reynolds stress closures.

With the reasoning above, in conjunction with the K equation, regarding the turbulent diffusion of Reynolds stress and turbulent kinetic energy the Schmidt number σ_k is taken as 1.0.

Modelling the destruction term

Consider isotropic homogeneous turbulence decaying in the absence of a mean strain rate field (see *e.g.* Hinze[41] or Batchelor[4]), where the K and ε equations reduce to

$$\begin{aligned} \frac{dK}{dt} &= -\varepsilon \\ \frac{d\varepsilon}{dt} &= T^\varepsilon - D^\varepsilon \end{aligned} \qquad (4.97)$$

If turbulence under these circumstances is assumed to decay in a self similar manner in the sense that the decay rates of K and ε at most differ by a factor

$$\frac{K/\frac{dK}{dt}}{\varepsilon/\frac{d\varepsilon}{dt}} = C_{\varepsilon 2} \tag{4.98}$$

then it follows that in order to mimic this behaviour the ε equation should be modelled as

$$\frac{d\varepsilon}{dt} = -C_{\varepsilon 2}\frac{\varepsilon^2}{K} \tag{4.99}$$

From the K and the modelled ε equations (4.97,4.99) the evolution of K can then be solved

$$K(t) = K_0\left\{1 + (C_{\varepsilon 2} - 1)\frac{\varepsilon_0}{K_0}t\right\}^{-n}, \qquad n \equiv \frac{1}{C_{\varepsilon 2} - 1}$$

where the 0-subscripts denote initial values. This type of power-law decay is well documented in the literature on wind tunnel experiments of grid generated turbulence, *e.g.* , in the study by Comte-Bellot & Corrsin[15] where exponents of $n = 1.2$ - 1.3 was found.

The rate of decay may be expected to depend on the Reynolds number which is one of the parameters that are relevant in isotropic turbulence. Assuming a self-similar decay behaviour of high Reynolds number turbulence, described by the simple model spectrum of figure 4.6, Comte-Bellot & Corrsin[15] and Reynolds[82] showed that $C_{\varepsilon 2}$ is related to the low wavenumber exponent a of the energy spectrum function. Evaluation of the kinetic energy from the model spectrum gives

$$K(t) = \int_0^\infty E(k,t)\,dk = \text{const} \times [\varepsilon(t)]^m, \qquad m = \frac{2(a+1)}{3a+5}$$

which is consistent with the first assumption (4.98) regarding the decay behaviour of K and ε if

$$C_{\varepsilon 2} = \frac{3a+5}{2(a+1)} \tag{4.100}$$

In the above discussion about the behaviour and modelling of decaying turbulence at high Reynolds numbers the viscous scales themselves, where the dissipation takes place, have been ignored. Similarity behaviour of the large scales was assumed and found to be supported by observations in both numerical simulations and wind tunnel experiments. While the aim was to derive a model equation for ε it is rather a model for the transfer of energy from large to small scales ($\varepsilon_f \equiv -\int_0^{k_i} T(k)\,dk$) that has been arrived at.

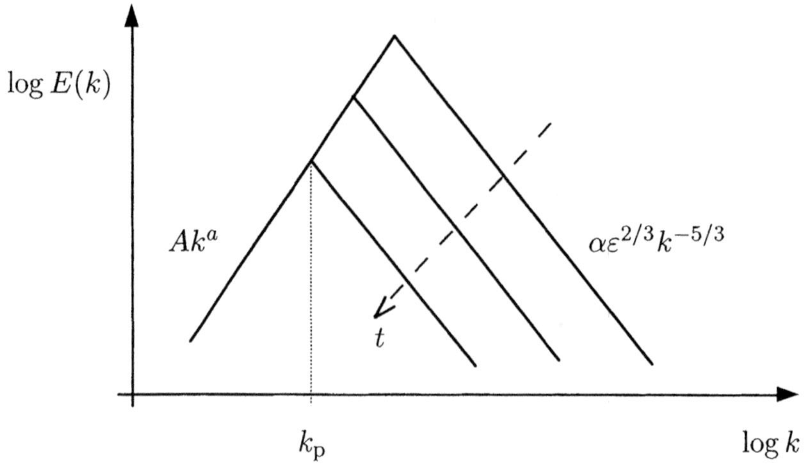

Figure 4.6. Self similar decay of high Reynolds number model spectrum where A is assumed to be constant and k_p decreases with time.

To further illustrate the benefit of devising a model equation for the energy drainage rather than modelling the equation for the actual viscous dissipation ($\varepsilon \equiv \nu\overline{u'_{i,l}u'_{i,l}}$) we may study the viscous destruction term

$$D^\varepsilon \equiv 4\nu^2 \int_0^\infty k^4 E(k)\, dk$$

appearing in the exact ε equation. By assuming a universal equilibrium spectrum like $E(k) \sim \varepsilon^{2/3}k^{-5/3}f(k\eta)$ it is readily shown that for large Re_T

$$D^\varepsilon \sim \sqrt{Re_T}\frac{\varepsilon^2}{K}$$

We have seen, though, that the net effect of $T^\varepsilon - D^\varepsilon$ is only weakly dependent on Re_T so we may conclude that also T^ε increases with the square root of the Reynolds number. Thus, instead of modelling two completely different presumably very large terms of an exact equation, a model equation has been constructed based on the concept of one of the most striking features of turbulence, namely that of a cascade of energy from large to small scales.

Empiricism shows that (4.99) is a good model also for low Reynolds number decaying turbulence, although somewhat higher values of the exponent n give best fit to data. An upper bound of n can be estimated by considering vanishingly low Re_T ("final period of decay") where the non-linear term becomes negligible and by assuming $E = Ak^a$ (A =constant) for the low wave numbers we get a decay exponent of $\frac{a+1}{2}$.

As seen above the large scale behaviour is of greatest relevance in isotropic turbulence. Saffman[92] suggested a value of $a = 2$ for the low wave number exponent, whereas *e.g.* Tennekes & Lumley[126] and Batchelor[4]) have suggested $a = 4$. Saffman argues that both are possible, depending on how the turbulence is generated. Recent large-eddy simulations (Chasnov[11]) have shown that decaying turbulence with a spectrum characterized by k^2 at low wave numbers to a good approximation reaches a similarity state. With $a = 2$ we would get the following bounds

$$5/3 < C_{\varepsilon 2} < 11/6 \qquad \text{or} \qquad 3/2 > n > 6/5$$

Most experiments on grid-generated turbulence fall into this interval. The 'final period of decay' must be regarded as an extreme situation far from what one would expect to find in real turbulence. Hence, an appropriate value of $C_{\varepsilon 2}$ should be close to the upper limit of the interval given above.

The importance of the structure of the large turbulence scales, presumably non-universal and generally unknown, for the energy drainage rate, however, points to an inherent weakness of one-point turbulence closures.

Modelling the ε production term

For Reynolds numbers in homogeneous turbulence high enough for "local isotropy" to prevail the small eddies that dominate the contribution to double-velocity-derivative correlations like $\langle u'_{i,l}u'_{j,k}\rangle$ are approximately isotropic, which also means that the mean flow induced production terms $\mathcal{P}^{\varepsilon(1)}$ (4.56) and $\mathcal{P}^{\varepsilon(2)}$ (4.57) are, to the same approximation, zero. It is, however, reasonable to assume that the rate of energy drainage ($\varepsilon_f \sim K^{3/2}/l$) is intensified as K is increased by production. Also, for moderate Re_T dissipative scales are not wholly equivalent to the smallest locally isotropic scales and $\mathcal{P}^{\varepsilon(1)} \neq 0$ and $\mathcal{P}^{\varepsilon(2)} \neq 0$ in general. A reasonable and simple assumption is to say that $\mathcal{P}^{\varepsilon(1)} + \mathcal{P}^{\varepsilon(2)} \sim \varepsilon\, P/K$, reflecting a coupling between energetic and dissipative scales. The third mean flow related production term $\mathcal{P}^{\varepsilon(3)}$ is negligibly small essentially everywhere in wall bounded shear flows (cf. *e.g.* Rodi & Mansour[87]).

The K-ε equations are then given by

$$\begin{aligned} \frac{dK}{dt} &= \mathcal{P} - \varepsilon \\ \frac{d\varepsilon}{dt} &= C_{\varepsilon 1}\frac{\mathcal{P}}{K}\varepsilon - C_{\varepsilon 2}\frac{\varepsilon^2}{K} \end{aligned} \qquad (4.101)$$

Homogeneous shear flow data may be used to assign $C_{\varepsilon 1}$ a numeric value. Homogeneous shear flows appear to reach an equilibrium state with K and ε growing in a manner so that the turbulence time scale K/ε approaches an approximately constant value and the normalized Reynolds stresses $\overline{u_i u_j}/K$

are nearly constant. From the K and ε equations an equation for K/ε can be derived

$$\frac{d(K/\varepsilon)}{dt} = C_{\varepsilon 2} - 1 - (C_{\varepsilon 1} - 1)\frac{\mathcal{P}}{\varepsilon}$$

Using $C_{\varepsilon 2} = 11/6 \approx 1.83$ and the shear flow data of Tavoularis & Corrsin [125], where $\mathcal{P}/\varepsilon \approx 1.8$ and $K/\varepsilon \approx$ constant, a numeric value of $C_{\varepsilon 1} = 1.46$ is obtained. These may be compared with the 'standard' values suggested by Launder *et al.* [55], $C_{\varepsilon 1} = 1.44$ and $C_{\varepsilon 2} = 1.92$.

We further note that there is no term associated with the mean flow second order derivative terms in the standard high Reynolds number formulation (4.91) of the ε-equation (cf. 4.55, 4.58). For wall-bounded shear-flows we know that $\partial^2 U/\partial y^2$ rapidly decreases as we move away from the wall. Appreciable effects are found only in the buffer region (and viscous sublayer for cases with a strong pressure gradient). The modelling of the rest of the production term is one of the weakest parts of the (4.91) formulation, but is difficult to remedy without extending the $K-\varepsilon$-closure concept.

Calibration of the Schmidt number σ_ε

The complex diffusion, or spatial redistribution, term in the exact transport equation for the (homogeneous part of the) dissipation is also here replaced by a simple gradient diffusion expression. This approach has proven to be reasonably successful and is used in all the modelled transport equations.

Consider the situation near a solid wall. Advection and pressure gradient terms are small up to the log-layer and after integration the mean flow equation may be written (with y as the wall-normal coordinate)

$$(\nu + \nu_T)\frac{\partial U}{\partial y} \approx u_\tau^2 \tag{4.102}$$

We next restrict our attention to the log-layer where the viscous stress is negligible and $\frac{\partial U}{\partial y} = \frac{u_\tau}{\kappa y}$. Inserting this together with the eddy-viscosity relation into (4.102) we get

$$K^2 \approx \frac{u_\tau \kappa}{C_\mu} y\varepsilon \tag{4.103}$$

For the log-layer we further assume that $\mathcal{P} \approx \varepsilon$, which gives that

$$\varepsilon \approx \nu_T \frac{\partial U}{\partial y} = C_\mu \frac{K^2}{\varepsilon}\left(\frac{u_\tau}{\kappa y}\right)^2 \tag{4.104}$$

Combining (4.103) and (4.104) we obtain the well known relations for K and ε in the loglayer

$$\begin{aligned} \varepsilon &\approx \frac{u_\tau^3}{\kappa y} \\ K &\approx \frac{u_\tau^2}{\sqrt{C_\mu}} \end{aligned} \tag{4.105}$$

Since K is constant and production is balanced by dissipation, the K equation is trivially satisfied. Using $\mathcal{P} \approx \varepsilon$ and the assumption of negligible advection we can write the ε equation as

$$0 = (C_{\varepsilon 1} - C_{\varepsilon 2}) \frac{\varepsilon^2}{K} + \frac{\partial}{\partial y} \left(\frac{\nu_T}{\sigma_\varepsilon} \frac{\partial \varepsilon}{\partial y} \right) \tag{4.106}$$

Combining (4.106) and (4.105) we finally obtain

$$\sigma_\varepsilon = \frac{\kappa^2}{\sqrt{C_\mu}} (C_{\varepsilon 2} - C_{\varepsilon 1}) \tag{4.107}$$

Hence, with the above 'standard' values $\kappa = 0.41, C_\mu = 0.09, C_{\varepsilon 2} = 1.92, C_{\varepsilon 1} = 1.44$ we would get $\sigma_\varepsilon \approx 1.2$ to be compared with the 'standard' value of 1.3. Alternatively, with $C_{\varepsilon 2} = 1.83, C_{\varepsilon 1} = 1.46$ we would get $\sigma_\varepsilon \approx 1.5$.

Near-wall treatment and low Reynolds number formulations

Still the most common way to treat wall boundary conditions in commercial CFD applications is to use so called log-layer boundary conditions. This means that the ε equation is retained in its high Reynolds number form (4.91) and the equations are integrated only down to the inner portion of the log-layer. In this way one strongly reduces numeric problems and the need for dense computational meshes in the areas where viscous diffusion is strong. One also avoids modification to the model to account for the low Reynolds numbers near the wall. If the first node, y_1 is chosen to lie in the (inner portion) of the log-layer (usually chosen in the range $30 < y_1^+ < 100$) we have

$$U(y_1) = u_\tau \left(\frac{1}{\kappa} \ln y_1^+ + B \right) \tag{4.108}$$

The values of K and ε are then given by (4.105).

Normally the friction velocityis not known, which means that some iterative process has to be used or simply determine u_τ as

$$u_\tau = \frac{\sqrt{K(y = y_1)}}{C_\mu^{1/4}}$$

and use

$$\frac{\partial K}{\partial y} = 0 \quad \text{at} \quad y = y_1$$

In separated regions or for flow near separation the above approach is highly questionable. This is also true in a number of other situations, *e.g.* in cases with strong effects of rotation, or for unsteady flows.

For flows where the existance universal wall functions is not established a low Reynolds number formulation of the turbulence model equations must be formulated. There are a number of low Re versions of the K-ε model suggested in the literature (cf. Patel *et al.* [76] or Rodi & Mansour [87]) which for two-dimensional shear layers can be written on the form

$$\frac{\mathrm{D}K}{\mathrm{D}t} = \frac{\partial}{\partial y}\left[\left(\nu + \frac{\nu_t}{\sigma_K}\right)\frac{\partial K}{\partial y}\right] + \nu_t\left(\frac{\partial U}{\partial y}\right)^2 - \varepsilon \tag{4.109}$$

$$\frac{\mathrm{D}\tilde{\varepsilon}}{\mathrm{D}t} = \frac{\partial}{\partial y}\left[\left(\nu + \frac{\nu_t}{\sigma_\varepsilon}\right)\frac{\partial \tilde{\varepsilon}}{\partial y}\right] + C_{\varepsilon 1} f_1 \frac{\tilde{\varepsilon}}{K}\nu_t\left(\frac{\partial U}{\partial y}\right)^2 - C_{\varepsilon 2} f_2 \frac{\tilde{\varepsilon}^2}{K} + E \tag{4.110}$$

$$\nu_t = C_\mu f_\mu \frac{K^2}{\tilde{\varepsilon}}, \qquad \tilde{\varepsilon} \equiv \varepsilon - D$$

The low Reynolds number corrections to the high Reynolds number model form are found in the damping functions f_μ, f_1 and f_2 and in the extra terms D and E. The various proposed models differ through the choice of these quantities. In the fully turbulent region away from the wall all three damping functions approach unity and the two extra terms approach zero. The damping function f_μ is introduced to make the model predict a shear stress $-\overline{u'v'} \sim y^3$ in agreement with the expected limiting behaviour. An often used form of D is such that

$$\tilde{\varepsilon} = \varepsilon - 2\nu\left(\frac{\partial K^{1/2}}{\partial y}\right)^2 \sim y^2 \text{ as } y \to 0$$

This redefined dissipation rate has the advantage of approaching zero at the wall (see section 4.5.5), as opposed to ε, whence $\tilde{\varepsilon}_{\mathrm{w}} = 0$ may be used as boundary condition. Away from the wall $\varepsilon \to \tilde{\varepsilon}$. The role of the extra term E (or alternatively f_1) is to increase the production of $\tilde{\varepsilon}$ near the wall. Yet, most models predict the maximum of ε away from the wall in disagreement with DNS data (cf. So *et al.* [111]). The damping functions are usually expressed as functions of either y^+ or $Re_{\tilde{\varepsilon}} \equiv K^2/\nu\tilde{\varepsilon}$, or both. For instance, in the Launder-Sharma model the low Reynolds number corrections are given by

$$f_\mu = \exp\left(\frac{-3.4}{(1 + Re_{\tilde{\varepsilon}}/50)^2}\right), \quad f_1 = 1.0, \quad f_2 = 1 - 0.3\exp(-Re_{\tilde{\varepsilon}}^2)$$

$$D = 2\nu \left(\frac{\partial K^{1/2}}{\partial y} \right)^2 , \quad E = 2\nu\nu_t \left(\frac{\partial^2 U}{\partial y^2} \right)^2$$

As boundary conditions at the wall most low Reynolds number K-ε models use

$$K = \tilde{\varepsilon} = 0 \quad \text{at} \quad y = 0$$

Exceptions are models that use (*e.g.* the Lam-Bremhorst model [53])

$$\varepsilon = \nu \frac{\partial^2 K}{\partial y^2} \quad \text{or} \quad \frac{\partial \varepsilon}{\partial y} = 0 \quad \text{at} \quad y = 0$$

There is no *a priori* reason why the latter condition should hold (see equation 4.77). Yet another boundary condition for ε was suggested by Lindberg [66] who used

$$\varepsilon^+ = \langle a_1^{+2} \rangle + \langle c_1^{+2} \rangle \approx 0.17$$

which is justifiable insofar one has reason to beleive that this value is reasonably constant with respect to flow conditions.

4.7.2. THE $K - \omega$ MODEL

The $K-\omega$ model has roots that go back to Kolmogorov[51] and was revived and extended by Saffman[93]. Much of its present popularity is owed to extensive work by Wilcox and co-workers, see *e.g.* the papers by Saffman & Wilcox[94] and Wilcox[136]. The model is also given an extensive presentation in the Wilcox[137] book. The proponents of the $K-\omega$ model claim that the main advantages of this model as compared to the $K-\varepsilon$-model lie in a more natural treatment of the near-wall region and in the treatment of the turbulent diffusion. These arguments and their validity will not be discussed in detail here.

The quantity ω has been given several different interpretations, but in essence it may be taken as an inverse time scale of the large eddies. The dissipation is now modelled through the following relation with ω

$$\varepsilon = C_\mu \omega K \tag{4.111}$$

The model, as described by Wilcox[136],[137], is defined by the modelled mean flow equation (4.88) (and continuity), the kinetic energy equation (4.89) with ε as above, and, for *Re*, the following ω-equation

$$\frac{\mathrm{D}\omega}{\mathrm{D}t} = 2\alpha S_{ij} S_{ij} - \beta\omega^2 + \frac{\partial}{\partial x_i} \left[\left(\nu + \frac{\nu_T}{\sigma_\omega} \right) \frac{\partial \omega}{\partial x_i} \right] \tag{4.112}$$

and the eddy-viscosity relation

$$\nu_T = K\omega \tag{4.113}$$

The model parameter values given by Wilcox[137] are:

$$C_\mu = 0.09, \quad \sigma_k = 2.0, \quad \sigma_\omega = 2.0, \quad \alpha = 5/9 \approx 0.56, \quad \beta = 3/40 = 0.075. \tag{4.114}$$

One may note, perhaps somewhat surprisingly, that the Schmidt number σ_k here is given a value twice that in the standard $K-\varepsilon$-model (the calibration of this parameter is an issue rather separate from the choice of the Z-quantity). It may be interesting to compare the formulation (4.112) of the ω-equation with that derived by inserting (4.111) into the modelled ε-equation (4.91). This gives an equation very similar to (4.112). The form of the production and destruction terms are identical, but the parameters would instead attain the values

$$\alpha = 2(C_{\varepsilon 1} - 1) = 0.88, \qquad \beta = C_\mu(C_{\varepsilon 2} - 1) \approx 0.083. \tag{4.115}$$

The turbulent'diffusion' term in the transformed ω equation reads

$$\frac{\partial}{\partial x_i}\left[\frac{\nu_T}{\sigma_\varepsilon}\frac{\partial \omega}{\partial x_i}\right] + \frac{\omega}{K}\frac{\partial}{\partial x_i}\left[\nu_T\left(\frac{1}{\sigma_\varepsilon} - \frac{1}{\sigma_k}\right)\frac{\partial K}{\partial x_i}\right] + \frac{2\nu_T}{K\sigma_\varepsilon}\frac{\partial K}{\partial x_i}\frac{\partial \omega}{\partial x_i}$$

and can be seen to involve terms that are not of gradient diffusion type. The original $K-\omega$ model is known to have problems near free-stream boundaries (see Wilcox[137]). To remedy this behaviour Menter[72] has proposed to include a cross-diffusion term that is equivalent to the last term of the above expression of the diffusion term in the $K-\omega$ model. This means that the similarity between the two models become even closer. A further modification that has been proposed is a formulation where the influence of the cross-diffusion term is weighted such that it only will influence the flow far away from walls. The generality of such formulations for complex flows can perhaps be doubted.

The $K-\omega$ model has been used extensively for boundary layer problems with different types of pressure gradients. For instance, in boundary layers with a local separation bubble the behaviour after reattachment has been found to be improved over typical $K-\varepsilon$ models. As seen from the transformation of the ε equation into an ω equation the qualitative difference between the two models are few, at least in their high Reynolds number formulations.

The near-wall behaviour of the two quantities is quite different though, and perhaps the main prediction differences noted between the two models can probably be ascribed to the differences in the near-wall treatment.

4.8. Differential Reynolds stress models for incompressible flow

Although eddy viscosity type of models have proven to work well in shear stress dominated wall-bounded flows of engineering interest there are many examples of situations in which they produce predictions that are far from the experimental results. Typical situations where this occurs are flows with sudden change in mean strain rate, flow in rotating or stratified fluids, flows in non-circular ducts exhibiting secondary flow patterns, flows along curved surfaces and flows with boundary layer separation. Models based on the transport equations for the Reynolds stress tensor (RST closures) have prerequisites for a better accommodation of the underlying physics of such flows.

We will here consider *classical* RST-models, by which we refer to closures that are based on the transport equations for R_{ij} and the total dissipation rate. Hence, the complete closure consists of the Reynolds averaged Navier-Stokes equations and the continuity equation, complemented by seven transport equations for the turbulence quantities.

The primary field variables are U_i, P, $\overline{u_i' u_j'}$ and ε. Geometrical constraints and the kinematic viscosity, ν, may enter the problem through a parametrical dependence. Aside from the primary closure variables themselves, dimensionless frame invariant groups of these are formed and used in the mathematical modelling of the various terms. Classical works in the area are Rotta[90] 1951, Launder, Reece & Rodi[55] 1975, and Lumley[68] 1978 and reviews on the subject are found in Reynolds[82] 1976, Reynolds[83] 1987, Launder[54] 1989, Groth[32] 1991 and Speziale[121] 1991.

In the following we will discuss the modelling of the individual terms in the RST-equations. Desirable closure properties inspired by empericism based on experiments and direct numerical simulation along with application of kinematic and other constraints for model development and validation will be discussed. Model formulations will often be presented and discussed in terms of non-dimensional anisotropy measures. The Reynolds stress anisotropy tensor, a_{ij}, was introduced in section (4.4). We may define a corresponding anisotropy measure for the dissipation rate tensor

$$e_{ij} \equiv \frac{\varepsilon_{ij}}{\varepsilon} - \frac{2}{3}\delta_{ij} \tag{4.116}$$

With these measures at hand it is illustrative to formulate the transport equations for the turbulence quantities in the following way

$$\begin{aligned} \frac{\mathrm{D}K}{\mathrm{D}t} &= \mathcal{P} - \varepsilon - \frac{\partial}{\partial x_m}\left(J_m - \nu\frac{\partial K}{\partial x_m}\right) + \overline{u_i' f_i'} \qquad (4.117) \\ \frac{\mathrm{D}\varepsilon}{\mathrm{D}t} &= \mathcal{P}^{\varepsilon(1)} + \mathcal{P}^{\varepsilon(2)} + \mathcal{P}^{\varepsilon(3)} + T^{\varepsilon} - D^{\varepsilon} \end{aligned}$$

$$-\frac{\partial}{\partial x_m}\left(J_m^\varepsilon - \nu\frac{\partial\varepsilon}{\partial x_m}\right) + 2\nu\overline{u'_{i,m}f'_{i,m}} \qquad (4.118)$$

$$\frac{\mathrm{D}\,a_{ij}}{\mathrm{D}t} = \mathcal{P}_{ij}^{(a)} + \frac{1}{K}\Pi_{ij} - \frac{\varepsilon}{K}(e_{ij} - a_{ij}) + \mathit{Diff}_{ij}^{(a)} + G_{ij}^{(a)} \qquad (4.119)$$

where the right hand sides of the K and ε equations are given by (4.40-4.42) and (4.56-4.61) and the anisotropy production, volume force generation or redistribution, and diffusion terms are defined as

$$\mathcal{P}_{ij}^{(a)} \equiv -\left(\frac{4}{3}S_{ij} + a_{ik}S_{jk} + a_{jk}S_{ik} - \left(a_{ij} + \frac{2}{3}\delta_{ij}\right)a_{lk}S_{lk}\right) - (a_{ik}\Omega_{jk} + a_{jk}\Omega_{ik})$$

$$G_{ij}^{(a)} \equiv \frac{1}{K}\left(\overline{u'_i f'_j} + \overline{u'_j f'_i} - \left(a_{ij} + \frac{2}{3}\delta_{ij}\right)\overline{u'_i f'_i}\right)$$

$$\mathit{Diff}_{ij}^{(a)} \equiv -\frac{1}{K}\frac{\partial}{\partial x_k}\left(J_{ijk} - \frac{2}{3}\delta_{ij}J_k\right) + \frac{a_{ij}}{K}\frac{\partial J_k}{\partial x_k} + \frac{\nu}{K}\left(\frac{\partial^2(Ka_{ij})}{\partial x_k\partial x_k} - a_{ij}\frac{\partial^2 K}{\partial x_k \partial x_k}\right)$$

With this formulation the two transport equations for the dimensional quantities, K and ε, are kept separate from the equation for the evolution of the anisotropy. The latter comprises five independent equations since $a_{ii} = 0$. It is noteworthy that $\mathcal{P}_{ij}^{(a)}$ and $\mathcal{P}$ are explicitly described in terms of K and a_{ij}, and hence need no modelling. One may also note that the only term in the K-equation that needs modelling in terms of the primary quantities is J_m.

Equation (4.119) illustrates the influence of the various effects on the anisotropy. Obviously, a mean strain or rotation rate will have direct influence through the production term. The second and third terms on the right hand side of (4.119) represent the redistributive effects of pressure-strain rate and dissipation. We note that the difference in anisotropy, $e_{ij} - a_{ij}$, will contribute to the net change of a_{ij}.

In the presence of a system rotation we have a Coriolis volume force ($f'_i = -2\epsilon_{ilm}\Omega_l^{\mathrm{s}}u'_m$) the K- and ε-equations are both left unchanged ($\langle u'_i f'_i = 0\rangle$) whereas a_{ij} is influenced by through the generation term $G_{ij}^{(a)}$ that is given by

$$G_{ij}^{(a)} = -2\Omega_l^{\mathrm{s}}\left(\epsilon_{jlm}a_{im} + \epsilon_{ilm}a_{jm}\right)$$

Hence, we see that at the RST-equation level effects of rotational mean flow and system rotation are directly accounted for to lowest order through the production and volume-force generation/redistribution terms.

4.8.1. THE DISSIPATION RATE TENSOR

The modelling of the dissipation rate tensor is suitably divided into two separate problems, firstly that of modelling the total dissipation rate, which mainly is an issue of modelling the terms in the ε-equation, and secondly that of modelling the dissipation rate anisotropy e_{ij} defined in (4.116). Supported by the Kolmogorov theory of small scale universal equilibrium which more or less postulates local isotropy for dissipating scales at high enough Reynolds numbers, the ε_{ij} is often modelled as being isotropic, *i.e.*

$$\varepsilon_{ij} = \tfrac{2}{3}\varepsilon\delta_{ij} \quad \text{or} \quad e_{ij} = 0$$

There are however a number of indications that this is not very well satisfied in flows with high rates of mean strain, shear or rotation at moderate Reynolds numbers. Experiments and direct numerical simulations (Sjögren & Johansson[108]) have shown that significant degrees of dissipation rate anisotropy can exist even at relatively high Reynolds numbers in the presence of large mean strain rates or in the relaxation from such straining.

Turbulence in the immediate vicinity of a solid wall boundary is an example of a low Reynolds number strongly anisotropic flow exhibiting strong anisotropy also in the dissipation rate tensor.

A possible approach to the modelling problem is to assume that the small scale contribution to the dissipation rate anisotropy is negligible. With S and L referring to small and large scales, respectively, we then have

$$\varepsilon_{ij} = \varepsilon^{(\mathrm{L})}(e_{ij}^{(\mathrm{L})} + \frac{2}{3}\delta_{ij}) + \frac{2}{3}\delta_{ij}\varepsilon^{(\mathrm{S})} \tag{4.120}$$

(see *e.g.* Naot, Shavit & Wolfshtein[74] or Hanjalic & Launder[40]). The total dissipation, hence, is split into two parts, where the relation between these can be taken as

$$\varepsilon = \frac{K^{3/2}}{l}\left(B_2 + B_1\frac{\nu}{K^{1/2}l}\right) \tag{4.121}$$

This can be seen as a correction to the normal high Reynolds number relation $\varepsilon \sim K^{3/2}/l$ between the ε and the integral scale l. We may obtain an estimate of B_1 by approximating the correlation function as $\mathcal{R}_{\alpha\alpha}(r\vec{e}_\alpha) = R_{\alpha\alpha}\exp(-r^2/2\lambda^2)$ (cf. Schumann & Pattersson[98]) in the low Re_T limit which yields $B_1 = 5\pi$. Assuming a model spectrum representing turbulence at high Reynolds numbers (see figure 4.6) one gets that $B_2 \to 0.31$ as $Re_T \to \infty$.

The model for $e_{ij}^{(\mathrm{L})}$ (here taken to equivalentto e_{ij}) does not influence the relation between K, l and ε since its trace is zero. Rotta[90] suggested an anisotropic relation of the above kind in which the large scale dissipation rate anisotropy was set equal to the Reynolds stress anisotropy. This is perhaps not the best choice. In fact, linear theory (RDT) shows that during the initial stages of homogeneous isotropic turbulence subjected to an arbitrary mean shear or strain the dissipation rate anisotropy is only half of that of the Reynolds stresses. On the other hand, the asymptotic relations (4.79) show that at a wall $e_{ij} = a_{ij}$ to leading order.

The idea of linking e_{ij} to a_{ij} is probably sound since these two quantities are determined by the same range of scales of the turbulent field. As an alternative to the Rotta proposal a possible candidate to such a model relationship might be the non-linear model by Hallbäck *et al.* [36] that was based on considerations of turbulence at moderate Reynolds numbers under imposed mean strain so that both the energetic scales and the dissipative scales where affected by the mean strain. The model was calibrated against RDT and tested against DNS-data of homogeneous turbulence at fairly low Reynolds numbers, and is given by

$$e_{ij} = \left[1 + \alpha\left(\frac{1}{2}II_a - \frac{2}{3}\right)\right] a_{ij} - \alpha\left(a_{ik}a_{kj} - \frac{1}{3}II_a\delta_{ij}\right) \tag{4.122}$$

where the model parameter (α) was set to 3/4 but could be allowed to depend on the Reynolds number and strain rate parameters.

Sjögren & Johansson[109] further developed this into a fouth order model that satisfies strong realizability, and both the RDT-condition and the relation $e_{ij} = a_{ij}$ at a solid wall

$$e_{ij} = \left(1 - \frac{1}{2}F\right) a_{ij}, \quad F = 1 - \frac{9}{8}\left(II_a - III_a\right). \tag{4.123}$$

This model was tested for channel flow and a number of homogeneous flows, and compared successfully with experimental data for the total dissipation rate anisotropy in homogeneous axisymmetric turbulence (Sjögren & Johansson[108]). Close to solid boundaries an assumption of isotropic dissipation rate would obviously violate the realizability condition. A near-wall model that satisfies the asymptotic relations (4.79) was also formulated by Launder and coworkers (see, *e.g.* , Launder[54]), although that model was considerably more complicated in that it *e.g.* involved the wall-normal vector.

If the small scale contribution (instead of being neglected) would be assumed to scale in the same manner as the small scale part of the total dissipation one would obtain a total anisotropy tensor as

$$e_{ij} = \left(1 + \frac{B_2}{B_1}\frac{K^{1/2}l}{\nu}\right)^{-1} e_{ij}^{(\mathrm{L})}(a_{kl}) \tag{4.124}$$

With the above values for B_1 and B_2 one would get a model not to different from that of Hanjalić & Launder[40]. This seems to be a fairly good approximation in low-Reynolds number channel flow (cf. Mansour *et al.* [69]) for the normal components of e_{ij}. However, it gives far too low degree of dissipation rate anisotropy in, *e.g.* , initially strained homogeneous turbulence relaxing towards isotropy (Sjögren & Johansson[108]). This would, in fact, suggest that a better alternative would be to use $e_{ij}^{(\mathrm{L})}$ alone as the estimate of the total dissipation rate anisotropy.

4.8.2. THE PRESSURE-STRAIN RATE TERM

The treatment of the pressure-strain rate correlation term is traditionally based on the formal solution of the Poisson equation for the pressure field (in the absence of body forces)

$$-\frac{1}{\rho}p'_{,kk} = 2U_{i,j}u'_{j,i} + u'_{i,j}u'_{j,i} - \overline{u'_{i,j}u'_{j,i}} \tag{4.125}$$

with the wall boundary condition

$$\frac{1}{\rho}p'_{,2} = \nu u'_{2,22} \tag{4.126}$$

where x_2 is the wall-normal coordinate. The linearity in p of both the Poisson equation and the boundary condition allows superposition of solutions and the pressure field is normally decomposed into three parts: the *rapid* and *slow* parts and the *Stokes* term

$$\begin{aligned}
\tfrac{1}{\rho}p'^{(\mathrm{r})}_{,kk} &= -2U_{i,j}u'_{j,i} && \text{wall b.c. } p'^{(\mathrm{r})}_{,2} = 0 \\
\tfrac{1}{\rho}p'^{(\mathrm{s})}_{,kk} &= -u'_{i,j}u'_{j,i} + \overline{u'_{i,j}u'_{j,i}} && \text{wall b.c. } p'^{(\mathrm{s})}_{,2} = 0 \\
p'^{(\mathrm{S})}_{,kk} &= 0 && \text{wall b.c. } p'^{(\mathrm{S})}_{,2} = \mu u'_{2,22}
\end{aligned} \tag{4.127}$$

Here, effects of the wall boundary are still present in the rapid and slow parts due to the homogeneous boundary conditions that are prescribed on the wall. The solution of the rapid part, for instance, can be written as

$$p'^{(\mathrm{r})}(\vec{x}) = \frac{1}{4\pi}\int_V 2U^{(1)}_{k,l}u'^{(1)}_{l,k}\,G(\vec{x},\vec{x}^{(1)})\,dV^{(1)}$$

and correspondingly for the slow term. The effect of the wall condition is included in the form of the Green's function $G(\vec{x},\vec{x}^{(1)})$. In the case of a plane wall boundary the Green's function takes the form (cf. Launder *et al.* [55] and Shih & Lumley[104])

$$G(\vec{x},\vec{x}^{(1)}) = \frac{1}{|\vec{x}^{(1)}-\vec{x}|} + \frac{1}{|\vec{x}^{(1)}-\vec{x}^{(*)}|}$$

where $\vec{x}^{(*)}$ is the mirror image point of $\vec{x}$.

In forming the expressions for the rapid and slow pressure-strain rate correlations, for the following modelling work, the by far most common approach is to impose three local homogeneity assumptions (cf. Chou[14]):

1) the boundary conditions for the rapid and slow parts of the pressure field are prescribed at infinity, and the Green's function assumes the special form $1/|\vec{x}^{(1)}-\vec{x}|$.
2) the mean velocity gradient field is expanded around the point $\vec{x}$ and only the zero order term is retained, thus allowing it to be extracted from of the Poisson integral.
3) two-point correlations expressed as functions of position $\vec{x}$ and separation $\vec{r}=\vec{x}^{(1)}-\vec{x}$, are assumed to vary slowly enough with the position $\vec{x}$ for the velocity-derivative correlations to be approximated by two-point velocity-correlation derivatives with respect to $\vec{r}$ only.

With these approximations the pressure-strain rate correlations become

$$\Pi_{ij}^{(r)}(\vec{x}) \approx -\frac{1}{2\pi}\frac{\partial U_q}{\partial x_p}\int\left\{\frac{\partial^2\overline{u_i' u_p'^{(1)}}}{\partial r_j \partial r_q} + \frac{\partial^2\overline{u_j' u_p'^{(1)}}}{\partial r_i \partial r_q}\right\}\frac{dV^{(1)}}{|\vec{x}^{(1)}-\vec{x}|} \tag{4.128}$$

$$\Pi_{ij}^{(s)}(\vec{x}) \approx -\frac{1}{4\pi}\int\left\{\frac{\partial^3\overline{u_i' u_p'^{(1)} u_q'^{(1)}}}{\partial r_j \partial r_p \partial r_q} + \frac{\partial^3\overline{u_j' u_p'^{(1)} u_q'^{(1)}}}{\partial r_i \partial r_p \partial r_q}\right\}\frac{dV^{(1)}}{|\vec{x}^{(1)}-\vec{x}|} \tag{4.129}$$

where (1)-indices are assigned to integration variables. All effects of the local inhomogeneity and the Stokes term are lumped into a *wall* term $\Pi_{ij}^{(w)}$.

Bradshaw *et al.* [6] used the DNS data of Kim, Moin & Moser[50] of turbulent channel flow (half-channel width $\delta^+ = 180$) to study the difference between the exact rapid pressure and the rapid pressure as obtained with the correct boundary condition but employing the second homogeneity assumption, and found that the results were indistinguishable down to $y^+ \approx 40$. The data also show that the Stokes term ($\Pi_{ij}^{(S)}$) is negligible for $y^+ > 20$. At the wall all elements of the pressure-strain rate tensor vanishes except Π_{12}, all parts of which are non-zero.

The 'exact' rapid and slow pressure-strain rate terms, based on the exact solution of (4.127) for the corresponding pressures, were evaluated from the DNS data [50] showing that for $10 < y^+ < 80$

$$|\Pi_{11}^{(s)}| > |\Pi_{11}^{(r)}|, \quad |\Pi_{22}^{(s)}| > |\Pi_{22}^{(r)}|, \quad |\Pi_{12}^{(s)}| > |\Pi_{12}^{(r)}| \text{ and } \quad |\Pi_{33}^{(s)}| < |\Pi_{33}^{(r)}|$$

Thus, the slow term dominates in the major part of this region except for the 33-element. Whether this holds for higher Reynolds numbers is still unclear. Beyond $y^+ \approx 80$ the pressure-strain correlation terms resemble those of a homogeneous turbulent shear flow where $\Pi^{(s)}_{22} \approx -\Pi^{(s)}_{11}$ and $\Pi^{(r)}_{33} \approx -\Pi^{(r)}_{11}$, *i.e.* the R_{22} component is supplied by energy mainly by the slow term whereas the rapid pressure strain dominates the transfer to the transverse component R_{33} (Rogers *et al.* [89]).

The rapid part Introducing the fourth rank tensor $\boldsymbol{M}$

$$M_{ijpq} \equiv -\frac{1}{8\pi K}\int \frac{\partial^2 \mathcal{R}_{ij}}{\partial r_p \partial r_q}\frac{dV}{|\vec{r}|} = \frac{1}{2K}\int \frac{k_p k_q}{k^2}\widehat{\mathcal{R}}_{ij} d^3k \tag{4.130}$$

the homogeneous approximation of the rapid term (4.128) can be written as

$$\Pi^{(r)}_{ij} = 4KU_{p,q}\{M_{iqpj} + M_{jqpi}\} \tag{4.131}$$

The problem is thus to model the dimensionless tensor $\boldsymbol{M}$. The $\boldsymbol{M}$-tensor satisfies two symmetry conditions, a continuity condition and the so called Green's condition

$$\begin{aligned} M_{ijpq} &= M_{jipq} \\ M_{ijpq} &= M_{ijqp} \\ M_{ijjq} &= 0 \\ M_{ijpp} &= \tfrac{R_{ij}}{2K} = \tfrac{1}{2}a_{ij} + \tfrac{1}{3}\delta_{ij} \end{aligned} \tag{4.132}$$

The second symmetry condition is obvious and the first one follows from the fact that the anti-symmetric part of the spectrum tensor integrates to zero.

The standard approach is to model $\boldsymbol{M}$ in terms of $\boldsymbol{a}$. A complete linear expansion satisfying the symmetry conditions of (4.132) is given by

$$\begin{aligned} M_{ijpq} = {} & A_1\,\delta_{ij}\delta_{pq} + A_2\,(\delta_{ip}\delta_{jq} + \delta_{iq}\delta_{jp}) + A_3\,\delta_{ij}a_{pq} \\ & + A_4\,a_{ij}\delta_{pq} + A_5\,(\delta_{ip}a_{jq} + \delta_{iq}a_{jp} + \delta_{jp}a_{iq} + \delta_{jq}a_{ip}) \end{aligned} \tag{4.133}$$

Imposing the last two conditions of (4.132) yields all of the coefficients expressed in terms of only one;

$$A_1 = \frac{4}{15}, \quad A_2 = -\frac{1}{15}, \quad A_3 = \frac{C_2}{2}, \quad A_4 = \frac{5+2C_2}{11}, \quad A_5 = -\frac{2+3C_2}{11}$$

Insertion of the linear ansatz (4.133) and the above coefficients into (4.131) yields the linear model of Launder *et al.* [55]

$$\begin{aligned}\Pi_{ij}^{(r)} &= \frac{4}{5}KS_{ij} + \frac{9C_2+6}{11}K(a_{ik}S_{jk} + a_{jk}S_{ik} - \frac{2}{3}a_{kl}S_{kl}\delta_{ij}) \\ &\quad + \frac{-7C_2+10}{11}K(a_{ik}\Omega_{jk} + a_{jk}\Omega_{ik}) \qquad (4.134)\end{aligned}$$

where $C_2 = 0.4$ was found to give best fit in calibration against the normal stresses in the nearly homogeneous shear flow of Champange *et al.* [9]. Calibration against RDT would give that $C_2 = 10/7$, as a consequence of which, as pointed out by Shih *et al.* [106], the mean rotation related term of (4.134) would vanish. The best compromise value, for a wide range of flow situations, seems to be somewhat higher than that proposed by Launder *et al.* A value of (or close to) 5/9 has been proposed by Lumley[68] and used in algebraic Reynolds stress modelling by Taulbee[123] and Wallin & Johansson[133].

A further truncated version of (4.134) (Launder *et al.* [55]) is given by

$$\begin{aligned}\Pi_{ij}^{(r)} &= \frac{4}{3}\gamma KS_{ij} + \gamma K(a_{ik}S_{jk} + a_{jk}S_{ik} - \frac{2}{3}a_{kl}S_{kl}\delta_{ij}) \\ &\quad + \gamma K(a_{ik}\Omega_{jk} + a_{jk}\Omega_{ik}) \\ &= -\gamma(\mathcal{P}_{ij} - \frac{2}{3}\mathcal{P}\delta_{ij}) \qquad (4.135)\end{aligned}$$

where $\gamma = 0.6$ was chosen in order to give the correct initial response to mean strain of isotropic turbulence ($\Pi_{ij}^{(r)} = 4/5KS_{ij}$). The simplified version (4.135), known as 'isotropization of production', does not satisfy the symmetry conditions of the $\boldsymbol{M}$-tensor.

The most general ansatz for $\boldsymbol{M}(\boldsymbol{a})$ satisfying the symmetry conditions consists of 15 tensorially independent terms, each multiplied by a scalar function coefficient. Insertion into (4.131) gives (Johansson & Hallbäck [44])

$$\begin{aligned}\frac{1}{K}\Pi_{ij}^{(r)} &= S_{pq}\,[Q_1\delta_{ip}\delta_{jq} + Q_2\,(a_{ip}\delta_{jq} + a_{jp}\delta_{iq}) \\ &\quad + (Q_3\delta_{ij} + Q_4a_{ij} + Q_5a_{ik}a_{kj})\,a_{pq} + Q_6a_{iq}a_{jp} \\ &\quad + (Q_7\delta_{ij} + Q_5a_{ij} + Q_8a_{ik}a_{kj})\,a_{pl}a_{ql}] \\ &\quad + \Omega_{pq}\,[Q_9\,(a_{ip}\delta_{jq} + a_{jp}\delta_{iq}) + Q_{10}a_{pk}\,(a_{jk}\delta_{iq} + a_{ik}\delta_{jq}) \\ &\quad + Q_{11}a_{pk}\,(a_{jk}a_{iq} + a_{ik}a_{jq})] \qquad (4.136)\end{aligned}$$

The scalar functions Q_α may depend on II_a, III_a and on Re_T. A dependence of any of the mean deformation rate parameters is less motivated since the $\boldsymbol{M}$-tensor does not respond directly to changes in the mean strain rate tensor. This is, hence, the reason for not incorporating $U_{i,j}$ in the tensorial ansatz for $\boldsymbol{M}$. The continuity and Green's conditions reduce the number of unknowns to seven independent scalar functions and determines the constant part of Q_1 to be 4/5, which is in exact agreement with the initial response of isotropic turbulence to an imposed high mean strain rate (Crow[17]).

All existing models for the rapid pressure-strain model, based on an expansion in a_{ij}, can be identified as subsets of (4.136). The unknown scalar functions must be expanded in their arguments, and the most natural approach is to assume a truncated Taylor expansion in II_a and III_a since the tensorial ansatz resembles a Taylor expansion. Finally the strong realizability condition can be imposed on the model. It is worth mentioning that no linear model can satisfy strong realizability. Sjögren & Johansson[109] analyzed the model expression (4.136) further and included a fifth order rapid pressure strain rate model into a composite Reynolds stress closure.

The above approach can give arbitrarily good agreement with RDT for irrotationally strain flows with increasing order of truncation of the Taylor expansion of the scalar functions (see Johansson & Hallbäck [44] or Lee[61]). For rotational mean deformations, however, the above expansion does not converge to the RDT result with increasing order of truncation. For turbulence subjected to rapid system rotation in the absence of mean strain ('pure rotation') a model of type (4.136) predicts undamped oscillations of the components of a_{ij} regardless of the choice of scalar functions, whereas the RDT-solution shows a damped oscillatory behaviour (see Mansour, Shih & Reynolds[70] or Johansson *et al.* [45]) towards a, not necessarily isotropic, equilibrium state set by the initial conditions. This illustrates a severe inherent limitation of classical RST-modelling.

The slow part The coupling between the pressure-strain rate term and the

two-point correlation spectrum, which integrated over wavenumber space gives the Reynolds stresses, is less clear for the slow term than for the rapid term. The justification for modelling the slow pressure-strain-rate correlation in terms of the Reynolds stresses is thus weaker than for the rapid part.

Even though not rigourously shown it is regarded well established wisdom, based on experience, that the slow pressure-strain rate term plays the role of redistributing energy among velocity components towards isotropy. The simplest possible way to model this tendency of isotropization mathematically is to apply dimensional analysis and assume that the pressure-strain rate is proportional to the deviation from isotropy as was suggested by Rotta[90];

$$\overline{\frac{p'^{(s)}}{\rho}\left(\frac{\partial u_i'}{\partial x_j}+\frac{\partial u_j'}{\partial x_i}\right)} = -C_{\mathrm{R}}\frac{K^{3/2}}{l}\left(\frac{R_{ij}}{K}-\frac{2}{3}\delta_{ij}\right) \tag{4.137}$$

In isotropic turbulence the pressure has, indeed, been shown to scale on the kinetic energy, $\sqrt{\overline{(p'/\rho)^2}} \sim K$ (Batchelor[4]). This is on the other hand far from true at a wall where $K=0$ but $p'^{(s)}_{\mathrm{rms}} \neq 0$, reflecting the non-local character of the pressure field. Usually the high Reynolds number estimate $l \sim K^{3/2}/\varepsilon$ is used to eliminate the turbulence length scale from (4.137) and reformulate the model as

$$\Pi_{ij}^{(s)} = -C_1 \varepsilon a_{ij} \tag{4.138}$$

where C_1 is known as the "Rotta constant", usually assigned a value in the range 1.5 - 1.8. This type of modelling has been adopted by most modellers although C_1 is in general not a constant but rather a function of the turbulence state.

Notable is that a linear model cannot satisfy the strong realizability condition unless the parameter C_1 is allowed to depend on scalar measures of the turbulence state, such as the two-componentality parameter F or the Reynolds number Re_T. However, if C_1 is forced to zero in the two-component limit all components of the slow pressure-strain rate tensor go to zero, which is not motivated from the real physical constraint. For this reason and in order to better mimic the observed behaviour of relaxing strongly anisotropic turbulence, a tensorially quadratic term is sometimes included;

$$\Pi_{ij}^{(s)} = -\varepsilon\left\{\beta_1 a_{ij} + \beta_2\left(a_{ik}a_{kj} - \frac{1}{3}II_a\delta_{ij}\right)\right\} \tag{4.139}$$

where the coefficients β_1 and β_2 may depend on II_a, III_a, Re_T, S^* and Ω^*. One may note that in axisymmetry the linear model is tensorially adequate since there $\Pi_{ij}^{(s)}$ and a_{ij}, each contains only one independent element. An ansatz of the type (4.139) requires that the principal axes of a_{ij} and $\Pi_{ij}^{(s)}$ are aligned. DNS data of Rogers *et al.* [89] show that this is approximately the case in homogeneous shear flow turbulence approaching a quasi-equilibrium state with the principal axes of both tensors rotated about 20° relative to the mean flow direction.

Sjögren & Johansson[109] constructed a nonlinear model based on the type of model (4.139) with

$$\beta_1 = f(Re_T)\beta_1^*(II_a, III_a), \quad \beta_2 = f(Re_T)\beta_2^*(II_a, III_a).$$

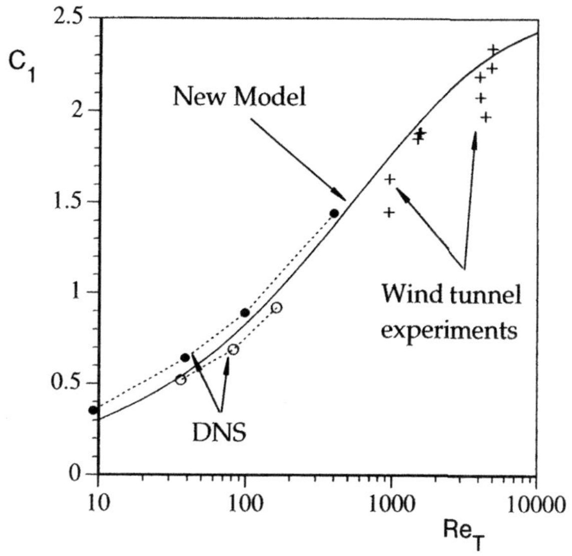

Figure 4.7. The Rotta parameter C_1 from DNS of axisymmetric relaxation ($\{\bullet\}$ and $\{\circ\}$) and according to the model equations (4.140) and (4.141) {solid line}. Included are also experimental results (+) based on direct measurements of the dissipation rate tensor ε_{ij} (T. Sjögren, private communication).

A strong dependence on the Reynolds number was inferred in the theoretical study by Weinstock [134], [135] with $C_1 \to 3.6$ as $Re_T \to \infty$. This type of behaviour was also found by Hallbäck & Johansson [37] in DNS with Re_T ranging over one and a half decade. Actually, the Re_T trend observed in those DNS can be captured if the finite-Reynolds-number corrected relation between ε and l (4.121) suggested by Rotta[90] is used to express the original Rotta proposal (4.137) on the form (4.138). If C_{R} is assumed constant then the "Rotta constant" C_1 varies with Re_T as

$$C_1 = C_{\mathrm{R}} f(Re_T) \tag{4.140}$$

where $f(Re_T)$ is given by

$$f(Re_T) \equiv \frac{K^{3/2}/l}{\varepsilon}, \; f(Re_T) \approx \sqrt{\frac{Re_T}{4B_1} + \left(\frac{B_2}{8B_1} Re_T\right)^2} - \frac{B_2}{8B_1} Re_T. \tag{4.141}$$

Good agreement with DNS and new experimental data (Sjögren & Johansson[108]) is obtained for $C_{\mathrm{R}} = 0.8$ ($B_2 = 0.31$), figure 4.7. For this choice of C_{R} the "Rotta constant" C_1 approaches $C_{\mathrm{R}}/B_2 \approx 2.6$ as $Re_T \to \infty$.

The recent experiments of Sjögren & Johansson[108] show clearly that non-linear models should be used for the slow presure strain rate. DNS-results also support this view.

An experimental finding in purely relaxing turbulence is also that the rate of relaxation depends on the initial state of anisotropy, being lower for positive values of the third invariant (III_a) (Choi[13] and Le Penven *et al.* [65]). Description of this feature also necessitates the use of a non-linear model.

Pressure reflection effects

The formal solution of the Poisson equation for the pressure field (section 4.8.2) in the case of a wall boundary indicate how one should modify the homogeneous model expression to account for pressure reflection effects. These contributions to the pressure strain are of the form (cf. Launder *et al.* [55] or Shih & Lumley[104])

$$\overline{\frac{p'}{\rho}\frac{\partial u_i'}{\partial x_j}}(\vec{x}) = -\frac{1}{4\pi}\int\left\{2\frac{\partial U_p^{(1)}}{\partial x_q^{(1)}}\overline{\frac{\partial u_q'^{(1)}}{\partial x_p^{(1)}}\frac{\partial u_i'}{\partial x_j}} + \overline{\frac{\partial u_p'^{(1)}}{\partial x_q^{(1)}}\frac{\partial u_q'^{(1)}}{\partial x_p^{(1)}}\frac{\partial u_i'}{\partial x_j}}\right\}\frac{dV^{(1)}}{|\vec{x}^{(1)}-\vec{x}^{(*)}|} \quad (4.142)$$

where $dV^{(1)}$ indicates that the integral is evaluated over all points $\vec{x}^{(1)}$ in a half space down to the wall and $\vec{x}^{(*)}$ is the image point of $\vec{x}$. At positions far from the wall the Green's function becomes small and the contribution of this part of the solution negligible.

While these integrals may appear similar to those already modeled in the homogeneous case they differ in the occurrence of the image point in the Green's function and in the fact that, due to strong inhomogeneity, the correlation functions cannot be assumed to depend on the separation vector alone. Hence, the integral of the rapid part does not satisfy the symmetry and Green's conditions. Launder *et al.* [55] suggested a model of the form

$$\Pi_{ij}^{(w)} = \left\{C_1^{(*)}\varepsilon a_{ij} + 4K\frac{\partial U_p}{\partial x_q}(M_{iqpj}^{(*)} + M_{jqpi}^{(*)})\right\} f\left(\frac{l}{x_2}\right) \quad (4.143)$$

where f is a damping function approaching zero away from the wall and $\boldsymbol{M}^{(*)}$ is a tensor similar to $\boldsymbol{M}$ with the exceptions mentioned above. Instead symmetry and zero trace conditions should be imposed directly on the pressure-strain rate tensor, which gives a model that contains three unknown parameters instead of one as in the homogeneous case. Their final model expression reads

$$\Pi_{ij}^{(w)} = \left\{0.125\varepsilon a_{ij} - 0.03(\overline{u_i'u_k'}\Omega_{jk} + \overline{u_j'u_k'}\Omega_{ik})\right\}\frac{K^{3/2}}{\varepsilon x_2} \quad (4.144)$$

In the log-layer (where $K^{3/2}/\varepsilon x_2 \approx 3$) the wall term gives the following contributions

$$\begin{aligned}\Pi_{11}^{(w)} &= 0.37\varepsilon a_{11} + 0.090SK(-a_{12})\\ \Pi_{22}^{(w)} &= 0.37\varepsilon a_{22} - 0.090SK(-a_{12})\\ \Pi_{12}^{(w)} &= 0.37\varepsilon a_{12} + 0.045SK(a_{11} - a_{22})\end{aligned}$$

In the log-layer $SK/\varepsilon \approx SK/\mathcal{P} \approx K/u_\tau^2 \approx 3.5$. The contribution to the transverse components given by the 'slow' term which is small since a_{33} is relatively small. From the above expressions we recognize a reduction of the effective Rotta constant by 0.4. The overall effect of the pressure reflexion is to reduce intercomponent energy transfer and thereby allow a higher degree of Reynolds stress anisotropy which is in agreement with the observed discrepancy between homogeneous shear flow data and near-wall turbulence data (table 4.1).

TABLE 4.1. Typical data of homogeneous equilibrium shear flow and log-layer near-wall data.

	a_{11}	a_{22}	a_{33}	a_{12}
hom. shear[a]	0.40	−0.29	−0.11	−0.28
log-layer[b]	0.51	−0.42	−0.09	−0.24

[a] Tavoularis & Corrsin [125]
[b] Table 1 [55]

A model similar to (4.143) which in addition also satisfies realizability was devised by Shih & Lumley [104] and a model containing the wall-normal unit vector has been suggested by Gibson & Launder [27].

There is a trend today to try to use model formulations that are free from the explicit appearance of the distance normal to the wall (x_2). Instead of the traditional use of pressure strain model formulations that contain wall reflexion terms there is an emerging use of realizable model formulations (cf. *e.g.* Launder & Li [59] and Sjögren & Johansson[109]). As the strongly anisotropic, close to two-component, state near a wall is approached these models tend to decrease the strength of the pressure strain correlation.

4.8.3. ROTATING CHANNEL FLOW – AND ILLUSTRATIVE EXAMPLE

To illustrate the differences in the inherent dynamics of the set of equations involved in eddy-viscosity based two-equation models and Reynolds stress closures, the flow in a channel rotating around the homogeneous spanwise axis, will here serve as an example. Typical for thisflow case is that one side of the channel will be 'stabilized' with a suppression of turbulence as a consequence, and the other side will 'destabilized' with enhanced turbulent activity as a consequence. This will lead to different wall shear stresses on the two walls and an asymmetric mean velocity profile.

Standard eedy-viscosity-based two equation models will leave the channel flow unaffected by the system rotation and thereby predict a symmetric profile as the non-rotating case. This displays the inability of such models to handle effects of rotation in an adequate way.

Differential Reynolds stress models on the other hand was shown by Launder, Tselepidakis & Younis 1987 [57] to capture the essential effects of the imposed rotation already with the standard linear forms (the Rotta return-to-isotropy and the isotropization-of-production pressure strain models, and isotropic dissipation rate tensor) of the models for the individual terms. For the near-wall treament the Gibson & Launder [27] model was used. The fact that rotation enters naturally in the individual components of the stress transport equation is clear form the contribution of the volume force term (*i.e.* the Corilolis term $f'_i = -2\epsilon_{ilm}\Omega^{\mathrm{s}}_l u'_m$).

$$\overline{u'_i f'_j} + \overline{u'_j f'_i} = -2\Omega^{\mathrm{s}}_k(\overline{u'_j u'_m}\epsilon_{ikm} + \overline{u'_i u'_m}\epsilon_{jkm}) \equiv G_{ij}$$

The Coriolis term itself does not produce nor reduce turbulence activity directly ($G_{ii} = 0$) but it acts to redistribute energy among the stress components;

$$G_{11} = 4\Omega^{\mathrm{s}}\overline{u'_1 u'_2}\,, \quad G_{22} = -4\Omega^{\mathrm{s}}\overline{u'_1 u'_2}\,, \quad G_{12} = -2\Omega^{\mathrm{s}}(\overline{u'_1 u'_1} - \overline{u'_2 u'_2})$$

At the 'pressure side' $\partial U/\partial y > 0$ and the shear stress $-\overline{u'_1 u'_2} > 0$, and energy is being transfered to the $\overline{u'_2 u'_2}$ component. The enhanced turbulent mixing gives a thinner and stronger shear layer near the wall and an enhanced turbulence production ($\mathcal{P}_{ij}$). On the other side of the channel the reverse scenario takes place. The effects of the system rotation also enters into the rapid pressure-strain rate term. Deriving a Poisson equation from equation (4.70) for the fluctuating velocity component in a rotating frame, we see that the 'homogeneous' expression for the rapid part of the pressure strain (4.131) should be altered to

$$\Pi^{(\mathrm{r})}_{ij} = 4K\,(U_{p,q} + \epsilon_{plq}\Omega^{\mathrm{s}}_l)\,\{M_{iqpj} + M_{jqpi}\} \tag{4.145}$$

In the 'isotropization-of-production' model (4.135) this gives that half the Coriolis term should be added to the production rate tensor

$$\Pi^{(\mathrm{r})}_{ij} = -\gamma(\mathcal{P}_{ij} + \frac{1}{2}G_{ij} - \frac{2}{3}\mathcal{P}\delta_{ij})$$

4.8.4. THE ε EQUATION IN RST CLOSURES

The modelled RST equations are usually complemented by a model equation for the dissipation rate. To the lowest order of refinement the latter equation is typically given by (see section 4.3)

$$\frac{\mathrm{D}\,\varepsilon}{\mathrm{D}t} = C_{\varepsilon 1}\frac{\mathcal{P}}{K}\varepsilon - C_{\varepsilon 2}\frac{\varepsilon^2}{K} - \frac{\partial}{\partial x_m}\left(J_m^{\varepsilon} - \nu\frac{\partial \varepsilon}{\partial x_m}\right) \tag{4.146}$$

where

$$\frac{\mathcal{P}}{K} = -a_{lm}S_{lm}$$

$$C_{\varepsilon 1} \approx 1.5\,, \quad C_{\varepsilon 2} \approx 1.9$$

and the transport term is again described by a gradient diffusion hypothesis. The Daly & Harlow[18] model reads

$$J_l^{\varepsilon} = -c_{\varepsilon}\frac{K}{\varepsilon}\overline{u_l u_m}\frac{\partial \varepsilon}{\partial x_m} \tag{4.147}$$

with $c_{\varepsilon} = 0.15$ (Launder *et al.* [55]).

4.8.5. WALL BOUNDARY CONDITIONS AND LOW REYNOLDS NUMBER FORMULATIONS

In analogy with the situation for two-equation models, much of the CFD applications of DRSM treated with commercial codes use log-layer conditions to replace the actual wall-boundary conditions. For Reynolds stress transport closures the values of the individual Reynolds stresses in the log-layer have to be specified, typically as

$$\langle {u'_1}^2\rangle \approx 5u_\tau^2, \quad \langle {u'_2}^2\rangle \approx u_\tau^2, \quad \langle {u'_3}^2\rangle \approx 2u_\tau^2, \quad \langle u'_1 u'_2\rangle \approx -u_\tau^2 \ \text{ or } \ \frac{\partial\langle u'_1 u'_2\rangle}{\partial y} = 0,$$

where the chosen numeric values of the coefficients are based on empiricism.

For RST closures that are to be integrated all the way down to wall it is quite clear that something has to be done about the slow part of the pressure strain tensor. At a wall all its components vanish, $\Pi_{ij}^{(\mathrm{s})} = 0$, except for $\Pi_{12}^{(\mathrm{s})}$ (and $\Pi_{32}^{(\mathrm{s})}$) which in general are non zero. Also, we have the problem of the treatment of a non-zero dissipation at the wall.

The wall-correction term of the pressure-strain correlation presented in section 4.8.2 often form the basis of low Reynolds number models. To account for the anisotropy of the dissipation rate tensor model formulations of the kind described in the discussion in section 4.8.1 are used.

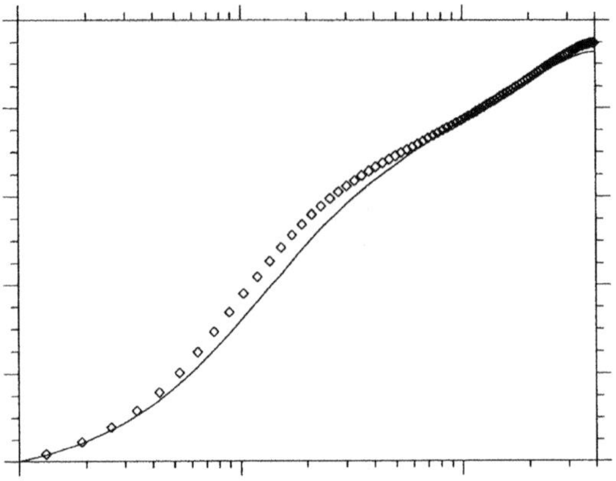

Figure 4.8. Comparison between the Sjögren & Johansson[109] DRSM prediction and DNS data (Kim *et al.* [50]) for the mean velocity in turbulent channel flow.

Corrections to the turbulence and pressure transport terms are usually neglected because at low Reynolds numbers the viscous diffusion is normally the dominates the transport. The composite modelling of the pressure strain and pressure diffusion of Shih & Mansour [105] is an exception to this.

A DRSM without wall-damping functions

The recently developed model of Sjögren & Johansson[109] is constructed to be integrated all the way to the wall but avoids completely damping functions and near-wall corrections. Instead it is based on non-linear models for the two parts of the pressure strain and the dissipation rate anisotropy, that all satisfy the condition of strong realizability. The model constants involved are mainly calibrated against a set of different homogeneous turbulent flows.

Figures 4.8 and 4.9 show comparisons between the model predictions and the DNS data of Kim *et al.* [50] for channel flow at a friction Reynolds number of 395. The two figures show the mean velocity and Reynolds shear stress, respectively. Considering the fact that no damping functions are used, and no explicit use is made of the wall normal vector or wall-normal distance, the agreement is more than satisfactory.

4.9. Algebraic Reynolds stress models

Standard two-equation models are still dominating in the context of industrial flow computations. In flows with strong effects of streamline curvature, adverse pressure gradients, flow separation or system rotation, such models fail to give accurate predictions. Turbulence models based on the transport

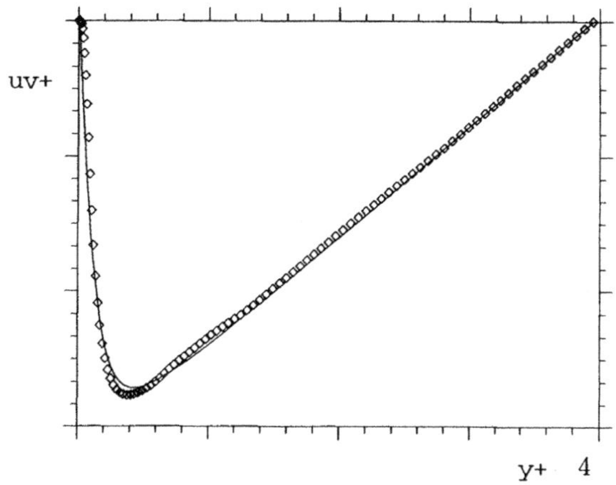

Figure 4.9. Comparison between the Sjögren & Johansson[109] DRSM prediction and DNS data (Kim *et al.* [50]) for the Reynolds shear stress in turbulent channel flow.

equations for the individual Reynolds stresses have the natural potential for dealing with, *e.g.* , the associated complex dynamics of intercomponent transfer. An intermediate level between this level and eddy viscosity models is represented by the algebraic Reynolds stress models (ARSM) where the Boussinesq hypothesis may be said to be replaced by an algebraic approximation of the transport equation for the Reynolds stress anisotropy tensor. A number of features from the Reynolds stress transport models are retained, but with a with a considerably higher degree of robustness in the numerics when explicit forms of the algebraic approximations are used. The implementations of such algebraic Reynolds stress models, referred to as EARSM, are also relatively simple extensions of codes for eddy-viscosity models. The class of problems suited for this type of models is of course more limited than for differential Reynolds stress models.

In constructing algebraic approximations of the Reynolds stress transport equations one obviously has to make rather drastic assumptions about the advection and difusion equations. A natural approach to this problem was suggested by Rodi[84], [85]) namely to assume that the advection minus the diffusion of the individual Reynolds stresses can be expressed as the product of the corresponding quantity for the kinetic energy, K, and the individual Reynolds stresses normalized by K. This results in an implicit relation between the stress components and the mean velocity gradient field that replaces the Boussinesq hypothesis. Since the algebraic Reynolds stress model is determined from the modeled Reynolds stress transport equation no additional model constants are needed and some of the basic behaviour and experiences of the particular RST model will be inherited.

A systematic way of describing this type of model can be based on taking the kinetic energy, a_{ij} and ε as primary quantities instead of $\overline{u_i'u_j'}$ and ε. We may then start by formulating the modelled transport equations for the scalar (and dimensional) quantities, in a manner similar to that in standard two-equation models.

$$\frac{\mathrm{D}K}{\mathrm{D}t}+\frac{\partial T_l^{(K)}}{\partial x_l} = \mathcal{P}-\varepsilon \tag{4.148}$$

$$\frac{\mathrm{D}\varepsilon}{\mathrm{D}t}+\frac{\partial T_l^{(\varepsilon)}}{\partial x_l} = \left(C_{\varepsilon 1}\mathcal{P}-C_{\varepsilon 2}f_\varepsilon\varepsilon\right)\frac{\varepsilon}{K}. \tag{4.149}$$

No direct modelling of the kinetic energy production term is needed here. An advantage over eddy-viscosity models is also that the transport terms here can be modelled by a gradient diffusion with a tensor diffusivity coefficient, such as in the model of Daly & Harlow[18]. Here, this gives

$$\begin{aligned} T_l^{(K)} &= -c_s'\frac{K}{\varepsilon}\overline{u_l u_m}\frac{\partial K}{\partial x_m} \\ T_l^{(\varepsilon)} &= -c_\varepsilon\frac{K}{\varepsilon}\overline{u_l u_m}\frac{\partial \varepsilon}{\partial x_m}. \end{aligned} \tag{4.150}$$

Launder *et al.* [55] recommend $c_s' = 0.25$ and $c_\varepsilon = 0.15$.

The dissipation of the turbulent kinetic energy, ε, is the most commonly used quantity for determining the turbulent length-scale but also other alternatives are possible. Wallin & Johansson[133] made comparisons with ω as the auxilliary quantity, and found some advantages for some boundary layer problems.

In the ARSM-approximation the Boussinesq hypothesis is, hence, replaced by a_{ij} transport equation where advection and diffusion terms are neglected (the advection term is indeed exactly zero for all stationary parallel mean flows, such as fully developed channel and pipe flows). This gives a local description of the anisotropy, but one should keep in mind that history effects enter through the K and ε equations. There is no direct influence of system rotation on these equations. The linearity of the Boussinesq hypothesis excludes any dependence of a_{ij} on the rotational (antisymmetric) part of the mean velocity gradient tensor. An ARSM approach here represents a systematic method to construct a non-linear stress relationship that includes effects of the rotational part of the mean velocity gradient tensor.

Since Coriolis terms enter explicitly in the Reynolds stress transport equations, also the ARSM will inheret this natural description of effects of system rotation (see Wallin & Johansson[133] and Sjögren[107]). In the following we will formulate the equations in a non-rotating system, where the ARSM-approximation may be written

$$\frac{\overline{u_i u_j}}{K}(\mathcal{P}-\varepsilon) = \mathcal{P}_{ij} - \varepsilon_{ij} + \Pi_{ij} \tag{4.151}$$

The dissipation rate tensor, ε_{ij}, and the pressure strain, Π_{ij}, need to be modeled whereas the production terms, $\mathcal{P}_{ij}$ and $\mathcal{P} = \mathcal{P}_{ii}/2$ do not need any modelling since they are explicit in the Reynolds stress tensor. The mean velocity gradient tensor $U_{i,j}$ will here be split into a mean strain and a mean rotation tensor, which normalized with the turbulent time-scale, $\tau \equiv K/\varepsilon$, are defined as S_{ij} and Ω_{ij} denote these tensors

$$\begin{aligned} S_{ij} &= \frac{\tau}{2}\left(U_{i,j} + U_{j,i}\right) \\ \Omega_{ij} &= \frac{\tau}{2}\left(U_{i,j} - U_{j,i}\right). \end{aligned} \tag{4.152}$$

The production term (normalized with ε) can be expressed as

$$\frac{\mathcal{P}_{ij}}{\varepsilon} = -\frac{4}{3}S_{ij} - \left(a_{ik}S_{kj} + S_{ik}a_{kj}\right) + a_{ik}\Omega_{kj} - \Omega_{ik}a_{kj}. \tag{4.153}$$

The resulting form of the ARSM assumption depends on the choice of the models for ε_{ij} and Π_{ij}. Anisotropic dissipation modelling in this context has been studied by Jongen, Mompean & Gatski[48]. In the following we treat only the case with an isotropic assumption for the dissipation rate tensor, and the Rotta model, Rotta[90], for the slow pressure strain

$$\Pi_{ij}^{(s)} - \varepsilon_{ij} = -c_1 \varepsilon a_{ij} - \frac{2}{3}\varepsilon\delta_{ij}. \tag{4.154}$$

In order to allow solution of the resulting implicit ARSM relation the pressure strain model has to be linear or quasi-linear, meaning that it must be tensorially linear in **a** but may depend nonlinearly on the production to dissipation ratio.

For the rapid pressure strain rate we will here choose the general linear model of Launder, Reece & Rodi[55], which for a non-rotating system normally is given by (4.134). In the following we will mainly use boldface notation, *e.g.* , $\mathbf{a}$, $\mathbf{S}$ and $\boldsymbol{\Omega}$, to denote second rank tensors, and $\mathbf{I}$ denotes the identity matrix. The inner product of two matrices is defined as $(\mathbf{SS})_{ij} \equiv (\mathbf{S}^2)_{ij} \equiv S_{ik}S_{kj}$ and $\mathrm{tr}\{\}$ denotes the trace.

Using this notation we may write the LRR model as,

$$\begin{aligned} \frac{\boldsymbol{\Pi}^{(r)}}{\varepsilon} &= \frac{4}{5}\mathbf{S} + \frac{9c_2+6}{11}\left(\mathbf{aS} + \mathbf{Sa} - \frac{2}{3}\mathrm{tr}\{\mathbf{aS}\}\mathbf{I}\right) \\ &+ \frac{7c_2 - 10}{11}\left(\mathbf{a}\boldsymbol{\Omega} - \boldsymbol{\Omega}\mathbf{a}\right). \end{aligned} \tag{4.155}$$

Also the SSG model by Speziale *et al.* [119] may be expressed in this form if it is linearized according to Gatski & Speziale[22]. The SSG pressure-strain model reads

$$\begin{aligned}\frac{\mathbf{\Pi}}{\varepsilon} = & -\left(\frac{C_1}{2}+\frac{C_1^*}{2}\frac{\mathcal{P}}{\varepsilon}\right)\mathbf{a}+\left(C_3-\frac{C_3^*}{2}II_a^{\frac{1}{2}}\right)\mathbf{S} \\ & + \frac{C_4}{2}\left(\mathbf{aS}+\mathbf{Sa}-\frac{2}{3}\mathrm{tr}\{\mathbf{aS}\}\mathbf{I}\right)-\frac{C_5}{2}\left(\mathbf{a\Omega}-\mathbf{\Omega a}\right) \\ & + \frac{C_2}{4}\left(\mathbf{a}^2-\frac{1}{3}II_a\mathbf{I}\right) \end{aligned} \tag{4.156}$$

where $II_a = \mathrm{tr}\{\mathbf{a}^2\}$ and the coefficients are

$$C_1 = 3.4, \quad C_1^* = 1.8, \quad C_2 = 4.2, \quad C_3 = \frac{4}{5},$$

$$C_3^* = 1.30, \quad C_4 = 1.25, \quad C_5 = 0.40. \tag{4.157}$$

The linearized SSG used by Gatski & Speziale[22] is then obtained by neglecting the quadratic anisotropy term and for the II_a invariant they used the equilibrium value predicted by the SSG model for two-dimensional homogeneous turbulence. This results in the following set of coefficients

$$C_1 = 3.4, \quad C_1^* = 1.8, \quad C_2 = 0, \quad C_3 = 0.36,$$

$$C_3^* = 0, \quad C_4 = 1.25, \quad C_5 = 0.40. \tag{4.158}$$

Gatski & Speziale[22] based their explicit algebraic Reynolds stress model (EARSM) on this linearized SSG pressure-strain model and implied an additional approximation in order to avoid the non-linearity in the ARSM equation system. The approximation was to use the asymptotic value for the production to dissipation ratio as a universal constant

$$\frac{\mathcal{P}}{\varepsilon} = \frac{C_{\varepsilon 2}-1}{C_{\varepsilon 1}-1}. \tag{4.159}$$

Various improvements on this model have subsequently been proposed by Gatski, Speziale and their co-workers.

The resulting ARSM for all quasi-linear pressure strain models may be written

$$N\mathbf{a} = -A_1\mathbf{S} + (\mathbf{a\Omega}-\mathbf{\Omega a}) - A_2\left(\mathbf{aS}+\mathbf{Sa}-\frac{2}{3}\mathrm{tr}\{\mathbf{aS}\}\right) \tag{4.160}$$

where

$$N = A_3 + A_4\frac{\mathcal{P}}{\varepsilon}. \tag{4.161}$$

TABLE 4.2. The coefficients in the general ARSM for different models.

	A_1	A_2	A_3	A_4
LRR	$\dfrac{88}{15\,(7c_2+1)}$	$\dfrac{5-9c_2}{7c_2+1}$	$\dfrac{11\,(c_1-1)}{7c_2+1}$	$\dfrac{11}{7c_2+1}$
Wallin & J. ($c_1, c_2 = 1.8, 5/9$)	1.20	0	1.80	2.25
Original LRR ($c_1, c_2 = 1.5, 0.4$)	1.54	0.37	1.45	2.89
linearized SSG	1.22	0.47	0.88	2.37
Gatski & Speziale	1.22	0.47	5.36	0

Note that the relation (4.160) is nonlinear since N contains the production to dissipation ratio which is equivalent to $\mathrm{tr}\{\mathbf{aS}\}$. The coefficients $A1 - A4$ are given in table 4.2 for a few different models.

The ARSM approximation of the a_{ij} transport equation with this approach is equivalent to neglecting the advective term (and diffusion) in the chosen coordinate system. Hence, the adequacy of the ARSM approach is coupled to the choice of a coordinate system where the neglection of advection terms in the a_{ij} equation can be justified.

4.9.1. EXPLICIT ALGEBRAIC REYNOLDS STRESS MODELS

Direct numerical solution of the implicit **a**-relation (4.160) is not a feasible alternative in complex flows. The difficulties are associated with the lack of diffusion or damping in the equation system. One instead wishes to construct an explicit **a**-relation

$$\mathbf{a} = \mathbf{a}(\mathbf{S}, \boldsymbol{\Omega}) \tag{4.162}$$

Such explicit algebraic Reynolds stress models, EARSM, where the Reynolds stresses are explicitly related to the mean flow field are much more numerically robust and have been found to have almost a negligible effect on the computational effort as compared to a standard two-equation model.

The most general form for **a** in terms of **S** and **Ω** consists of ten tensorially independent groups to which all higher order tensor combinations can be reduced with the aid of the Caley-Hamilton theorem.

$$\begin{aligned}
\mathbf{a} &= \beta_1 \mathbf{S} \\
&+ \beta_2 \left(\mathbf{S}^2 - \frac{1}{3} II_S \, \mathbf{I} \right) + \beta_3 \left(\boldsymbol{\Omega}^2 - \frac{1}{3} II_\Omega \, \mathbf{I} \right) + \beta_4 \left(\mathbf{S}\boldsymbol{\Omega} - \boldsymbol{\Omega}\mathbf{S} \right) \\
&+ \beta_5 \left(\mathbf{S}^2\boldsymbol{\Omega} - \boldsymbol{\Omega}\mathbf{S}^2 \right) + \beta_6 \left(\mathbf{S}\boldsymbol{\Omega}^2 + \boldsymbol{\Omega}^2\mathbf{S} - \frac{2}{3} IV \, \mathbf{I} \right) \\
&+ \beta_7 \left(\mathbf{S}^2\boldsymbol{\Omega}^2 + \boldsymbol{\Omega}^2\mathbf{S}^2 - \frac{2}{3} V \, \mathbf{I} \right) + \beta_8 \left(\mathbf{S}\boldsymbol{\Omega}\mathbf{S}^2 - \mathbf{S}^2\boldsymbol{\Omega}\mathbf{S} \right) \\
&+ \beta_9 \left(\boldsymbol{\Omega}\mathbf{S}\boldsymbol{\Omega}^2 - \boldsymbol{\Omega}^2\mathbf{S}\boldsymbol{\Omega} \right) + \beta_{10} \left(\boldsymbol{\Omega}\mathbf{S}^2\boldsymbol{\Omega}^2 - \boldsymbol{\Omega}^2\mathbf{S}^2\boldsymbol{\Omega} \right) \qquad (4.163)
\end{aligned}$$

The β coefficients may be functions of the five independent invariants of $\mathbf{S}$ and $\boldsymbol{\Omega}$, which can be written as

$$II_S = \mathrm{tr}\{\mathbf{S}^2\}, \quad II_\Omega = \mathrm{tr}\{\boldsymbol{\Omega}^2\}, \quad III_S = \mathrm{tr}\{\mathbf{S}^3\},$$

$$IV = \mathrm{tr}\{\mathbf{S}\boldsymbol{\Omega}^2\}, \quad V = \mathrm{tr}\{\mathbf{S}^2\boldsymbol{\Omega}^2\}. \qquad (4.164)$$

Also other scalar parameters may be involved.

In two-dimensional mean flows, there are only three independent tensor groups, *e.g.* , the $\beta_{1,2,4}$ groups, and two independent invariants, II_S and II_Ω.

The determination of the ten β coefficients in general three-dimensional mean flows is very complex for the general form of a quasi-linear pressure strain rate model. In a true EARSM approach the values of these coeficients should be derived from the corresponding differential Reynolds stress models. Among pioneering work with this approach we may mention Pope[78], Taulbee[123] and Gatski & Speziale[22]. The alternative to calibrate the β:s from some chosen set of 'basic flows' (Shih *et al.* [101]) could be referred to as a nonlinear eddy-viscosity approach.

In the following we will present results only for the LRR model with $c_2 = 5/9$ (Taulbee[123], Wallin & Johansson[133]) for which the implicit a-relation reduces to

$$N\mathbf{a} = -\frac{6}{5}\mathbf{S} + (\mathbf{a}\boldsymbol{\Omega} - \boldsymbol{\Omega}\mathbf{a}) \qquad (4.165)$$

where N is closely related to the production to dissipation ratio ($\mathcal{P}/\varepsilon = -\mathrm{tr}\{\mathbf{aS}\}$),

$$N = c_1' + \frac{9}{4}\frac{\mathcal{P}}{\varepsilon} \qquad (4.166)$$

and

$$c_1' = \frac{9}{4}(c_1 - 1)\,. \qquad (4.167)$$

A complicating factor is the nonlinearity of equation (4.165). Wallin & Johansson[133] solved the nonlinear system of equations in the form of a linear system of (five) equations complemented by a nonlinear scalar equation for $\mathcal{P}/\varepsilon$. For two-dimensional mean flows, Johansson & Wallin[46] and Girimaji[28][29] independently showed that this equation has a closed and fully explicit solution that can be expressed in a compact form.

The removal of the need for ad-hoc relations for $\mathcal{P}/\varepsilon$ represents a substantial improvement for this type of modelling. A constant $\mathcal{P}/\varepsilon$ gives wrong asymptotic behaviour for large strain rates, also noticed by Speziale & Xu[120], while the fully consistent solution of the non-linear equation system automatically fulfills the correct asymptotic behaviour. Also in the very near-wall region, the correct asymptotic behaviour significantly improves the predictions.

Following Wallin & Johansson[133] the procedure to solve the equation (4.165) is the following: First, the general form for the anisotropy, equation (4.163), is inserted into the simplified ARSM equation (4.165) where N is not yet determined. The resulting linear equation system for the β coefficients is then solved by using the fact that higher order tensor groups can be reduced with the aid of Cayley-Hamilton theorem where the ten groups in the general form (4.163) forms a complete basis. The β coefficients are now functions of the production to dissipation ratio, $\mathcal{P}/\varepsilon$, or N. The final step is to formulate the non-linear scalar equation for N or $\mathcal{P}/\varepsilon$. We will refer to this as the WJ model.

4.9.2. THE WJ MODEL FOR two-dimensional MEAN FLOWS

For two-dimensional mean flows the solution is reduced to only two non-zero coefficients β_1, β_4, and the anisotropy tensor can be expressed as

$$\mathbf{a} = -\frac{6}{5}\frac{1}{N^2 - 2II_\Omega}\left(N\mathbf{S} + \mathbf{S}\boldsymbol{\Omega} - \boldsymbol{\Omega}\mathbf{S}\right) \tag{4.168}$$

It is clearly seen that the denominator, $N^2 - 2II_\Omega$, cannot become singular since II_Ω is always negative. The non-linear equation for N in two-dimensional mean flow can be derived by introducing the solution of $\mathbf{a}$ for two-dimensional mean flow in the definition of N. The resulting equation is cubic and can be solved in a closed form with the solution for the positive root being

$$N = \begin{cases} \dfrac{c_1'}{3} + \left(P_1 + \sqrt{P_2}\right)^{1/3} + \operatorname{sign}\left(P_1 - \sqrt{P_2}\right) \mid P_1 - \sqrt{P_2} \mid^{1/3} & , P_2 \geq 0 \\ \dfrac{c_1'}{3} + 2\left(P_1^2 - P_2\right)^{1/6} \cos\left(\dfrac{1}{3}\arccos\left(\dfrac{P_1}{\sqrt{P_1^2 - P_2}}\right)\right) & , P_2 < 0 \end{cases} \tag{4.169}$$

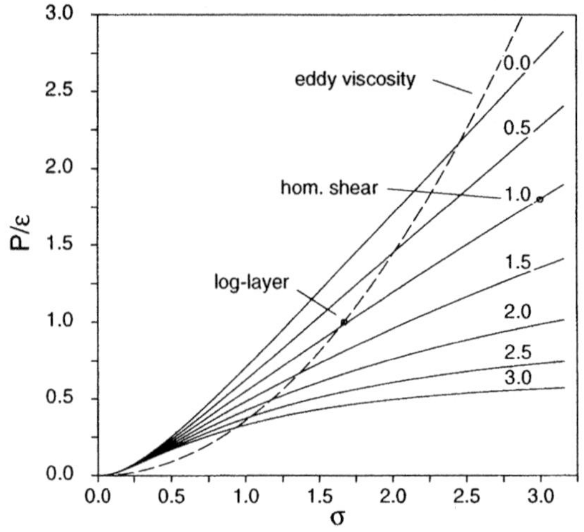

Figure 4.10. Production to dissipation ratio versus strain rate σ for different rotation ratios ω/σ. The current model (——) compared to an eddy viscosity model (---).

with

$$P_1 = \left(\frac{c_1'^2}{27} + \frac{9}{20}II_S - \frac{2}{3}II_\Omega\right)c_1', \quad P_2 = P_1^2 - \left(\frac{c_1'^2}{9} + \frac{9}{10}II_S + \frac{2}{3}II_\Omega\right)^3. \tag{4.170}$$

N remains real and positive for all possible values of II_S and II_Ω. The production to dissipation ratio may then be found from (4.166).

Figure 4.10 illustrates the behaviour of the solution for $\mathcal{P}/\varepsilon$. We note that it is zero for all cases with $\sigma = 0$, *i.e.* irrespective of the value of ω, where σ and ω are defined as $\sigma \equiv \sqrt{II_S/2}$ and $\omega \equiv \sqrt{II_\Omega/2}$. The model also has the attractive feature of reproducing a monotonously decreasing $\mathcal{P}/\varepsilon$ ratio with increasing influence of rotation. For all parallel shear flows $\sigma = \omega$.

Homogeneous shear flow is a classical corner stone case for calibration of turbulence models. Tavoularis & Corrsin[125] has experimentally shown that the asymptotic value of $SK/\varepsilon \approx 6$ corresponding to $\sigma = \omega = 3$. In the experiments the production to dissipation ratio was found to be approximately 1.8, marked as a circle in figure 4.10. The WJ model exactly replicates that result. Furthermore, in the log-layer of a boundary layer we know that the production balances the dissipation rate, $\mathcal{P} = \varepsilon$, which is obtained with the WJ model when the strain rate $\sigma = 1.69$. This is within the range of σ values found in the log-layer of the DNS data for channel flow (Kim[49]). This is also consistent with an effective $C_\mu = 0.09$ which gives $\sigma = 1.67$, also marked as a circle in the figure.

TABLE 4.3. The computed anisotropy in the log-layer using the current model assuming balance between turbulence production and dissipation compared to channel DNS data (Kim 1989).

	a_{12}	a_{11}	a_{22}	a_{33}	σ
DNS	−0.29	0.34	−0.26	−0.08	1.65
Current model	−0.30	0.25	−0.25	0.00	1.69

TABLE 4.4. The anisotropy in asymptotic homogeneous shear flow using the current model with $\sigma = 3.0$ compared to measurements by [125].

	a_{12}	a_{11}	a_{22}	a_{33}	$\mathcal{P}/\varepsilon$
expr	−0.30	0.40	−0.28	−0.12	1.8
Current model	−0.30	0.31	−0.31	0.00	1.8

The anisotropies predicted with the WJ model for the two cases are compared in tables 4.3 and 4.4. Most important is to correctly predict the a_{12} anisotropy since this is the only component of the anisotropy tensor that contributes to the turbulent production in parallel flows. The a_{22} component is also important since this is the only term that contributes to the turbulent diffusion term. The tables show that a_{12} and a_{22} are well predicted for the two different cases. The a_{11} and a_{33} components are, however, not as well predicted due to the simplification of setting $c_2 = 5/9$ since this implies that $a_{33} = 0$.

For moderate to large strain rates (in parallel flows) the WJ model predicts an a_{12} anisotropy that is nearly constant. This range includes values relevant for the log-layer and the asymptotic homogeneous shear flow. This also means that it also avoids unrealizable results for large strain rates, unlike standard eddy viscosity models. Bradshaw's assumption, that is adopted by [71] in the SST model, forces the a_{12} anisotropy to be constant for $\mathcal{P}/\varepsilon$ ratios greater than unity, which gives that $\beta_1 \sim 1/\sigma$. This is hence fulfilled for the WJ model in the limit of large strain rates.

An example where the benefits of a direct, explicit solution of the production to dissipation ratio is further illustrated is rotating plane channel flow. For this case the effective II_Ω in the rotating system is decreased on one side and increased on the other, which results naturally in a decreased production on the stabilized side and an increased production on the destabilized side.

4.9.3. THE WJ MODEL FOR THREE-DIMENSIONAL MEAN FLOW

For general three-dimensional mean flows the WJ model gives the following β-coefficients

$$\begin{aligned} \beta_1 &= -\frac{N\left(2N^2 - 7II_\Omega\right)}{Q} & \beta_3 &= -\frac{12N^{-1}IV}{Q} & & \\ \beta_4 &= -\frac{2\left(N^2 - 2II_\Omega\right)}{Q} & \beta_6 &= -\frac{6N}{Q} & \beta_9 &= \frac{6}{Q} \end{aligned} \tag{4.171}$$

where all the other coefficients are identically zero. The denominator

$$Q = \frac{5}{6}\left(N^2 - 2II_\Omega\right)\left(2N^2 - II_\Omega\right) \tag{4.172}$$

is also here clearly seen always to remain positive since II_Ω always is negative.

The non-linear equation for N or the corresponding equation for $\mathcal{P}/\varepsilon$ is *e.g.* obtained by introducing the above solution (4.171) for **a** into the definition of N. The resulting equation is of sixth order. Wallin & Johansson[133] proposed an approximative solution of this equation in terms of a linear expansion around the solution for the corresponding cubic equation valid for two-dimensional flows.

Fully developed turbulent flow in a circular pipe rotating around its length axis is an interesting case, since it represents a three-dimensional flow that can be described with only one spatial coordinate in a cylindrical coordinate system. If the flow is laminar, the tangential velocity, U_θ, varies linearly with the radius, r, like a rigid body rotation. In turbulent flow, on the other hand, the tangential velocity is nearly parabolic (Imao *et al.* [42]), which cannot be described with an eddy viscosity turbulence model. Wallin & Johansson[133], among others, have demonstrated that cubic terms are needed in the EARSM to capture this feature. One may note that such terms are not present in twodimensional formulations of EARSM and not in most of the EARSM:s proposed in the literature.

When EARSM is used with integration all the way down to solid walls near-wall damping functions are needed in a similar fashion as with standard two-equation models. Wallin & Johansson[133] showed, however, that a correct asymptotic behaviour for large strain rates which is possible to achieve with EARSM improves the near-wall situation substantially over that for eddy-viscosity models. They showed by comparison with DNS channel flow data that a simple van Driest damping function was quite sufficient to obtain excellent agreement for the shear stress anisotropy. For further details the reader is referred to that paper.

4.9.4. COMPRESSIBLE EARSM

The WJ model also includes a version for compressible turbulent flow in which the equations were expressed by use of Favre averages and compressibility of the mean flow was accounted for, but no direct compressibility corrections were included. Hence, no attempt was made to include effects of pressure dilation, compresssible contributions to the dissipation etc. In wall bounded flows with Mach numbers below 5 compressibility effects due to turbulent fluctuations may be neglected and the effect of compressibility enters into the problem essentially only through the mean flow compressibility (Fridriech [21]). Friedrich[21] concludes, on the other hand, that compressibility effects due to turbulent fluctuations may be important in hypersonic, high Mach number, wall bounded flows and in mixing layers at rather moderate convective Mach numbers.

In this approach, the stress anisotropy and the turbulent kinetic energy must first be redefined as

$$a_{ij} \equiv \frac{\overline{\rho u_i u_j}}{\rho K} - \frac{2}{3}\delta_{ij} \qquad\qquad K \equiv \frac{\overline{\rho u_i u_i}}{2\rho}$$

where ρ is the local mean density of the fluid. The trace of the strain is not zero for compressible flow. We may instead use a somewhat redefined normalized strain rate tensor

$$S_{ij} \equiv \frac{\tau}{2}\left(U_{i,j} + U_{j,i}\right) - \frac{\mathcal{D}}{3}\delta_{ij} \tag{4.173}$$

where the normalized dilatation of the mean flow is defined as $\mathcal{D} \equiv \tau U_{k,k}$.

With the redefinition of $\mathbf{S}$ to have zero trace the incompressible solution process may be used. The redefinition also implicates, however, that the S_{33} component is non-zero for two-dimensional mean flow in the $x - y$ plane and that the simplifications for two-dimensional mean flow are not strictly valid for compressible flow.

The general linear model of the LRR rapid pressure strain rate model for incompressible flow does not have zero trace in compressible flow, so the model needs to be generalized according to [131] and reads

$$\begin{aligned} \Pi_{ij}^{(r)} &= -\frac{c_2+8}{11}\left(\mathcal{P}_{ij} - \frac{2}{3}\mathcal{P}\delta_{ij}\right) - \frac{30c_2-2}{55}\rho K\left(U_{i,j} + U_{j,i} - \frac{2}{3}U_{l,l}\delta_{ij}\right) \\ &- \frac{8c_2-2}{11}\left(D_{ij} - \frac{2}{3}\mathcal{P}\delta_{ij}\right) \end{aligned} \tag{4.174}$$

where $\mathcal{P}_{ij} = -\overline{\rho u_i u_k}U_{j,k} - \overline{\rho u_j u_k}U_{i,k}$ and $D_{ij} = -\overline{\rho u_i u_k}U_{k,j} - \overline{\rho u_j u_k}U_{k,i}$. The incompressible models for the slow pressure strain and the dissipation tensor can be used also here. The general ARSM for compressible flow can now be expressed in the same form as the incompressible one with a modified c_1'. With $c_2 = 5/9$we obtain

$$\left(c_1 - 1 - \frac{2}{3}\mathcal{D} - \mathrm{tr}\{\mathbf{aS}\}\right)\mathbf{a} = -\frac{8}{15}\mathbf{S} + \frac{4}{9}(\mathbf{a\Omega} - \mathbf{\Omega a}). \qquad (4.175)$$

The solution of the simplified compressible ARSM equation is the same as the incompressible solution except for the definition of the c_1' coefficient, which for the compressible case is

$$c_1' = \frac{9}{4}\left(c_1 - 1 - \frac{2}{3}\mathcal{D}\right) \qquad (4.176)$$

bearing in mind that S_{ij} here is defined by (4.175).

For two-dimensional mean flows Wallin & Johansson[133] derived an approximation of the compressible ARSM equation by use of an expansion around a solution obtained by use of the two-dimensional simplification valid in incompressible flow

$$\mathbf{a} = \mathbf{a}^{2\mathrm{D}} - \frac{4}{15\left(c_1 - 1 - \mathrm{tr}\{\mathbf{a}^{2\mathrm{D}}\mathbf{S}^{2\mathrm{D}}\}\right)}\mathcal{D}\left(\mathbf{I}^{2\mathrm{D}} - \frac{2}{3}\mathbf{I}\right) + \mathcal{O}\left(\mathcal{D}^2\right), \qquad (4.177)$$

where $S_{ij}^{2\mathrm{D}} \equiv (\tau/2)(U_{i,j} + U_{j,i}) - \mathcal{D}\delta_{ij}^{2\mathrm{D}}/2$ and $\mathbf{a}^{2\mathrm{D}}$ is the zeroth order solution

$$\mathbf{a}^{2\mathrm{D}} = \beta_1 \mathbf{S}^{2\mathrm{D}} + \beta_4\left(\mathbf{S}^{2\mathrm{D}}\mathbf{\Omega} - \mathbf{\Omega}\mathbf{S}^{2\mathrm{D}}\right) \qquad (4.178)$$

obtained as described for the incompressible case, with the only difference being that c_1' is now given by (4.176).

In Wallin & Johansson[133] the compressible EARSM together with near wall formulations was tested on a shock boundary layer interaction where a turbulent boundary layer at Mach 5 on a flat plate interacts with an oblique shock from a shock generator above the plate. The shock strength was sufficient to cause a boundary layer separation. The corresponding experiment was carried out by Schülein *et al.* [96]. The computations were performed with the EARSM based on both a $K-\varepsilon$ model and a $K-\omega$ (Wilcox[137]) model as the platform. The size of the separated region is underestimated by the two original eddy viscosity models whereas an essentially correct separation length was obtained for the two EARSM:s. The skin friction behaviour downstream reattachment was significantly better with the $K-\omega$-based EARSM.

References

1. Aupoix, B., Cousteix, J. and Liandrat, J. 1989. MIS: A way to derive the dissipation equation. In *Turbulent Shear Flow*, vol. 6 (ed. J.-C. André), Springer.
2. Aronson, D. and Löfdahl, L. 1993. The plane wake of a cylinder: Measurements and inferences on turbulence modelling. To appear in *Phys. Fluids*.
3. Bardina, J., Ferziger, J.H. and Rogallo, R.S. 1985. Effect of rotation on isotropic turbulence; computation and modelling. *J. Fluid Mech.* **154**, 321-336.
4. Batchelor, G.K. 1953. *The theory of homogeneous turbulence*, Cambridge.
5. Batchelor, G.K. and Proudman, I. 1954. The effect of rapid distortion of a fluid in turbulent motion. *Quart. J. Mech. Appl. Math.* **7**, 83-103.
6. Bradshaw, P., Mansour, N.N. and Piomelli, U. 1987. On local approximations of the pressure-strain term in turbulence models. In *Proc. of 1987 Summer Program, Report CTR-S87*, Stanford University, USA.
7. Brasseur, J.G. and Corrsin, S. 1987. Spectral evolution of the Navier-Stokes equations for low order couplings of Fourier modes. In *Advances in Turbulence 2*, Berlin, Germany.
8. Cambon, C., Jacquin, L. and Lubrano, J.L. 1992. Toward a new Reynolds stress model for rotating turbulent flows. *Phys. Fluids* **4**, 812-824.
9. Champagne, F.H., Harris, V.G. and Corrsin, S. 1970. Experiments on nearly homogeneous turbulent shear flow. *J. Fluid Mech.* **41**, 81-139.
10. Chasnov, J.R. 1993. Computation of the Loitsianski integral in decaying isotropic turbulence. *Phys. Fluids* **5**, 5279.
11. Chasnov, J.R. 1994. Similarity states of passive scalar transport in isotropic turbulence. *Phys. Fluids* **6**, 1036.
12. Chien, K.Y. 1982. Predictions of Channel and Boundary-Layer Flows with a Low-Reynolds-Number Turbulence Model. *AIAA J.* **20**, 33–38.
13. Choi, K.-S. 1983. A study of the return to isotropy of homogeneous turbulence. Ph. D. Thesis, Cornell University.
14. Chou, P.Y. 1945. On velocity correlations and the solutions of the equations of turbulent motion. *Quart. of Applied Math.* **3**, 38-54.
15. Comte-Bellot, G. and Corrsin, S. 1966. The use of a contraction to improve the isotropy of grid-generated turbulence. *J. Fluid Mech.* **25**, 657-682.
16. Craft, T.J., Launder, B.E. and Leschziner, M.A. 1993. On the prediction of turbulent flow in spirally fluted tubes. In *Refined flow modelling and turbulence measurements, 5th Int. Symposium*, eds Coantic et al, p327, Presses de l'école nationale des Ponts et chaussées.
17. Crow, S.C. 1968. Viscolelastic properties of fine-grained incompressible turbulence. *J. Fluid Mech.* **33**, 1-20.
18. Daly, B.J. and Harlow, F.H. 1970. Transport equations in turbulence. *Int. Phys. Fluids* **13**, 2634–2649.
19. Durbin, P.A. and Speziale, C.G. 1994. Realizability of second-moment closure via stochastic analysis. *J. Fluid Mech.* **280**, 395407.
20. Ekander, H. 1988. Prediction of curved and rotating channel flow with an improved algebraic Reynolds stress model. Trita-Mek-88-06, Dept. of Mech., Royal Institute of Technology, Stockholm, Sweden.
21. Friedrich, R. 1996. Compressibility effects due to turbulent fluctuations. Oral presentation at Sixth European Turbulence Conference, Lausanne, July 1996 and published as chapter IV of **Critical Review of Prediction Methods for Compressible Turbulent Attached and Separated Flows**, Ed. Ingemar Lindblad, FFA TN 1995-43.
22. Gatski, T.B and Speziale C.G. 1993. On explicit algebraic stress models for complex turbulent flows. *J. Fluid Mech.* **254**, 59–78.
23. Gence, J.N. and Mathieu, J. 1979. On the application of successive plane strains to grid-generated turbulence. *J. Fluid Mech.* **93**, 501-513.

24. Gence, J.N. and Mathieu, J. 1980. The return to isotropy of a homogeneous turbulence having been submitted to two successive plane strains. *J. Fluid Mech.* **101**, 555-566.
25. George, W.K. and Hussein, H.J. 1991. Locally axisymmetric turbulence. *J. Fluid Mech.* **233**, 1-23.
26. Germano, M. 1986. A proposal for a redefinition of the turbulent stresses in the filtered Navier-Stokes equations. *Phys. Fluids* **29**, 2323-2324.
27. Gibson, M.M. and Launder, B.E. 1978. Ground effects on pressure fluctuations in the atmospheric boundary layer. *J. Fluid Mech.* **86**, 491-511.
28. Girimaji, S.S. 1995. Fully-explicit and self-consistent algebraic Reynolds stress model *ICASE Report No. 95-82.*
29. Girimaji, S.S. 1996. Improved algebraic Reynolds stress model for engineering flows In **Engineering Turbulence Modelling and Experiments 3** pp 121–129. Eds W. Rodi and G. Bergeles. Elsevier. Science B.V.
30. Grant, H.L., Stewart, R.W. and Molliet, A. 1962. Turbulence spectra from a tidal channel. *J. Fluid Mech.* **12**, 241-268.
31. Groth, J. 1991. Description of the pressure effects in the Reynolds stress transport equations. *Phys. Fluids A* **3**, 2276-2277.
32. Groth, J. 1991. On the modelling of homogeneous turbulence. PhD Thesis, Dept. of Mech., Royal Institute of Technology, Stockholm, Sweden.
33. Groth, J., Hallbäck M. and Johansson, A.V. 1989. Measurement and modeling of anisotropic turbulent flows. In *Advances in Turbulence 2*, eds H.-H. Fernholz and H.E. Fiedler, p84, Springer.
34. Groth, J., Hallbäck M. and Johansson, A.V. 1990. A non-linear model for the dissipation rate term in Reynolds stress models. In *Engineering Turbulence Modelling and Experiments*, eds W. Rodi and E.N. Ganić, p93, Elsevier.
35. Hallbäck, M. Groth, J. and Johansson, A.V. 1989. A Reynolds stress closure for the dissipation in anisotropic turbulent flows. In *Proc. of the Seventh Symposium on Turbulent Shear Flows*, Stanford University, USA.
36. Hallbäck, M. Groth, J. and Johansson, A.V. 1990. An algebraic model for non-isotropic turbulent dissipation rate in Reynolds stress closures. *Phys. Fluids A* **2**, 1859-1866.
37. Hallbäck, M. and Johansson, A.V. 1992. Modelling of pressure-strain in Homogeneous turbulence. In *Advances in Turbulence 4*, TU Delft, Holland.
38. Hallbäck, M 1993. Development and critical evaluation of turbulence models through direct numerical simulations of homogeneous turbulence. Trita-Mek-93-2, Dept. of Mech., Royal Institute of Technology, Stockholm, Sweden.
39. Hanjalić, K, and Launder, B.E. 1972. A Reynolds stress model of turbulence and its application to thin shear flows. *J. Fluid Mech.* **52**, 609-638.
40. Hanjalić, K, and Launder, B.E. 1976. Contribution towards a Reynolds-stress closure for low-Reynolds-number turbulence. *J. Fluid Mech.* **74**, 593-610.
41. Hinze, J.O. 1975. *Turbulence* Second Edition, McGraw-Hill.
42. Imao, S., Itoh, M. and Harada, T. 1996. Turbulent characteristics of the flow in an axially rotating pipe *J. Heat and Fluid Flow* **17**, 444–451.
43. Jacquin, L., Leuchter, O., Cambon, C. and Mathieu, J. 1990. Homogeneous turbulence in the presence of rotation. *J. Fluid Mech.* **220**, 1-52.
44. Johansson, A.V. and Hallbäck, M. 1994. Modelling of rapid pressure strain in Reynolds-stress closures. *J. Fluid Mech.* **269**, 143-168.
45. Johansson, A.V., Hallbäck, M. and Lindborg, E. 1994. Modelling of rapid pressure strain in Reynolds-stress closures – difficulties associated with rotational mean flows. *Appl. Sci. Res.* **53**, 119-137.
46. Johansson, A.V. and Wallin, S. 1996. A new explicit algebraic Reynolds stress model. Proc. Sixth European Turbulence Conference, Lausanne, July 1996, Ed. P. Monkewitz, 31–34.
47. Jones, W.P. and Launder, B.E. 1972. The prediction of laminarization with a two-

equation model of turbulence.. *Int. J of Heat and Mass Transfer* **15**, 310–314..

48. Jongen, T., Mompean, G. and Gatski, T.G. 1998. Accounting for Reynolds stress and dissipation rate anisotropies in inertial and noninertial frames. *Phys. Fluids* **10**, 674–684.
49. Kim, J. 1989. On the structure of pressure fluctuations in simulated turbulent channel flow. *J. Fluid Mech.* **205**, 421–451.
50. Kim, J., Moin, P. and Moser, R. 1987. Turbulence statistics in fully developed channel flow at low Reynolds number. *J. Fluid Mech.* **177**, 133-166.
51. Kolmogorov, A.N. 1942. Equations of turbulent motion of an incompressible fluid.. Izvestia Academy of Sciences, USSR; Physics, 6, 56-58..
52. Kovasznay, L.S.G. 1953. Turbulence in supersonic flow. *J. Aeronaut. Sci.* **20**, 657–674.
53. Lam, C.K.G. and Bremhorst, K.A. 1981. Modified form of K-ε model for predicting wall turbulence. *ASME, J. Fluids Engineering* **103**, 456-460.
54. Launder, B.E. 1989. Second-moment closure: present ... and future?. *Int. J. Heat and Fluid Flow* **10**, 282-300.
55. Launder, B.E., Reece, G.J. and Rodi, W. 1975. Progress in the development of a Reynolds-stress turbulence closure.. *J. Fluid Mech.* **41**, ,. 537–566.
56. Launder, B.E. and Spalding, D.B. 1972. **Mathematical models of turbulence**, Academic Press..
57. Launder, B.E., Tselepidakis, D.P. and Younis, B.A. 1987. A second-moment closure study of rotating channel flow. *J. Fluid Mech.* **183**, 63-75.
58. Launder, B.E.and Reynolds, W.C. 1983. Asymptotic near-wall stress dissipation rates in a turbulent flow. *Phys. Fluids A* **26**, 1157.
59. Launder, B.E. and Li, S.-P. 1993. On the elimination of wall-topography parameters from second-moment closure. *Phys. Fluids* **6**, 999-1006.
60. Lee, M.J. 1989. Distortion of homogeneous turbulence by axisymmetric strain and dilatation. *Phys. Fluids A* **1**, 1541-1557.
61. Lee, M.J. 1990. A contribution toward rational modeling of the pressure-strain-rate correlation. *Phys. Fluids A* **2**, 630-633.
62. Lee, M.J., Kim, J. and Moin, P. 1989. Structure of turbulence at high shear rate. *J. Fluid Mech.* **216**, 561-583.
63. Lee, M.J. and Reynolds, W.C. 1985. Numerical Experiments on the Structure of Homogeneous Turbulence. Report No. TF-24, Dept. of Mech. Eng., Stanford Univ., USA.
64. Lele, S.K. 1994. Compressibility effects on turbulence. *Ann. Rev. Fluid Mech.* **26**, 211-254.
65. Le Penven, L., Gence, J.N. and Comte-Bellot, G. 1985. On the approach to isotropy of homogeneous turbulence: effect of the partition of kinetic energy among the velocity components. In *Frontiers in Fluid Mechanics*, eds Davis and Lumley, Springer Verlag.
66. Lindberg, P.Å. 1994. Near-wall turbulence models for 3D boundary layers. *Appl. Sci. Res.* **53**, 139-162.
67. Lindborg, E., Hallbäck, M. and Burden, A.D. 1992. EDQNM and DNS – comparative calculations. In *Advances in Turbulence 4*, TU Delft, Holland.
68. Lumley, J.L. 1978. Computational modeling of turbulent flows. *Adv. Appl. Mech.* **18**, 123-177.
69. Mansour, N.N., Kim, J. and Moin, P. 1988. Reynolds-stress and dissipation-rate budgets in a turbulent channel flow. *J. Fluid Mech.* **194**, 15-44.
70. Mansour, N.N., Shih, T.-H. and Reynolds, W.C. 1991. The effects of rotation on initially anisotropic homogeneous flows. *Phys. Fluids A* **3**, 2421-2425.
71. Menter, F.R. 1993. Zonal Two Equation $k-\omega$ Turbulence Models for Aerodynamic Flows. *AIAA 93-2906*
72. Menter, F.R. 1992 Improved Two-Equation $k-\omega$ Turbulence Model for Aerodynamic Flows. *NASA TM-103975*

73. Morris, P.J. 1984. Modeling the pressure redistribution terms. *Phys. of Fluids* **27**, 1620-1623.
74. Naot, D., Shavit, A. and Wolfstein, M. 1973. Two-point correlation model and the redistribution of Reynolds stresses. *Phys. Fluids* **16**, 738-743.
75. Pao, Y.H. 1965. Structure of turbulent velocity and scalar fields at large wavenumbers. *Phys. of Fluids* **8**, 1063-1075.
76. Patel, V.C., Rodi, W. and Scheurer G. 1985. Turbulence Models for near-wall and low Reynolds number flows: a review. *AIAA J.* **23**, 1308-1319.
77. Pearson, J.R.A. 1959. The effect of uniform distortion on weak homogeneous turbulence. *J. Fluid Mech.* **5**, 274-288.
78. Pope, S.B. 1975. A more general effective-viscosity hypothesis *J. Fluid Mech.* **72**, 331–340.
79. Pope, S.B. 1985. Pdf methods for turbulent reacting flows. *Prog. Energy Combust. Sci.* **11**, 119-192.
80. Prandtl, L. 1925. ber die ausgebildete Turbulenz.. *ZAMM* **5**, 136–139.
81. Prandtl, L. 1945. ber ein neues Formelsystem fr die ausgebildete Turbulenz.. Nachr. Akad. Wiss. Gttingen. Math.-Phys. Kl. 1945, 6–19.
82. Reynolds, W.C. 1976. Computation of turbulent flows. *Ann. Rev. Fluid Mech.* **8**, 183-208.
83. Reynolds, W.C. 1987. Fundamentals of turbulence for turbulence modelling and simulation. *Lecture Notes for von Karman Institute*, AGARD-CP-93, NATO.
84. Rodi, W. 1972. The prediction of free turbulent boundary layers by use of a two equation model of turbulence. *Ph.D. thesis, University of London.*
85. Rodi, W. 1976. A new algebraic relation for calculating the Reynolds stresses. *Z. angew. Math. Mech.* **56**, T219–221.
86. Rodi, W. 1980. *Turbulence models and their application in hydrolics - a state of the art review*, Int. Association for Hydraulic Research, Delft, The Netherlands.
87. Rodi, W and Mansour, N.N. 1993. Low Reynolds number K-ε modelling with the aid of direct simulation data. *J. Fluid Mech.* **250**, 509-529.
88. Rogallo, R.S. 1981. Numerical Experiments in Homogeneous Turbulence. NASA-TM81315, Ames Research Center, USA.
89. Rogers, M.M., Moin, P. and Reynolds, W.C. 1986. The structure and modeling of the hydrodynamic and passive scalar fields in homogeneous turbulence shear flow. Rep. no. TF-25, Stanford University, California 94305.
90. Rotta, J 1951. Statistische theorie nichthomogener turbulenz I. *Zeitschrift für Physik* **129**, 547-572.
91. Rubinstein, R. and Barton, J.M. 1992. Renormalization group analysis of the Reynolds stress transport equation. *Phys. Fluids* **4**, 1759-1766.
92. Saffman, P.G. 1967. The large-scale structure of homogeneous turbulence. *J. Fluid Mech.* **27**, 581-593.
93. Saffman, P.G. 1970. A model for inhomogeneous turbulent flow.. *Proc. Roy. Soc. London A* **317**, 417-433..
94. Saffman, P.G. and Wilcox, D.C. 1974. Turbulence-model predictions for turbulent boundary layers.. *AIAA Journal* **12**, ,. 541–546.
95. Sarkar, S. and Speziale, C.G. 1990. A simple nonlinear model for the return to isotropy in turbulence. *Phys. Fluids A* **2**, 84-93.
96. Schülein, E., Krogmann, P. and Stanewsky, E. 1996. Documentation of Two-Dimensional Impinging Shock/ Turbulent Boundary Layer Interaction Flow. DLR Report: DLR IB 223 - 96 A 49.
97. Schumann, U. 1977. Realizability of Reynolds-stress models. *Phys. Fluids* **20**, 721-725.
98. Schumann, U. and Pattersson, G.S. 1978. Numerical study of pressure and velocity fluctuations in nearly isotropic turbulence. *J. Fluid Mech.* **88**, 685-709.
99. Schumann, U. and Pattersson, G.S. 1978. Numerical study of the return of axisymmetric turbulence to isotropy. *J. Fluid Mech.* **88**, 711-735.

100. Shabbir, A. and Shih, T.H. 1992. Critical Assessment of Reynolds Stress Turbulence Models Using Homogeneous Flows NASA TM 105954, ICOMP-92-24, CMOTT-92-12.
101. Shih, T.H., Zhu, J. and Lumley, J.L. 1992. A Realizable Reynolds Stress Algebraic Equation Model NASA TM 105993, ICOMP-92-27, CMOTT-92-14.
102. Shih, T.-H. 1996. Constitutive relations and realizability of single-point turbulence closures Chapter 4 of **Turbulence and Transition Modelling**, Eds. M. Hallback, D.S. Henningson, A.V. Johansson and P.H. Alfredsson. Kluwer.
103. Shih, T.-H. and Lumley, J.L. 1993. Remarks on turbulent constitutive relations *Math. Comput. Modelling*, **18**, 9–16.
104. Shih, T.-H. and Lumley, J.L. 1986. Second-order modeling of near-wall turbulence. *Phys. of Fluids* **29**, 971-975.
105. Shih, T.-H. and Mansour, N.N. 1990. Modeling of near-wall turbulence. In *Engineering Turbulence Modelling and Experiment*, eds W.Rodi and E.N. Ganić, Elsevier.
106. Shih, T.-H., Reynolds, W.C. and Mansour, N.N. 1990. A spectrum model for weakly anisotropic turbulence. *Phys. Fluids A* **2**, 1500-1502.
107. Sjögren, T. 1997. Development and Validation of Turbulence Models Through Experiment and Computation Doctoral thesis, Dept. of Mechanics, KTH, Stockholm, Sweden.
108. Sjögren, T. and Johansson, A.V. 1998. Measurement and modelling of homogeneous axisymmetric turbulence. *J. Fluid Mech.* **374**, 59-90.
109. Sjögren, T. and Johansson, A.V. 1998. Development and calibration of algebraic non-linear models for terms in the Reynolds stress transport equations. *Submitted to Phys. Fluids* ,
SochJPF
110. So, R.M.C., Lai, Y.G. and Zhang, H.S. 1991. Second-Order Near-wall Turbulence Closures: A Review. *AIAA J.* **29**, 1819–1835.
111. So, R.M.C., Zhang, H.S. and Speziale, C.G. 1991. Near-wall modeling of the dissipation rate equation. *AIAA J.* **29**, 2069-2076.
112. So, R.M.C., Lai, Y.G., Zhang, H.S. and Hwang, B.C. 1991. Second-order near-wall turbulence closures: a review. *AIAA J.* **29**, 1819-1835.
113. Spalart, P. and Allmaras, S.R. 1992. A one-equation model for aerodynamic flows. AIAA 92-439.
114. Spencer, A.J.M. and Rivlin, R.S. 1959. The theory of matrix polynomials and its application to the mechanics of isotropic continua *Arch. Rat. Mech. Anal.* **2**, 309–336.
115. Speziale, C.G. 1981. Some interesting properties of two-dimensional turbulence. *Phys. Fluids* **24**, 1425-1427.
116. Speziale, C.G. 1987. On non-linear K-l and K-ε models of turbulence. *J. Fluid Mech.* **178**, 459-475.
117. Speziale, C.G., Mansour, N.N. and Rogallo, R.S. 1987. The decay of isotropic turbulence in a rapidly rotating frame. Report CTR-S87, Ames Research Center/ Stanford University, USA.
118. Speziale, C.G. 1988. Turbulence modeling in noninertial frames of reference. *Theoret. Comput. Fluid Dynamics* **1**, 3-19.
119. Speziale, C.G., Sarkar, S. and Gatski, T.B. 1991. Modelling the pressure-strain correlation of turbulence: an invariant dynamical systems approach. *J. Fluid Mech.* **227**, 245–272.
120. Speziale, C.G. and Xu, X.-H. 1996. Towards the development of second-order closure models for nonequilibrium turbulent flows *Int. J. Heat and Fluid Flow* **17**, 238–244.
121. Speziale, C.G. 1991. Analytical methods for the development of Reynolds-stress closures in turbulence. *Annu. Rev. Fluid Mech.* **23**, 107-157.
122. Spina, E.F., Smits, A.J. and Robinson, S.K. 1994. The physics of supersonic tur-

bulent boundary layers. *Ann. Rev. Fluid Mech.* **26**, 287–319.
123. Taulbee, D.B. 1992. An improved algebraic Reynolds stress model and corresponding nonlinear stress model. *Phys. Fluids A* **4**, 2555–2561.
124. Taulbee, D.B., Sonnenmeier, J.R. and Wall, K.M. 1994. Stress relation for three-dimensional turbulent flows *Phys. Fluids* **6**, 1399–1401.
125. Tavoularis, S. and Corrsin, S. 1981. Experiments in nearly homogeneous turbulent shear flow with a uniform mean temperature gradient. Part I. *J. Fluid Mech.* **104**, 311–347.
126. Tennekes, h. and Lumley, J.L. 1972. *A first course in turbulence*, MIT.
127. Townsend, A.A. 1954. The uniform distortion of homogeneous turbulence. *Quart. J. Mech. Appl. Math.* **7**, 104-127.
128. Townsend, A.A. 1976. *The structure of turbulent shear flow*, (2nd ed., Cambridge University Press.
129. Tucker, H.J. and Reynolds, A.J. 1968. The distortion of turbulence by irrotational plane strain. *J. Fluid Mech.* **32**, 657-673.
130. van Driest 1951. Turbulent boundary layers in compressible fluids. *J. Aeronaut. Sci.* **18**, 145-160.
131. Vandromme, D. 1992. Introduction to the Modeling of Turbulence *von Karman Institute for Fluid Dynamics, Lecture Series 1991-02.*
132. Wallin, S. and Johansson, A.V. 1996. A new explicit algebraic Reynolds stress turbulence model including an improved near-wall treatment. Proc. Flow Modeling and Turbulence Measurements VI, Tallahassee FL, Chen, Shih, Lienau and Kung (eds), 399–406.
133. Wallin, S. and Johansson, A.V. 1997. A complete explicit algebraic Reynolds stress turbulence model for incompressible and compressible turbulent flows. FFA TN 1997-51 (and submitted for publication).
134. Weinstock, J. 1981. Theory of pressure-strain-rate correlation for Reynolds-stress turbulence closures. Part 1. Off-diagonal element. *J. Fluid Mech.* **105**, 369-396.
135. Weinstock, J. 1982. Theory of the pressure-strain rate. Part 2. Diagonal elements. *J. Fluid Mech.* **116**, 1-29.
136. Wilcox, D.C. 1988. Reassessment of the scale determining equation for advanced turbulence models.. *AIAA Journal* **26**, 1299–1310..
137. Wilcox, D.C. 1994. Simulation of Transition with a Two-Equation Turbulence Model. *AIAA J.* **32**, 247–255.
138. Yoshizawa, A. 1995. Simplified statistical approach to complex turbulent flows and ensemble-mean compressible turbulence modelling.. *Phys. Fluids* **7**, 3105–3117.

Chapter 5

MODELLING OF TURBULENCE IN COMPRESSIBLE FLOWS

R. FRIEDRICH
Lehrstuhl für Fluidmechanik
TU München, Boltzmannstr. 15
85748 Garching, Germany

5.1. Introduction

The first attempts at turbulence modelling date back to the nineteenth century. They provided 'predictions' for a narrow class of incompressible flow problems. At the end of the twentieth century a similar situation (in some sense) prevails with respect to compressible turbulence modelling, insofar as our understanding of physical mechanisms is too limited to form a sufficiently broad basis for general modelling. However, due to advances in incompressible turbulence modelling the level at which predictions are made is higher and there are good prospects for faster progress than at those times. The optimistic view is cherished by recent achievements of direct numerical simulation (DNS) which has improved our knowledge of compressibility effects considerably during the last decade.

Let us begin with a rough classification of compressible turbulent flows into:

I) Flows with vanishing compressibility effects due to turbulent fluctuations and
II) Flows in which such effects play a role.

Type I flows are assumed to follow Morkovin's hypothesis in it's weak form (Morkovin [63]) which states that thermodynamic pressure and total temperature fluctuations are negligible for small turbulent Mach numbers implying negative density-temperature correlations. The hypothesis has led to the so–called strong Reynolds analogy (SRA) and is in line with the Van

Driest transformation (Van Driest [92]) which collapses velocity profiles of compressible turbulent boundary layers onto the incompressible law of the wall (Fernholz & Finley [28], Huang & Coleman [43]). Compressibility effects therefore manifest themselves in terms of mean density variations and can be modelled by straightforward adaptations of classical incompressible models. Besides boundary layers with zero or weak pressure gradient and freestream Mach numbers less than 5, mixing layers with convective Mach numbers less than 1 are commonly considered as examples of type I flows (Bradshaw [14]). It is also expected (although not confirmed at present) that type II flows in which fluctuations of the thermodynamic pressure become important, are encountered at hypersonic speeds. Unfortunately, direct numerical simulation data are not yet available to clarify this issue. A closer look at DNS results for different classes of flows, however, unveils the lack of subtlety of such a classification. Coleman *et al.* 's [19] DNS of supersonic fully developed flow in a channel with cooled walls *e.g.* shows that, although compressibility effects due to turbulent fluctuations are unimportant, the strong Reynolds analogy in it's form for nonadiabatic flows (Gaviglio [35], Rubesin [70]) does not apply. A more general representation of the analogy was therefore derived by Huang *et al.* [44] and shown to match the DNS data. Recent direct simulations of annular mixing layers with convective Mach numbers ranging from M_c=0.1 to 1.8 by Freund *et al.* [29] indicate that pressure fluctuations are subordinate to temperature and density fluctuations (related to their mean values, respectively) only for $M_c < 0.2$. For higher values of M_c, Morkovin's hypothesis for adiabatic flows does not apply. A third example where pressure fluctuations are non-negligible, is shock isotropic turbulence interaction. Based on DNS and linear theory Mahesh *et al.* [58] found considerable deviation from Morkovin's hypothesis (in it's weak form) behind shocks, although the deviations were seen to decrease with the downstream distance. To be more specific, total temperature fluctuations are generated immediately behind the shock as a result of shock oscillation and are convected into the far field. A second important finding of their work is that upstream entropy fluctuations lead to higher amplification rates of turbulent kinetic energy and vorticity across the shock than pure vortical fluctuations. This result should be of great value in explaining the interaction between shocks and strongly cooled boundary layers.

These examples show that a conclusive classification scheme for compressible turbulent flows is difficult to find at present, especially as long as our knowledge of compressibility effects is not complete.

The paper starts from the conservation laws for ideal gases in section 1, discusses the molecular transport coefficients for high-speed flow and the coupling between vorticity and dilatation transport. In section 2 the basic equations are statistically averaged and transport equations for unknown single-point correlations like the turbulent stress, the pressure variance and the turbulent heat flux are derived. Homogeneous shear flow is discussed in detail in terms of these equations, since this flow is fundamental for the development of turbulence models. Section 3 concentrates on compressibility effects due to turbulent fluctuations as derived from direct numerical simulation. The importance of linear mechanisms is emphasized in homogeneous isotropic, respectively sheared turbulence, and assumptions in the derivation of models for explicit compressibility terms are discussed. Compressibility effects in supersonic channel flow (as a typical example of wall-bounded turbulence) are summarized, before models for explicit compressibility terms are shown to fail. Section 4, finally contains a short (by no means exhaustive) survey and applications of most frequently used turbulence models which are all variable-density extensions of their incompressible counterparts.

5.1.1. EQUATIONS OF MOTION

Turbulent flows of compressible polyatomic gases for which the continuum hypothesis is considered valid are governed by the following set of conservation equations:

Mass:

$$\frac{\partial \rho}{\partial t} + \frac{\partial \rho u_j}{\partial x_j} = 0, \tag{5.1}$$

Momentum:

$$\frac{\partial \rho u_i}{\partial t} + \frac{\partial}{\partial x_j}(\rho u_i u_j) = \frac{\partial \sigma_{ij}}{\partial x_j} = -\frac{\partial p}{\partial x_i} + \frac{\partial \tau_{ij}}{\partial x_j}, \tag{5.2}$$

Energy:

$$\frac{\partial \rho E}{\partial t} + \frac{\partial}{\partial x_j}(\rho E u_j) = -\frac{\partial q_j}{\partial x_j} + \frac{\partial}{\partial x_j}(u_i(\tau_{ij} - p\delta_{ij})). \tag{5.3}$$

Body forces are assumed small in high-speed flows. ρ, u_i, p denote the density, velocity and pressure. E is the total energy which comprises the internal energy e and the kinetic energy per unit mass:

$$E = e + \frac{1}{2}u_i u_i. \tag{5.4}$$

According to Fourier's law, the heat flux by conduction, q_j, is related to the temperature gradient:

$$q_j = -k\frac{\partial T}{\partial x_j}. \tag{5.5}$$

For Newtonian fluids the stress tensor, τ_{ij}, is proportional to the rate-of-strain tensor s_{ij}:

$$\tau_{ij} = 2\mu\left(s_{ij} - \frac{1}{3}s_{kk}\delta_{ij}\right) + \mu_v s_{kk}\delta_{ij}, \tag{5.6}$$

$$s_{ij} = \frac{1}{2}\left(\frac{\partial u_i}{\partial x_j} + \frac{\partial u_j}{\partial x_i}\right) \quad . \tag{5.7}$$

Introducing the deviatoric part of the s_{ij}-tensor, namely:

$$s_{ij}^D = s_{ij} - \frac{1}{3}s_{kk}\delta_{ij} \tag{5.8}$$

which describes the pure straining motion without change of volume, eq. (5.6) reads:

$$\tau_{ij} = 2\mu s_{ij}^D + \mu_v s_{kk}\delta_{ij}. \tag{5.9}$$

The set of conservation equations is not yet complete. We have to add equations of state which relate the thermodynamic variables. The assumption of a thermally perfect gas, viz:

$$p = \rho R T \tag{5.10}$$

allows to describe even relaxation effects in the molecular translational and rotational degrees of freedom. Eq. (5.10) implies a caloric equation of state of the form:

$$e = e(T) \quad . \tag{5.11}$$

We will use the equations

$$de = c_v(T)dT \tag{5.12}$$

and

$$e = c_v T \tag{5.13}$$

interchangeably. The latter defines the calorically perfect gas and assumes constant specific heat at constant volume. The conservation of momentum and energy is controlled by the following molecular transport coefficients:

– the shear or dynamic viscosity μ

– the bulk viscosity μ_v
– the heat conductivity k.

They all depend on temperature alone. Sutherland's formula for the shear viscosity is valid in a range of temperatures, between $200K$ and $1200K$:

$$\frac{\mu}{\mu_0} = \left(\frac{T}{T_0}\right)^{\frac{3}{2}} \frac{T_0 + S_0}{T + S_0} \tag{5.14}$$

The coefficients are given in table 5.1 for 3 gases.

TABLE 5.1. Sutherland constants for dynamic viscosity, valid in the range $200K < T < 1200K$.

Gas	T_0(K)	S_0(K)	$10^5 \times \mu_0$ (Pa s)
N_2	273.1	106.6	1.665
O_2	273.1	138.8	1.921
CO_2	273.1	222.2	1.370

Bertolotti [10] has derived a new temperature dependence for the bulk viscosity of the form:

$$\frac{\mu_v(T)}{\mu(T)} = \left(\frac{\mu_v}{\mu}\right)_{T=293.3K} \exp\left(\frac{T - 293.3}{1940}\right) \quad , \tag{5.15}$$

and has demonstrated its damping effect on the instability of laminar boundary layers at M=4.5 (especially on Mack's second mode). It has to be noted, that the bulk viscosity is not a physical property of gases. It is rather an approximation designed to model the effect of rotational energy relaxation[1].

The pressure p at a point in a moving fluid is defined in a mechanical way as the mean normal stress with sign reversed, *i.e.* $p = -\sigma_{ii}/3$ (Batchelor [7]). The thermodynamic pressure, to which it is related, depends on two state variables, say ρ and e. The internal energy involves all molecular energies, *i.e.* translational and rotational *e.g.* If the rotational modes have relaxation times of the order of several collision intervalls and make a significant contribution to the internal energy e, then μ_v has to take care of the lag in the adjustment of the mechanical pressure to the continually changing values in ρ and e in a motion involving volume changes. The

[1]Vibrational energy relaxation cannot be treated in such an approximate way. A relaxation equation must be solved.

bulk viscosity is usually regarded as non-negligible in situations like the attenuation of high frequency sound waves or the structure of shock waves. Considering Bertolotti's findings [10], it is to be expected that μ_v also plays a considerable role in high-speed compressible turbulence.

The heat conductivity is likewise affected by the state of the internal energy of the molecules and follows a similar Sutherland law (Bertolotti [10]) obtained from a best fit to experimental data:

$$\frac{k}{k_0} = \left(\frac{T}{T_0}\right)^{3/2} \frac{T_0 + S_{k0}}{T + S_{k0}}. \tag{5.16}$$

The coefficients are contained in table 5.2.

TABLE 5.2. Sutherland constants for heat conductivity, valid in the range $200K < T < 1200K$.

Gas	$T_0(K)$	$S_{k0}(K)$	$10^2 \times k_0 \left(\frac{W}{mK}\right)$
N_2	273.1	166.6	2.440
O_2	273.1	222.2	2.480
CO_2	273.1	2222.2	1.455

For some situations an energy equation is needed in terms of the enthalpy h, where $h = e + p/\rho$, or of the stagnation (or total) enthalpy; $H = h + 1/2\ u_i u_i$. The balance equation for the total enthalpy is:

$$\rho \frac{DH}{Dt} = \frac{\partial p}{\partial t} - \frac{\partial q_j}{\partial x_j} + \frac{\partial}{\partial x_j}(u_i \tau_{ij}), \tag{5.17}$$

with the material derivative

$$\frac{D}{Dt} = \frac{\partial}{\partial t} + u_j \frac{\partial}{\partial x_j}. \tag{5.18}$$

Subtracting the kinetic energy equation from (5.17) leads to the enthalpy equation:

$$\rho \frac{Dh}{Dt} = \frac{Dp}{Dt} - \frac{\partial q_j}{\partial x_j} + \tau_{ij} \frac{\partial u_i}{\partial x_j}. \tag{5.19}$$

The last term on the right-hand-side is the dissipation rate per unit volume, ϕ. Using the symbol d for the dilatation,

$$d = \frac{\partial u_j}{\partial x_j}, \tag{5.20}$$

we express the dissipation rate in a form which shows that it is always positive, viz:

$$\phi = \tau_{ij}\frac{\partial u_i}{\partial x_j} = \tau_{ij}s_{ij} = 2\mu s_{ij}^D s_{ij}^D + \mu_v d^2. \tag{5.21}$$

The bulk viscosity provides additional dissipation. From (5.19) a temperature equation may be derived introducing the caloric state equation in the form

$$\frac{Dh}{Dt} = c_p\frac{DT}{Dt} \tag{5.22}$$

appropriate for moving fluids with $c_p = c_p(T)$.

For thermally and at the same time calorically perfect gases (from now on referred to as perfect gases), the enthalpy equation (5.19) can also be converted into an equation for the pressure alone. Using the gas law one gets

$$h = c_pT = \frac{\gamma}{\gamma-1}\frac{p}{\rho} \quad , \quad \gamma = c_p/c_v \quad . \tag{5.23}$$

and from (5.19) and the continuity equation:

$$\frac{Dp}{Dt} = -\gamma p\frac{\partial u_j}{\partial x_j} + (\gamma-1)\left(\phi - \frac{\partial q_j}{\partial x_j}\right). \tag{5.24}$$

p is thus a measure of internal energy which is altered reversibly during compression and expansion processes and irreversibly by dissipation and heat conduction.

Finally, with the help of the Gibbs fundamental equation:

$$Tds = dh - dp/\rho \tag{5.25}$$

an entropy balance equation is obtained from (5.19):

$$\rho T\frac{Ds}{Dt} = -\frac{\partial q_j}{\partial x_j} + \phi \tag{5.26}$$

or in the more informative form:

$$\rho\frac{Ds}{Dt} = \frac{\partial}{\partial x_j}\left(\frac{k\frac{\partial T}{\partial x_j}}{T}\right) + \frac{1}{T}\left(\phi + \frac{k}{T}\left(\frac{\partial T}{\partial x_j}\right)^2\right) \tag{5.27}$$

which shows that the entropy irreversibly increases within the flow field due to friction and heat conduction. The transfer of heat across a control surface (first term on the rhs) can otherwise increase or decrease the entropy. Walls that inhibit the heat conduction across their surfaces are called adiabatic. Setting all molecular transport coefficients to zero defines isentropic flow:

$$\frac{Ds}{Dt} = 0. \tag{5.28}$$

5.1.2. TRANSPORT OF DILATATION AND VORTICITY

In high-speed non-reactive flows dilatation is a measure of compressibility, in the sense that volume changes are caused by changes in the pressure (Lele [54]). Following Thompson [90] we express density changes in the continuity equation (5.1) by changes in the pressure and the entropy. Then, introducing the speed of sound, c, for thermally and calorically perfect gases, by:

$$c^2 = \left(\frac{\partial p}{\partial \rho}\right)_s = \gamma p/\rho, \tag{5.29}$$

we obtain:

$$d = \frac{\partial u_j}{\partial x_j} = -\frac{1}{\gamma p}\frac{Dp}{Dt} + \frac{1}{c_p}\frac{Ds}{Dt}. \tag{5.30}$$

With the help of the gas law (5.10), the kinetic energy equation, the enthalpy and entropy equations (5.19), (5.26) one finally gets:

$$d = -\frac{1}{c^2}\left\{\frac{1}{\rho}\frac{\partial p}{\partial t} - \frac{D}{Dt}\left(\frac{1}{2}u_i u_i\right) + \frac{u_i}{\rho}\frac{\partial \tau_{ij}}{\partial x_j} - \frac{1}{\rho}(\gamma - 1)\left(\phi - \frac{\partial q_j}{\partial x_j}\right)\right\}. \tag{5.31}$$

The first term on the rhs of (5.31) represents acoustic effects. It generates compressibility when the time scale of the pressure 'oscillations' is comparable to the local acoustic time scale. The second term is of the order of the Mach number squared and states that high-speed flows are typically compressible flows. The remaining terms generally don't lead to compressibility effects. Only in extreme situations might excessive heat transfer rates cause considerable volume changes. For reactive flows, eq. (5.31) must be supplemented by diffusion effects, heat release and changes in the molecular weight of the gas mixture. Supersonic combustion is a situation where chemical and compressibility effects strongly interact. The modelling of correlations which involve dilatation fluctuations is then extremely complicated.

Dilatation transport:

A transport equation for d is easily derived, taking the divergence of the momentum equation (5.2):

$$\frac{Dd}{Dt} = -\frac{\partial u_i}{\partial x_j}\frac{\partial u_j}{\partial x_i} - \frac{1}{\rho}\frac{\partial^2 p}{\partial x_i \partial x_i} + \frac{1}{\rho^2}\frac{\partial \rho}{\partial x_i}\frac{\partial p}{\partial x_i} + \frac{\partial}{\partial x_i}\left(\frac{1}{\rho}\frac{\partial \tau_{ij}}{\partial x_j}\right). \tag{5.32}$$

It is useful to express the first term on the rhs by the rate-of-strain tensor and the vorticity vector ω_i, defined by:

$$\omega_i = \epsilon_{ijk}\frac{\partial u_k}{\partial x_j} = \epsilon_{ijk} r_{kj}, \tag{5.33}$$

where ϵ_{ijk} is the alternating unit tensor and

$$r_{ij} = \frac{1}{2}\left(\frac{\partial u_i}{\partial x_j} - \frac{\partial u_j}{\partial x_i}\right) \tag{5.34}$$

the rate-of-rotation tensor. The following intermediate steps

$$\begin{aligned} \frac{\partial u_i}{\partial x_j}\frac{\partial u_j}{\partial x_i} &= (s_{ij} + r_{ij})(s_{ji} + r_{ji}) \\ &= s_{ij}s_{ij} - \frac{1}{2}\omega_i\omega_i = s_{ij}^D s_{ij}^D + \frac{1}{3}d^2 - \frac{1}{2}\omega_i\omega_i \end{aligned} \tag{5.35}$$

lead to the final form of the dilatation transport equation:

$$\begin{aligned} \frac{Dd}{Dt} &= -s_{ij}^D s_{ij}^D - \frac{1}{3}d^2 + \frac{1}{2}\omega_i\omega_i - \frac{1}{\rho}\frac{\partial^2 p}{\partial x_i \partial x_i} \\ &\quad + \frac{1}{\rho^2}\frac{\partial \rho}{\partial x_i}\frac{\partial p}{\partial x_i} + \frac{\partial}{\partial x_i}\left(\frac{1}{\rho}\frac{\partial \tau_{ij}}{\partial x_j}\right). \end{aligned} \tag{5.36}$$

Obviously, pure straining motions and volume changes act in the same direction. They decrease the magnitude of d in expansion zones, whereas any vortical motions directly increase the level of d and vice versa in compression zones. The pressure field acts on d via its Laplacian and via the dot product between density and pressure gradients and, finally, d is controlled by viscous effects. For incompressible isothermal flow (5.36) reduces to the well-known Poisson equation for the pressure:

$$\frac{\partial u_i}{\partial x_j}\frac{\partial u_j}{\partial x_i} = -\frac{1}{\rho}\frac{\partial^2 p}{\partial x_i \partial x_i}, \tag{5.37}$$

which underlines the fact that p is no longer a state variable, but is completely determined by the velocity field. This change in role of the pressure also reflects the difficulties in adequately modelling correlation functions involving pressure fluctuations.

Vorticity transport:

Taking the curl of the momentum equation (5.2) provides the vorticity transport equation for a compressible fluid in the form:

$$\frac{D\omega_i}{Dt} = \omega_j s_{ij} - \omega_i d - \epsilon_{ijk}\left(\frac{1}{\rho^2}\frac{\partial p}{\partial x_j}\frac{\partial \rho}{\partial x_k} - \frac{\partial}{\partial x_j}\left(\frac{1}{\rho}\frac{\partial \tau_{kl}}{\partial x_l}\right)\right). \tag{5.38}$$

The first term on the rhs changes vorticity by stretching, contracting or tilting of vortex lines. It is this term which increases vorticity fluctuations in turbulent flows while kinetic energy is transferred from large to small scales in a cascade process, until the loss of vorticity by viscosity compensates the gain by stretching at the smallest scales. In compressible flows two extra effects appear, namely the increase in vorticity in compression zones ($d < 0$) or a corresponding decrease in expansion zones and a change due to the baroclinic torque term (3rd term on the rhs). If pressure and density gradients are not parallel, the pressure force does not pass through the center of gravity of the fluid particle and a moment about this center exists which changes ω_i (Smits and Dussauge [81]). The baroclinic term is zero for barotropic flows, for which the pressure is a function of density alone (isentropic flow of thermally and calorically perfect gases; *e.g.*). Baroclinic effects are discussed by Mahesh *et al.* [58] in the context of shock/turbulence interaction. Finally, we emphasize the explicit non-linear coupling between dilatation and vorticity, which contributes to the complexity of compressible turbulent flows and is reflected in the two transport equations (5.36) and (5.38).

5.2. Averaged equations

5.2.1. DEFINITION OF AVERAGES

For compressible flows it is common practice to work with two different averages simultaneously, the Reynolds-average, denoted by a bar and the Favre- or mass-weighted average, characterized by a tilde. Density and pressure are mostly written in terms of Reynolds-averages and fluctuations, viz:

$$\rho = \bar{\rho} + \rho'$$

$$p = \bar{p} + p' \tag{5.39}$$

whereas temperature, internal energy and velocity are split into

$$\begin{aligned} T &= \tilde{T} + T'', \\ e &= \tilde{e} + e'', \\ u_i &= \tilde{u}_i + u_i''. \end{aligned} \tag{5.40}$$

Mass-weighted averages are defined as

$$\bar{\rho}\tilde{T} = \overline{\rho T}, \qquad etc. \tag{5.41}$$

implying

$$\overline{\rho T''} = 0, \tag{5.42}$$

but

$$\overline{T''} \neq 0 \tag{5.43}$$

in general. The mean of the Reynolds-fluctuation, however, vanishes. It remains to state, how mean quantities can be obtained in the computation (DNS/LES) or the experiment. One way is by ensemble averaging over a large number of realizations:

$$\bar{\rho}^e \equiv \lim_{N\to\infty} \frac{1}{N} \sum_{n=1}^{N} \rho_n. \tag{5.44}$$

Another is by time-averaging over a finite time-interval τ which is large enough to cover all turbulent time scales but small with respect to the statistical unsteadiness of the flow:

$$\bar{\rho}^t = \frac{1}{\tau} \int_0^{\tau} \rho(t+\theta)d\theta \tag{5.45}$$

For stationary turbulence, τ may go to infinity, in principle. If a computed flow is homogeneous in one or two or even all three directions, spatial averaging in these directions is common. We assume that all mean values coincide in the special case of stationary and homogeneous turbulence (ergodic hypothesis), and that there is usually a way to obtain stable statistical values in general flow situations. We simply denote such statistical quantities by the overbar, $\overline{}$, or the tilde, $\widetilde{}$, and do not care anymore how they have been obtained. We further assume that the averaging procedure commutes with differentiation, is linear and preserves constants.

5.2.2. AVERAGED CONSERVATION EQUATIONS

The conservation equations not only contain products of ρ and u_i *e.g.* , but also products of p and u_i or of μ and gradients of u_i etc. When averaging these products, it proves in general more convenient to use Favre variables and their fluctuations in terms resulting from convection and Reynolds-averages and -fluctuations in the remaining terms (see Huang *et al.* [44]). The averaged mass, momentum and total energy equations are:
Mass:

$$\frac{\partial \bar{\rho}}{\partial t} + \frac{\partial \bar{\rho}\widetilde{u_j}}{\partial x_j} = 0, \tag{5.46}$$

Momentum:

$$\frac{\partial \bar{\rho}\widetilde{u_i}}{\partial t} + \frac{\partial}{\partial x_j}(\bar{\rho}\widetilde{u_i}\widetilde{u_j}) = -\frac{\partial}{\partial x_j}\overline{\rho u_i'' u_j''} - \frac{\partial \bar{p}}{\partial x_i} + \frac{\partial \overline{\tau_{ij}}}{\partial x_j}, \tag{5.47}$$

Energy:

$$\frac{\partial \bar{\rho}\tilde{E}}{\partial t} + \frac{\partial}{\partial x_j}(\bar{\rho}\widetilde{u_j}\tilde{E}) =$$
$$-\frac{\partial}{\partial x_j}\overline{\rho u_j'' E''} - \frac{\partial \overline{q_j}}{\partial x_j} + \frac{\partial}{\partial x_j}\left(\overline{u_i}(\overline{\tau_{ij}} - \bar{p}\delta_{ij}) + \overline{u_i'\tau_{ij}'} - \overline{u_j' p'}\right). \tag{5.48}$$

Several terms in these equations need some discussion. The mean total energy, $\tilde{E}$, *e.g.* contains the kinetic energy of the mean motion and the turbulent kinetic energy K:

$$\tilde{E} = \tilde{e} + \frac{1}{2}\widetilde{u_i}\widetilde{u_i} + \underbrace{\frac{1}{2}\widetilde{u_i'' u_i''}}_{K}. \tag{5.49}$$

Consequently, a fluctuation E'' is defined as

$$E'' = e'' + \widetilde{u_i} u_i'' + \frac{1}{2}u_i'' u_i'' - K. \tag{5.50}$$

The energy flux term (See the 1st term on the rhs of (5.48)) can then be written as:

$$\overline{\rho u_j'' E''} = \overline{\rho u_j'' e''} + \widetilde{u_i}\overline{\rho u_i'' u_j''} + \frac{1}{2}\overline{\rho u_i'' u_i'' u_j''}, \tag{5.51}$$

i.e. as the sum of the turbulent heat flux, the work done by the Reynolds stress tensor, $\overline{\rho u_i'' u_j''}$, and the turbulent transport or diffusion.

The mean viscous stress, $\overline{\tau_{ij}}$, is

$$\overline{\tau_{ij}} = 2\overline{\mu}\overline{s_{ij}^D} + \overline{\mu_v}\ \overline{s_{kk}}\delta_{ij} + 2\overline{\mu' s_{ij}^{D'}} + \overline{\mu_v' s_{kk}'}\delta_{ij}. \tag{5.52}$$

It contains correlations betweeen viscosity fluctuations, $\mu' = \mu(T')$ and fluctuations of the rate-of-strain tensor:

$$s_{ij}' = \frac{1}{2}\left(\frac{\partial u_i'}{\partial x_j} + \frac{\partial u_j'}{\partial x_i}\right). \tag{5.53}$$

Similarly, the mean conductive heat flux is:

$$\overline{q_j} = -\bar{k}\frac{\partial \bar{T}}{\partial x_j} - \overline{k'\frac{\partial T'}{\partial x_j}}. \tag{5.54}$$

Consistent definitions of τ_{ij}' and q_j', in the sense that $\overline{\tau_{ij}'} = 0$ and $\overline{q_j'} = 0$, are:

$$\begin{aligned} \tau_{ij}' &= 2\mu' s_{ij}^{D'} + \mu_v' s_{kk}'\delta_{ij} \\ &\quad -2\overline{\mu' s_{ij}^{D'}} - \overline{\mu_v' s_{kk}'}\delta_{ij} \\ &\quad +2\overline{\mu} s_{ij}^{D'} + \overline{\mu_v} s_{kk}'\delta_{ij} \\ &\quad +2\mu'\overline{s_{ij}^D} + \mu_v'\overline{s_{kk}}\delta_{ij}, \end{aligned} \tag{5.55}$$

$$q_j' = -k'\frac{\partial T'}{\partial x_j} + \overline{k'\frac{\partial T'}{\partial x_j}} - k'\frac{\partial \bar{T}}{\partial x_j} - \bar{k}\frac{\partial T'}{\partial x_j}. \tag{5.56}$$

Subtracting the transport equation for the kinetic energy of the mean motion, $\frac{1}{2}\widetilde{u_i}\widetilde{u_i}$, from the total energy balance equation (5.48), leads to the balance equation for the mean internal energy $\bar{\rho}\tilde{e}$:

$$\begin{aligned} &\frac{\partial \bar{\rho}\tilde{e}}{\partial t} + \frac{\partial}{\partial x_j}(\bar{\rho}\widetilde{u_j}\tilde{e}) = \\ &-\frac{\partial}{\partial x_j}\overline{\rho u_j'' e''} - \frac{\partial \overline{q_j}}{\partial x_j} - \bar{p}\bar{d} - \overline{p'd'} + \overline{\tau_{ij}}\ \overline{s_{ij}} + \overline{\tau_{ij}' s_{ij}'}. \end{aligned} \tag{5.57}$$

In the averaged conservation equations above, the following (single-point) correlations appear that require closure:

- the turbulent (or Reynolds) stresses, $\overline{\rho u_i'' u_j''} = \bar{\rho}\widetilde{u_i'' u_j''}$
- the turbulent heat fluxes, $\overline{\rho u_j'' e''} = \bar{\rho}\widetilde{u_j'' e''}$

– the turbulent mass flux, $\overline{u_i''} = -\overline{\rho' u_i'}/\bar{\rho}$
– the pressure-velocity correlation, $\overline{p' u_j'}$
– the velocity triple correlation, $\overline{\rho u_i'' u_i'' u_j''}$
– the pressure dilatation, $\overline{p' d'} = \overline{p' \frac{\partial u_j'}{\partial x_j}}$
– the turbulent dissipation rate, $\bar{\rho}\epsilon = \overline{\tau_{ij}' s_{ij}'}$
– the transport by viscous stresses, $\overline{\tau_{ij}' u_i'}$
– viscosity rate-of-strain correlations. *e.g.* $\overline{\mu' s_{ij}'}$
– and the heat conductivity temperature gradient correlation, $\overline{k' \partial T' / \partial x_j}$

The turbulent heat flux and the pressure velocity correlation can be combined for perfect gases to:

$$\begin{aligned} &\overline{\rho u_j'' e''} + \overline{p' u_j'} = \overline{\rho u_j'' (e'' + p'/\rho)} = \\ &\overline{\rho c_p u_j'' T''} + R\tilde{T}\overline{\rho' u_j''} = \overline{\rho c_p u_j'' T''} - \bar{p}\overline{u_j''}. \end{aligned} \tag{5.58}$$

Note that $\overline{p' u_j'} = \overline{p' u_j''}$ and $\bar{p} = \bar{\rho} R \tilde{T}$. The turbulent mass flux $\overline{u_i''}$ describes the difference between Reynolds and Favre-averaged velocities

$$\overline{u_i} - \tilde{u}_i = \overline{u_i''} = -\overline{\rho' u_i'}/\bar{\rho} = -\overline{\rho u_i'}/\bar{\rho}, \tag{5.59}$$

and is one of the explicit compressibility terms.

We follow Huang *et al.* [44] and split the turbulent dissipation rate into a total number of five terms. From the definition of τ_{ij}' in eq. (5.55) we first get:

$$\bar{\rho}\epsilon = \overline{\tau_{ij}' s_{ij}'} = \bar{\rho}(\epsilon_1 + \epsilon_2 + \epsilon_3), \tag{5.60}$$

where

$$\bar{\rho}\epsilon_1 = 2\bar{\mu}\overline{s_{ij}^{D'} s_{ij}^{D'}} + \bar{\mu}_v \overline{s_{kk}' s_{jj}'} \tag{5.61}$$

$$\bar{\rho}\epsilon_2 = 2\overline{\mu' s_{ij}^{D'} s_{ij}^{D'}} + \overline{\mu_v' s_{kk}' s_{jj}'} \tag{5.62}$$

$$\bar{\rho}\epsilon_3 = 2\overline{\mu' s_{ij}^{D'}} \cdot \overline{s_{ij}^{D}} + \overline{\mu_v' s_{kk}'} \cdot \overline{s_{jj}}. \tag{5.63}$$

The quantity $\bar{\rho}\epsilon_1$ is the most important one among these and can be expressed as the sum of a solenoidal (ϵ_s), a dilatational (ϵ_d) and an inhomogeneous term (ϵ_I):

$$\bar{\rho}\epsilon_1 = \bar{\rho}(\epsilon_s + \epsilon_d + \epsilon_I) \tag{5.64}$$

with

$$\bar{\rho}\epsilon_s = \bar{\mu}\overline{\omega_i'\omega_i'}, \tag{5.65}$$

$$\bar{\rho}\epsilon_d = \left(\frac{4}{3}\bar{\mu} + \overline{\mu_v}\right)\overline{s_{kk}'s_{jj}'}, \tag{5.66}$$

$$\bar{\rho}\epsilon_I = 2\bar{\mu}\left(\frac{\partial^2}{\partial x_i \partial x_j}\overline{u_i'u_j'} - 2\frac{\partial}{\partial x_i}\overline{u_i's_{jj}'}\right). \tag{5.67}$$

$\bar{\rho}\epsilon_s$ has the same form as in incompressible turbulence and is usually obtained from a transport equation. The compressible or dilatational dissipation rate, $\bar{\rho}\epsilon_d$, is an explicit compressibility term for which several algebraic models have been proposed in the past. The inhomogeneous term vanishes for homogeneous turbulence. The turbulent dissipation rate $\bar{\rho}\epsilon$ not only appears in the internal energy balance equation, but also in the turbulence kinetic energy equation which will be derived next.

5.2.3. TURBULENT STRESS TRANSPORT EQUATIONS

The steps needed in order to derive a transport equation for the turbulent (or Reynolds) stress follow from the time-derivative of

$$\overline{\rho u_i u_j} = \bar{\rho}\widetilde{u_i}\widetilde{u_j} + \overline{\rho u_i'' u_j''}, \tag{5.68}$$

viz:

$$\begin{aligned}\frac{\partial}{\partial t}\overline{\rho u_i'' u_j''} &= \overline{\frac{\partial}{\partial t}(\rho u_j)u_i + \frac{\partial}{\partial t}(\rho u_i)u_j - \frac{\partial \rho}{\partial t}u_i u_j} \\ &\quad - \left(\frac{\partial}{\partial t}(\bar{\rho}\widetilde{u_j})\widetilde{u_i} + \frac{\partial}{\partial t}(\bar{\rho}\widetilde{u_i})\widetilde{u_j} - \frac{\partial \bar{\rho}}{\partial t}\widetilde{u_i}\widetilde{u_j}\right).\end{aligned} \tag{5.69}$$

Substituting from equations (5.1), (5.2), (5.46) and (5.47) and rearranging we obtain the turbulent stress transport equation:

$$\begin{aligned}&\frac{\partial}{\partial t}(\bar{\rho}\widetilde{u_i''u_j''}) + \frac{\partial}{\partial x_k}(\widetilde{u_k}\bar{\rho}\widetilde{u_i''u_j''}) = \\ &\bar{\rho}P_{ij} + \bar{\rho}\Pi_{ij}^D + \bar{\rho}\Pi_{ij}^{DL} + M_{ij} - \bar{\rho}\epsilon_{ij} - \frac{\partial}{\partial x_k}(\bar{\rho}D_{ijk}^T) + \frac{\partial}{\partial x_k}D_{ijk}^V\end{aligned} \tag{5.70}$$

in which $\bar{\rho}P_{ij}$ represents the production rate tensor, $\bar{\rho}\Pi_{ij}^D$ the deviatoric part of the pressure-strain rate tensor; $\bar{\rho}\Pi_{ij}^{DL}$ the pressure-dilatation term (which appears as a consequence of subtracting this term out of the pressure-strain term), M_{ij} the mass flux variation, $\bar{\rho}\epsilon_{ij}$ the turbulent dissipation rate tensor and $\bar{\rho}D^T_{ijk}$, D^V_{ijk}, the turbulent, respectively viscous diffusion terms. These terms are defined by:

$$P_{ij} = -\widetilde{u_i''u_k''}\frac{\partial \tilde{u}_j}{\partial x_k} - \widetilde{u_j''u_k''}\frac{\partial \tilde{u}_i}{\partial x_k}, \tag{5.71}$$

$$\bar{\rho}\Pi^D_{ij} = \overline{p'\left(\frac{\partial u_i'}{\partial x_j} + \frac{\partial u_j'}{\partial x_i}\right)} - \frac{2}{3}\overline{p'd'}\delta_{ij} \quad , \tag{5.72}$$

$$\rho\Pi^{DL}_{ij} = \frac{2}{3}\overline{p'd'}\delta_{ij}, \tag{5.73}$$

$$M_{ij} = \overline{u_i''}\left(\frac{\partial \bar{\tau}_{jk}}{\partial x_k} - \frac{\partial \bar{p}}{\partial x_j}\right) + \overline{u_j''}\left(\frac{\partial \bar{\tau}_{ik}}{\partial x_k} - \frac{\partial \bar{p}}{\partial x_i}\right), \tag{5.74}$$

$$\bar{\rho}\epsilon_{ij} = \overline{\tau'_{ik}\frac{\partial u_j'}{\partial x_k}} + \overline{\tau'_{jk}\frac{\partial u_i'}{\partial x_k}}, \tag{5.75}$$

$$\bar{\rho}D^T_{ijk} = -(\overline{\rho u_i''u_j''u_k''} + \overline{p'(u_i'\delta_{jk} + u_j'\delta_{ik})}), \tag{5.76}$$

$$D^V_{ijk} = \overline{\tau'_{ik}u_j'} + \overline{\tau'_{jk}u_i'}. \tag{5.77}$$

With the help of the fluctuating viscous stress tensor according to eq. (5.55) the dissipation rate tensor, $\bar{\rho}\epsilon_{ij}$, can be expressed in the form:

$$\begin{aligned}
\bar{\rho}\epsilon_{ij} \quad = \quad & \bar{\mu}\left(2\overline{\frac{\partial u_i'}{\partial x_k}\frac{\partial u_j'}{\partial x_k}} + \overline{\frac{\partial u_i'}{\partial x_k}\frac{\partial u_k'}{\partial x_j}} + \overline{\frac{\partial u_j'}{\partial x_k}\frac{\partial u_k'}{\partial x_i}}\right) \\
& +2\left(\bar{\mu}_v - \frac{2}{3}\bar{\mu}\right)\overline{s'_{ij}s'_{kk}} \\
& +\overline{\mu'\left(2\frac{\partial u_i'}{\partial x_k}\frac{\partial u_j'}{\partial x_k} + \frac{\partial u_i'}{\partial x_k}\frac{\partial u_k'}{\partial x_j} + \frac{\partial u_j'}{\partial x_k}\frac{\partial u_k'}{\partial x_i}\right)} \\
& +2\overline{\left(\mu_v' - \frac{2}{3}\mu'\right)s'_{ij}s'_{kk}} + 2\left(\overline{\mu'\frac{\partial u_i'}{\partial x_k}}\overline{s_{jk}} + \overline{\mu'\frac{\partial u_j'}{\partial x_k}}\overline{s_{ik}}\right) \\
& +2\overline{(\mu_v' - \frac{2}{3}\mu')s'_{ij}}\overline{s_{kk}}.
\end{aligned} \tag{5.78}$$

It can be verified that the trace of this tensor is twice the dissipation rate, $\bar{\rho}\epsilon$, defined in (5.60), *i.e.*

$$\bar{\rho}\epsilon = \frac{1}{2}\bar{\rho}\epsilon_{jj}. \tag{5.79}$$

For later reference, we split the dominant part of the dissipation rate tensor (namely the first line of eq. (5.78) which does not involve correlations with viscosity fluctuations) into solenoidal, dilatational and inhomogeneous parts, in analogy to eq. (5.64). We get:

$$\begin{aligned} \bar{\rho}\epsilon^1_{ij} &= \bar{\mu}\left(2\overline{\frac{\partial u'_i}{\partial x_k}\frac{\partial u'_j}{\partial x_k}} + \overline{\frac{\partial u'_i}{\partial x_k}\frac{\partial u'_k}{\partial x_j}} + \overline{\frac{\partial u'_j}{\partial x_k}\frac{\partial u'_k}{\partial x_i}}\right) + 2\left(\overline{\mu_v} - \frac{2}{3}\bar{\mu}\right)\overline{s'_{ij}s'_{kk}} \\ &= \bar{\rho}(\epsilon^s_{ij} + \epsilon^d_{ij} + \epsilon^I_{ij}) \end{aligned} \tag{5.80}$$

where

$$\bar{\rho}\epsilon^s_{ij} = 2\bar{\mu}(2\overline{r'_{ik}r'_{jk}} + \overline{s'_{ik}r'_{jk}} + \overline{s'_{jk}r'_{ik}}), \tag{5.81}$$

$$\bar{\rho}\epsilon^d_{ij} = 2\left(\overline{\mu_v} + \frac{4}{3}\bar{\mu}\right)\overline{s'_{ij}s'_{kk}}, \tag{5.82}$$

$$\begin{aligned} \bar{\rho}\epsilon^I_{ij} &= 2\bar{\mu}\left(\frac{\partial^2}{\partial x_j \partial x_k}\overline{u'_i u'_k} + \frac{\partial^2}{\partial x_i \partial x_k}\overline{u'_j u'_k} - \frac{\partial}{\partial x_i}\overline{u'_j s'_{kk}}\right. \\ &\quad \left. - \frac{\partial}{\partial x_j}\overline{u'_i s'_{kk}} - 2\frac{\partial}{\partial x_k}\overline{u'_k s'_{ij}}\right). \end{aligned} \tag{5.83}$$

Note that contracting the solenoidal part leads to

$$\bar{\rho}\epsilon^s_{jj} = 4\bar{\mu}\overline{r'_{ij}r'_{ij}} = 2\bar{\mu}\overline{\omega'_i\omega'_i} \tag{5.84}$$

which is twice $\bar{\rho}\epsilon_s$.

The balance equation for the turbulent kinetic energy, $\bar{\rho}K = \bar{\rho}\,\widetilde{u''_i u''_i}/2$, is the trace of eq. (5.70) multiplied by $\frac{1}{2}$:

$$\begin{aligned} \frac{\partial}{\partial t}(\bar{\rho}K) &+ \frac{\partial}{\partial x_j}(\widetilde{u_j}\bar{\rho}K) = -\bar{\rho}\widetilde{u''_i u''_j}\frac{\partial \widetilde{u_i}}{\partial x_j} + \overline{p'd'} \\ &+ \overline{u''_i}\left(\frac{\partial \overline{\tau_{ij}}}{\partial x_j} - \frac{\partial \bar{p}}{\partial x_i}\right) - \overline{\tau'_{ij}s'_{ij}} \\ &- \frac{\partial}{\partial x_j}\left(\bar{\rho}\,\frac{\widetilde{(u''_i u''_i u''_j)}}{2} + \overline{p'u'_j} + \overline{\tau'_{ij}u'_i}\right). \end{aligned} \tag{5.85}$$

Comparison of the K-equation with the balance equation (5.57) for the mean internal energy, $\bar{\rho}\tilde{e}$, shows that K and $\tilde{e}$ exchange energy via the following terms:

– pressure dilatation, $\overline{p'd'}$
– turbulent dissipation rate, $\bar{\rho}\epsilon = \overline{\tau'_{ik}s'_{ik}}$
– mass flux variation, $\overline{u''_i}\left(\frac{\partial \overline{\tau_{ik}}}{\partial x_k} - \frac{\partial \bar{p}}{\partial x_i}\right)$.

In eq. (5.57) the mass flux variation is contained implicitely and appears, if the Favre-averaged velocity is used to express the dissipation rate by the mean velocity field and the work done by compression or expansion, viz:

$$
\begin{aligned}
&-\bar{p}\bar{d} + \overline{\tau_{ij}}\,\overline{s_{ij}} = \\
&-\bar{p}\left(\frac{\partial \widetilde{u_j}}{\partial x_j} + \frac{\partial \overline{u''_j}}{\partial x_j}\right) + \overline{\tau_{ij}}\left(\frac{\partial \widetilde{u_i}}{\partial x_j} + \frac{\partial \overline{u''_i}}{\partial x_j}\right) = \\
&-\bar{p}\frac{\partial \widetilde{u_j}}{\partial x_j} + \overline{\tau_{ij}}\frac{\partial \widetilde{u_i}}{\partial x_j} + \frac{\partial}{\partial x_j}\left(-\bar{p}\overline{u''_j} + \overline{\tau_{ij}}\,\overline{u''_i}\right) + \overline{u''_i}\left(\frac{\partial \bar{p}}{\partial x_i} - \frac{\partial \overline{\tau_{ij}}}{\partial x_j}\right)
\end{aligned}
\tag{5.86}
$$

On the other hand, turbulent kinetic energy is extracted from the mean motion via the production term $-\bar{\rho}\widetilde{u''_i u''_k}\partial \widetilde{u_i}/\partial x_k$. The whole energy exchange in compressible turbulent flows is illustrated in an instructive diagram by Lele [54].

The viscous diffusion term

$$
D^V_{iij}/2 = \overline{\tau'_{ij}u'_i} \tag{5.87}
$$

can be simplified by neglecting fluctuations of viscosity (see *e.g.* Gatski [33]). The complete and simplified expressions are:

$$
\begin{aligned}
\overline{\tau'_{ij}u'_i} &= 2\bar{\mu}\overline{s'_{ij}u'_i} + \left(\overline{\mu_v} - \frac{2}{3}\bar{\mu}\right)\overline{s'_{ii}u'_j} + 2\overline{\mu' s^{D'}_{ij} u'_i} + \overline{\mu'_v s'_{ii} u'_j} \\
&\quad + 2\overline{\mu' u'_i}\cdot \overline{s^D_{ij}} + \overline{\mu'_v u'_j}\cdot \overline{s_{ii}} \\
&\approx \bar{\mu}\left(\frac{1}{2}\frac{\partial}{\partial x_j}\overline{u'_i u'_i} + \frac{\partial}{\partial x_i}\overline{u'_i u'_j} - \frac{5}{3}\overline{s'_{ii}u'_j}\right) + \overline{\mu_v}\,\overline{s'_{ii}u'_j}.
\end{aligned}
\tag{5.88}
$$

Between correlations involving Favre fluctuations and those involving Reynolds fluctuations there are the following relations:

$$
\widetilde{u''_i u''_j} + \overline{u''_i}\cdot\overline{u''_j} = \overline{u'_i u'_j} + \overline{\rho' u'_i u'_j}/\bar{\rho}, \tag{5.89}
$$

$$
\widetilde{s''_{ii}u''_j} - \overline{\rho' s'_{ii}}\,\overline{u''_j}/\bar{\rho} = \overline{s'_{ii}u'_j} + \overline{\rho' s'_{ii} u'_j}/\bar{\rho}. \tag{5.90}
$$

5.2.4. TRANSPORT EQUATIONS FOR THE PRESSURE VARIANCE AND THE TURBULENT HEAT FLUX

There is a third equation in which the pressure-dilatation correlation appears as an explicit compressibility term, the pressure variance equation.

We will see later (section 5.3.2) that a simplified version of this equation forms the basis for Zeman's pressure-dilatation model. Mostly algebraic models are being used to treat the turbulent heat flux. Since it is advantageous to solve a heat flux transport equation, it will be derived in this subsection.

Pressure variance transport equation

This equation is obtained via the following procedure:

$$\frac{\partial}{\partial t}\overline{p'^2} = 2\overline{p'\frac{\partial p}{\partial t}}. \tag{5.91}$$

We keep in mind that eq. (5.24) for the pressure involves the assumption of perfect gases. Then, multiplying eq. (5.24) by $2p'$, using Favre as well as Reynolds splitting where desirable and averaging, we get:

$$\begin{aligned}\frac{\partial}{\partial t}\overline{p'^2} \quad + \quad & \widetilde{u_j}\frac{\partial}{\partial x_j}\overline{p'^2} = -2\overline{p'u_j''}\frac{\partial \bar{p}}{\partial x_j} - 2\gamma\overline{p'^2}\frac{\partial \widetilde{u_j}}{\partial x_j} \\ & -2\gamma\bar{p}\,\overline{p'\frac{\partial u_j''}{\partial x_j}} - (2\gamma-1)\overline{p'^2\frac{\partial u_j''}{\partial x_j}} - \frac{\partial}{\partial x_j}\overline{u_j''p'^2} \\ & +2(\gamma-1)\left(\overline{p'\tau_{ij}'}\frac{\partial \widetilde{u_i}}{\partial x_j} + \overline{p'\frac{\partial u_i''}{\partial x_j}}\overline{\tau_{ij}} + \overline{p'\frac{\partial u_i''}{\partial x_j}\tau_{ij}'} - \overline{p'\frac{\partial q_j'}{\partial x_j}}\right)\end{aligned} \tag{5.92}$$

The equation is discussed by Sarkar [72] and Lele [54]. Pressure fluctuations are produced when the mean flow has strong pressure gradients or thin compression zones like shocks. In the special case of a homogeneous turbulence field, the third term on the rhs (the pressure-dilatation term) is the dominant term. DNS data show that the pressure dilatation term is weakly positive (and oscillatory) for isotropic turbulence and predominantly negative (and again oscillatory) for homogeneous shear turbulence (Sarkar [72]). An explanation for the different signs is given by Sarkar *et al.* [74]. These few comments already show the limitations of a model which is based on a direct relation between the time evolution of $\overline{p'^2}$ and the pressure dilatation, since algebraic models for $\partial\overline{p'^2}/\partial t$ can hardly predict the sign change.

Turbulent heat flux transport equation

We present equations for both $\bar{\rho}\widetilde{e''u_i''}$ and $\bar{\rho}\widetilde{h''u_i''}$ and note that these equations can be converted one into the other for perfect gases. The relation guiding the derivation of a transport equation for $\bar{\rho}\widetilde{e''u_i''}$ is:

$$\frac{\partial}{\partial t}\overline{\rho e'' u_i''} = \overline{u_i \frac{\partial}{\partial t}(\rho e) + e\frac{\partial}{\partial t}(\rho u_i) - e u_i \frac{\partial \rho}{\partial t}} - \left(\widetilde{u_i}\frac{\partial}{\partial t}(\bar{\rho}\tilde{e}) + \tilde{e}\frac{\partial}{\partial t}(\bar{\rho}\widetilde{u_i}) - \tilde{e}\widetilde{u_i}\frac{\partial \bar{\rho}}{\partial t}\right), \tag{5.93}$$

A similar relation holds for $\partial\overline{\rho h'' u_i''}/\partial t$. After combining the corresponding equations we obtain the transport equation for $\bar{\rho}\widetilde{e'' u_i''}$ in the form:

$$\begin{aligned} \frac{\partial}{\partial t}\overline{\rho e'' u_i''} &+ \widetilde{u_j}\frac{\partial}{\partial x_j}\overline{\rho e'' u_i''} = -\overline{\rho u_i'' u_j''}\frac{\partial \tilde{e}}{\partial x_j} \\ &- \overline{\rho e'' u_j''}\frac{\partial \widetilde{u_i}}{\partial x_j} - (\overline{\rho e'' u_i''} + \overline{p u_i''})\frac{\partial \widetilde{u_j}}{\partial x_j} + \overline{u_i'' \tau_{jk}}\frac{\partial \widetilde{u_j}}{\partial x_k} \\ &- \frac{\partial}{\partial x_j}\overline{\rho e'' u_i'' u_j''} - \overline{p u_i''\frac{\partial u_j''}{\partial x_j}} + \overline{u_i'' \tau_{jk}\frac{\partial u_j''}{\partial x_k}} \\ &- \overline{e''\frac{\partial p}{\partial x_i}} + \overline{e''\frac{\partial \tau_{ij}}{\partial x_j}} - \overline{u_i''\frac{\partial q_j}{\partial x_j}}. \end{aligned} \tag{5.94}$$

In order to avoid the appearance of too many terms, the instantaneous pressure and the viscous stress tensor have been kept in the correlations. A turbulent heat flux is generated due to mean temperature gradients (first term on the rhs), due to mean shear and mean dilatation. In compressible homogeneous turbulence there is no turbulent heat flux because there is no mean temperature gradient (Blaisdell *et al.* [11]).

The transport equation for $\overline{\rho h'' u_i''}$ reads:

$$\begin{aligned} \frac{\partial}{\partial t}\overline{\rho h'' u_i''} + \widetilde{u_j}\frac{\partial}{\partial x_j}\overline{\rho h'' u_i''} &= -\overline{\rho u_i'' u_j''}\frac{\partial \tilde{h}}{\partial x_j} \\ -\overline{\rho h'' u_j''}\frac{\partial \widetilde{u_i}}{\partial x_j} - \overline{\rho h'' u_i''}\frac{\partial \widetilde{u_j}}{\partial x_j} &+ \overline{u_i'' \tau_{jk}}\frac{\partial \widetilde{u_j}}{\partial x_k} \\ -\frac{\partial}{\partial x_j}\overline{\rho h'' u_i'' u_j''} + \widetilde{u_j}\overline{u_i''\frac{\partial p}{\partial x_j}} &+ \overline{u_i'' u_j''\frac{\partial p}{\partial x_j}} + \overline{u_i''\frac{\partial p}{\partial t}} \\ +\overline{u_i'' \tau_{jk}\frac{\partial u_j''}{\partial x_k}} - \overline{h''\frac{\partial p}{\partial x_i}} &+ \overline{h''\frac{\partial \tau_{ij}}{\partial x_j}} - \overline{u_i''\frac{\partial q_j}{\partial x_j}}. \end{aligned} \tag{5.95}$$

Similar production terms appear as in eq. (5.94). The turbulent heat fluxes and the pressure velocity correlation are related by:

$$\overline{\rho h'' u_i''} = \overline{\rho e'' u_i''} + \overline{p u_i''} \tag{5.96}$$

which becomes for perfect gases:

$$\overline{\rho h'' u_i''} = \frac{\gamma}{\gamma - 1} \overline{p u_i''}. \tag{5.97}$$

In this case the turbulent heat flux equations (5.94) and (5.95) can also be considered as transport equations for the pressure velocity correlation. Since $\overline{\rho e'' u_i''}$ vanishes in homogeneous turbulence, $\overline{\rho h'' u_i''}$ and $\overline{p u_i''}$ are zero as well. It is certainly wrong to conclude from this fact that in general inhomogeneous flows these terms are of minor importance. In strongly compressible flows with shocks, *e.g.* , these terms are important and have to be modelled properly (Gatski [33]).

5.2.5. TRANSPORT EQUATIONS FOR HOMOGENEOUS SHEAR FLOW

Homogeneous shear flow has been frequently used in the past to test turbulence models and to tune model constants. Direct numerical simulation has provided insight into the physics of compressible turbulent shear flow and has helped to develop pressure-dilatation and compressible dissipation rate models. It is therefore useful to discuss the characteristics of this flow. We first derive conditions under which homogeneous turbulent flows can exist.

Necessary conditions for homogeneity

Homogeneous turbulence has statistics of fluctuating quantities which are independent of space. This definition allows for non-zero mean velocity gradients. Blaisdell *et al.* [11] have inspected transport equations for $\overline{u_i'' u_j''}$ (a quantity which is unusual in turbulence modelling), for $\overline{\rho'^2}$ and $\overline{p'^2}$ in order to find necessary conditions under which initially homogeneous turbulence remains homogeneous.

Starting from transport equations for u_i'' and ρ', one concentrates on terms involving velocities only. Then, from the identities

$$\frac{\partial u_i''}{\partial t} = \frac{1}{\rho}\left(\frac{\partial}{\partial t}(\rho u_i) - u_i \frac{\partial \rho}{\partial t}\right) - \frac{1}{\bar{\rho}}\left(\frac{\partial}{\partial t}(\bar{\rho}\tilde{u}_i) - \tilde{u}_i \frac{\partial \bar{\rho}}{\partial t}\right), \tag{5.98}$$

$$\frac{\partial \rho'}{\partial t} = \frac{\partial \rho}{\partial t} - \frac{\partial \bar{\rho}}{\partial t}, \tag{5.99}$$

one finds:

$$\frac{\partial u_i''}{\partial t} = -\tilde{u}_j \frac{\partial u_i''}{\partial x_j} - u_j'' \frac{\partial \tilde{u}_i}{\partial x_j} - u_j'' \frac{\partial u_i''}{\partial x_j} + \frac{1}{\bar{\rho}} \frac{\partial}{\partial x_j} \overline{\rho u_i'' u_j''} + \ldots, \tag{5.100}$$

$$\begin{aligned}\frac{\partial \rho'}{\partial t} &= -\rho' \frac{\partial \widetilde{u_j}}{\partial x_j} - \bar{\rho} \frac{\partial u_j''}{\partial x_j} - \rho' \frac{\partial u_j''}{\partial x_j} \\ &\quad - u_j'' \frac{\partial \bar{\rho}}{\partial x_j} - \widetilde{u_j} \frac{\partial \rho'}{\partial x_j} - u_j'' \frac{\partial \rho'}{\partial x_j}. \end{aligned} \tag{5.101}$$

Averaging the combined transport of $u_j'' \partial u_i'' / \partial t$ and $u_i'' \partial u_j'' / \partial t$ as well as the transport of $2\rho' \partial \rho' / \partial t$ and applying homogeneity, gives rise to:

$$\frac{\partial}{\partial t} \overline{u_i'' u_j''} = -\overline{u_i'' u_k''} \frac{\partial \widetilde{u_j}}{\partial x_k} - \overline{u_j'' u_k''} \frac{\partial \widetilde{u_i}}{\partial x_k} + \overline{u_i'' u_j'' \frac{\partial u_k''}{\partial x_k}} + \ldots \tag{5.102}$$

$$\frac{\partial}{\partial t} \overline{\rho'^2} = -2\overline{\rho'^2} \frac{\partial \widetilde{u_j}}{\partial x_j} - 2\bar{\rho} \overline{\rho' \frac{\partial u_j''}{\partial x_j}} - 2\overline{\rho' u_j''} \frac{\partial \bar{\rho}}{\partial x_j} - \overline{\rho'^2 \frac{\partial u_j''}{\partial x_j}}. \tag{5.103}$$

Now, Blaisdell *et al.* [11] argue that due to homogeneity, correlation functions can only depend on time. Since the left hand side of equations (5.102), (5.103) is solely a function of time, the same must be valid for the right hand side, especially for each of the rhs terms. Thus from (5.102):

$$\frac{\partial \widetilde{u_i}}{\partial x_j} = A_{ij}(t) \tag{5.104}$$

which corresponds to a linear mean velocity field

$$\widetilde{u_i}(\vec{x}, t) = A_{ij}(t) x_j. \tag{5.105}$$

The time-dependent constant of integration is not considered because it can be treated as a time-dependent body force, which is not of interest to us. Similarly, eq. (5.103) provides the condition

$$\bar{\rho} = \bar{\rho}(t), \tag{5.106}$$

since $\overline{\rho' \partial u_j'' / \partial x_j}$ is generally non-zero. Restrictions on $\bar{p}$ can be derived from eq. (5.92). From the third term on the rhs one gets (since $\overline{p'd'}$ is non-zero):

$$\bar{p} = \bar{p}(t). \tag{5.107}$$

From the averaged equation of state for perfect gases:

$$\bar{p} = \bar{\rho} R \tilde{T}\,. \tag{5.108}$$

One finally concludes that for spatially uniform mean density and pressure the temperature must also be uniform is space, *i.e.* $\tilde{T} = \tilde{T}(t)$.

Mean flow characteristics of homogeneous shear turbulence

Taking equations (5.104), (5.106) and (5.107) into account the mean mass and momentum transport equations (5.46), (5.47) reduce to:

$$\frac{d\bar{\rho}}{dt} + \bar{\rho}A_{jj}(t) = 0, \tag{5.109}$$

$$\frac{\partial \widetilde{u_i}}{\partial t} + \widetilde{u_k}\frac{\partial \widetilde{u_i}}{\partial x_k} = 0. \tag{5.110}$$

Differentiation of (5.110) with respect to x_j provides a nonlinear coupled set of ordinary differential equations to be satisfied by the mean velocity gradient tensor A_{ij}:

$$\frac{dA_{ij}}{dt} + A_{ik}A_{kj} = 0. \tag{5.111}$$

These equations which apply to any homogeneous turbulence field further simplify in the case of homogeneous shear flow with constant shear rate in x_2-direction. We assume a mean momentum transport in x_1-direction:

$$\widetilde{u_i}(\vec{x},t) = (\widetilde{u_1}(x_2,t),0,0) \tag{5.112}$$

and obtain from (5.109) and (5.111):

$$\bar{\rho} = const, \tag{5.113}$$

$$A_{12}(t) = \frac{d\widetilde{u_1}}{dx_2} = S = const. \tag{5.114}$$

The mean internal energy balance (5.57) reduces to

$$\bar{\rho}\frac{d\tilde{e}}{dt} = -\overline{p'd'} + \bar{\mu}S^2 + \bar{\rho}\epsilon \tag{5.115}$$

if fluctuations of viscosity are neglected in the mean flow dissipation rate (2nd term on the rhs). The turbulence kinetic energy (TKE) balance takes the simple form:

$$\bar{\rho}\frac{dK}{dt} = -\bar{\rho}\widetilde{u_1''u_2''}S + \overline{p'd'} - \bar{\rho}\epsilon. \tag{5.116}$$

The sum of $\tilde{e}$ and K increases in time due to TKE-production and mean flow dissipation. Since $\overline{p'd'}$ is predominantly negative, $\tilde{e}$ grows in time. DNS data indicate an exponential growth of K for long times, *i.e.* $dK/dt \sim K$ and, since the production and $\overline{p'd'}$ terms do not balance, ϵ grows in time as well. The mean shear rate introduces directionality into the turbulence structure. Hence, the Reynolds stress tensor is homogeneous, but anisotropic:

$$\widetilde{u_i''u_j''} = \begin{pmatrix} \widetilde{u_1''^2} & \widetilde{u_1''u_2''} & 0 \\ \widetilde{u_1''u_2''} & \widetilde{u_2''^2} & 0 \\ 0 & 0 & \widetilde{u_3''^2} \end{pmatrix} , \qquad \widetilde{u_1''^2} \neq \widetilde{u_2''^2} \neq \widetilde{u_3''^2} \,. \tag{5.117}$$

Blaisdell *et al.* [11] provide a conclusive argument that the mass flux $\overline{\rho u_i'}$ vanishes in homogeneous turbulence. This means that Reynolds and Favre averaged velocities coincide. In the TKE dissipation rate, $\bar{\rho}\epsilon$, the inhomogeneous term vanishes. If we further exclude viscosity fluctuations, $\bar{\rho}\epsilon$ takes the simple form:

$$\bar{\rho}\epsilon = \bar{\mu}\overline{\omega_i'\omega_i'} + \left(\frac{4}{3}\bar{\mu} + \overline{\mu_v}\right)\overline{d'^2}. \tag{5.118}$$

Turbulent stress transport in homogeneous shear flow

Although the homogeneity condition removes several unknown terms from the full transport equations, compressible homogeneous shear turbulence is still today a flow case which is hard to predict accurately. One reason is certainly the difficulty associated with pressure strain modelling. From equations (5.70) - (5.83) we obtain the transport equations for the four Reynolds stresses:

$$\bar{\rho}\frac{d}{dt}\left(\frac{1}{2}\widetilde{u_1''^2}\right) = -\bar{\rho}\widetilde{u_1''u_2''}S + \overline{p'\frac{\partial u_1'}{\partial x_1}} - \bar{\rho}\left(\epsilon_{11}^S + \epsilon_{11}^d\right), \tag{5.119}$$

$$\bar{\rho}\frac{d}{dt}\left(\frac{1}{2}\widetilde{u_2''^2}\right) = \overline{p'\frac{\partial u_2'}{\partial x_2}} - \bar{\rho}\left(\epsilon_{22}^S + \epsilon_{22}^d\right), \tag{5.120}$$

$$\bar{\rho}\frac{d}{dt}\left(\frac{1}{2}\widetilde{u_3''^2}\right) = \overline{p'\frac{\partial u_3'}{\partial x_3}} - \bar{\rho}\left(\epsilon_{33}^S + \epsilon_{33}^d\right), \tag{5.121}$$

$$\bar{\rho}\frac{d}{dt}\left(\widetilde{u_1''u_2''}\right) = -\bar{\rho}\widetilde{u_2''^2}S + \overline{p'\left(\frac{\partial u_1'}{\partial x_2} + \frac{\partial u_2'}{\partial x_1}\right)} - \bar{\rho}\left(\epsilon_{12}^S + \epsilon_{12}^d\right). \tag{5.122}$$

Only the streamwise component and the shear stress obtain energy directly via production. The remaining components are fed by redistribution of energy. As shown by Blaisdell *et al.* [11] the ordering of the normal stresses in compressible homogeneous shear turbulence is the same as in the incompressible case, *i.e.* the streamwise component is the most energetic, followed by the spanwise component, with the component in the shear direction being the least energetic. So far, the only explicit differences to the incompressible case are the appearance of dilatational dissipation rates $\bar{\rho}\epsilon_{ij}^d$

and the fact that the pressure strain terms sum up to a non-zero pressure-dilatation term.

Compressibility parameters

The TKE equation (5.116) can be used to derive two independent compressibility parameters which are needed to quantify effects of compressibility on the turbulence structure. To this end, we nondimensionalize the K-equation choosing $1/S$ as the time scale, two different length scales l_0, λ_0 (an integral scale l_0 and a Taylor microscale λ_0), u_0 as a fluctuating velocity scale, $\bar{\rho} u_0^2$ as a characteristic pressure fluctuation and μ_0 as a reference viscosity. The index '0' refers to constant initial values of these quantities. Then the nondimensional TKE equation reads:

$$M_{go}\frac{dK^*}{dt^*} = -M_{go}\overline{\rho^* u_1''^* u_2''^*} - M_{to}\overline{p'^* d'^*} - \frac{M_{to}}{Re_{to}}\left(\frac{l_0}{\lambda_0}\right)^2 \epsilon^*, \tag{5.123}$$

with the:

– gradient Mach number:

$$M_{go} = Sl_0/\overline{c_0} \tag{5.124}$$

– turbulent Mach number:

$$M_{to} = u_0/\overline{c_0} \tag{5.125}$$

– turbulent Reynolds number:

$$Re_{to} = \bar{\rho} l_0 u_0/\mu_0 \tag{5.126}$$

as free parameters. The star $*$ indicates non-dimensional quantities. (It has this meaning only in eq. (5.123)).

The turbulent Reynolds number Re_{to} drops out of the list of nondimensional parameters, if a simple estimate of l_0/λ_0, as provided by Tennekes and Lumley [89], is assumed to hold, namely

$$l_0/\lambda_0 = O\left(Re_{to}^{1/2}\right). \tag{5.127}$$

This is also found if u_0^3/l_0 is used to scale ϵ. Sarkar [73] has found the gradient Mach number, M_{go}, to be the key parameter in explaining the stabilizing effect of compressibility on the TKE growth rate. We will come back to this point later. Earlier Blaisdell *et al.* [11] had already used a gradient Mach number which they called shear rate Mach number and which differs from Sarkar's definition only in the specification of the integral length scale. Durbin and Zeman [23] had introduced a distortion Mach number in their RDT analysis that can be interpreted as the mean Mach number change across an 'eddy'. The expression 'distortion Mach number' was then adopted by Jacquin *et al.* [45], Cambon *et al.* [17] and Simone *et al.* [80] to parameterize rapidly sheared and strained compressible homogeneous turbulence. It is important to note that M_{go} and M_{to} are two independent parameters since their ratio Sl_0/u_0 can be changed via l_0, u_0.

There is some ambiguity in the definition of l_0. While Sarkar [73] chooses

$$l_0 = \int_{-\infty}^{\infty} \left.(\widetilde{u_1''(x_2)u_1''(x_2+r_2)})\right|_{t=0} dr_2 \;/\; \left.\widetilde{u_1''^2(x_2)}\right|_{t=0}, \qquad (5.128)$$

Blaisdell *et al.* [11] define

$$l_0 = \int_{-\infty}^{\infty} \left.(\widetilde{u_1''(x_2)u_2''(x_2+r_2)})\right|_{t=0} dr_2 \;/\; \left.\widetilde{u_1''u_2''(x_2)}\right|_{t=0}. \qquad (5.129)$$

In view of the fact that the 2-2 components of the Reynolds stress tensor, of the pressure strain tensor and of the compressibe dissipation rate tensor are the most affected by compressibility (Blaisdell *et al.* [11]), a suitable alternative to the definitions (5.128), (5.129) would be:

$$l_0 = \int_{-\infty}^{\infty} \left.(\widetilde{u_2''(x_2)u_2''(x_2+r_2)})\right|_{t=0} dr_2 \;/\; \left.\widetilde{u_2''^2(x_2)}\right|_{t=0} \qquad (5.130)$$

A supporting argument for this definition follows from the linearized inviscid equations of motion. For homogeneous shear turbulence there exists a direct link between u_2' and d' in Fourier space (Friedrich and Bertolotti [30]). On the other hand, the linear inviscid momentum balance in x_2-direction,

$$\left(\frac{\partial}{\partial t} + \bar{u}_1 \frac{\partial}{\partial x_1}\right) u_2' = -\frac{1}{\bar{\rho}}\frac{\partial p'}{\partial x_2}, \qquad (5.131)$$

reflects the coupling between u_2- and p-fluctuations. Pressure fluctuations are found to be strongly affected by compressibility effects in plane and annular mixing layers (Vreman *et al.* [95], Freund *et al.* [29]). The suppression of the pressure strain rate correlation in the stress equation for the shear direction (x_2) is explained primarily by the reduction of pressure fluctuations due to compressibility. In fact, Freund *et al.* [29] demonstrate a significant decrease of the correlation length of velocity fluctuations in the shear direction, which supports a definition of l_0 according to (5.130).

5.3. Compressibility effects due to turbulent fluctuations and modelling of explicit compressibility terms

Most of our present knowledge about compressibility effects due to turbulent fluctuations (intrinsic effects of compressibility) stems from direct numerical simulations and rapid distortion analysis. In this chapter we will therefore first inspect the linear equations of motion in order to gain some insight into the physics of compressible turbulence, then summarize the most important recent findings of DNS before we discuss turbulence models derived from DNS data.

5.3.1. HOMOGENEOUS ISOTROPIC TURBULENCE

The statistical equations for decaying isotropic turbulence are obtained from those for homogeneous shear turbulence setting the mean shear rate equal to zero. Without loss of generality the mean velocity can be assumed zero. The mean density is constant, thus:

$$\bar{\rho} = const.\,, \quad \tilde{u}_i = 0. \tag{5.132}$$

The turbulent stress tensor is isotropic, with

$$\widetilde{u_1''^2} = \widetilde{u_2''^2} = \widetilde{u_3''^2} = 2K/3. \tag{5.133}$$

The time evolutions of the mean internal energy, $\bar{\rho}\tilde{e}$, and of the turbulent kinetic energy, $\bar{\rho}K$, are determined by $\overline{p'd'}$ and $\bar{\rho}\epsilon$ alone, consequently:

$$\tilde{e} + K = const. \tag{5.134}$$

While K decays, $\tilde{e}$ increases in time, along with $\bar{p}$ and $\tilde{T}$. The decay of K entrains that of $\bar{\rho}\epsilon$ and $\overline{p'd'}$ after initial transients. Normalized density, pressure and temperature variances decay as well. Integral and Taylor microscales grow after some initial decay.

Linear analysis of turbulent fluctuations

We assume fluctuations of density, pressure and temperature to be small with respect to their mean values, and velocity fluctuations to be small compared to the mean speed of sound, $\bar{c}_0$. The latter is equivalent to assuming low turbulent Mach number, M_{to}. There are two characteristic velocities in compressible isotropic turbulence, namely $u_0 = (2K_0/3)^{1/2}$ and $\overline{c_0}$. Together with an integral length scale l_0 they define two timescales:

- the eddy-turnover time, $\tau_t = l_0/u_0$
- and the acoustic time, $\tau_a = l_0/\overline{c_0}$.

The ratio of these two time scales is the turbulent Mach number:

$$M_{to} = \tau_a/\tau_t. \tag{5.135}$$

Low M_{to} means that the two time scales are disparate. In this situation we may linearize the equations of motion with respect to fluctuations. From (5.1), (5.2), (5.24) we then obtain for decaying isotropic turbulence:

$$\frac{\partial \rho'}{\partial t} + \bar{\rho} d' = 0, \tag{5.136}$$

$$\frac{\partial u_i'}{\partial t} = -\frac{1}{\bar{\rho}}\frac{\partial p'}{\partial x_i} + \bar{\nu}\frac{\partial^2 u_i'}{\partial x_j \partial x_j} + (\bar{\nu}/3 + \bar{\mu}_v/\bar{\rho})\frac{\partial d'}{\partial x_i}, \tag{5.137}$$

$$\frac{\partial p'}{\partial t} = -\gamma \bar{p} d' + (\gamma - 1)\bar{k}\frac{\partial^2 T'}{\partial x_j \partial x_j}. \tag{5.138}$$

An alternative energy equation follows from the entropy balance, viz:

$$\frac{\partial s'}{\partial t} = \frac{\bar{k}}{\bar{\rho}\bar{T}}\frac{\partial^2 T'}{\partial x_j \partial x_j}. \tag{5.139}$$

From this coupled set of equations we now derive wave equations for p', ρ', d' and diffusion equations for ω_i' and s'. Thereby we use the linearized perfect gas relation

$$p' = R(\rho' \bar{T} + T' \bar{\rho}) \tag{5.140}$$

and the linearized state relation $p' = p'(\rho', s')$ to express time-derivatives

$$\frac{\partial p'}{\partial t} = \gamma R \bar{T}\frac{\partial \rho'}{\partial t} + \frac{\bar{p}}{c_v}\frac{\partial s'}{\partial t} \tag{5.141}$$

or the Laplacian:

$$\frac{\partial^2 p'}{\partial x_j \partial x_j} = \gamma R \bar{T}\frac{\partial^2 \rho'}{\partial x_j \partial x_j} + \frac{\bar{p}}{c_v}\frac{\partial^2 s'}{\partial x_j \partial x_j}. \tag{5.142}$$

Equation (5.141) is valid on an acoustic time scale, τ_a, in which the time variations of $\bar{T}$ and $\bar{p}$ due to viscous effects can be neglected. (5.142) is straightforward due to spatial homogeneity of $\bar{T}, \bar{p}$.

Now, applying the Laplacian to eq. (5.136), differentiating (5.138), (5.139) with respect to time, taking the divergence of eq. (5.137), introducing (5.140) and combining these results, leads to the following wave equation for the pressure fluctuations:

$$\frac{\partial^2 p'}{\partial t^2} - \bar{c}^2 \frac{\partial^2 p'}{\partial x_j \partial x_j} = \gamma \left(\frac{4}{3}\bar{\nu} + \bar{\mu}_v/\bar{\rho} \right) \frac{\partial^2}{\partial x_j \partial x_j} \left(\frac{\partial p'}{\partial t} \right) + \left(1 - \frac{4}{3}\overline{Pr} - \frac{c_p \bar{\mu}_v}{\bar{k}} \right) \frac{\bar{p}}{\bar{c}_v} \frac{\partial^2 s'}{\partial t^2}. \quad (5.143)$$

The first term on the rhs accounts for sound absorption at high frequencies. $\overline{Pr} = \bar{\mu} c_p/\bar{k}$ is the mean Prandtl number. Equations, similar to (5.143) are obtained for ρ' and d':

$$\frac{\partial^2 \rho'}{\partial t^2} - \bar{c}^2 \frac{\partial^2 \rho'}{\partial x_j \partial x_j} = \left(\frac{4}{3}\bar{\nu} + \bar{\mu}_v/\bar{\rho} \right) \frac{\partial^2}{\partial x_j \partial x_j} \left(\frac{\partial \rho'}{\partial t} \right) + \frac{\bar{p}}{c_v} \frac{\partial^2 s'}{\partial x_j \partial x_j}, \quad (5.144)$$

$$\frac{\partial^2 d'}{\partial t^2} - \bar{c}^2 \frac{\partial^2 d'}{\partial x_j \partial x_j} = \left(\frac{4}{3}\bar{\nu} + \bar{\mu}_v/\bar{\rho} \right) \frac{\partial^2}{\partial x_j \partial x_j} \left(\frac{\partial d'}{\partial t} \right) - (\gamma - 1)\bar{T} \frac{\partial^2}{\partial x_j \partial x_j} \left(\frac{\partial s'}{\partial t} \right). \quad (5.145)$$

Taking the curl of eq. (5.137) and replacing the Laplacian of T' with the help of (5.140) and (5.142) gives rise to diffusion equations for the velocity and entropy fluctuations:

$$\frac{\partial \omega_i'}{\partial t} = \bar{\nu} \frac{\partial^2 \omega_i'}{\partial x_j \partial x_j}, \quad (5.146)$$

$$\frac{\partial s'}{\partial t} = \frac{\bar{\nu}}{\overline{Pr}} \left(\frac{\partial^2 s'}{\partial x_j \partial x_j} + \frac{R}{\bar{p}} \frac{\partial^2 p'}{\partial x_j \partial x_j} \right). \quad (5.147)$$

The conclusions we draw from these equations as long as viscosity plays a role, are:

- pressure (density and dilatation) fluctuations are coupled with entropy fluctuations. The coupling between p' and s' is via viscosity and therefore weak,
- only vorticity fluctuations develop independently,
- the wave character of p', ρ', d' and the diffusive character of ω_i', s' are obvious.

If we neglect viscosities, we note that:

- the vorticity and entropy fluctuations are frozen (this corresponds to Taylor's hypothesis used in experiments),

– the pressure and dilatation fluctuations follow a pure wave equation,
– density fluctuations are still coupled with entropy fluctuations.

Vorticity, pressure and entropy fluctuations have been called 'modes' by Kavasznay [47]) and Morkovin [63], and are since then referred to as the vorticity mode, the acoustic mode and the entropy mode of compressible turbulence. In isotropic turbulence these three modes are decoupled.

The decoupling is a result of the missing of mean shear or of gradients of mean thermodynamic variables. It has the important consequence that compressibility effects studied in DNS depend more on the initial conditions than on the turbulent Mach number (Blaisdell *et al.* [12]). Therefore DNS data of isotropic turbulence can in general not be used to validate turbulence models for the compressible dissipation rate and the pressure-dilatation correlation. We come back to the effect of initial conditions in chapter 5.3.1.

In order to demonstrate the coupling between the acoustic pressure mode and the velocity field, we apply Helmholtz's decomposition of u_i' into incompressible, $u_i^{s\prime}$, and compressible parts, $u_i^{d\prime}$:

$$u_i' = u_i^{s\prime} + u_i^{d\prime} \tag{5.148}$$

where

$$\frac{\partial u_i^{s\prime}}{\partial x_i} = 0, \quad \epsilon_{ijk}\frac{\partial u_k^{d\prime}}{\partial x_j} = 0. \tag{5.149}$$

Furthermore, we follow Simone [79] and use the Craya-Herring formalism (Craya [20], Cambon [16]) to express the Fourier transform of u_i' in an orthonormal frame of reference with bases $(e_i^{(1)}, e_i^{(2)}, e_i^{(3)})$ normal and parallel to the wavevector $\vec{k}$. Defining a pair of three-dimensional Fourier transforms by

$$\hat{A}(\vec{k},t) = \frac{1}{(2\pi)^3}\int\limits_{V(\vec{x})} A(\vec{x},t)exp(-i\vec{k}\cdot\vec{x})d\vec{x}, \tag{5.150}$$

$$A(\vec{x},t) = \int\limits_{V(\vec{k})} \hat{A}(\vec{k},t)exp(i\vec{k}\cdot\vec{x})d\vec{k}, \tag{5.151}$$

(where the caret denotes the Fourier coefficient and $i^2 = -1$), the Fourier transform of the velocity fluctuation[2] reads:

[2]For simplification we omit the dash in Fourier space

$$\hat{u}_i(\vec{k},t) = \underbrace{\hat{\varphi}^{(1)}(\vec{k},t)e_i^{(1)}(\vec{k}) + \hat{\varphi}^{(2)}(\vec{k},t)e_i^{(2)}(\vec{k})}_{\hat{u}_i^s} + \underbrace{\hat{\varphi}^{(3)}(\vec{k},t)e_i^{(3)}(\vec{k})}_{\hat{u}_i^d}. \quad (5.152)$$

The first two terms on the rhs of (5.152) represent the Fourier transform of the solenoidal velocity fluctuation $u_i^{s\prime}$; the third is the compressible (or dilatational) contribution, aligned with $\vec{k}$. $\hat{u}_i^s$ lies in the plane $(\vec{e}^{(1)}, \vec{e}^{(2)})$ perpendicular to $\vec{k}$. It is possible to fix the coordinates in that plane by introducing a polar axis $\vec{n}$ (Herring [41]) either according to symmetries of the mean flow (if any) or the statistical properties of the fluctuating field, viz:

$$\vec{e}^{(1)} = \frac{\vec{k} \times \vec{n}}{|\vec{k} \times \vec{n}|}, \quad \vec{e}^{(2)} = \vec{e}^{(3)} \times \vec{e}^{(1)}, \quad \vec{e}^{(3)} = \vec{k}/|\vec{k}|. \quad (5.153)$$

The Fourier transforms of vorticity and dilatation fluctuations have the form:

$$\hat{\omega}_i = \epsilon_{ijk} i k_j \hat{u}_k = ik\left(\hat{\varphi}^{(1)} e_i^{(2)} - \hat{\varphi}^{(2)} e_i^{(1)}\right), \quad (5.154)$$

$$\hat{d} = ik_i \hat{u}_i = ik\hat{\varphi}^{(3)}, \quad k = |\vec{k}|. \quad (5.155)$$

Only the incompressible modes $(\hat{\varphi}^{(1)}, \hat{\varphi}^{(2)})$ contribute to Kovasznay's vorticity mode, while the compressible mode $\hat{\varphi}^{(3)}$ alone determines the dilatation fluctuation. Fourier transforming the inviscid equations of motion (5.136) - (5.138), leads to:

$$\frac{\partial \hat{\rho}}{\partial t} + \bar{\rho} ik\hat{\varphi}^{(3)} - 0, \quad (5.156)$$

$$\frac{\partial \hat{u}_i}{\partial t} = -\frac{1}{\bar{\rho}} ik_i \hat{p}, \quad (5.157)$$

$$\frac{\partial \hat{p}}{\partial t} = -\gamma \bar{p} ik\hat{\varphi}^{(3)}, \quad (5.158)$$

Introducing (5.152) into (5.157) and multiplying (5.157) scalarly with $e_i^{(j)} (j = 1, 2, 3)$, one obtains a set of equations determining the behaviour of the incompressible and compressible modes $\hat{\varphi}^{(3)}, \hat{\varphi}^{(4)} = i\hat{p}/(\bar{\rho}\bar{c})$:

$$\frac{\partial \hat{\varphi}^{\alpha}}{\partial t} = 0, \quad \alpha = 1, 2, \quad (5.159)$$

$$\frac{\partial \hat{\varphi}^{(3)}}{\partial t} + \bar{c}k\hat{\varphi}^{(4)} = 0, \quad (5.160)$$

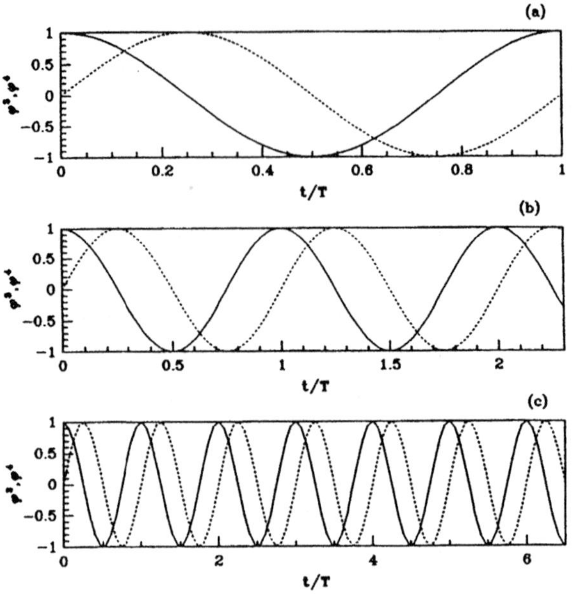

Figure 5.1. Time evolution of compressible modes in acoustic equilibrium (isotropic turbulence, increasing wavenumber from top to bottom). Taken from Simone [79] by permission.

$$\frac{\partial \hat{\varphi}^{(4)}}{\partial t} - \bar{c}k\hat{\varphi}^{(3)} = 0. \tag{5.161}$$

Cambon *et al.* [17] and Simone [79] call this the purely acoustic regime of isotropic turbulence, where energy is exchanged between the dilatational velocity and the pressure at a frequency $\bar{c}k$, while the solenoidal component $u_i^{s\prime}$ is frozen. Differentiating equations (5.160) and (5.161) with respect to time and combining them gives rise to the two wave equations in Fourier space:

$$\frac{\partial^2 \hat{\varphi}^{(3)}}{\partial t^2} + \bar{c}^2 k^2 \hat{\varphi}^{(3)} = 0, \tag{5.162}$$

$$\frac{\partial^2 \hat{\varphi}^{(4)}}{\partial t^2} + \bar{c}^2 k^2 \hat{\varphi}^{(4)} = 0, \tag{5.163}$$

which, together with eq. (5.159), demonstrate the independence of all modes and *e.g.* the fact that the pressure fluctuations have no influence on the solenoidal velocity component. The analytical solution of eqs. (5.162), (5.163) is given by Simone [79] in the form:

$$\hat{\varphi}^{(3)}(\vec{k},t) = \hat{\varphi}^{(3)}(\vec{k}_0,0)\cos\omega t - \hat{\varphi}^{(4)}(\vec{k}_0,0)\sin\omega t, \tag{5.164}$$

$$\hat{\varphi}^{(4)}(\vec{k},t) = \hat{\varphi}^{(3)}(\vec{k}_0,0)\sin\omega t + \hat{\varphi}^{(4)}(\vec{k}_0,0)\cos\omega t, \tag{5.165}$$

with $\omega = \bar{c}k = \bar{c}_0 k_0$. k_0 is the magnitude of the wavevector at time t_0. $\vec{k}_0$ plays the same role in wavenumber space as the Lagrangian coordinate in physical space. In isotropic turbulence $\vec{k}$ is independent of time, *i.e.* $\vec{k} = \vec{k}_0$. Figure 5.1 shows the time evolution of the compressible modes $\hat{\varphi}^{(3)}$, $\hat{\varphi}^{(4)}$ for initial values between 0 and 1, corresponding to acoustical equilibrium[3], *i.e.* to a situation in which the kinetic energy of the dilatational velocity mode equals the potential energy of the pressure mode:

$$\frac{1}{2}\overline{u_i^{d'} u_i^{d'}} = \frac{1}{2}\overline{p'p'}/(\bar{\rho}\bar{c})^2. \tag{5.166}$$

A measure for acoustical equilibrium or non-equilibrium had earlier been introduced by Sarkar *et al.* [74] in the form of the energy partition parameter

$$F = \gamma\bar{\rho}\bar{p}\frac{\overline{u_i^{d'} u_i^{d'}}}{\overline{p'p'}}. \tag{5.167}$$

One easily verifies that eq. (5.166) corresponds to $F = 1$. Sarkar *et al.* [74] had shown, based on DNS data of isotropic turbulence that $F = 1$ even holds for high turbulent Mach numbers ($M_{to} = 0.5$) and had used the value $F = 1$ in deriving a model for the compressible dissipation rate (see section 3.1.3).

Importance of initial conditions

Blaisdell *et al.* [11, 12] emphasize that initial conditions strongly influence the time evolution of correlation functions which makes it difficult to use simulations of decaying isotropic turbulence to evaluate compressible turbulence models. This difficulty is removed when a mean shear rate is introduced (See section 5.3.2).

The authors have performed several DNS to demonstrate the effect of initial conditions. They varied the initial rms-values of density and temperature, of turbulent Mach number and of the fraction χ of compressible kinetic energy, defined as:

$$\chi = \overline{u_i^{d'} u_i^{d'}}/(2K). \tag{5.168}$$

Table 3 contains the simulation parameters of six of the simulations. The

[3] A 'strong form' of acoustical equilibrium was used to specify initial conditions for fig. 5.1, by assuming that the energy balance occurs at each wavenumber.

TABLE 5.3. Initial parameters for DNS of isotropic turbulence. Taken from Blaisdell *et al.* [11].

Case	iga96	igb96	idc96	ie96	ifd96	ife96
$\rho_{rms,0}$	0.1	0.016	0	0.15	0.15	0
$T_{rms,0}$	0.1	0.016	0	0.15	0.15	0
χ_o	0.1	0.816	0	0.25	0.25	0
M_{to}	0.05	0.11	0.3	0.3	0.7	0.7
Re_{to}	90.6	360	160	160	90	91

number 96 appearing in the case identifications refers to the 96^3 grid used in these simulations. The initial spectra are the same in all cases. Cases iga96 and igb96 have been chosen to test how well ideas from linear acoustics could predict the evolution of turbulence. While the turbulent Mach numbers are low in both cases, the equilibrium parameters F_0 are very different. Case iga96 has an F_0 of $1/40$, corresponding to comparatively large pressure fluctuations and case igb96 is characterized by $F_0 \approx 40$ and large dilatational velocity fluctuations. Cases idc96 and ie96 have moderate turbulent Mach numbers and differ in the initial fluctuations of density, temperature and dilatational velocity. Cases ifd96 and ife96 have been selected to show the effect of higher Mach number. The turbulent Reynolds numbers $Re_{to} = 2\bar{\rho}K/(\epsilon\bar{\mu})$ do not vary much from case to case and need therefore not be discussed.

Figure 5.2 shows the nondimensional variances of pressure and density fluctuations for the cases iga96 (top) and igb96 (bottom). Simulation iga96 starts with remarkable density and pressure fluctuations. Energy is transferred from pressure to dilatational velocity fluctuations in a highly oscillatory way. The transfer process reaches an acoustical equilibrium state after one eddy-turnover time. Simulation igb96 on the other hand produces pressure and density fluctuations from low levels, in close agreement with an isentropic process. These results demonstrate the existence of linear acoustic mechanisms in compressible turbulence as described in the previous subsection.

Figure 5.3 aims at illustrating the dependence of compressibility effects on initial conditions. It shows the time evolution of the dilatational fraction χ_ϵ of the turbulent dissipation rate:

$$\chi_\epsilon = \frac{\epsilon_d}{\epsilon} = \frac{\epsilon_d/\epsilon_s}{1+\epsilon_d/\epsilon_s}, \tag{5.169}$$

for the 4 remaining flow cases of table 3. The top figure contains the

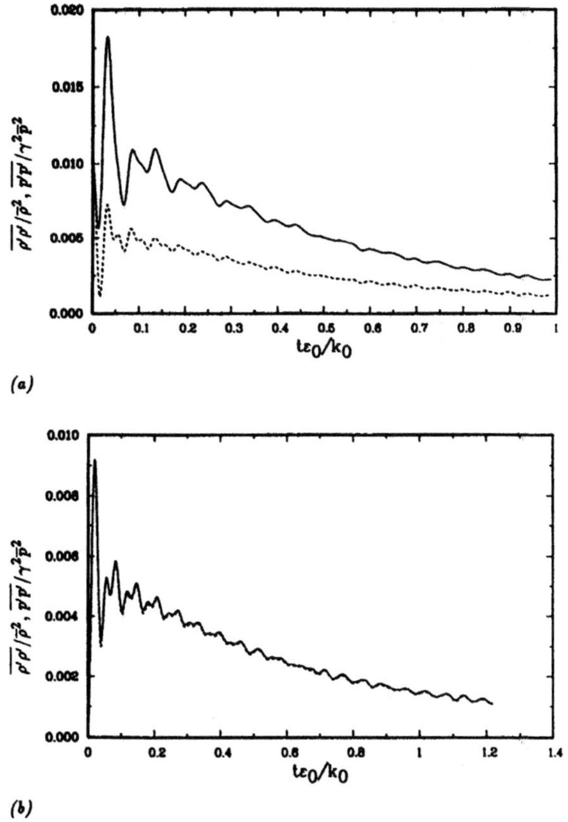

Figure 5.2. Evolution of normalized density and pressure variances. Blaisdell et al's [11] DNS of isotropic turbulence. Top: $F_0 = 1/40$, large initial pressure fluctuations. Bottom: $F_0 = 40$, large initial dilatational velocity fluctuations. By permission of the authors.

$M_{to} = 0.3$ cases, the bottom figure the $M_{to} = 0.7$ cases. The values of χ_ϵ differ from case to case. It is obvious that χ_ϵ not only depends on M_t, but also on the initial conditions. The behaviour of χ_ϵ is also consistent with linear analysis, since the coupling between acoustical and vortical modes is weak. Strong coupling could only be due to nonlinearities and would lead to curves which approach each other. Of course, nonlinear vortex interactions are weak for low Reynolds number turbulence. Blaisdell *et al.* [11] could however show that the dependence on initial conditions persists at higher Reynolds numbers.

The authors conclude that simulations of isotropic turbulence cannot be used to validate turbulence models. Recently Ristorcelli and Blaisdell [68] have designed consistent initial conditions for DNS. Tests for decaying isotropic turbulence show the natural development of the flow during the

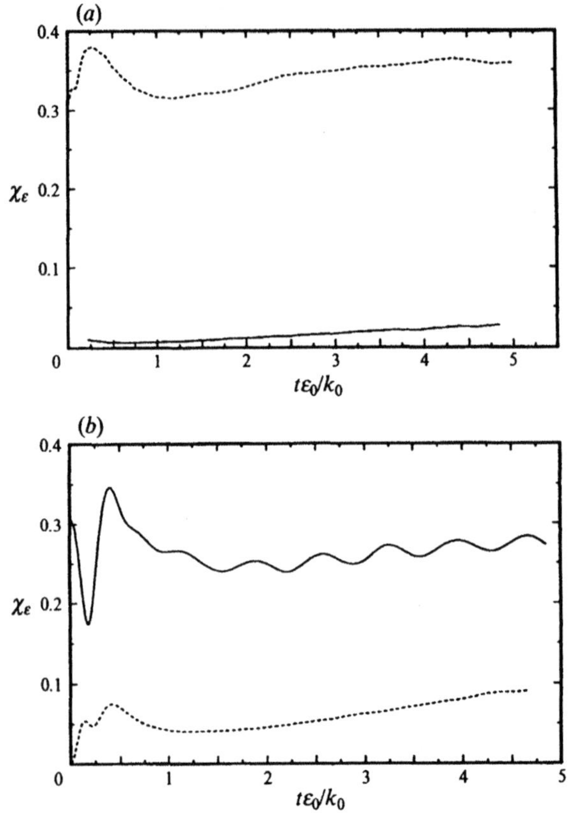

Figure 5.3. Influence of initial conditions on time evolution of dilatational fraction of turbulent dissipation rate. Modes are uncoupled, therefore, curves do not approach each other. Taken from Blaisdell *et al.* [12] by permission.

initial phase.

Turbulence models for the compressible dissipation rate

Model proposed by Sarkar et al. :

Sarkar *et al.* 's [74] model is based on ideas from linear acoustics and uses assumptions such as

$$F = 1, \quad \overline{p'd'} \approx 0. \tag{5.170}$$

Based on an asymptotic analysis for low turbulent Mach numbers M_t and on DNS data, the authors show that isotropic turbulence, in the state of acoustic equilibrium, is characterized by (5.170). F=1 is a good approximation even at $M_t = 0.5$. Using the definitions (5.65), (5.66) for ϵ_s, ϵ_d and $\bar{\mu}_v = 0$, the dilatational fraction χ_ϵ of the dissipation rate reads

$$\chi_\epsilon = \frac{4\overline{d'^2}/3}{\overline{\omega'_i\omega'_i} + 4\overline{d'^2}/3} \tag{5.171}$$

Then defining compressible and incompressible Taylor microscales λ_d, λ_s:

$$\lambda_d^2 = \overline{u_i^{d'}u_i^{d'}}/\overline{d'^2}, \tag{5.172}$$

$$\lambda_s^2 = \overline{u_i^{s'}u_i^{s'}}/\overline{\omega'_i\omega'_i} \tag{5.173}$$

and using χ, the fraction of compressible kinetic energy, given by eq. (5.168), χ_ϵ takes the form:

$$\chi_\epsilon = \frac{\chi}{\chi + \frac{3}{4}\left(\frac{\lambda_d}{\lambda_s}\right)^2(1-\chi)}, \tag{5.174}$$

from which ϵ_d/ϵ_s is obtained:

$$\frac{\epsilon_d}{\epsilon_s} = \frac{4\lambda_s^2\chi}{3\lambda_d^2(1-\chi)}. \tag{5.175}$$

For weakly compressible turbulence

$$\lambda_d/\lambda_s = O(1) \tag{5.176}$$

is a permissible assumption. In compressed turbulence, as *e.g.* in shock-turbulence interactions, relation (5.176) certainly makes no sense in general. Introducing (5.176) into (5.175) leads to:

$$\frac{\epsilon_d}{\epsilon_s} = \beta_1\chi + O(\chi^2), \tag{5.177}$$

where $\beta_1 = O(1)$. Now, from the definition of F

$$F = \gamma\bar{\rho}\bar{p}\frac{\overline{u_i^{d'}u_i^{d'}}}{\overline{p'p'}} = \frac{(\gamma\bar{p})^2}{\overline{p'p'}}\chi M_t^2, \tag{5.178}$$

the order of magnitude relation

$$\overline{p'p'} = O(\bar{\rho}^2u_{rms}^4) = O(\bar{\rho}^2K^2) \tag{5.179}$$

and $F = 1$, the authors conclude

$$\chi = O(M_t^2). \tag{5.180}$$

From (5.177) they obtain the model

$$\epsilon_d = \epsilon_s \alpha_1 M_t^2. \tag{5.181}$$

The model constant α_1 is specified in the following way: Decaying isotropic turbulence is governed by the transport equations:

$$\bar{\rho} c_v \frac{d\tilde{T}}{dt} = -\overline{p'd'} + \epsilon_s + \epsilon_d, \tag{5.182}$$

$$\bar{\rho} \frac{dK}{dt} = \overline{p'd'} - \epsilon_s - \epsilon_d, \tag{5.183}$$

$$\frac{d\epsilon_s}{dt} = -c_{\epsilon 2} \frac{\epsilon_s^2}{K}. \tag{5.184}$$

The equations for $\tilde{e}$ and K are still exact. Only the last equation is modelled. $c_{\epsilon 2}$ is set to 1.83, which is the proper value for decaying incompressible isotropic turbulence at high Reynolds number. With the help of eq. (5.183) and the definition $M_t^2 = 2K/(\gamma R\tilde{T})$, the equation for $\tilde{T}$ is converted into an equation for M_t^2. Setting $\overline{p'd'}$ to zero the set of equations for M_t^2, K, ϵ_s is solved for different initial conditions and different values of α_1. The best agreement between modelled and directly simulated turbulence decay was achieved for $\alpha_1 = 1$. For homogeneous shear turbulence, however, Sarkar [72] suggested the value $\alpha_1 = 0.5$.

Zeman's model:

Zeman [100] has formulated a model assuming the existence of shock-like structures in the flow. On the basis of a stochastic shocklet model he inferred the parametric expression

$$\frac{\epsilon_d}{\epsilon_s} = c_z f(M_t, F). \tag{5.185}$$

$c_z = 0.75$. The so-called shocklet dissipation function $f(M_t, F)$ is an integral functional of a pdf (m, F) of the fluctuating Mach number m. F is the kurtosis of m, *i.e.* $F = \overline{m^4}/\left(\overline{m^2}\right)^2$ and characterizes the departure from Gaussianity. From DNS of homogeneous turbulence, F was estimated as $F \approx 4$ and $f(M_t, F)$ approximated as

$$f(M_t) = \left[1 - exp\left(-\frac{1}{2}(\gamma + 1)(M_t - M_{to})^2/\Lambda^2\right)\right] \mathcal{H}(M_t - M_{to}), \tag{5.186}$$

TABLE 5.4. Parameters of Zeman's dissipation model

	Λ	M_{to}
free shear flows	0.6	$0.1(2/(\gamma+1))^{1/2}$
boundary layers	0.66	$0.25(2/(\gamma+1))^{1/2}$

where γ is the ratio of specific heats, $\mathcal{H}(x)$ the Heaviside step function and M_{to} a threshold value, below which shocklets cannot occur. The parameters Λ, M_{to} have different values for free and wall bounded turbulence, see table 5.4. The function $f(M_t)$ is taken from Wilcox [99].

Fauchet's model:

Based on a new two-point closure for weakly compressible isotropic turbulence, Fauchet [26], Fauchet *et al.* [27] derived a model for the compressible dissipation rate valid for low turbulent Mach numbers:

$$\frac{\epsilon_d}{\epsilon_s} = c_\epsilon M_t^4 ln\left(Re_L\right) Re_L^{-1} \quad , \qquad \text{for } M_t < 0.1 \tag{5.187}$$

M_t and Re_L are defined as follows:

$$M_t = (2K)^{1/2}/\overline{c_o} \quad , \qquad Re_L = (2K)^{1/2} L/\nu_o. \tag{5.188}$$

The integral scale L is computed from the energy spectrum $E(k)$:

$$L = \frac{\pi}{8K}\int_0^\infty \frac{E(k)}{k} dk \quad . \tag{5.189}$$

In contrast to Sarkar's model where ϵ_d/ϵ_s grows with M_t^2 and contrary to results obtained with EDQNM models by Bertoglio *et al.* [9] where ϵ_d/ϵ_s is found to scale as M_t^2 (in perfect agreement with Sarkar), Fauchet's model predicts a behaviour which coincides with Ristorcelli's [65] pseudo-sound theory. Moreover, Fauchet is able to confirm his model by LES results of forced isotropic turbulence with a subgrid-scale model of the Chollet-Lesieur type.

5.3.2. HOMOGENEOUS SHEAR TURBULENCE

We quickly recall the mean flow characteristics of homogeneous shear turbulence, discussed in section 5.2.5, before we turn to the linear inviscid equations for the turbulent fluctuations. The mean density and mean shear rate are space-time constants, viz:

$$\overline{\rho} = \text{const.}, \qquad \frac{d\tilde{u}_1}{dx_2} = S = \text{const.} \tag{5.190}$$

The mean pressure and mean temperature are homogeneous in space, but grow in time due to dissipative effects.

Linear inviscid analysis of turbulent fluctuations

From equations (5.1), (5.2) and (5.24) we derive the linear inviscid equations of motion for homogeneous shear turbulence:

$$\frac{D\rho'}{Dt} + \overline{\rho} d' = 0, \tag{5.191}$$

$$\frac{Du_i'}{Dt} + u_2' S \delta_{i1} + \frac{1}{\overline{\rho}} \frac{\partial p'}{\partial x_i} = 0, \tag{5.192}$$

$$\frac{Dp'}{Dt} + \gamma \overline{p} d' = 0. \tag{5.193}$$

The material derivative along mean flow trajectories has the form:

$$\frac{D}{Dt} = \frac{\partial}{\partial t} + S x_2 \frac{\partial}{\partial x_1}. \tag{5.194}$$

Blaisdell *et al.* [11, 12] have discussed these equations in a transformed coordinate system which moves with the mean velocity. They have Fourier transformed the vorticity equations in order to show the coupling between vorticity and acoustic modes due to the presence of mean shear. It is this coupling which ensures that measures of compressibility become independent of their initial conditions. We follow Simone [79], Simone *et al.* [80] and Fourier transform equations (5.191) - (5.194) using the Craya-Herring reference frame in Fourier space with bases $(e_i^{(1)}, e_i^{(2)}, e_i^{(3)})$ normal and parallel to the wavevector $\vec{k}$. The polar vector $\vec{n}$ is chosen parallel to the direction of mean shear, *i.e.* the x_2-direction. The base vectors then have the form:

$$\begin{aligned} e_i^{(1)} &= (\epsilon_{i12} k_1 - \epsilon_{i23} k_3)/k' \quad , \\ e_i^{(2)} &= (k_2 k_i - k^2 n_i)/(k k') \quad , \\ e_i^{(3)} &= k_i / k \quad . \end{aligned} \tag{5.195}$$

ϵ_{ijk} is the alternating unit tensor and k' the magnitude of $\vec{k} \times \vec{n}$:

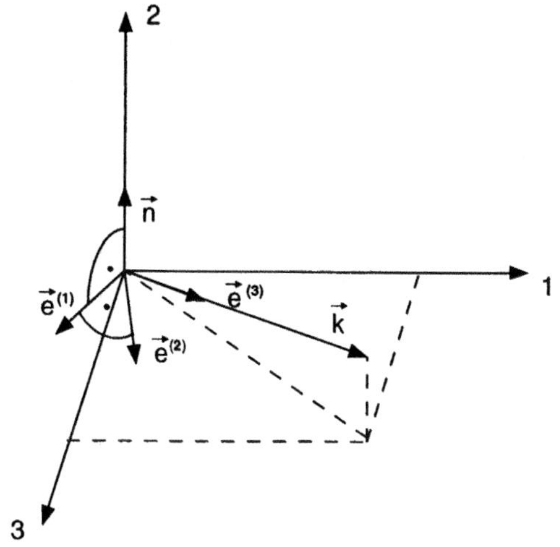

Figure 5.4. Craya-Herring local reference frame in wavenumber space. The polar vector $\vec{n}$ is parallel to the mean shear direction. The compressible mode is parallel to $\vec{k}$. The vorticity vector lies in the plane perpendicular to $\vec{k}$.

$$\left|\vec{k} \times \vec{n}\right| = k' = (k_1^2 + k_3^2)^{1/2}. \tag{5.196}$$

Figure 5.4 shows the Craya-Herring reference frame. The compressible velocity mode is parallel to the wavevector $\vec{k}$. The incompressible modes lie in the plane perpendicular to $\vec{k}$. This plane also contains the vorticity mode. Before transforming the linear equations we discuss the mean flow trajectories in physical and Fourier space. The material derivative in eq. (5.194) corresponds to a partial derivative with respect to time at fixed Lagrangian coordinate X_j:

$$\frac{D}{Dt} = \left(\frac{\partial}{\partial t}\right)_{\vec{X}} \quad . \tag{5.197}$$

In homogeneous shear flow Eulerian and Lagrangian coordinates are related by:

$$x_i = x_i\left(X_j, t_0, t\right) \quad , \qquad x_i\left(X_j, t_0, t\right) = F_{ij}\left(t_0, t\right) X_j \quad . \tag{5.198}$$

x_i is the position of a fluid particle moving with the mean flow at time t, and X_j denotes its position at time t_0, *i.e.* $X_j = x_j(t_0)$. The gradient displacement tensor is defined by the mean shearing motion:

$$F_{ij}(t_0,t) = \frac{\partial x_i(X_j,t_0,t)}{\partial X_j} = \delta_{ij} + St\delta_{i1}\delta_{j2} \quad . \tag{5.199}$$

The fluid particle coordinates are therefore:

$$x_1 = X_1 + StX_2 \quad , \qquad x_2 = X_2 \quad , \qquad x_3 = X_3 \quad . \tag{5.200}$$

A moving coordinate system in physical space corresponds to a time-dependent wave vector $\vec{k}(t)$. $\vec{k}(t)$ represents the position of a fluid particle in Fourier space at time t which at time t_0 had the position $\vec{K}$ (see Lesieur [55]). The wavenumbers are obtained from:

$$k_i = F_{ji}^{-1}(t_0,t)K_j \quad . \tag{5.201}$$

The tensor F_{ji}^{-1} is the inverse of the transpose of F_{ij} given in eq. (5.199). The components of k_i are:

$$k_1 = K_1 \quad , \qquad k_2 = K_2 - K_1 St \quad , \qquad k_3 = K_3 \quad . \tag{5.202}$$

One easily verifies the wave conservation law (Cambon [15]):

$$k_i x_i = K_i X_i \quad . \tag{5.203}$$

The Fourier transform of the material derivative, eq. (5.194), is denoted by the symbol $\frac{\mathcal{D}}{\mathcal{D}t}$:

$$\frac{\mathcal{D}}{\mathcal{D}t} = \frac{\partial}{\partial t} - k_1 S \frac{\partial}{\partial k_2} \quad . \tag{5.204}$$

The linear inviscid equations of motion finally take the spectral form:

$$\frac{\mathcal{D}\hat{\rho}}{\mathcal{D}t} + ik\overline{\rho}\hat{\varphi}^{(3)} = 0 \quad , \tag{5.205}$$

$$\frac{\mathcal{D}\hat{u}_i}{\mathcal{D}t} + \hat{u}_2 S\delta_{i1} + \frac{ik_i}{\overline{\rho}}\hat{p} = 0 \quad , \tag{5.206}$$

$$\frac{\mathcal{D}\hat{p}}{\mathcal{D}t} + ik\gamma\overline{p}\hat{\varphi}^{(3)} = 0 \quad . \tag{5.207}$$

The transformed velocity fluctuation, $\hat{u}_i$, consists of a solenoidal and a compressible part, as defined in eq. (5.152). By scalar multiplication of eq. (5.206) with $e_i^{(j)}$ $(j = 1,2,3)$, Simone [79] gets the mode equations. The derivation needs some care since the base vectors depend on $\vec{k}$ and only implicitly on time. Eq. (5.206) is equivalent to the component relations:

$$\frac{\mathcal{D}\hat{\varphi}^{(1)}}{\mathcal{D}t} + S\frac{k_3}{k}\hat{\varphi}^{(2)} - S\frac{k_2 k_3}{kk'}\hat{\varphi}^{(3)} = 0 \quad , \tag{5.208}$$

$$\frac{\mathcal{D}\hat{\varphi}^{(2)}}{\mathcal{D}t} - S\frac{k_1 k_2}{k^2}\hat{\varphi}^{(2)} + S\frac{k_1}{k'}\hat{\varphi}^{(3)} = 0 \quad , \tag{5.209}$$

$$\frac{\mathcal{D}\hat{\varphi}^{(3)}}{\mathcal{D}t} - 2S\frac{k_1 k'}{k^2}\hat{\varphi}^{(2)} + S\frac{k_1 k_2}{k^2}\hat{\varphi}^{(3)} + \bar{c}k\hat{\varphi}^{(4)} = 0 \quad . \tag{5.210}$$

The mode $\hat{\varphi}^{(4)}$ replaces the pressure:

$$\hat{\varphi}^{(4)} = i\frac{\hat{p}}{\bar{\rho}\,\bar{c}} \quad . \tag{5.211}$$

One immediately recognizes the differences to the case of isotropic turbulence. The incompressible modes $\hat{\varphi}^{(1)}$, $\hat{\varphi}^{(2)}$ are no longer frozen, but get energy from the dilatational mode $\hat{\varphi}^{(3)}$. There is even a coupling among the incompressible modes as a result of the mean shearing motion. For $S = 0$ the equations (5.159), (5.160) are recovered. The compressible velocity mode on the other hand is fed by the acoustic pressure mode $\hat{\varphi}^{(4)}$. Analytical solutions of this set of equations are not available. Simone *et al.* [80], however, discuss two special solutions. The first is the *solenoidal limit*, given by $\hat{\varphi}^{(3)} = 0$. From (5.208) and (5.209) one then gets:

$$\frac{\mathcal{D}\hat{\varphi}^{(1)}}{\mathcal{D}t} + S\frac{k_3}{k}\hat{\varphi}^{(2)} = 0 \quad , \tag{5.212}$$

$$\frac{\mathcal{D}\hat{\varphi}^{(2)}}{\mathcal{D}t} - S\frac{k_1 k_2}{k^2}\hat{\varphi}^{(2)} = 0 \quad . \tag{5.213}$$

Taking the wavenumber relations (5.202) into account, the second equation can be directly integrated. The exact solution of (5.212) is then found. Simone *et al.* [80] show that the modes $k\hat{\varphi}^{(2)}$ and $\hat{\varphi}^{(1)}$ are directly linked to the variables $\nabla^2 u_2$ and ω_2 which are typically used in the stability analysis of parallel incompressible shear flow (Orr–Sommerfeld equations).
The second special case is the so–called *pressure–released limit.* In this case pressure fluctuations are unable to draw energy from the dilatational velocity field. The mathematical model for this limit is obtained by removing the $\hat{\varphi}^{(4)}$ mode from the equations. The most simple way to study this case is in the form of eq. (5.192). One has:

$$\frac{Du_1'}{Dt} + u_2'S = 0 \quad , \qquad \frac{Du_2'}{Dt} = 0 \quad , \qquad \frac{Du_3'}{Dt} = 0 \quad , \tag{5.214}$$

which means that the vertical and spanwise velocity fluctuations are only advected with the mean velocity $\tilde{u_1}$. The integration along particle trajectories yields the solution:

$$\begin{aligned} u_1'(x_i,t) &= u_1'(X_i,t_0) - St\, u_2'(X_i,t_0) \quad , \\ u_2'(x_i,t) &= u_2'(X_i,t_0) \quad , \\ u_3'(x_i,t) &= u_3'(X_i,t_0) \quad . \end{aligned} \tag{5.215}$$

It immediately leads to the following expression for the turbulent kinetic energy, provided the turbulence field is initially isotropic:

$$\frac{K(t)}{K(0)} = 1 + \frac{(St)^2}{3} \quad . \tag{5.216}$$

A term linear in St does not appear because the Reynolds shear stress vanishes in isotropic turbulence. In the case where the initial turbulence field is not completely isotropic, Simone [79] obtains:

$$\frac{K(t)}{K(0)} = 1 + \frac{\overline{u_2'(0)u_2'(0)}}{K(0)}(St)^2 - 2St\frac{\overline{u_1'(0)u_2'(0)}}{K(0)} \quad . \tag{5.217}$$

This is an important result of rapid-distortion theory (RDT)[4], because the pressure–released regime constitutes an upper bound to the TKE amplification. This will be shown later in comparison to DNS data.
Another aspect that can be deduced from the linear equations, concerns the special role of the velocity fluctuations in the direction of mean shear. The 2-component of the momentum equation (5.192) and the energy equation (5.193) reflect the direct coupling between u_2- and pressure fluctuations on one side and pressure and dilatation fluctuations on the other side:

$$\frac{Du_2'}{Dt} + \frac{1}{\bar{\rho}}\frac{\partial p'}{\partial x_2} = 0 \quad , \tag{5.218}$$

$$\frac{Dp'}{Dt} + \gamma\bar{p}d' = 0 \quad . \tag{5.219}$$

The Fourier transforms of these relations are:

$$\frac{\mathcal{D}\hat{u}_2}{\mathcal{D}t} + \frac{1}{\bar{\rho}}ik_2\hat{p} = 0 \quad , \tag{5.220}$$

$$\frac{\mathcal{D}\hat{p}}{\mathcal{D}t} + \gamma\bar{p}\hat{d} = 0 \quad . \tag{5.221}$$

[4]The combination of linear solutions with statistical averaging is frequently referred to as rapid-distortion theory (RDT).

Operating $\mathcal{D}/\mathcal{D}t$ on eq. (5.220) while considering the wave number relations (5.202) and introducing (5.221) gives:

$$\frac{\mathcal{D}^2\hat{u}_2}{\mathcal{D}t^2} + \frac{k_1}{k_2}S\frac{\mathcal{D}\hat{u}_2}{\mathcal{D}t} = ik_2\bar{c}^2\hat{d} \quad . \tag{5.222}$$

This second-order hyperbolic equation for the transformed velocity fluctuation in the direction of mean shear contains the dilatation as 'forcing term'. It provides a direct link between compressibility effects (through $\hat{d}$) and the u_2-fluctuations. A second equation is available in the form of the wave equation for $\hat{d}$:

$$\frac{\mathcal{D}^2\hat{d}}{\mathcal{D}t^2} + k^2\bar{c}^2\hat{d} = -4k_1k_2S\frac{\hat{p}}{\bar{\rho}} = -4ik_1S\frac{\mathcal{D}\hat{u}_2}{\mathcal{D}t} \quad , \tag{5.223}$$

where

$$\frac{\mathcal{D}^2}{\mathcal{D}t^2} = \frac{\partial^2}{\partial t^2} - 2k_1S\frac{\partial^2}{\partial t\partial k_2} + (k_1S)^2\frac{\partial^2}{\partial k_2^2} \quad . \tag{5.224}$$

Equations (5.222) and (5.223) have to be solved for suitable initial conditions. From these equations it can be concluded that intrinsic effects of compressibility in the form of dilatation fluctuations at low turbulent Mach number (the acoustic time scale is small with respect to the turbulent time scale) should scale with the u_2-fluctuations. In some sense this result supports the definition of an integral length scale based on the transverse two-point correlation of u_2-fluctuations to be used in the definition of M_g, see eq. (5.130).

Some findings based on DNS of homogeneous shear turbulence

The extensive work of Blaisdell *et al.* [11] is full of useful information. We concentrate on a few findings and conclusions only.

Independence of initial conditions:

The authors have performed five direct simulations with initial turbulent Mach numbers of 0.5, but different rms-values of density- and temperature fluctuations and varying fractions of compressible kinetic energy, χ. Fig. 5.5 shows the time evolution of χ_ϵ, the dilatational fraction of the dissipation rate. Depending on initial conditions this quantity goes through different transient phases, but finally settles down and becomes independent of its initial value. Other measures of compressibility such as χ and $\overline{\rho'^2}$ show a similar behaviour. This independence from the initial conditions is due to the coupling between vorticity and dilatation fluctuations and means that

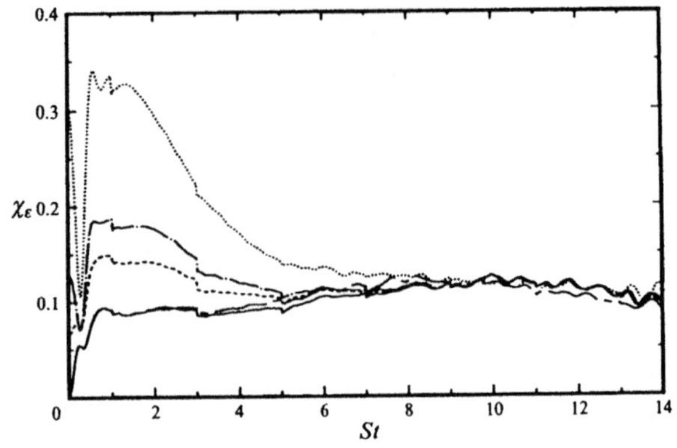

Figure 5.5. The dilatational fraction of the dissipation rate gets independent of initial conditions in homogeneous shear turbulence. Taken from Blaisdell *et al.* [12] by permission.

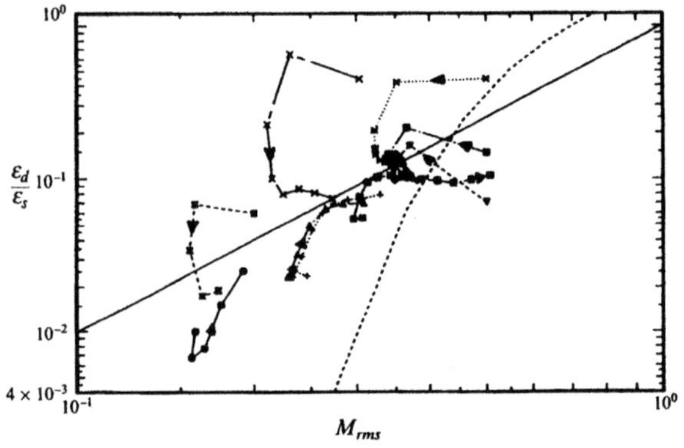

Figure 5.6. Comparison of Sarkar *et al.* 's [74] ——— , and Zeman's [100] model - - - - for compressible dissipation rate with DNS data of Blaisdell *et al.* [11] for homogeneous shear flow. Taken from Blaisdell *et al.* [11] by permission.

algebraic turbulence models for explicit compressibility terms may be able to capture compressibility effects.
Test of models for explicit compressibility terms:

Models for ϵ_d proposed by Sarkar *et al.* [74] and Zeman [100] and discussed in section 5.3.1 have been examined by Blaisdell *et al.* [11, 12]. The results of eleven DNS with different initial turbulent Mach numbers and rms-fluctuations are displayed in figure 5.6 and compared with Sarkar's and Zeman's models. Depending on χ_0 the simulations have different initial

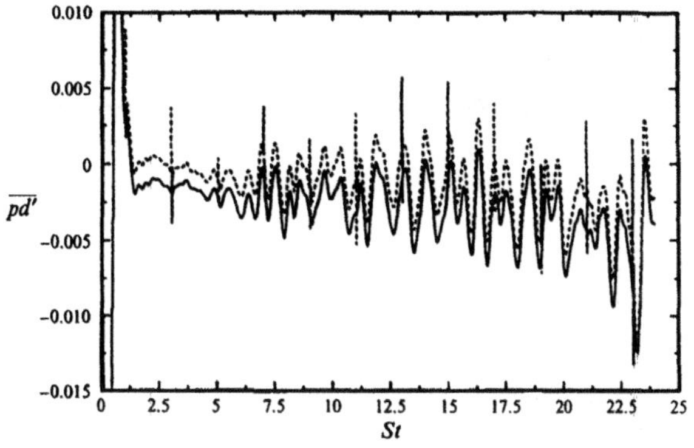

Figure 5.7. Time evolutions of $\overline{p'd'}$ and $d\overline{p'^2}/dt/(2\gamma\overline{p})$. Confirmation of linear theory. Taken from Blaisdell *et al.* [12] by permission.

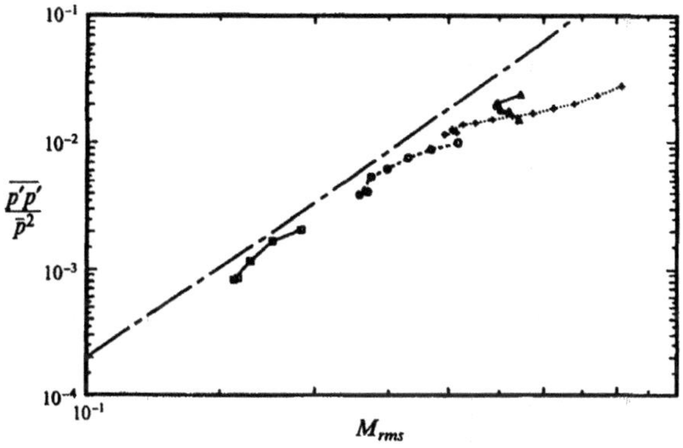

Figure 5.8. Test of Zeman's pressure-dilatation model. Pressure variance versus turbulent Mach number. Taken from Blaisdell *et al.* [12] by permission.

values of ϵ_d/ϵ_s and progress according to the arrows. Only the final values should be compared with the model results. The solid/dashed curves represent the results of Sarkar resp. Zeman. Obviously, Sarkar's model provides proper prediction, at least up to M_t around 0.3, whereas Zeman's model predicts a much faster increase of ϵ_d/ϵ_s with M_t [5] and does not match so well the results of the simulations.

A model for the pressure-dilatation correlation proposed by Zeman [101], relates $\overline{p'd'}$ to $\overline{p'^2}$ via the relation

[5]Blaisdell *et al.* use an rms Mach number instead of M_t which takes fluctuations in the sound speed into account. Both quantities do not differ noticeably.

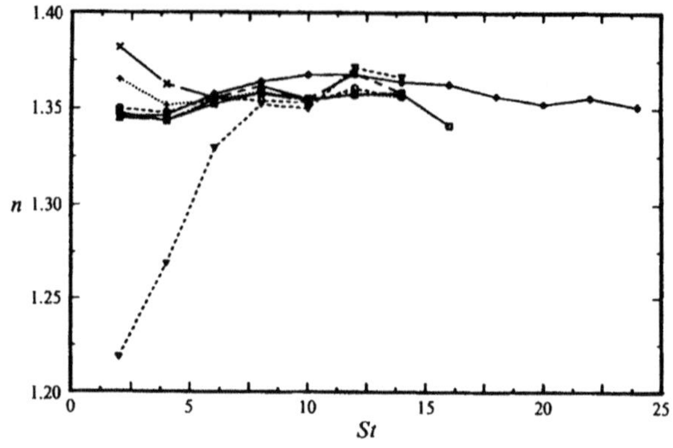

Figure 5.9. Development of mean polytropic coefficient. Independence of initial conditions. n close to γ. Taken from Blaisdell *et al.* [12] by permission.

$$\overline{p'\frac{\partial u_i'}{\partial x_i}} = -\frac{1}{2\gamma\bar{p}}\frac{d\overline{p'^2}}{dt} \tag{5.225}$$

which can be derived from the linear equation (5.193) (cf. also the discussion in section 5.2.4). The validity of this relation has been checked by Blaisdell *et al.* [11, 12] using a highly resolved DNS (sha 192) with 192^3 grid points for an initial M_t of 0.4 and vanishing initial values of ρ_{rms}, T_{rms}, χ. Figure 5.7 presents the time evolutions of $\overline{p'd'}$ and of the rhs of eq. (5.225). The model (discussed below) assumes the time variation of $\overline{p'^2}$ to scale with M_t^4 at low M_t and with M_t^2 at higher M_t. The DNS data follow an M_t^2 scaling much earlier than Zeman's model, cf. figure 5.8. This means that the model constants or the fitting function should be modified.

'Mean' polytropic coefficient:

In turbulence modelling one has usually to assume some correlation among fluctuations of thermodynamic variables. In his one-equation model Rubesin [69] has used the following relation

$$\frac{p'}{\bar{p}} = n\frac{\rho'}{\bar{\rho}} = \frac{n}{n-1}\frac{T''}{\tilde{T}}. \tag{5.226}$$

The polytropic coefficient n is γ in isentropic flow, $n = 1$ in isothermal and $n = 0$ in isobaric flow. In turbulent flow, a local quantity n is not well defined. Blaisdell *et al.* [11] therefore define an average polytropic coefficient

$$n = \left(\frac{\overline{p'^2}/\overline{p}^2}{\overline{\rho'^2}/\overline{\rho}^2} \right)^{1/2} \tag{5.227}$$

and demonstrate for homogeneous shear turbulence its independence on initial conditions after some transient behaviour and that it tends to a value slightly less than γ. This means that density, temperature and pressure fluctuations follow nearly isentropic processes. Figure 5.9 shows results for n according to (5.227) for seven different DNS of sheared turbulence.

This behaviour is in complete contrast to what is found in supersonic channel flow (see Huang *et al.* [44]) where n obviously follows an isobaric process. An explanation for this behaviour is not easily found. It is however interesting to note that density-temperature correlations in the core region of the channel indicate that the flow there shows a tendency towards isentropic behaviour.

Structure of solenoidal and dilatational velocity fields:

Based on a Helmholtz decomposition of the fluctuating velocity vector Blaisdell *et al.* [11] have investigated the different structure of the solenoidal and dilatational velocity fields. The Reynolds stresses $R_{ij} = \overline{\rho u_i' u_j'}$ are written as the sum of solenoidal, dilatational and cross terms

$$R_{ij} = R_{ij}^s + R_{ij}^d + R_{ij}^c \quad , \tag{5.228}$$

where

$$R_{ij}^s = \overline{\rho u_i^{s\prime} u_j^{s\prime}} \tag{5.229}$$

$$R_{ij}^d = \overline{\rho u_i^{d\prime} u_j^{d\prime}} \tag{5.230}$$

$$R_{ij}^c = \overline{\rho u_i^{s\prime} u_j^{d\prime}} + \overline{\rho u_j^{s\prime} u_i^{d\prime}} \tag{5.231}$$

The anisotropy tensors of the 's' and 'd' components are:

$$b_{ij}^s = R_{ij}^s / R_{kk}^s - \delta_{ij}/3 \quad , \quad b_{ij}^d = R_{ij}^d / R_{kk}^d - \delta_{ij}/3 \quad . \tag{5.232}$$

Each of them possesses three scalar invariants. The first, being the trace, is zero. The other two are (omitting the superscripts):

$$II = -\frac{1}{2} b_{ij} b_{ji} \quad , \quad III = \frac{1}{3} b_{ij} b_{jk} b_{ki} \quad .$$

They must lie within the regions bordered by

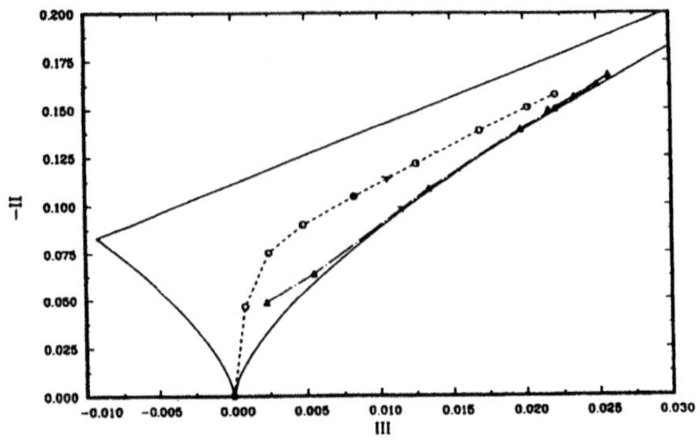

Figure 5.10. Solenoidal (○) and dilatational (△) Reynolds stress anisotropy tensors for Blaisdell *et al.* 's DNS sha192. Taken from Blaisdell *et al.* [11] by permission.

$$
\begin{aligned}
II &= -\frac{3}{2^{2/3}}(-III)^{2/3} \quad , \\
II &= -3III - \frac{1}{9} \quad , \\
II &= -\frac{3}{2^{2/3}}(III)^{2/3}
\end{aligned} \tag{5.233}
$$

in the (III, II) plane and form an invariant map first introduced by Lumley [57]. This map is presented in figure 5.10 for Blaisdell et al's DNS sha192. One notices that b^d_{ij} lies very close to the curve which corresponds to axisymmetric expansion and thus indicates that $u_i^{d\prime}$ tends to be planar in contrast to $u_i^{s\prime}$. A velocity field associated with sound waves can indeed be planar.

TKE growth rate and Reynolds stress anisotropy:

DNS data of Blaisdell *et al.* [11, 12] and Sarkar *et al.* [74] for compressible homogeneous shear flow had indicated an exponential growth of TKE at a rate considerably below that for incompressible turbulence, cf. figure 5.11. This stabilizing effect of compressibility was, until recently, attributed to the presence of explicit compressibility terms like the compressible dissipation rate ϵ_d and the pressure-dilatation correlation $\overline{p'd'}$ in the K-equation (5.116). Indeed ϵ_d acts as a sink and $\overline{p'd'}$ is predominantly negative. It was Sarkar [73] who provided new insight into the stabilizing effect of compressibility. His findings were later confirmed by Simone *et al.* [80] and for mixing layers by Vreman *et al.* [95]. Sarkar [73] performed two DNS series

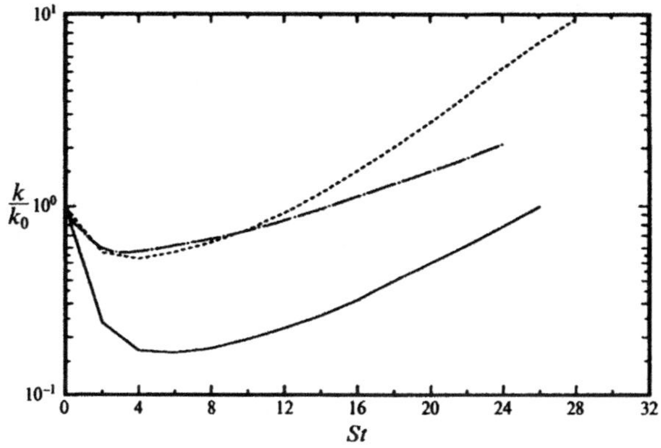

Figure 5.11. Comparison between compressible and incompressible time evolutions of K. ———, – – – – incompressible homogeneous shear flow. —·—· sha192 shows reduced growth rate. Taken from Blaisdell *et al.* [12] by permission.

A, B of homogeneous shear flow in which he varied the initial values M_{go} and M_{to} individually from run to run in order to demonstrate for large nondimensional times St that

- the reduction in the growth rate of TKE as M_{go}, respectively M_{to}, increaseses, is primarily due to the reduced level of turbulence production, while the response of explicit dilatational terms is much less pronounced
- the stabilizing effect of compressibility is substantially larger in the DNS series A than in series B. In series A, M_{go} is increased via the shear rate S while $\bar{c}_o$ and M_{to} are kept constant. In series B, on the other hand, M_{to} was increased and M_{go} was kept constant.

The different behaviour is a consequence of the fact that M_{go} and M_{to} are two independent parameters since their ratio Sl_o/u_o can be varied via l_o, u_o. We recall that the nondimensionalized TKE equation (5.123) had already indicated the relevance of M_{go} with respect to the production term. The DNS parameters of Sarkar's series A are given in Table 5.5.

$Re_{\lambda o}$ is the Reynolds number based on the velocity scale u_o and on Taylor's microscale $\lambda_o = u_o/\overline{(\omega_i'\omega_i')_o}^{1/2}$. Sarkar [73] writes the K-equation (5.116) as an evolution equation for the growth rate $\Lambda = (1/SK)\,dK/dt$, viz:

$$\Lambda = -2b_{12}\left(1 - \frac{\epsilon_s}{P}\right) - \frac{\epsilon_d - \overline{p'd'}/\overline{\rho}}{SK} = -2b_{12}\left(1 - X_\epsilon\right) \quad , \qquad (5.234)$$

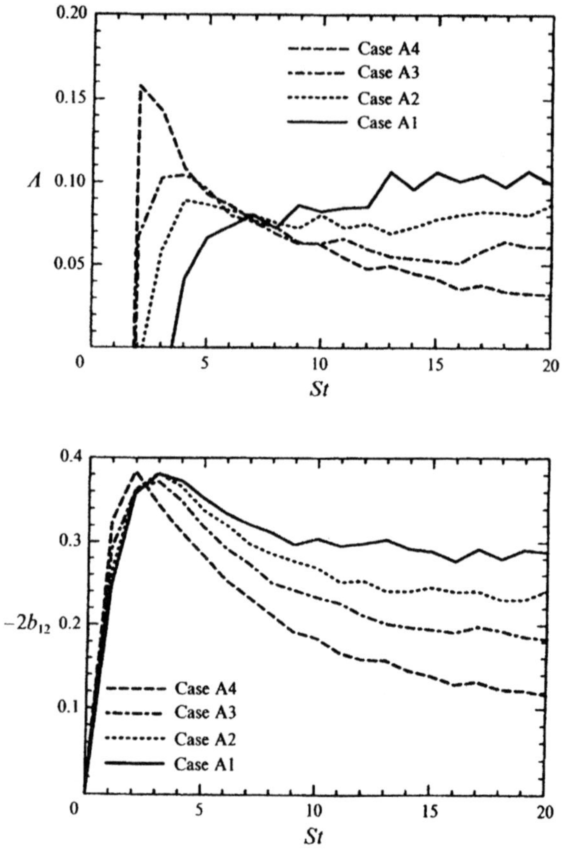

Figure 5.12. Growth rate reduction (top) by increased M_{g0} is primarily due to reduced non-dimensional production (bottom). Taken from Sarkar [73] by permission.

TABLE 5.5. Parameters of Sarkar's [73] series A simulations.

Case	M_{go}	M_{to}	$Re_{\lambda o}$	Pr_o	$(SK/\epsilon)_o$
A1	0.22	0.40	14	0.7	1.8
A2	0.44	0.40	14	0.7	3.6
A3	0.66	0.40	14	0.7	5.4
A4	1.32	0.40	14	0.7	10.8

where $b_{12} = \widetilde{u_1' u_2'}/(2K)$ is the Reynolds shear stress anisotropy (cf. eq. (5.232)) and P the production term. The quantity

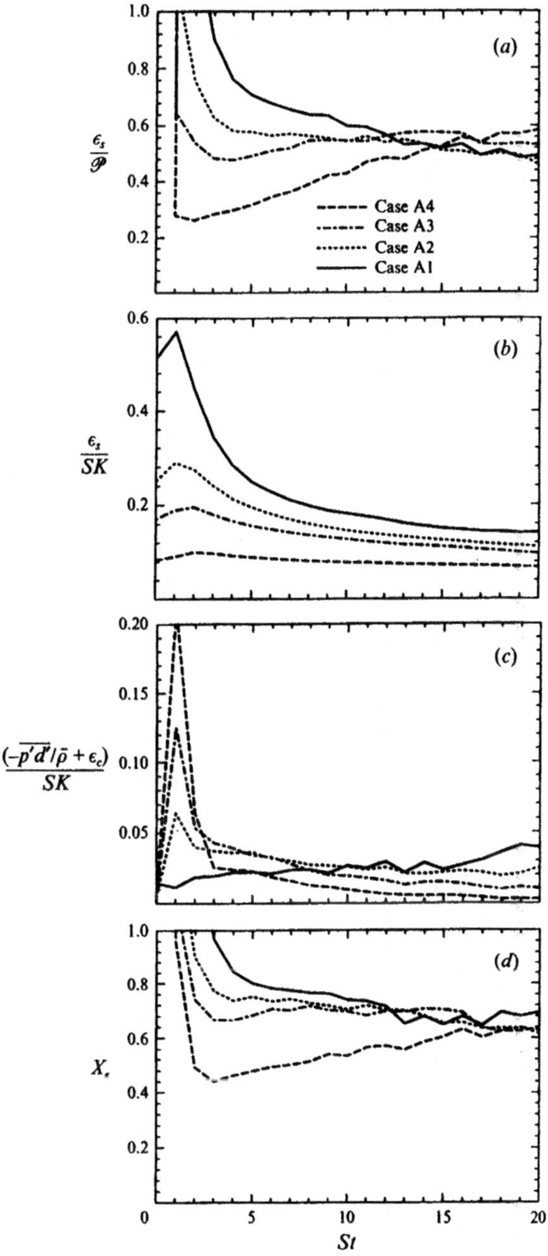

Figure 5.13. Evolution of ϵ_s, of explicit compressibility terms and of the sum of all terms (except b_{12}). The effect of M_{g0} is for all terms lower than for b_{12}. Taken from Sarkar [73] by permission.

$$X_\epsilon = \left(\epsilon_s + \epsilon_d - \overline{p'd'}/\overline{\rho}\right)/P \tag{5.235}$$

combines all terms other than the production term. Figure 5.12 contains the time evolutions of Λ and $-2b_{12}$. We note that these quantities attain asymptotic values at large St which decrease as M_{go} increases. The non-dimensional production term, $P/(SK) = -2b_{12}$, is strongly inhibited when the gradient Mach number is increased. Figure 5.13 shows the evolution of ϵ_s in two different non-dimensionalizations, the evolutions of the explicit compressibility terms and of X_ϵ. One recognizes that the long-time behaviour ($St = 20$) of all these terms is much less affected by the gradient Mach number than the turbulent production ($-2b_{12}$). The effect of M_{go} on the dilatational terms is small, but non-negligible. It is against intuition that the sum $\left(-\overline{p'd'}/\overline{\rho} + \epsilon_d\right)$ is decreasing instead of increasing with M_{go}. From X_ϵ at ($St = 20$) Sarkar [73] concludes that the large reduction in the value of Λ is almost wholly due to the large reduction in the magnitude of the Reynolds shear stress anisotropy b_{12}. He further comments, based on a Helmholtz decomposition of the velocity field, that the solenoidal part of b_{12} is responsible for the reduced production rate. The effect of compressibility on the b_{11} and b_{22} components of the anisotropy tensor is presented in figure 5.14. There is a systematic increase in the magnitude of these anisotropies from case A1 to case A4 and it may be concluded that the pressure-strain correlation tensor in the homogeneous shear flow is significantly changed due to compressibility.

Pressure strain correlation tensor:

The trace of the pressure strain correlation tensor is the pressure-dilatation correlation $\overline{p'd'}$. Sarkar [72] showed that this correlation, obtained from spatial averaging of instantaneous turbulence fields, oscillates in time in the case of homogeneous shear flow. For low to moderate initial turbulent Mach numbers the oscillations take place at fast (acoustic) time scales, but do not contribute in a time-integrated sense to the evolution of K. By decomposing the fluctuating pressure into incompressible and compressible parts,

$$p' = p^{I'} + p^{C'} \quad , \tag{5.236}$$

Sarkar [72] was able to attribute the oscillations in $\overline{p'd'}$ to the compressible component $\overline{p^{C'}d'}$ which does not need modelling due to its negligible integrated effect. The model he derived for $\overline{p^{I'}d'}$ will be discussed below.

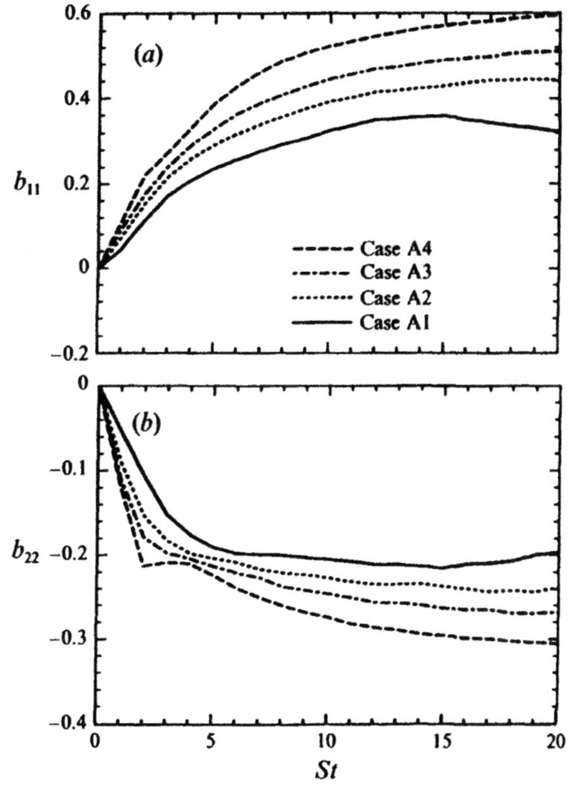

Figure 5.14. Effect of M_{g0} on streamwise and transverse Reynolds stress anisotropies. Taken from Sarkar [73] by permission.

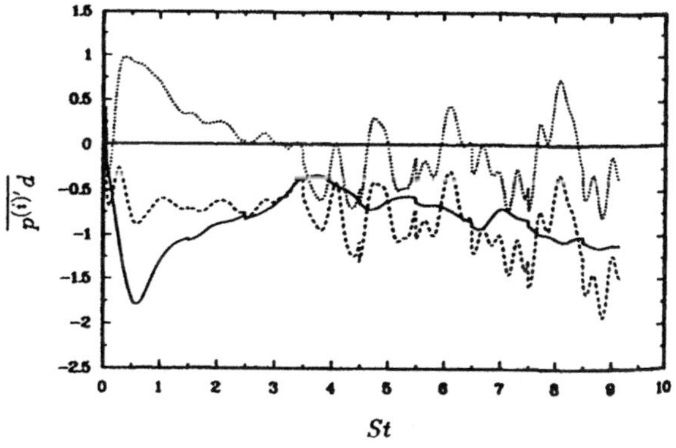

Figure 5.15. Time evolution of $\overline{p'd'}$ (– – – –) and its compressible (· · · · · ·) and incompressible (———) contributions. Taken from Blaisdell and Sarkar [13] by permission.

Figure 5.15 illustrates the time evolution of $\overline{p'd'}$ and of its compressible and incompressible contributions for a flow with the parameters: $M_{to} = 0.4$, $(2SK/\epsilon)_o = 13.6$, $Re_{to} = 441$, $\rho_{rms,o} = 0$, $\chi_o = 0$. $p_{rms,o}$ was obtained from a Poisson equation.

The motivation for the decomposition (5.236) was derived from the Poisson equation for p'. In incompressible turbulence such a Poisson equation forms the basis for the derivation of pressure-strain models. Sarkar's analysis starts from the following equation for the instantaneous pressure:

$$\frac{\partial^2 p}{\partial x_i \partial x_i} = -\frac{\partial^2}{\partial x_i \partial x_j}(\rho u_i u_j) + \frac{\partial^2 \rho}{\partial t^2} + \frac{\partial^2 \tau_{ij}}{\partial x_i \partial x_j} \quad , \tag{5.237}$$

which is the divergence of the momentum equation (5.2) combined with the continuity equation (5.1). Splitting each quantity into a mean and a fluctuation,

$$u_i = \overline{u}_i + u_i' \quad , \quad p = \overline{p} + p' \quad , \quad \rho = \overline{\rho} + \rho' \quad , \quad \tau_{ij} = \overline{\tau}_{ij} + \tau_{ij}' \quad , \tag{5.238}$$

substituting these into (5.237), and subtracting the mean of this relation gives after some algebra:

$$\begin{aligned} p'_{,jj} = & - 2\overline{u}_{i,j}\left(\overline{\rho}u_j'\right)_{,i} - 2\overline{u}_{i,i}\left(\overline{\rho}u_j'\right)_{,j} - 2\overline{u}_{i,ij}\left(\overline{\rho}u_j'\right) - \left(\overline{\rho}u_i'u_j'\right)_{,ij} \\ & - \rho'\left(\overline{u}_{i,i}\right)^2 - \rho'\overline{u}_{i,j}\overline{u}_{j,i} - 2\overline{u}_{i,j}\left(\rho'u_j'\right)_{,i} - 2\overline{u}_{i,i}\left(\rho'u_j'\right)_{,j} \\ & - 2\overline{u}_{i,ij}\left(\rho'u_j'\right) - \left(\rho'u_i'u_j'\right)_{,ij} + \frac{D^2\rho'}{Dt^2} + \tau'_{ij,ij} \quad . \end{aligned} \tag{5.239}$$

For convenience, commas are used to denote spatial derivatives. The first four terms on the right-hand side of this equation depend explicitly on $\overline{\rho}$, terms five to ten depend on ρ' and its gradients. Term eleven contains unsteadiness and mean convection of ρ' and the last term describes explicit viscous effects. Now, the first four terms are similar to the source terms for incompressible constant density flow. All remaining terms vanish in that case. Therefore, the fluctuating 'incompressible' pressure is associated with the first four source terms and satisfies the Poisson equation:

$$p^{I'}_{,jj} = -2\overline{u}_{i,j}\left(\overline{\rho}u_j'\right)_{,i} - 2\overline{u}_{i,i}\left(\overline{\rho}u_j'\right)_{,j} - 2\overline{u}_{i,ij}\left(\overline{\rho}u_j'\right) - \left(\overline{\rho}u_i'u_j'\right)_{,ij} \quad . \tag{5.240}$$

Evaluating this equation, Sarkar obtains $p^{I'}$. Since the DNS provides p', the compressible part is the difference of p' and $p^{I'}$. That way the incompressible and compressible contributions to $\overline{p'd'}$ were obtained.

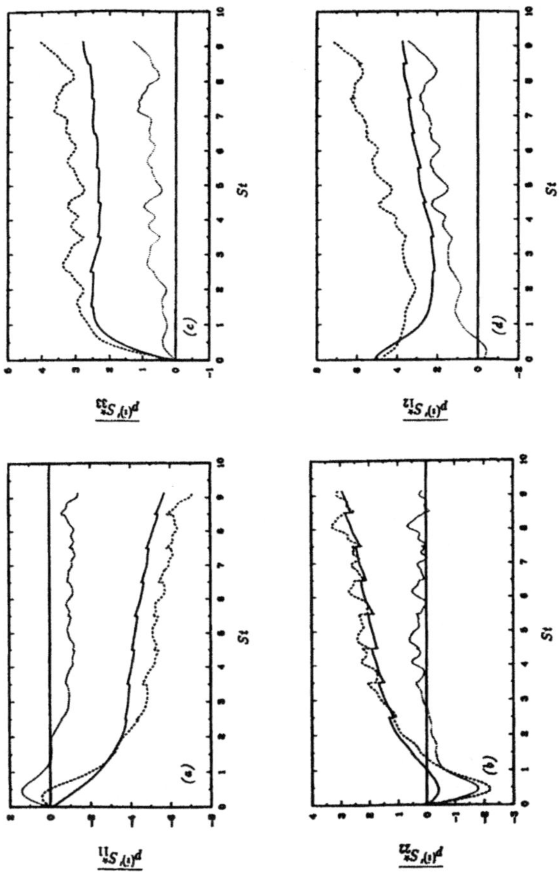

Figure 5.16. Time evolutions of the four deviatoric pressure-strain correlations, split as in fig. 1b. Taken from Blaisdell and Sarkar [13] by permission.

The split pressure fields $p^{I'}$, $p^{C'}$ were later used by Blaisdell and Sarkar [13] to decompose the pressure strain terms. They discuss the deviatoric part of the pressure strain tensor

$$\begin{aligned} \overline{p's^{*\prime}_{ij}} &= \overline{p'\left(s'_{ij} - \frac{1}{3}s'_{kk}\delta_{ij}\right)} \qquad {}^{6)} \qquad (5.241) \\ &= \overline{p^{I'}\left(s'_{ij} - \frac{1}{3}s'_{kk}\delta_{ij}\right)} + \overline{p^{C'}\left(s'_{ij} - \frac{1}{3}s'_{kk}\delta_{ij}\right)} \quad , \end{aligned}$$

based on DNS data for the flow parameters listed above, which correspond

[6]The star (∗) replaces the (D) used in equation (5.8) to denote the deviatoric deformation tensor.

to a high M_{to} / low M_{go} case. Figure 5.16 contains the time evolutions of the four relevant pressure strain components, each of them split as indicated in equation (5.241). The discontinuities in the curves are due to the remeshing process in the numerical simulations. The following observations were made:

- The normal components of the three pressure strain tensors $\overline{p' s^{*\prime}_{ij}}$, $\overline{p^{I'} s^{*\prime}_{ij}}$ and $\overline{p^{C'} s^{*\prime}_{ij}}$ behave in a way similar to $\overline{p'd'}$, $\overline{p^{I'}d'}$ and $\overline{p^{C'}d'}$. The compressible parts are oscillatory and have much smaller magnitudes than the incompressible parts, consequently their contributions to the time evolutions of the Reynolds stresses are smaller.
- While all the 1-1 components are negative, all the 2-2 and 3-3 components are positive. The compressible 2-2 component is the weakest among these normal components.
- The compressible part of the off-diagonal (1-2) component is, however, not small and thus contributes significantly to the time evolution of $\overline{\rho u_1' u_2'}$. Its magnitude reaches 75 % of that of the incompressible part. Therefore, $\overline{p^{C'} s_{12}'}$ needs modelling. Its positive sign indicates a reduction of the magnitude of $\overline{\rho u_1' u_2'}$ (cf. eq. (5.122)).

From correlation coefficients of the decomposed pressure fluctuations Blaisdell and Sarkar [13] conclude that the two pressure fields $p^{I'}$ and $p^{C'}$ are not statistically independent, which makes a proper independent modelling of $\overline{p^{I'} s^{*\prime}_{ij}}$ and $\overline{p^{C'} s^{*\prime}_{ij}}$ difficult. Splitting the incompressible field $p^{I'}$ into rapid and slow parts

$$p^{I'} = p^{R'} + p^{s'} \quad , \tag{5.242}$$

moreover shows that the compressible pressure $p^{C'}$ is more closely associated with the rapid incompressible pressure $p^{R'}$ than with its slow part $p^{s'}$.

The role of linear processes in explaining structural changes due to intrinsic compressibility:

One of several important issues of the work by Simone *et al.* [80] is that linear rapid-distortion theory (RDT) is capable of predicting the TKE growth rate and the Reynolds shear stress anisotropy in surprisingly close agreement with DNS data. The authors solve the rapid distortion equations (5.207) to (5.210) numerically and perform direct simulations for a turbulent Mach number of 0.25 and various gradient Mach numbers. In order to allow for comparison, they select two flow cases which are close to Sarkar's cases A3 and A4 (See table 5.5). Their distortion Mach number (M_d) definition uses the large-eddy lengthscale $(2K)^{3/2}/\epsilon$ instead of an integral lengthscale based on a two-point velocity correlation. Otherwise it

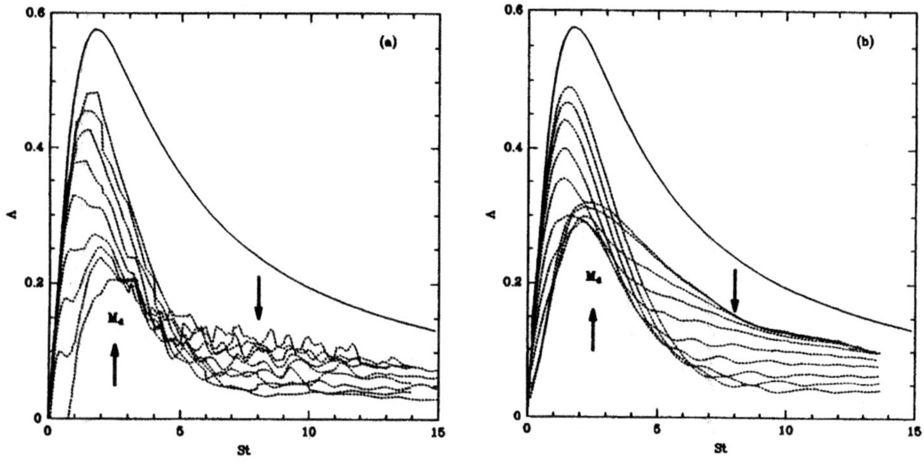

Figure 5.17. Comparison between DNS (left) and RDT (right) for TKE growth rate. Solid line: pressure-released limit. Destabilization for $St < 4$ and stabilization at later times. Taken from Simone *et al.* [80] by permission.

coincides with the definition of M_g. Therefore, their values for M_{do} are by a factor of more than 6 larger than Sarkar's values for M_{go}. All the computations (RDT and DNS) start from initial conditions which correspond to a good approximation to isotropic turbulence in acoustic equilibrium. The following findings may be listed:

- For small non-dimensional times ($St < 4$) compressibility acts destabilizing. Only at later times ($St > 4$) Sarkar's stabilizing effect of $M_{go}(M_{do})$ is observed. This 'crossover' behaviour which appears only in sheared (not in compressed) homogeneous turbulence is explained based on a semi-analytical analysis. The growth rate Λ, defined in equation (5.234) is used to show these effects.
- The destabilizing/stabilizing effect of M_{do} on the growth rate is nearly completely due to the behaviour of b_{12}, the non-dimensional production term.
- A Helmholtz decomposition of the velocity field shows that the dilatational part of the shear stress anisotropy, b_{12}^d is practically independent of M_{do}, while the solenoidal part b_{12}^s is dramatically decreased (with growing M_{do}) over the entire range of St. The source for these structural changes of the solenoidal velocity fluctuations is the feedback of the dilatational disturbances upon the solenoidal field (cf. eqs. (5.208), (5.209)).

Figure 5.17 shows the time evolution of the TKE growth rate as obtained from DNS data (left) and RDT (right). The solid lines indicate the pressure-

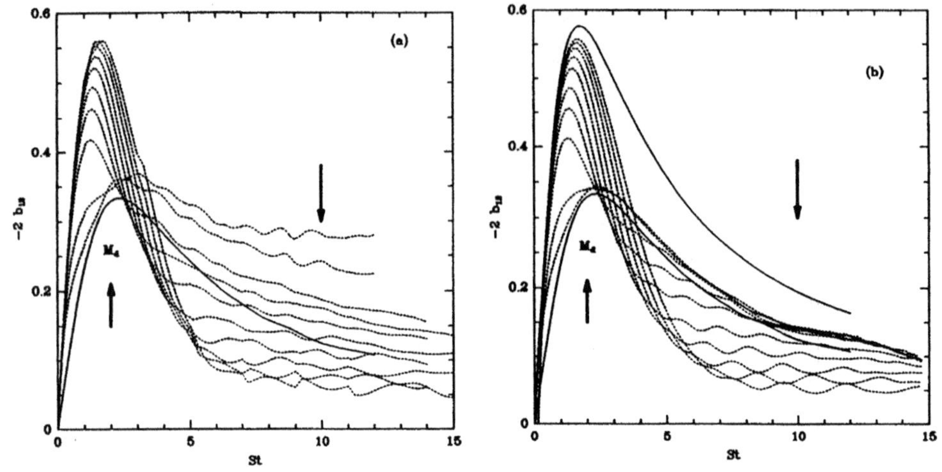

Figure 5.18. Comparison between DNS (left) and RDT (right) for b_{12}. Incompressible and pressure-released limits (higher curve). Taken from Simone *et al.* [80] by permission.

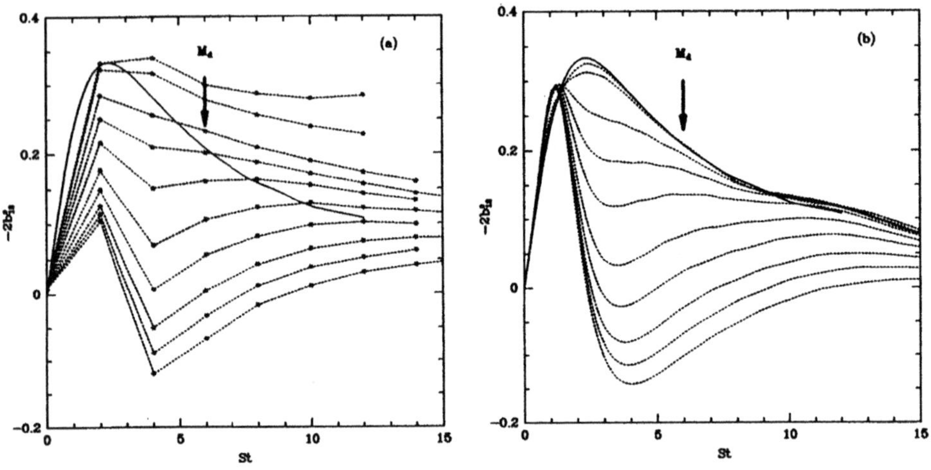

Figure 5.19. Behaviour of solenoidal b_{12}-component. Left: DNS, right: RDT. This component is strongly affected by M_d. Taken from Simone *et al.* [80] by permission.

released limit which was discussed at the end of section 5.3.2 (eqs. (5.214)-(5.217)) and is given by

$$\Lambda^{pr} = \frac{2St}{3 + (St)^2} \quad . \tag{5.243}$$

This result follows from eq. (5.216) for initially isotropic turbulence with the normalized TKE, $K(0) = 1$. It represents an upper bound, found for initial $M_{do} \gg 1$. All curves exhibit the 'crossover' feature which is due to

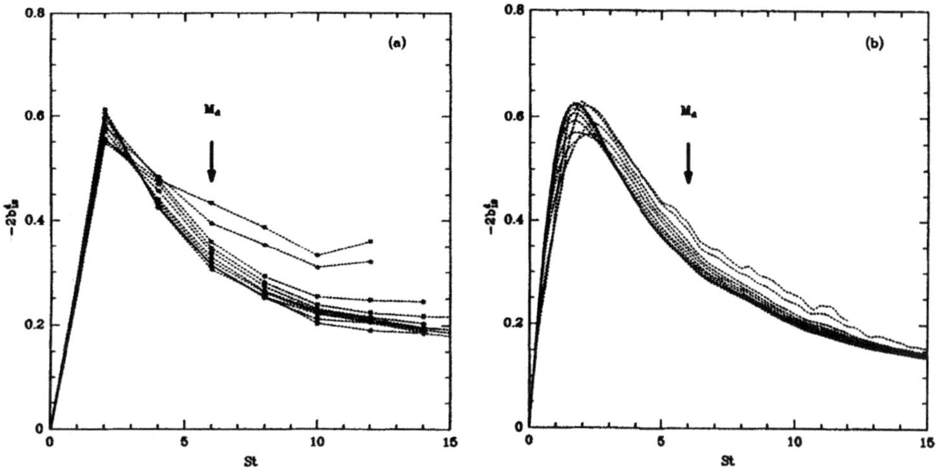

Figure 5.20. Behaviour of dilatational b_{12}-component. Left: DNS, right: RDT. It is essentially unaffected. Taken from Simone *et al.* [80] by permission.

the coupling of solenoidal and dilatational velocity fields.
The DNS and RDT histories of b_{12} on left and right of figure 5.18 not only confirm Sarkar's [73] observation that the non-dimensional production term in eq. (5.234) is primarily responsible for the exponential growth rate behaviour, but demonstrate that it can be extended to the regime of destabilization ($St < 4$). The right figure presenting RDT results contains two solid curves. The upper represents the pressure-released limit and the lower the RDT result for incompressible flow. Both form upper and lower bounds for the regime in which an increase in $M_{do}(M_{go})$ acts destabilizing and the agreement between DNS and RDT is striking.
Histories of the solenoidal and dilatational contributions to the Reynolds shear stress anisotropy, b_{ij}^s and b_{ij}^d, are shown in figures 5.19 and 5.20. While the solenoidal component is dramatically decreased with increasing M_{do} over the entire range of St, the dilatational part is essentially unaffected by compressibility. The structural changes are therefore almost exclusively felt in the solenoidal field. RDT predicts almost independence of b_{ij}^d on M_d. The large-St limits of both terms are certainly not properly captured by the RDT, because it does not treat nonlinear effects.
Summarizing, the overall good agreement between DNS and RDT has to be emphasized; which means that much of the structural effect of compressibility can be fairly well reproduced by solving linear equations. This remarkable result should be used for the development of improved turbulence models.

Dissipation rate tensor:

The work of Blaisdell *et al.* [11] also provides information concerning the dissipation rate anisotropy. Unfortunately, only one flow case is considered, so that the Mach number dependence of ϵ_{ij} and of its solenoidal and dilatational components ϵ_{ij}^s, ϵ_{ij}^d (cf. eqs. (5.81), (5.82)) can only be grasped from comparison with the incompressible case. A few observations pertinent to Blaisdell *et al.* 's run sha92 are:

- In the transport equations for the four Reynolds stresses (5.119 - 5.122) only ϵ_{22}^d is relevant and comparable in size to ϵ_{22}^s. All other components of ϵ_{ij}^d are negligibly small. ϵ_{12}^s becomes small as the simulation progresses (in agreement with the incompressible case).
- The solenoidal dissipation rate anisotropy tensor, $d_{ij}^s = \epsilon_{ij}^s/\epsilon_{kk}^s - \delta_{ij}/3$ is highly aligned with the Reynolds stress anisotropy tensor, b_{ij}. The same is true for incompressible turbulence. The dilatational tensor, d_{ij}^d, is not aligned with b_{ij}.
- However, d_{ij}^d tends to be aligned with b_{ij}^d. A physically interesting result, but difficult to use for modelling of ϵ_{ij}^d.

These results show that the dilatational dissipation rate tensor is strongly anisotropic. So, besides altering the structure of the Reynolds stress anisotropy directly (via inviscid mode coupling) and of the pressure strain correlation tensor, intrinsic compressibility could lead to further modification of the Reynolds stress tensor via structural changes of the dissipation rate tensor.

Pressure-dilatation models

Zeman's model and extensions:

Zeman's [101] model is based on the argument that for low turbulent Mach number homogeneous isotropic and sheared turbulence the pressure-dilatation correlation and the time derivative of the pressure variance are closely connected. The corresponding relation had already been given in eq. (5.225) and figure 5.7 had shown the excellent agreement between this relation and DNS data. Introducing relation (5.225) into the K-equation for homogeneous isotropic turbulence and neglecting viscosity fluctuations gives:

$$\frac{d}{dt}\left[\frac{1}{2}\left(\overline{\rho u_i'' u_i''} + \frac{1}{\gamma \overline{p}} \ \overline{p'^2}\right)\right] = -\overline{\rho}\left(\epsilon_s + \epsilon_d\right) \quad . \tag{5.244}$$

The term in the square brackets is the total energy of turbulence. In a state of acoustic non-equilibrium kinetic and potential energies are exchanged. Assuming the rhs of eq. (5.244) to be closed, a model is needed for the pressure variance. Zeman postulates that $\overline{p'^2}$ relaxes to an equilibrium value on an acoustic time scale τ_a:

$$\frac{d\overline{p'^2}}{dt} = -\frac{\overline{p'^2} - p_e^2}{\tau_a} \quad , \tag{5.245}$$

where $\tau_a = 0.13\,(2K/\epsilon_s)\,M_t$.

The equilibrium value p_e^2 is related to K and M_t via the empirical model:

$$\frac{p_e^2}{2\overline{\rho}^2 K \overline{c}^2} = \frac{1}{\gamma^2 M_t^2}\frac{p_e^2}{\overline{p}^2} = \frac{\alpha M_t^2 + \beta M_t^4}{1 + \alpha M_t^2 + \beta M_t^4} = f\left(M_t^2\right) \tag{5.246}$$

This model combines two assumptions: (1) the equilibrium ratio of compressible to solenoidal kinetic energy is $\overline{u_i^{d\prime} u_i^{d\prime}} / \overline{u_i^{s\prime} u_i^{s\prime}} = \alpha M_t^2 + \beta M_t^4$, and (2) in a state of acoustic equilibrium, the compressible potential and kinetic energies are equal, *i.e.*

$$p_e^2 \approx \overline{\rho}^2 \overline{c}^2 \overline{u_i^{d\prime} u_i^{d\prime}} \quad , \tag{5.247}$$

(cf eq. (5.167)). The model (5.246) allows for a M_t^4 behaviour for small M_t in agreement with the acoustic analysis of Sarkar *et al.* [75] and a M_t^2 variation at higher Mach numbers. The complete model reads:

$$\overline{p'd'} = -\frac{1}{2\overline{\rho}^2\overline{c}^2}\frac{\overline{p'^2} - \gamma^2\overline{p}^2 M_t^2 \cdot f\left(M_t^2\right)}{\tau_a} \quad , \tag{5.248}$$

where $\tau_a = 0.13\,(2K/\epsilon_s)\,M_t$ and $f\left(M_t^2\right)$ is defined in eq. (5.246). The constants α, β were originally given as $\alpha = 1$, $\beta = 2$. Later Durbin and Zeman [23] extended the model to account for one-dimensional compression using RDT:

$$\overline{p'd'} = -\frac{1}{\overline{\rho}^2\overline{c}^2}\left(\frac{1}{2}\frac{d\overline{p'^2}}{dt} + \frac{\gamma - 1}{2}\overline{p'^2}\frac{\partial \tilde{u}_i}{\partial x_i}\right) \quad . \tag{5.249}$$

The first term on the rhs is a slow relaxation term (see (5.245)) and the second a rapid compression term. The coefficient in front of the rapid term was slightly modified by Zeman and Coleman [103]. In an attempt to provide a pressure-dilatation model for boundary layers in quasi-equilibrium, Zeman [102] analyzed the pressure variance equation in more detail and obtained:

$$\overline{p'd'} = 2f_\rho\left(M_t\right)\overline{c}^2\widetilde{u_i''u_j''}\frac{K}{\epsilon}\frac{1}{\overline{\rho}}\frac{\partial\overline{p}}{\partial x_i}\frac{\partial\overline{p}}{\partial x_j} \quad , \tag{5.250}$$

where

$$f_\rho\left(M_t\right) = 0.02\left(1 - exp\left(-M_t^2/0.2\right)\right) . \tag{5.251}$$

Aupoix et al's model:

Instead of using linear acoustics and scaling relations, Aupoix, Blaisdell, Reynolds and Zeman [100] derive a transport equation for $\overline{p'd'}$ assuming that the fluid behaves as a perfect gas, and model it. For homogeneous turbulence the model reads:

$$\frac{d\overline{p'd'}}{dt} = -c_1\overline{\rho}\frac{M_t^2}{\tau}\frac{dK}{dt} - c_2\frac{1}{\tau}\overline{p'd'} \quad , \tag{5.252}$$

with

$$\tau = K^{3/2}/\left(\epsilon\overline{c}\right) \quad , \quad c_1 = 0.25 \quad , \quad c_2 = 0.2 \quad . \tag{5.253}$$

Comparison of model predictions with DNS data for homogeneous isotropic and sheared turbulence shows that the model works well for sheared flows. Isotropic flows with various initial acoustics allowed for fair predictions only due to the sensitivity of these flows to initial conditions.

Sarkar's model:

Sarkar [72] derived a model for homogeneous flows based on a Poisson equation for the pressure fluctuations, a decomposition into incompressible and compressible pressure fields and the fact that the time-integrated contribution of the latter is negligible. Homogeneous flows have no density gradients. Equation (5.240), therefore, simplifies to:

$$p_{,jj}^{I'} = -2\overline{\rho}\,\overline{u}_{i,j}u_{j,i}' - 2\overline{\rho}\,\overline{u}_{i,i}u_{j,j}' - \overline{\rho}\left(u_i'u_j'\right)_{,ij} \quad . \tag{5.254}$$

As is done in incompressible modelling, the first two rhs terms define a rapid and the last term, a slow part, cf. the defining equation (5.242). The rapid pressure fluctuation satisfies the Poisson equation:

$$p^{R\prime}_{,jj} = -2\overline{\rho}\,\overline{u}_{i,j}u'_{j,i} - 2\overline{\rho}\,\overline{u}_{i,i}u'_{j,j} \tag{5.255}$$

and the slow fluctuation obeys

$$p^{s\prime}_{,jj} = -\overline{\rho}\left(u'_i u'_j\right)_{,ij} \quad . \tag{5.256}$$

Homogeneous flow is periodic in all spatial directions and thus allows to solve eqs. (5.255) and (5.256) exactly by Fourier transforms

$$k^2\hat{p}^R = 2i\,\overline{\rho}\,k_i\overline{u}_{i,j}\hat{u}_j + 2i\,\overline{\rho}\,k_j\overline{u}_{i,i}\hat{u}_j \quad , \tag{5.257}$$

$$k^2\hat{p}^s = -\overline{\rho}k_ik_j\widehat{u_iu_j} \quad . \tag{5.258}$$

Multiplying these relations by the complex conjugate (denoted by $*$) of the Fourier transformed dilatation fluctuation, $\hat{d}^* = -ik_m\hat{u}^*_m$, and integrating over all wavenumbers, gives:

$$\overline{p^{R\prime}d'} = 2\overline{\rho}\,\overline{u}_{i,j}\int\frac{k_ik_m}{k^2}E_{jm}d\vec{k} + 2\overline{\rho}\,\overline{u}_{i,i}\int\frac{k_jk_m}{k^2}E_{jm}d\vec{k} \quad , \tag{5.259}$$

$$\overline{p^{s\prime}d'} = \overline{\rho}\int\frac{k_ik_jk_m}{k^2}\left(i\widehat{u_iu_j}\hat{u}^*_m - i\widehat{u_iu^*_j}\hat{u}_m\right)d\vec{k} \quad , \tag{5.260}$$

where

$$E_{jm} = \left(\hat{u}_j\hat{u}^*_m + \hat{u}_m\hat{u}^*_j\right)/2 \tag{5.261}$$

denotes the spectrum of the Reynolds stress tensor $\overline{u'_ju'_m}$.

Sarkar [72] has evaluated the integrals in the above equations using scaling arguments, truncated Taylor-series expansions and an order of magnitude analysis to obtain:

$$\overline{p'd'} \approx \overline{p^{I\prime}d'} = 2\alpha_2M_t\overline{\rho}\,\overline{u}_{i,j}b_{ij}K + \alpha_3\overline{\rho}\epsilon_sM_t^2 + \frac{16}{3}\alpha_4M_t^2\overline{\rho}\,\overline{u}_{i,i}K \quad , \tag{5.262}$$

where b_{ij} is the Reynolds stress anisotropy tensor. The model coefficients are $\alpha_2 = 0.15$, $\alpha_3 = 0.2$. The last coefficient has not yet been calibrated for flows with homogeneous compression. The second term on the rhs of (5.262) accounts for the slow pressure part. Sarkar shows that the model provides good agreement with DNS data of homogeneous shear turbulence when used in SOC predictions with the SSG pressure-strain model and his ϵ_d-model. However, Sarkar also mentions that eq. (5.262) forms a reasonable approximation for weakly inhomogeneous flows *without* walls. In wall-bounded flows he expects $\overline{p'd'}$ to be smaller because the rms rapid pressure is a smaller fraction of $2\overline{\rho}K$ and the wall normal velocity fluctuation is damped.

El Baz and Launder's model:

Although this model is not derived from observations related to DNS of homogeneous shear flow, it is included in this subsection for later reference. The model is empirical and was designed in order to account for compressibility effects in mixing layers. An attempt was made to treat the influence of dilatational terms on the pressure-strain correlation within a Reynolds stress modelling framework. Only the rapid part of this correlation was modified with respect to the model for incompressible flow. The trace of this term (denoted by ϕ_{kk2}) vanishes in incompressible flows. In compressible flows it is generally nonzero and provides a finite pressure-dilatation correlation which is of the form (see El Baz and Launder [25]):

$$\overline{p'd'} = 3\overline{\rho}{M_t}^2\left(\frac{4}{3}K\frac{\partial\tilde{u}_i}{\partial x_i} + \widetilde{u_i''u_j''}\frac{\partial\tilde{u}_i}{\partial x_j}\right). \tag{5.263}$$

Applications of the model to predict simple-stream and two-stream mixing layers with sub- and supersonic convective Mach numbers provided broadly satisfactory agreement with available experiments.

Ristorcelli's model:

Ristorcelli [66] has conducted a small M_t singular perturbation expansion of the compressible Navier-Stokes equations about a mean state corresponding to homogeneous turbulence without mean dilatation. The first order expansion of the continuity equation provides a diagnostic relation for the dilatation fluctuation of the form:

$$-\gamma d' = \frac{\partial p_1'}{\partial t} + {v_j}'\frac{\partial p_1'}{\partial x_j} \tag{5.264}$$

where ${v_j}'$ and ${p_1}'$ denote the incompressible velocity and pressure fluctuations. Ristorcelli calls ${p_1}'$ the 'pseudo-pressure' in order to distinguish it from the pressure associated with the acoustic problem. Multiplying (5.264) by ${p_1}'$, averaging and applying the homogeneity condition leads to:

$$-2\gamma \overline{p_1' d'} = \frac{d\overline{{p_1'}^2}}{dt} \quad . \tag{5.265}$$

This relation is similar to that used by Zeman [101] with the difference that the 'pressure-dilatation' formed with the incompressible part of p' (cf. also Sarkar [72]) now equals the time rate of change of the variance of the incompressible pressure. The advantage is that a model results which is consistent with the low ${M_t}^2$ asymptotics. While Zeman's model indicates $\overline{p'^2} \to 0$ as ${M_t}^2 \to 0$, Ristorcelli's model guarantees that $\overline{p_1'^2}$ is a function of K in the incompressible limit.

Without going into details of the model derivation it may be said that Ristorcelli [67] uses the theory of incompressible homogeneous isotropic turbulence to obtain an exact expression for the time rate of change of $\overline{{p_1}'^2}$. In the case of homogeneous sheared turbulence, scaling assumptions lead to:

$$\overline{p'd'} = -\chi_{pd} M_t^2 \epsilon_s \left(\frac{-\widetilde{u_i'' u_j''}}{\epsilon_s} \frac{\partial \tilde{u}_i}{\partial x_j} - 1 \right) . \tag{5.266}$$

The model coefficient χ_{pd} depends on Kolmogorov's scaling parameter α and the relative strain rate $(\widetilde{s_{ij}}\widetilde{s_{ij}})^{\frac{1}{2}} K/\epsilon_s$. The result (5.266) is similar to that of Aupoix *et al.* [5] and of Sarkar [72]. All three models allow for a change in sign from isotropic to sheared turbulence. This change in sign depends on M_t for Sarkar's model. In the case of Aupoix *et al.* 's and Ristorcelli's models it simply depends on whether the K-production exceeds dissipation. For more details of a model comparison we refer to Ristorcelli [67].

5.3.3. COMPRESSIBLE CHANNEL FLOW

Fully developed supersonic channel flow has been investigated by Coleman, Kim and Moser [19], using DNS. As for intrinsic compressibility effects, channel flow is very similar to boundary layer flow along cooled walls. In this respect the various aspects discussed in this chapter also apply to corresponding compressible boundary layers.

In order to achieve supersonic flow in a channel, the heat produced by dissipation has to be removed via wall cooling. In reality, the flow is driven by a mean pressure gradient which implies mean density and temperature

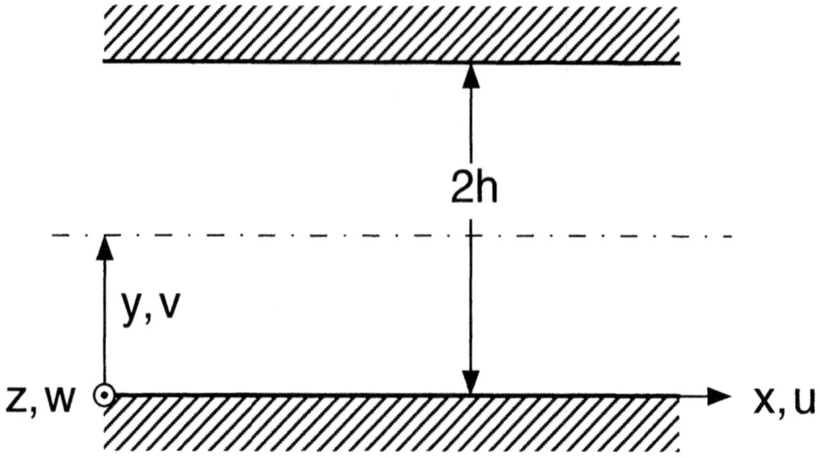

Figure 5.21. Channel geometry and coordinate system

gradients in axial direction. With the aim of preserving streamwise homogeneity in the simulation, Coleman *et al.* [19] have replaced the mean axial pressure gradient by an axially uniform body force. This force was adjusted so that the total mass flux through the channel remained constant. The channel geometry and the coordinate systems are given in figure 5.21.

Statistical quantities were obtained by averaging over time and streamwise (x) and spanwise (z) directions. Free parameters of the flow are the Mach number M, based on the bulk velocity u_m and the wall sound speed c_w, and the Reynolds number Re, based on the bulk density ρ_m, bulk velocity, channel half-width h and wall viscosity. The Prandtl number Pr was assumed constant, like the ratio of specific heats, γ. The dynamic viscosity varied as the 0.7th power of the temperature. The bulk quantities are defined by:

$$\rho_m = \int_0^1 \overline{\rho} d(y/h), \qquad \rho_m u_m = \int_0^1 \overline{\rho}\widetilde{u} d(y/h). \tag{5.267}$$

The walls were cooled and kept at constant temperature T_w. The wall shear stress, τ_w, and the heat flux, q_w, through the wall are results of the computation. They define the friction velocity

$$u_\tau^2 = \frac{\tau_w}{\rho_w}, \tag{5.268}$$

(which characterizes the level of the velocity fluctuations) and a typical temperature fluctuation T_τ:

$$T_\tau = \frac{|q_w|}{\rho_w c_p u_\tau}, \tag{5.269}$$

with the heat flux through the wall:

$$q_w = -k_w \left(\frac{\partial \overline{T}}{\partial y}\right)_w. \tag{5.270}$$

Mean flow properties

Statistically steady fully developed compressible channel flow driven by a mean body force $\overline{\rho}\widetilde{f_x}$, guarantees

$$\frac{\partial \overline{\rho}}{\partial x} = \frac{\partial \widetilde{u}}{\partial x} = 0 \tag{5.271}$$

and as a result of the mass balance:

$$\widetilde{v} = 0, \tag{5.272}$$

as long as the walls are impermeable. (5.272) also means that $\overline{v} = \overline{v''}$. From the averaged momentum equation (5.47) in axial direction, we obtain:

$$0 = \frac{\partial}{\partial y}(\overline{\tau_{xy}} - \overline{\rho}\widetilde{u''v''}) + \overline{\rho}\widetilde{f_x}. \tag{5.273}$$

Using the symmetry and wall boundary conditions

$$\begin{aligned} \overline{\tau_{xy}} &= 0, & \widetilde{u''v''} &= 0, & \text{at } y &= h \\ \overline{\tau_{xy}} &= \tau_w, & \widetilde{u''v''} &= 0, & \text{at } y &= 0 \end{aligned} \tag{5.274}$$

and integrating (5.273) from the wall to the symmetry plane gives:

$$\tau_w = \rho_m \widetilde{f_x} h. \tag{5.275}$$

The effective mean pressure gradient

$$-\frac{\partial \overline{p}}{\partial x} = \overline{\rho}\widetilde{f_x} = \frac{\tau_w}{h}\frac{\overline{\rho}}{\rho_m} \tag{5.276}$$

therefore depends on y through $\overline{\rho}$. This behaviour contrasts with that in a real pressure-driven channel flow, where $\partial\overline{p}/\partial x = -\tau_w/h$. In fact the authors missed to replace $-\partial\overline{p}/\partial x$ by $\widetilde{f_x}$ alone. As a result of eq. (5.276), the total shear stress τ_T in the channel behaves as:

$$\tau_T = \overline{\tau_{xy}} - \overline{\rho}\widetilde{u''v''} = \tau_w \left(1 - \int_0^{y/h} \overline{\rho}/\rho_m d\left(y/h\right) \right), \tag{5.277}$$

i.e. it does not vary linearly, as in actual channel flow. The overall effect, however, is so small in the two computed flow cases, that it has no impact on the conclusions drawn with respect to compressibility effects.

The energy balance (cf. Huang *et al.* [44]) provides further useful insight into the flow. The mean total enthalpy conservation equation reads:

$$\frac{\partial}{\partial x}\overline{\rho u H} + \frac{\partial}{\partial y}\overline{\rho v H} = \frac{\partial}{\partial y}\left(\overline{u_i \tau_{iy}} - \overline{q_y}\right). \tag{5.278}$$

Since the mean effective pressure appears in the first term on the lhs, this term is non-zero. The mean flux of total enthalpy in axial direction, namely

$$\begin{aligned} \overline{\rho u H} &= \overline{\rho}\tilde{u}\left(\tilde{e} + \frac{1}{2}\tilde{u}^2\right) + \tilde{u}\overline{p} + \overline{u''p} + \tilde{u}\left(\overline{\rho u_i'' u_i''/2} + \overline{\rho u''^2}\right) \\ &\quad + \overline{\rho u''\left(e'' + u_i'' u_i''/2\right)}, \end{aligned} \tag{5.279}$$

has the axial gradient:

$$\frac{\partial}{\partial x}\overline{\rho u H} = \tilde{u}\frac{\partial \overline{p}}{\partial x}. \tag{5.280}$$

Similarly, one finds:

$$\frac{\partial}{\partial y}\overline{\rho v H} = \frac{\partial}{\partial y}\left(\overline{v''p} + \tilde{u}\overline{\rho u''v''} + \overline{\rho v''\left(e'' + u_i'' u_i''/2\right)}\right), \tag{5.281}$$

$$\frac{\partial}{\partial y}\left(\overline{u_i \tau_{iy}} - \overline{q_y}\right) = \frac{\partial}{\partial y}\left(\overline{u}\,\overline{\tau_{xy}} + \overline{v}\,\overline{\tau_{yy}} + \overline{u_i' \tau_{iy}'} - \overline{q_y}\right). \tag{5.282}$$

With these results, the total enthalpy balance (5.278) can be integrated from the wall to the symmetry plane, to give

$$q_w = \int_0^h \tilde{u}\frac{\partial \overline{p}}{\partial x} dy = -\tau_w u_m \,. \tag{5.283}$$

This result implies that all correlations appearing in equations (5.281), (5.282) vanish at the wall and in the symmetry plane, which can be easily shown. Eq. (5.283) states that, due to the negative pressure gradient, the heat leaves the channel and its flux increases with the Mach number $M =$

u_m/c_w. It corresponds to the work done by the mean pressure gradient in one half of the channel. See Huang *et al.* [44] for a detailed discussion of the total energy balance within the channel.

Effects of compressibility

DNS results of Coleman *et al.* [19] will be discussed in this subsection which show the effect of compressibility. This will also include a comparison with incompressible channel flow data. The two flow cases of Coleman *et al.* [19] to which we refer, are shown in table 6.

TABLE 5.6. Parameters of fully developed supersonic channel flow used in DNS by Coleman *et al.* [19].

Case	M	Re	$\bar{T}_c/T_w$	Pr	γ
A	1.5	3000	1.38	0.7	1.4
B	3.0	4880	2.47	0.7	1.4

The bulk Mach and Reynolds numbers

$$M = u_m/c_w \quad , \qquad Re = \rho_m u_m h/\mu_w \tag{5.284}$$

increase from case A to B. In order to minimize differences due to Reynolds number effects, the bulk Reynolds number has been selected such that the local Reynolds numbers $\overline{\rho}\,\overline{u}y/\overline{\mu}$ are as similar as possible. Instead of fixing the wall temperature, the ratio of centerline to wall temperature is specified. A few global quantities computed by Huang *et al.* [44] are of interest and listed in table 5.7.

TABLE 5.7. Computed global parameters.

Case	Re_c	h^*	u_τ/u_m
A	2760	151	0.0545
B	2871	150	0.0387

The centerline Reynolds numbers of both cases

$$Re_c = \overline{\rho_c}\,\overline{u_c}h/\overline{\mu_c} \tag{5.285}$$

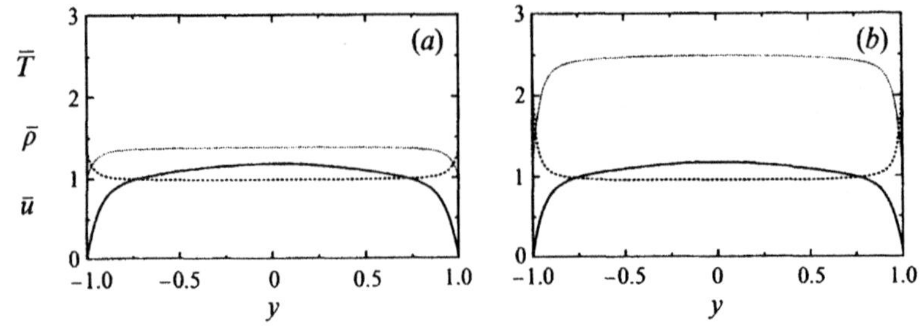

Figure 5.22. Profiles of mean density, temperature and velocity. Left: $M = 1.5$, right: $M = 3$. Taken from Coleman *et al.* [19] by permission.

are very similar. The centerline values h^* of a newly defined wall coordinate

$$y^* = \overline{\rho}\left(\tau_w/\overline{\rho}\right)^{1/2} y/\overline{\mu} \tag{5.286}$$

(a sort of local Reynolds number) are practically the same. As Huang *et al.* [44] state, y^* is perhaps the best among three possible definitions of wall coordinates to collapse all compressible channel flow data. For the log law presentation the coordinate

$$y^+ = \rho_w y u_\tau / \mu_w \tag{5.287}$$

will be used, where $u_\tau = (\tau_w/\rho_w)^{1/2}$.

Mach number effects:

The direct simulations of Coleman *et al.* [19] and the further analysis of their data by Huang *et al.* [44] provide complete insight into compressibility effects. Before we go into more details, we list up those quantities the magnitudes of which increase with increasing bulk Mach number M:

- the heat flux at the wall, q_w
- the near–wall gradients of $\overline{\rho}$ and $\overline{T}$
- the rms-fluctuations of ρ and T
- the turbulent Mach number M_t
- the compressible dissipation rate ϵ_d/ϵ_s and terms associated with $\overline{u''}$, $\overline{v''}$.

The following figures illustrate some of these effects. Figure 5.22 shows profiles of $\overline{\rho}$, $\overline{T}$, $\overline{u}$. The near–wall variations are rapid and qualitatively different from those found in adiabatic–wall boundary layers (See Fernholz and Finley [28], Dussauge *et al.* [24]). In the present case the walls are

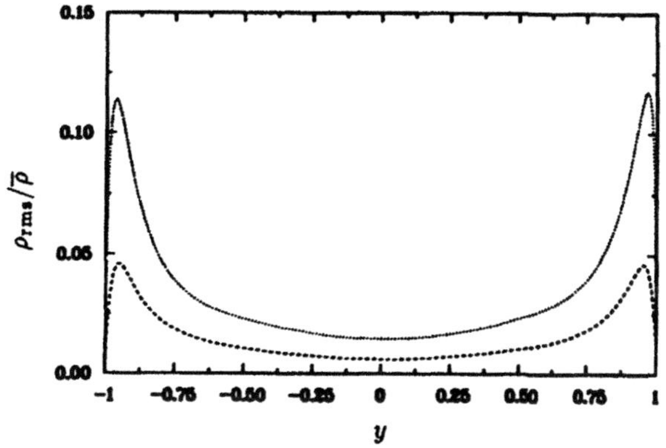

Figure 5.23. Profiles of rms density fluctuations: – – – – , Case A; · · · · · · , Case B. Taken from Coleman *et al.* [19] by permission.

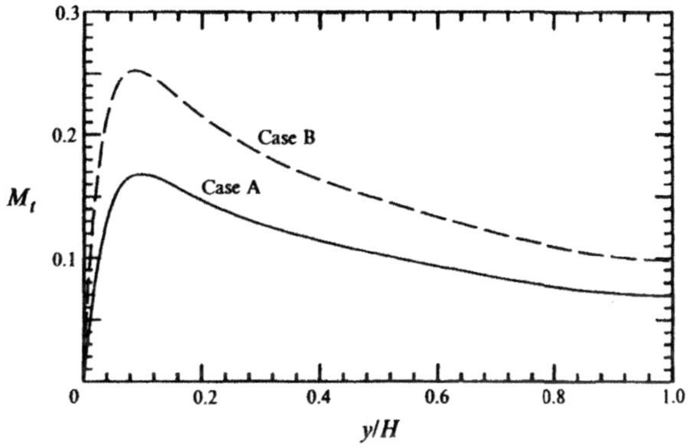

Figure 5.24. Profiles of turbulent Mach number for cases A, B. Taken from Huang *et al.* [44] by permission.

colder than the bulk of the flow and allow the heat generated by dissipation to be transferred out of the channel. The mean density and temperature variations, however, are such that the mean flow is approximately isobaric. This can be concluded from the relation

$$\frac{\partial \overline{p}}{\partial y} \approx \overline{\rho}\frac{\partial \overline{T}}{\partial y} + \overline{T}\frac{\partial \overline{\rho}}{\partial y} \tag{5.288}$$

and the fact that the pressure gradient is everywhere smaller than 0.5% of the first term on the rhs, taken at the wall. Profiles of the rms density fluctuations and of $M_t = K^{1/2}/\bar{c}$ are presented in figures 5.23 and 5.24. Near the

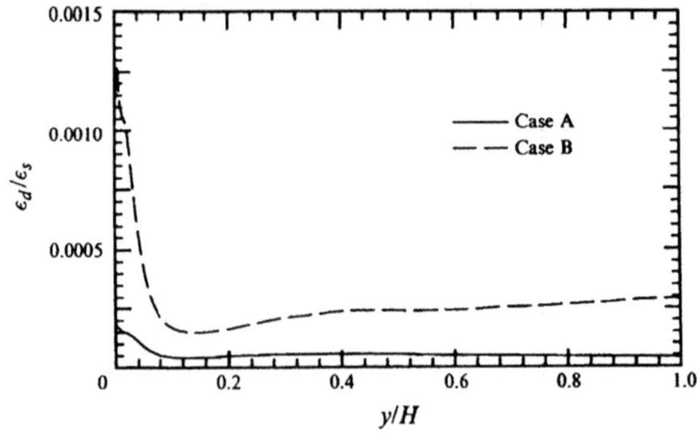

Figure 5.25. Profiles of compressible dissipation rate for cases A, B. Taken from Huang *et al.* [44] by permission.

wall the density fluctuations are greater that 10% of $\overline{\rho}$. The turbulent Mach numbers are fairly large. The temperature fluctuations (not shown) reach similarly high values. From joint probability density functions which reflect negative $\rho - T$ and positive $u - T$ correlations near the wall, Coleman *et al.* [19] conclude that these fluctuations are primarily a result of solenoidal passive mixing across a mean gradient rather than of acoustic nature. The maximum rms total temperature fluctuation is 20% of the mean value in the $M = 3$ case. The compressible dissipation rate profiles are depicted in figure 5.25. $\epsilon_d \,/\, \epsilon_s$ increases with M, but is never significantly larger than 10^{-3}. The paper by Huang *et al.* [44] also discusses the different contributions to ϵ (cf. eqs. (5.61) - (5.67)). It turns out that correlations involving viscosity fluctuations are non-negligible close to the wall. ϵ_3 even reaches maximum values of about 16% of ϵ, deep in the viscous wall region. $\epsilon_I \,/\, \epsilon$ on the other hand does not exceed values of 2,5%. The pressure-dilatation correlation $\overline{p'd'}$ changes its sign from negative to positive values very close to the wall. It however, never exceeds 1% of the solenoidal dissipation rate. Consequently, the explicit compressibility terms ϵ_d and $\overline{p'd'}$ are negligible in supersonic channel flow.

The question arises whether the flow satisfies Morkovin's hypothesis. It will be shown below, that on the whole it does.

Near-wall scaling:

Many of the channel flow quantities, when properly scaled, collapse onto their values for incompressible flow. This fact is well-known for boundary layer flow. In a careful evaluation of compressible turbulent boundary layer

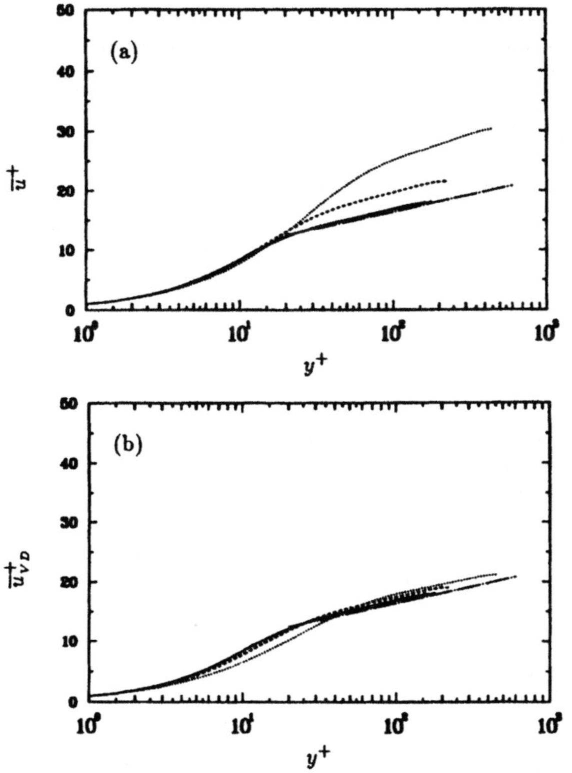

Figure 5.26. Comparison of mean velocity profiles for cases A, B with that for incompressible flow (Kim, Moin and Moser [48]) in wall units (top) and with Van Driest transformation (bottom). Taken from Coleman *et al.* [19] by permission.

data, Fernholz and Finley [28] had shown that the incompressible law of the wall is preserved when the velocity profile is transformed according to

$$\overline{u_{VD}} = \int_0^{\overline{u}} (\overline{\rho}/\rho_w)^{\frac{1}{2}} d\overline{u} \,. \tag{5.289}$$

This is Van Driest's [92] transformation. The density-weighted velocity was found to satisfy the log-law

$$\overline{u_{VD}}^+ = \frac{1}{\kappa} ln y^+ + C \,, \tag{5.290}$$

with $\kappa = 0.41$ and $C = 5.2$, see also Huang and Coleman [43]. The superscript + denotes wall units given by τ_w, ρ_w, μ_w. The wall coordinate y^+ was already defined in (5.287) and $\overline{u}^+ = \overline{u} \,/\, u_\tau = \overline{u} \,/\, (\tau_w \,/\, \rho_w)^{1/2}$. The

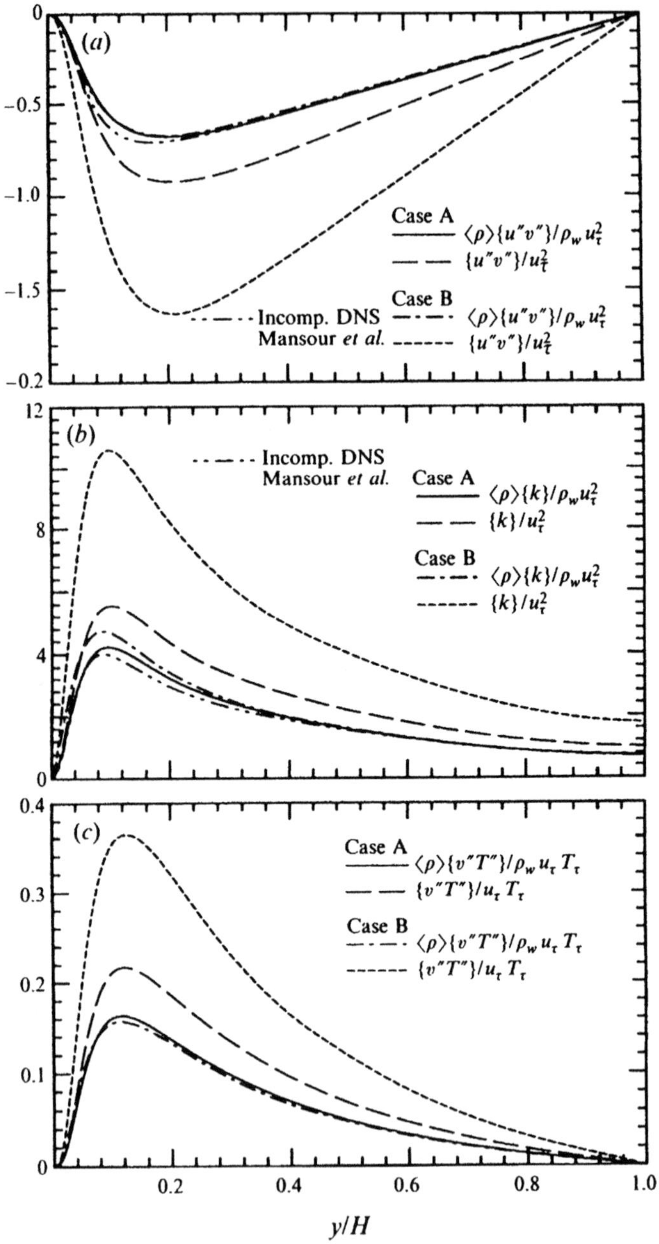

Figure 5.27. Effect of density scaling as suggested by Morkovin [63] for turbulent shear stress (a), TKE (b) and turbulent heat flux (c). Taken from Huang *et al.* [44] by permission.

same scaling holds for channel flow. Figure 5.26 confirms the validity of the Van Driest transformation in Coleman's supersonic channel. A similar collapse of curves is achieved for the Reynolds stress, the TKE and the turbulent heat flux, when these quantities are scaled with $\rho_w u_\tau^2$ and $\rho_w u_\tau T_\tau$, respectively, see figure 5.27. This scaling had been suggested by Morkovin [63]. The figure also contains incompressible channel flow data of Mansour *et al.* [59]. As Huang *et al.* [44] point out, the incomplete collapse of the density-scaled variables is due to Reynolds number differences.

These few examples clearly show that in supersonic channel flow the major Mach number effect leads to rapid mean property variations, whereas intrinsic compressibility effects are negligible. These findings seem to be supported by DNS data of super- and hypersonic boundary layers obtained by Guo and Adams [36] for Mach numbers ranging from 3 to 6 along hot adiabatic walls. They found negligibly small values for $\overline{p'd'}$ and $\overline{\rho}\epsilon_d$ compared to $\overline{\rho}\epsilon_s$.

Testing models for explicit compressibility terms

Based on the observation that in supersonic boundary layers along adiabatic walls pressure and total temperature fluctuations are small compared to the mean values of pressure and total temperature, Morkovin [63] formulated the strong Reynolds analogy (SRA):

$$\frac{T_{rms}}{\widetilde{T}} \approx (\gamma - 1) M^2 \frac{u_{rms}}{\widetilde{u}} . \tag{5.291}$$

This relation is sometimes used in the form:

$$\frac{T'}{\overline{T}} \approx -(\gamma - 1) M^2 \frac{u'}{\overline{u}} . \tag{5.292}$$

The SRA was found to be inadequate in cases with non-zero heat flux at the wall. Extensions of the analogy were therefore proposed by Gaviglio [35] and Rubesin [70]. Huang *et al.* [44] have derived a more general representation of an SRA for non-adiabatic flows which matches the DNS results very well and reads:

$$\frac{T'/\overline{T}}{(\gamma - 1) M^2 u'/\overline{u}} \approx \frac{1}{Pr_t} \frac{\partial \overline{T}}{\partial \overline{u}} \frac{c_p}{\overline{u}} \approx \frac{\widetilde{v''T''}}{\widetilde{u''v''}} \frac{c_p}{\overline{u}} . \tag{5.293}$$

The turbulent Prandtl number Pr_t is defined by

$$Pr_t = \frac{\widetilde{u''v''} \frac{\partial \widetilde{T}}{\partial y}}{\widetilde{v''T''} \frac{\partial \widetilde{u}}{\partial y}} . \tag{5.294}$$

Relation (5.293) is also used to derive a model for the mean Favre-averaged fluctuations, such as $\overline{u_i''}$ and $\overline{T''}$. The authors follow Rubesin [70] and assume that the fluid behaves in a polytropic manner

$$\frac{p'}{\overline{p}} = n\frac{\rho'}{\overline{\rho}} = \frac{n}{n-1}\frac{\rho T'}{\overline{\rho}\tilde{T}} . \tag{5.295}$$

Substituting (5.293) into (5.295) they obtain a relation for $\rho' \,/\, \overline{\rho}$ which finally leads to the following model for any variable $\overline{f''}$:

$$\overline{f''} = -\frac{\overline{\rho' f''}}{\overline{\rho}} \approx -\frac{1}{n-1}\frac{1}{\tilde{T}}\frac{\widetilde{v''T''}}{\widetilde{u''v''}}\frac{\overline{\rho u' f''}}{\overline{\rho}} = -\frac{1}{n-1}\frac{1}{\tilde{T}}\frac{\widetilde{v''T''}}{\widetilde{u''v''}}\frac{\overline{\rho u'' f''}}{\overline{\rho}} \tag{5.296}$$

Since supersonic channel flow is nearly isobaric, Huang *et al.* [44] have chosen $n = 0$ (for an isobaric process) in relation (5.296) to obtain an excellent agreement for $\overline{u''}$, $\overline{v''}$, $\overline{T''}$ with DNS data. The result of this comparison is presented in figure 5.28.

It remains to compare models for the pressure-dilatation (discussed in subsection 5.3.2) with DNS data. Figure 5.29 shows the result of Huang *et al.* . Predictions with models of Sarkar (eq. (5.262)) and El Baz and Launder (eq. (5.263)) were contrasted with DNS data. It is obvious that these models grossly overestimate the effects and should therefore not be used in their present form for wall flow predictions. Finally, figure 5.30 presents a plot of the compressible dissipation rate versus the turbulent Mach number. It shows that $\epsilon_d \,/\, \epsilon_s$ does not depend on M_t alone. Moreover, the values are much below those of predictions.

If one questions why *e.g.* Sarkar's pressure-dilatation model fails, one has to remember that it was designed based on assumptions for homogeneous flows which have no gradients of mean thermodynamic variables. Supersonic channel flow, on the other hand, has important mean property variations. We will demonstrate later that Sarkar's $\overline{p'd'}$-model was quite successful in predictions of homogeneous shear flow. This shows that the search for universal models which are capable of predicting both wall-bounded and free turbulent flows equally well, is most probably of little success. The development of different models for different classes of flow problems promises to be more useful.

One may also question, why the level of explicit compressibility terms is so low in wall-bounded turbulence. There is no conclusive answer at present to this question. However, there are some facts which point to the impermeability constraint as a possible reason. This constraint keeps the velocity fluctuations in the direction of main shear (the wall normal direction) at a low level. For homogeneous shear flow it could be shown that

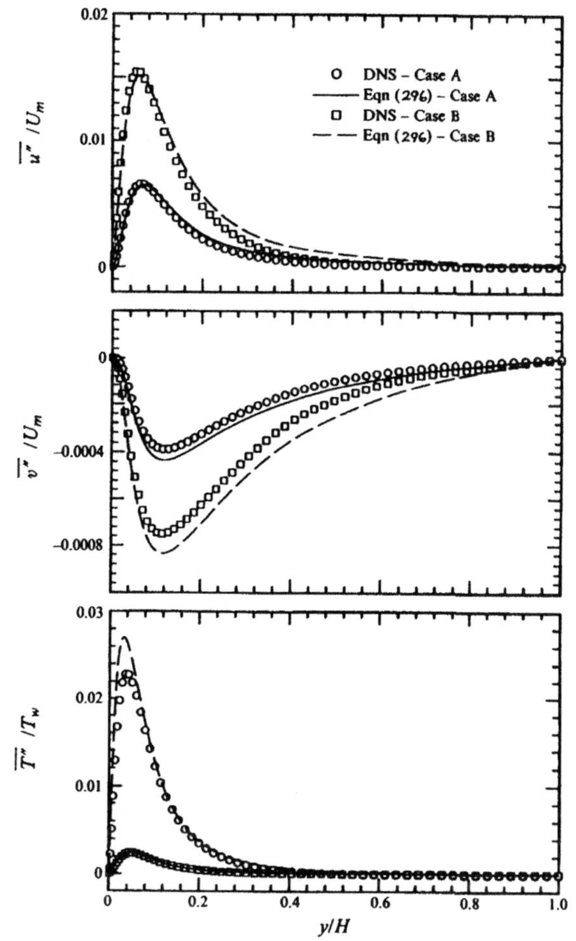

Figure 5.28. Testing models for the mass flux. Taken from Huang *et al.* [44] by permission.

there is a direct link between this velocity component and the dilatation (cf. subsection 5.3.2). The idea that intrinsic compressibility is controlled by the level of the velocity fluctuations in the shear direction gets some support from a linear stability analysis of Coleman's channel data with non-zero amplitudes of the wall normal velocity perturbation (Friedrich and Bertolotti [30]). Work is in progress in Munich to further investigate this aspect.

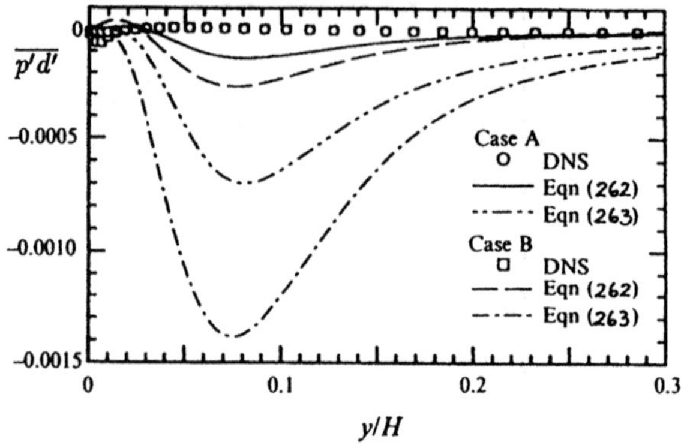

Figure 5.29. Comparison of pressure dilatation models of Sarkar [72] and El Baz and Launder [25] with DNS data for supersonic channel flow. Taken from Huang *et al.* [44] by permission.

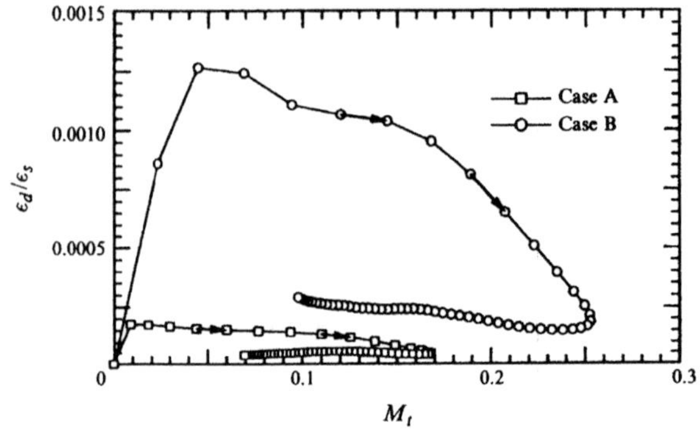

Figure 5.30. Dilatational dissipation rate versus turbulent Mach number. Taken from Huang *et al.* [44] by permission.

5.4. Transport equation models

From the previous discussion it appears that most of the wall bounded supersonic flows are unaffected by intrinsic compressibility effects. This means that generalizations of Morkovin's strong Reynolds analogy (SRA) apply, giving support to variable density extensions of incompressible turbulence models as proper tools to predict these flows.

The situation is more complex for free turbulent flows. DNS of annular mixing layers by Freund *et al.* [29] showed *e.g.* that 'acoustic modes' cannot be neglected when the convective Mach number exceeds values of 0.3, *i.e.* Morkovin's hypothesis that $p'/\overline{p} \ll \rho'/\overline{\rho}$ is then invalid. Changes in the turbulence structure of homogeneous sheared flow due to compressibility effects (M_g) had already been discussed. The prediction of such effects must fail with variable density extensions of models designed for incompressible flows. It needs completely new second-order-closure models, especially new models for the pressure-strain correlation (or the velocity pressure-gradient correlation) and the dissipation rate tensor. The development of these new tools has to be supported by DNS.

In this chapter we give a brief (by no means exhaustive) overview over some transport equation models, starting with two-equation models, because we feel this is the minimum level one should choose to predict complex flows (involving separation *e.g.*). For a discussion of algebraic and one-equation models we refer to the book of Wilcox [99], to the review articles of Knight [49] and Haase *et al.* [38]. A hierarchy of turbulence models for aerodynamic applications is discussed by Marvin and Huang [60] and Bardina *et al.* [6]. Various turbulence models have been tested within the ECARP project (see Haase *et al.* [38]), the ETMA project (see Dervieux *et al.* [21]) and in a recent ERCOFTAC Workshop organized by Batten *et al.* [8]. Finally, we refer to lecture notes of Speziale [84] and Gatski [33] on compressible turbulence closure models, and the valuable comparison of two-equation models by Viala [93].

In what follows, we discuss adaptations of models to wall-bounded flows only.

5.4.1. EDDY VISCOSITY BASED MODELS

The models discussed here use transport equations for two scalar properties of turbulence, namely the turbulence kinetic energy K and its solenoidal dissipation rate $\overline{\rho}\epsilon_s$ or specific dissipation rate ω. They are low-Reynolds number models which are integrated to the wall and do not need any wall functions. The Reynolds stresses are modelled in terms of the eddy viscosity μ_t as follows:

$$-\overline{\rho}\widetilde{u_i''u_j''} = \mu_t \left(\frac{\partial \widetilde{u_i}}{\partial x_j} + \frac{\partial \widetilde{u_j}}{\partial x_i} - \frac{2}{3}\frac{\partial \widetilde{u_k}}{\partial x_k}\delta_{ij} \right) - \frac{2}{3}\overline{\rho}K\delta_{ij} \tag{5.297}$$

For further use we define the quantity

$$\widetilde{s_{ij}} = \frac{1}{2}\left(\frac{\partial \widetilde{u_i}}{\partial x_j} + \frac{\partial \widetilde{u_j}}{\partial x_i} \right) . \tag{5.298}$$

Eq. (5.297) is called Boussinesq approximation.

The Launder-Sharma model

This so-called (K, ϵ) model is widely known and extensively used. Based on fundamental work of Jones and Launder [46] it was proposed by Launder and Sharma in [52]. The eddy viscosity, defined as

$$\mu_t = c_\mu f_\mu \overline{\rho} K^2 / \epsilon_s, \tag{5.299}$$

is scaled with the mean density $\overline{\rho}$, the turbulent velocity scale $K^{1/2}$ and the length scale $K^{3/2}/\epsilon_s$. c_μ is the square of the structure parameter $-\widetilde{u''v''}/K$ in the log-region of a boundary layer. The damping function, f_μ, aims at attenuating the eddy viscosity close to the wall, so that viscous and wall-blocking effects are properly taken into account, see *e.g.* Launder [50] and Durbin [22]. It is expressed in terms of a turbulent Reynolds number Re_t, defined below. The modelled transport equations have the form:

$$\begin{aligned} \frac{\partial (\overline{\rho} K)}{\partial t} + \frac{\partial (\overline{\rho} \widetilde{u_j} K)}{\partial x_j} &= -\overline{\rho} \widetilde{u_i'' u_j''} \widetilde{s_{ij}} - \overline{\rho} \epsilon_s^* \\ &+ \frac{\partial}{\partial x_j} \left((\overline{\mu} + \mu_t / \sigma_K) \frac{\partial K}{\partial x_j} \right) + \phi_K, \end{aligned} \tag{5.300}$$

$$\begin{aligned} \frac{\partial (\overline{\rho} \epsilon_s^*)}{\partial t} + \frac{\partial (\overline{\rho} \widetilde{u_j} \epsilon_s^*)}{\partial x_j} &= -c_{\epsilon 1} \frac{\epsilon_s^*}{K} \overline{\rho} \widetilde{u_i'' u_j''} \widetilde{s_{ij}} - c_{\epsilon 2} f_2 \overline{\rho} \frac{\epsilon_s^{*2}}{K} \\ &+ \frac{\partial}{\partial x_j} \left((\overline{\mu} + \mu_t / \sigma_\epsilon) \frac{\partial \epsilon_s^*}{\partial x_j} \right) + \phi_\epsilon. \end{aligned} \tag{5.301}$$

It must be emphasized that ϵ_s^* is not the true solenoidal dissipation rate. It approaches this quantity only in the fully turbulent wall regime. It is defined as:

$$\epsilon_s^* = \epsilon_s - 2\overline{\nu} \left(\frac{\partial \sqrt{K}}{\partial y} \right)^2, \tag{5.302}$$

where y is the wall normal coordinate. The second term on the rhs of eq. (5.302) is the wall value of ϵ_s. Consequently the wall value for ϵ_s^* is zero. There are two source terms ϕ_K and ϕ_ϵ in the transport equations. The first one, ϕ_K, is artificial. It makes sure, that in the K-transport there is a non-zero dissipation rate as the computation approaches the wall, *i.e.* for $y^+ < 10$. Hence

$$\phi_K = 2\overline{\rho}\, \overline{\nu} \left(\frac{\partial \sqrt{K}}{\partial y} \right)^2. \tag{5.303}$$

The term ϕ_ϵ represents an extra production in order to obtain the proper asymptotic behaviour in the 'dissipation' equation. It has the form:

$$\phi_\epsilon = 2\frac{\overline{\mu}\mu_t}{\overline{\rho}}\left(\frac{\partial^2\tilde{u}}{\partial y^2}\right)^2 . \tag{5.304}$$

$\tilde{u}$ is the mean velocity parallel to the wall. The near wall damping functions are:

$$\begin{aligned} f_\mu &= \exp\left(-3.4/\left(1+0.02\mathrm{Re}_t\right)^2\right), \\ f_2 = 1-0.3\exp\left(-\mathrm{Re}_t{}^2\right), &\qquad \mathrm{Re}_t = \overline{\rho}K^2/\left(\overline{\mu}\epsilon_s\right), \end{aligned} \tag{5.305}$$

and the model constants have the values:

$$\begin{aligned} c_\mu &= 0.09, \qquad c_{\epsilon 1} = 1.45, \qquad c_{\epsilon 2} = 1.92 \\ \sigma_K &= 1.0, \qquad \sigma_\epsilon = 1.3, \qquad \mathrm{Pr}_t = 0.9. \end{aligned} \tag{5.306}$$

Boundary conditions for K and ϵ_s^* to be used at solid walls are:

$$K = 0, \qquad \epsilon_s^* = 0 \qquad \text{at } y = 0. \tag{5.307}$$

For wall-bounded flows, the model gives good agreement with experimental results for zero and small mean pressure gradients, but is less accurate for large adverse pressure gradients (cf. Wilcox [99]). The model predictions are insensitive to freestream values of the turbulence. Due to its low-Reynolds number formulation it needs fine grid spacing near solid walls.

Wilcox's (K, ω)-model

This model transports the specific dissipation rate

$$\omega = \epsilon_s/\left(\beta^* K\right). \tag{5.308}$$

The eddy viscosity is

$$\mu_t = \alpha^*\frac{\overline{\rho}K}{\omega}. \tag{5.309}$$

We present a formulation of the model discussed by Wilcox [99] in which the asymptotically correct wall behaviour of K is achieved by taking into account low Reynolds number effects. The modelled transport equations read:

$$\frac{\partial(\overline{\rho}K)}{\partial t}+\frac{\partial(\overline{\rho}\tilde{u}_jK)}{\partial x_j} = -\overline{\rho}\widetilde{u_i''u_j''}\tilde{s}_{ij}-\beta^*\overline{\rho}\omega K + \frac{\partial}{\partial x_j}\left((\overline{\mu}+\sigma^*\mu_t)\frac{\partial K}{\partial x_j}\right) \quad , \qquad (5.310)$$

$$\frac{\partial(\overline{\rho}\omega)}{\partial t}+\frac{\partial(\overline{\rho}\tilde{u}_j\omega)}{\partial x_j} = -\alpha\frac{\omega}{K}\overline{\rho}\widetilde{u_i''u_j''}\tilde{s}_{ij}-\beta\overline{\rho}\omega^2 + \frac{\partial}{\partial x_j}\left((\overline{\mu}+\sigma\mu_t)\frac{\partial \omega}{\partial x_j}\right) \quad . \qquad (5.311)$$

This model does not need extra source terms to treat the near-wall region. The numerical wall boundary conditions require the specification of the distance from the wall to the first point off the wall. At no-slip surfaces the boundary conditions to be used, are:

$$K=0 \quad , \quad \omega = 10\frac{6\overline{\mu}}{\beta\overline{\rho}(y_1)^2} \quad , \quad y_1^+<1 \quad . \qquad (5.312)$$

y_1 is the distance of the first point away from the wall. The model constants and low Reynolds number corrections are:

$$\beta=3/40 \quad , \quad \sigma=0.5 \quad , \quad \sigma^*=0.5 \quad , \quad Pr_t=0.9 \quad ,$$

$$\alpha^* = \frac{\alpha_o^*+Re_t/R_K}{1+Re_t/R_K} \quad ,$$
$$\alpha = \frac{5}{9\alpha^*}\frac{\alpha_o+Re_t/R_\omega}{1+Re_t/R_\omega} \quad ,$$
$$\beta^* = \frac{9}{100}\frac{5/18+(Re_t/R_\beta)^4}{1+(Re_t/R_\beta)^4} \quad , \qquad (5.313)$$

where

$$\alpha_o^*=\beta/3\,, \quad \alpha_o=1/10\,, \quad R_\beta=8\,, \quad R_K=6\,, \quad R_\omega=27/10\,. \quad (5.314)$$

The turbulence Reynolds number is given by

$$Re_t = \frac{\overline{\rho} K}{\omega \overline{\mu}} \quad . \tag{5.315}$$

For boundary layer flows the following freestream values are recommended:

$$\omega_\infty > f \frac{u_\infty}{L} \quad , \quad \mu_{t\infty} < 10^{-2} \mu_{tmax} \quad , \quad K_\infty = \frac{\mu_{t\infty}}{\rho_\infty} \omega_\infty \quad , \tag{5.316}$$

where L is the approximate length of the computational domain and u_∞ the characteristic freestream velocity. A value of $f = 10$ should be used for this factor of proportionality.
Computations with this model reveal a sensitivity of the results to freestream conditions. On the other hand, the model proves superior in numerical stability to (K, ϵ) models, especially in the viscous sublayer. In the logarithmic region, the model gives good agreement with experimental results (See Bardina *et al.* [6]).

Menter's zonal (K, ω) *models*

Menter [61] has designed two different versions of the (K, ω) model. His so-called baseline (BSL) model combines Wilcox's (K, ω) model for the inner region of a boundary layer (up to roughly its half thickness) with a standard (K, ϵ_s) model in the outer region in order to avoid the dependence of the (K, ω) model on freestream conditions. The dissipation rate (ϵ_s) equation is transformed into an ω-equation using relation (5.308). This transformation produces an additional term in the ω-equation containing the scalar product of the K- and ω-gradients ('cross-diffusion' term). The original (K, ω) model is then multiplied by a blending function F_1 and the transformed (K, ϵ_s) model by a function $(1 - F_1)$ and both are finally added together to give the BSL model. Any constant of the BSL model is a blend of the constants of the original and the transformed models. The blending function F_1 is designed to have a value of one in the near wall region and of zero far from the wall. The blending takes place in the wake region of the boundary layer.
In order to improve predictions of separated flows, Menter has proposed a second model, the so-called shear-stress transport (SST) model. It consists in a modification of the eddy viscosity formula of the BSL model such that transport effects due to Reynolds shear stress are incorporated. The eddy viscosity is defined as

$$\mu_t = \frac{a \overline{\rho} K}{max\,(a\omega, \Omega F_2)} \quad . \tag{5.317}$$

a is the above mentioned structure parameter with a magnitude of 0.31. Ω is the mean vorticity $\partial\tilde{u}/\partial y$ and F_2 an auxiliary function of the wall distance y.

The boundary conditions to be used are the same as for the (K, ω) model.

The performance of the SST model has been tested by Menter and Rumsey [62] for an axisymmetric transonic shock-wave/boundary layer interaction at $M = 0.925$. Comparison with experimental data for the wall pressure distribution shows significant improvement of this model over the BSL- and the standard (K, ω) model.

Marvin and Huang [60] note the interesting aspect that Menter's modification of the eddy viscosity is in accord with ideas used in nonlinear eddy viscosity models where the coefficient c_μ is not a constant but a function of the dimensionless strain and vorticity invariants.

Modifications to account for mean dilatation

The modelled transport equations for the dissipation rate or the specific dissipation rate discussed above, do not account for changes in integral length scales that result from mean dilatation effects in the flow field. Speziale and Sarkar [87] were the first to propose a modification in the transport equation for the solenoidal dissipation rate. It is of the form:

$$\frac{\partial\left(\bar{\rho}\epsilon_s\right)}{\partial t} + \frac{\partial\left(\bar{\rho}\tilde{u}_j\epsilon_s\right)}{\partial x_j} = - c_{\epsilon 1}\frac{\epsilon_s}{K}\bar{\rho}\widetilde{u_i''u_j''}\left(\frac{\partial\tilde{u}_i}{\partial x_j} - \frac{1}{3}\frac{\partial\tilde{u}_k}{\partial x_k}\delta_{ij}\right) - \frac{4}{3}\bar{\rho}\epsilon_s\frac{\partial\tilde{u}_k}{\partial x_k} - c_{\epsilon 2}\bar{\rho}\frac{\epsilon_s^2}{K} + \frac{\partial}{\partial x_j}\left(\left(\bar{\mu} + \frac{\mu_t}{\sigma_\epsilon}\right)\frac{\partial\epsilon_s}{\partial x_j}\right) , \quad (5.318)$$

where $\mu_t = c_\mu\bar{\rho}K^2/\epsilon_s$, $c_\mu = 0.09$, $c_{\epsilon_1} = 1.44$, $c_{\epsilon_2} = 1.83$, $\sigma_\epsilon = 1.3$.

This is a high Reynolds number form of the dissipation rate equation which was recently again suggested by Speziale [84] and Gatski [33]. The production of dissipation term (first term on the rhs) contains the deviatoric strains. The second term accounts for mean dilatation effects. It must be noted that this equation like all the other 'ϵ'-equations is inappropriate in the log layer of wall-bounded flows, since the predicted slope of the Van Driest velocity does not equal $1/\kappa$ $(\kappa = 0.41)$. This fact was pointed out by Huang *et al.* [43]. The (K, ω) model on the other hand has the proper behaviour (Wilcox [99]). So, one might consider a blending of eq. (5.318) with the ω-equation, similar to Menter's [61] suggestion.

5.4.2. ALGEBRAIC STRESS MODELS

These models are extensions of the Boussinesq approximation (5.297). They form a very useful compromise between two-equation and full Reynolds stress transport models. The idea is to add higher order terms (h.o.t) to the stress-strain relationship (5.297) which contain nonlinear products of strain and vorticity tensors:

$$\bar{\rho}\widetilde{u_i'' u_j''} = -2\mu_t(\widetilde{s_{ij}} - \frac{1}{3}\widetilde{s_{kk}}\delta_{ij} + \text{h.o.t}) + \frac{2}{3}\bar{\rho}K\delta_{ij} \tag{5.319}$$

Since in all these models the eddy viscosity, μ_t, is related to K and its dissipation rate, ϵ_s, the idea basically is, to construct an anisotropic eddy viscosity with strain-dependent coefficients. The concept of including the higher-order terms in the above relationship starts from the two hypotheses to be made in the Reynolds stress transport eq. (5.70) and its trace (5.85):

$$2\bar{\rho}K\frac{Db_{ij}}{Dt} \equiv \frac{D}{Dt}(\bar{\rho}\widetilde{u_i'' u_j''}) - \frac{\widetilde{u_i'' u_j''}}{K}\frac{D(\bar{\rho}K)}{Dt} = 0 \tag{5.320}$$

$$\frac{\partial}{\partial x_k}(\bar{\rho}D_{ijk}^T + D_{ijk}^V) = 0 \tag{5.321}$$

Combining equations (5.70) and (5.85) under the assumptions (5.320), (5.321) provides an implicit relation containing the Reynolds stress anisotropy tensor b_{ij} and the pressure strain correlation tensor. A methodology to obtain explicit solutions of such a relation was proposed by Pope [64] for incompressible flow and later used by Gatski and Speziale [34], Shih [77] and recently by Wallin and Johansson [96]. For compressible flows, Speziale [84] and Gatski [33] propose an explicit quadratic constitutive relationship of the form:

$$\begin{aligned}\bar{\rho}\widetilde{u_i'' u_j''} &= \frac{2}{3}\bar{\rho}K\delta_{ij} - 2\mu_t^*\Big[\widetilde{s_{ij}} - \frac{1}{3}\widetilde{s_{kk}}\delta_{ij} + \alpha_2\frac{K}{\epsilon_s}(\widetilde{s_{ik}}\widetilde{r_{kj}} + \widetilde{s_{jk}}\widetilde{r_{ki}}) \\ &\quad - 2\alpha_3\frac{K}{\epsilon_s}(\widetilde{s_{ik}}\widetilde{s_{kj}} - \frac{1}{3}\widetilde{s_{kl}}\widetilde{s_{kl}}\delta_{ij})\Big]\end{aligned} \tag{5.322}$$

where

$$\mu_t^* = \bar{\rho}c_\mu^* K^2/\epsilon_s \tag{5.323}$$

and

$$c_\mu^* = c_\mu^*(II_{\tilde{s}}, II_{\tilde{r}}), \qquad \alpha_2 = \alpha_2(II_{\tilde{s}}, II_{\tilde{r}}), \qquad \alpha_3 = \alpha_3(II_{\tilde{s}}, II_{\tilde{r}}) \tag{5.324}$$

are functions of the invariants of the mean strain rate and rotation rate tensors

$$\widetilde{s_{ij}} = \frac{1}{2}(\widetilde{u}_{i,j} + \widetilde{u}_{j,i}), \qquad \widetilde{r_{ij}} = \frac{1}{2}(\widetilde{u}_{i,j} - \widetilde{u}_{j,i}). \tag{5.325}$$

For further details, the reader is referred to Abid *et al.* [2, 1]. The new model by Wallin and Johansson [96] is discussed in A. Johansson's lecture.

5.4.3. REYNOLDS STRESS TRANSPORT

Turbulence models based on the complete transport equation (5.70) for the Reynolds stress tensor contain closure assumptions for the pressure strain rate correlation, the turbulent and viscous diffusion terms, the dissipation rate and the mass flux variation. A model for the latter has already been mentioned in subsection 5.3.3.

Pressure strain rate modelling

This is an important task because the term can be comparable in magnitude to the production term or the dissipation rate (See DNS data of simple shear flows *e.g.*). The discussion in subsection 5.3.2 has also shown that intrinsic compressibility effects which are non-negligible in homogeneous shear flow, modify this term and thus affect the Reynolds stress anisotropy. Therefore, simple density extensions of existing models may do a good job in predicting wall-bounded turbulence (provided they are satisfactory in the corresponding incompressible cases), however new pressure strain models are needed to account for intrinsic compressibility effects. A step in this direction has been done by El Baz and Launder [25] with application to mixing layers. In his thesis, Vreman [94] models the growth rate reduction in the mixing layer due to reduced pressure fluctuations by introducing a damping factor which acts on the rapid part of the axial pressure-strain component and depends on the convective Mach number. The approach is successful, but cannot be generalized easily. Based on RDT and DNS data for homogeneous axially compressed turbulence Cambon *et al.* [17] have proposed a model which is capable of predicting the variations in the Reynolds stress anisotropies observed in the computation, viz:

$$\Pi_{ij}^{D} = (\Pi_{ij}^{R})_{inc} \cdot exp(-M_d/c_{\Delta m}). \tag{5.326}$$

The Launder-Reece-Rodi [51] model is used to express the rapid part of the incompressible pressure strain correlation. M_d is the distortion Mach number and $c_{\Delta m}$ a model constant. Along similar lines Uhlmann [91] has modified the rapid part of Π_{ij}

$$\Pi_{ij}^R = \Pi_{ij}^R\Big|_{LRR} \cdot exp(-M_d^2), \qquad \Pi_{ij}^S = \Pi_{ij}^S\Big|_{LRR} \tag{5.327}$$

(LRR stands for Launder-Reece-Rodi). His computations of homogeneous shear turbulence show still deficiencies in the shear stress anisotropy.

Modelling the deviatoric part of the pressure strain rate tensor reflects the attempt to distinguish between structural and energetic effects. In fact the dilatational part, $\overline{\rho}\Pi_{ij}^{DL}$, which has been subtracted out of the full term in eq. (5.70) accounts for the energy exchange between the pressure fluctuations, the internal and turbulent kinetic energies. Almost all available models are variable density extensions of classical incompressible models. According to their dependency on the Reynolds stress anisotropy tensor, b_{ij}, they are called linear, quadratic or cubic. Their general form is:

$$\overline{\rho}\Pi_{ij}^D = \overline{\rho}\epsilon_s A_{ij}(\mathbf{b}) + \overline{\rho}K M_{ijkl}(\mathbf{b})\frac{\partial \widetilde{u_k}}{\partial x_l}. \tag{5.328}$$

The second-order tensor $A_{ij}(\mathbf{b})$ is associated with the slow relaxation of the turbulence toward isotropy and the fourth-order tensor $M_{ijkl}(\mathbf{b})$ is associated with the rapid response of the turbulence to imposed mean velocity gradients. Speziale *et al.* [86] have provided exact interpretations of each of these terms for homogeneous flow.

As an example, we present the Speziale, Sarkar and Gatski [88] SSG model. These authors regroup three terms in eq. (5.70) and model the deviatoric part of the pressure-gradient velocity correlation. We note the following equivalence of terms:

$$\overline{\rho}\Pi_{ij}^D + \overline{\rho}\Pi_{ij}^{DL} + \frac{\partial}{\partial x_k}(\overline{\rho}D_{ijk}^T) = -\Pi_{ij} + \frac{2}{3}\overline{p'\frac{\partial u_k'}{\partial x_k}}\delta_{ij} - \frac{\partial}{\partial x_k}C_{ijk}. \tag{5.329}$$

The lhs-terms are from eq. (5.70). The terms on the rhs are defined by Speziale *et al.* [88] as:

$$\Pi_{ij} = \overline{u_i'\frac{\partial p'}{\partial x_j}} + \overline{u_j'\frac{\partial p'}{\partial x_i}} - \frac{2}{3}\overline{u_k'\frac{\partial p'}{\partial x_k}}\delta_{ij}. \tag{5.330}$$

$$C_{ijk} = \overline{\rho u_i'' u_j'' u_k''} + \frac{2}{3}\overline{p' u_k'}\delta_{ij} \tag{5.331}$$

Π_{ij} is the deviatoric pressure gradient velocity correlation and C_{ijk} the turbulent diffusion term. In the case of homogeneous turbulence, $C_{ijk,k}$ vanishes and the remaining terms take the form

$$-\Pi_{ij} + \frac{2}{3}\overline{p'\frac{\partial u'_k}{\partial x_k}}\delta_{ij} = \overline{p'(\frac{\partial u'_i}{\partial x_j} + \frac{\partial u'_j}{\partial x_i})} \tag{5.332}$$

which equals the sum of $\overline{\rho}\Pi_{ij}^D + \overline{\rho}\Pi_{ij}^{DL}$, used in eq. (5.70).

The SSG model for Π_{ij} has the form:

$$\begin{aligned}
\Pi_{ij} &= \overline{\rho}\Bigg[\left(c_1\epsilon_s - c_1^*\widetilde{u_i''u_j''}\frac{\partial \widetilde{u}_i}{\partial x_j}\right)b_{ij} \\
&\quad - c_2\epsilon_s\left(b_{ik}b_{kj} - \frac{1}{3}II_b\delta_{ij}\right) \\
&\quad - \left(c_3 - c_3^* II_b^{1/2}\right)K\left(\widetilde{s_{ij}} - \frac{1}{3}\widetilde{s_{kk}}\delta_{ij}\right) \\
&\quad - c_4K\left(b_{ik}\widetilde{s_{jk}} + b_{jk}\widetilde{s_{ik}} - \frac{2}{3}b_{kl}\widetilde{s_{kl}}\delta_{ij}\right) \\
&\quad - c_5K\left(b_{ik}\widetilde{r_{jk}} + b_{jk}\widetilde{r_{ik}}\right)\Bigg],
\end{aligned} \tag{5.333}$$

where

$$\begin{aligned}
&c_1 = 3.4, \qquad c_1^* = 1.8, \qquad c_2 = 4.2, \\
&c_3 = 4/5, \qquad c_3^* = 1.3, \qquad c_4 = 1.25, \\
&c_5 = 0.4, \qquad II_b = b_{ij}b_{ij}.
\end{aligned} \tag{5.334}$$

It is the compressible version of its incompressible counterpart presented in the review article of Gatski [32].

The turbulent third order diffusion term is modelled by Speziale and Sarkar [87] using a gradient hypothesis (see also Speziale [83]):

$$\overline{\rho u_i''u_j''u_k''} = -\frac{2}{3}c_s\overline{\rho}\frac{K^2}{\epsilon}\left(\frac{\partial}{\partial x_k}\widetilde{u_i''u_j''} + \frac{\partial}{\partial x_j}\widetilde{u_i''u_k''} + \frac{\partial}{\partial x_i}\widetilde{u_j''u_k''}\right), \tag{5.335}$$

where $c_s = 0.11$. The dissipation rate is taken as the sum of the solenoidal and compressible parts:

$$\epsilon = \epsilon_s + \epsilon_d \tag{5.336}$$

and for ϵ_d Sarkar's model (eq. (5.181)) is used. ϵ_s is determined from eq. (5.318) in which μ_t/σ_ϵ is replaced by

$$c_\epsilon\overline{\rho}\frac{K}{\epsilon_s}\widetilde{u_i''u_j''}, \tag{5.337}$$

with $c_\epsilon = 0.15$.

Further terms requiring modelling are the pressure-velocity correlation, the mass and turbulent heat flux, and the viscous diffusion term. Speziale and Sarkar [87] suggest to model the pressure-velocity correlation in terms of the mass and turbulent heat flux

$$\overline{p'u'_i} = -\overline{\rho}R\widetilde{T}\overline{u''_i} + \overline{\rho}R\widetilde{u''_i T''}. \tag{5.338}$$

In this relation the pressure fluctuation is assumed to be a thermodynamic quantity. Moreover the specific heats c_p, c_v are taken as constants. It is unclear at present whether this model behaves properly in wall bounded turbulence with negligible intrinsic compressibility effects. Finally, gradient transport models are proposed for

$$\overline{u''_i} = \frac{c_\mu}{\overline{\rho}\sigma_\rho}\frac{K^2}{\epsilon}\frac{\partial\overline{\rho}}{\partial x_i}, \qquad c_\mu = 0.09, \sigma_\rho = 0.5, \tag{5.339}$$

$$\widetilde{u''_i T''} = -\frac{c_\mu}{\sigma_T}\frac{K^2}{\epsilon}\frac{\partial\widetilde{T}}{\partial x_i}, \qquad \sigma_T = 0.7, \tag{5.340}$$

and the turbulent dissipation rate is assumed isotropic

$$\overline{\rho}\epsilon_{ij} = \frac{2}{3}\overline{\rho}\epsilon\delta_{ij}. \tag{5.341}$$

An asymptotic near-wall analysis in which viscosity fluctuations and some higher order correlations are neglected provides the following approximation for the viscous diffusion term:

$$D^V_{ijk} \approx \overline{\mu}\left(\frac{\partial}{\partial x_k}\widetilde{u''_i u''_j} + \frac{\partial}{\partial x_j}\widetilde{u''_i u''_k} + \frac{\partial}{\partial x_i}\widetilde{u''_j u''_k}\right). \tag{5.342}$$

The pressure-dilatation correlation has been assumed zero for reasons pertinent to homogeneous isotropic turbulence. This assumption is in accord with DNS data for wall-bounded turbulence. Applying the Speziale, Sarkar [87] model to boundary layer flow, it would then be appropriate to neglect ϵ_d as well.

5.4.4. HEAT FLUX TRANSPORT

Like many other compressible turbulence models, Speziale and Sarkar's [87] model uses a gradient-diffusion assumption for the turbulent heat flux. In compressible flow the temperature is not a passive scalar as in incompressible flow. It is directly coupled with the velocity field via density and pressure. More sophisticated heat flux modelling, therefore, seems appropriate. For incompressible and compressible flows Sommer *et al.* [82] have proposed a gradient-diffusion model which removes the constant turbulent Prandtl number assumption via a turbulent heat diffusivity κ_T which is computed from transport equations for the temperature variance $\widetilde{T''^2}$ and its dissipation rate ϵ_T. The model is:

$$\widetilde{u_i''T''} = -\kappa_T \frac{\partial \widetilde{T}}{\partial x_i} \tag{5.343}$$

$$\kappa_T = c_\lambda f_\lambda K \left(\frac{K}{\epsilon}\right)^{1/2} \left(\frac{\widetilde{T''^2}}{\epsilon_T}\right)^{1/2} \quad , \tag{5.344}$$

where $c_\lambda = 0.11$ and the near-wall damping function f_λ depends on the wall distance and turbulent Reynolds number:

$$f_\lambda = f_{PT} \cdot 0.1 Re_t^{-1/4} + \left(1 - exp\left(-y^+/45\right)\right)^2 \quad , \tag{5.345}$$

$$f_{PT} = exp\left(-(Re_t/80)^2\right) \quad , \quad Re_t = \overline{\rho} K^2 / \left(\overline{\mu}\epsilon\right) \quad . \tag{5.346}$$

The model (5.344) contains two different time or length scales which allow to account for thermal and flow effects with the same importance. The coefficient c_λ is selected such that the turbulent Prandtl number Pr_t has a value of 0.89 in the near-wall log-layer. For more details, especially the modelled transport equations, we refer to Sommer *et al.* [82] and to Viala [93].

Instead of assuming a gradient transport process and variable Prandtl number, one may solve the transport equation for the turbulent heat flux (eqs. (5.94), (5.95)) directly. This approach has been followed by Huang and Coakley [42]. They have derived an algebraic relation from the heat flux transport equation by using assumptions associated with the development of algebraic stress models. The relation has the form:

$$-\widetilde{u_k''T''} \frac{\partial \widetilde{u}_i}{\partial x_k} - c_{1T} \widetilde{u_i''T''} \frac{\epsilon}{K} + c_{2T} \widetilde{u_k''T''} \frac{\partial \widetilde{u}_i}{\partial x_k} = \widetilde{u_i''u_k''} \frac{\partial \widetilde{T}}{\partial x_k}, \tag{5.347}$$

with the model coefficients $c_{1T} = 3$, $c_{2T} = 0.5$. Solving this equation by direct inversion provides the heat flux components. Tests of the model in conjunction with two Reynolds stress models and comparison with a gradient transport model did not lead to clearly improved predictability. Further testing seems to be necessary.

An approach based on the solution of the modelled transport equations for the turbulent heat fluxes, $\overline{\rho}\widetilde{u_i''h''}$, was undertaken by Le Ribault and

Friedrich [53]. Most of the model assumptions are adaptations of corresponding incompressible models to account for property variations.

5.4.5. APPLICATIONS

In this section we discuss the behaviour of a few variable density extensions of pressure strain models and of explicit compressibility terms in computations of homogeneous shear flow. Computations of compressible boundary layers and of shock/boundary-layer interactions are finally presented for various turbulence models.

Homogeneous shear flow

The difficulty to predict compressible homogeneous shear turbulence reliably is demonstrated by Speziale *et al.* [85]. They used DNS data of Blaisdell *et al.* [12] to critically evaluate variable density extensions of the pressure-strain models of Launder, Reece and Rodi [51], of Fu, Launder and Tselepidakis [31] and of Speziale, Sarkar and Gatski [88]. We refer to these as the LRR, FLT and SSG models. Furthermore they tested compressible dissipation rate and pressure dilatation models of Sarkar *et al.* [74, 72] and Zeman [100, 101].

The effect of $\overline{p'd'}$ and ϵ_d on the time evolution of K is shown in figures 5.31 and 5.32. Without use of models for these terms, all pressure-strain models predict a much too rapid TKE growth. Adding models for $\overline{p'd'}$ and ϵ_d (either Sarkar *et al.* 's or Zeman's) provides good agreement with DNS data of Blaisdell *et al.* [12] at least for the FLT and SSG models. The linear LRR model still overpredicts the growth rate of K. Similar effects are observed in the evolution of ϵ_s (not shown). All models, however, fail to describe the Reynolds stress anisotropies, as presented in figure 5.33. This shows that variable density extensions of the LRR, FLT and SSG models are unable to predict the dramatic changes in the Reynolds stress anisotropies that arise from intrinsic compressibility effects.

Shock/boundary layer interaction

Interactions of shocks with turbulence fields are quite complex and depend strongly on the specific conditions upstream of the shock. In general, a shock produces (or increases existing) Reynolds stress anisotropy. Since it also modifies the length scales, changes in the dissipation rate anisotropies must be expected. At present it is unclear how these effects can be incorporated into Reynolds stress closures. Recent DNS data of Adams [3, 4] on shock/boundary layer interaction provide valuable physical insight into the

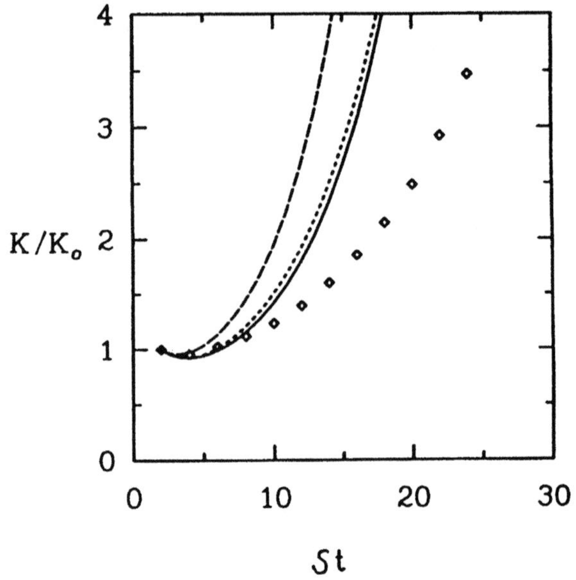

Figure 5.31. Comparison of model predictions for TKE (without explicit compressibility terms) with DNS data of Blaisdell *et al.* [12] for homogeneous shear turbulence. (— — —) Launder, Reece and Rodi [51] model; (- - - -) Fu, Launder and Tselepidakis [31] model; (———) SSG model; ◇ DNS data. Taken from Speziale *et al.* [85] by permission.

K-budget *e.g.* (the Reynolds stress budgets are not discussed but can be extracted from the data base) and thus prove useful in guiding the development of turbulence models.

In what follows we discuss applications of various models ranging from algebraic to SOC models. The first example is Settles *et al.* 's 24° adiabatic compression ramp at $M_\infty = 2.84$ and $Re_\delta = 1.69 \cdot 10^6$. The high ramp angle causes separation and a steep pressure rise well upstream of the corner. A separation shock merges with a recompression shock (resulting from compression waves at reattachment) downstream. Figure 5.34 shows computations with the standard (K, ω)-model performed by Haidinger [39] and Haidinger and Friedrich [40]. They demonstrate that the early pressure rise and the skin friction decay cannot be predicted with a two-equation model. The use of models for the explicit compressibility terms (ϵ_d, $\overline{p'd'}$) does not cure that problem. Figure 5.35 contains a comparison of results obtained with the Baldwin-Lomax, the (K, ω)- and a Reynolds stress transport model based on the SSG pressure-strain model and isotropic dissipation rates deduced from the ω-equation. The (K, ω) result is intermediate between the other two results. The Baldwin-Lomax model computes reattachment of the flow too far downstream. Only the Reynolds stress transport model

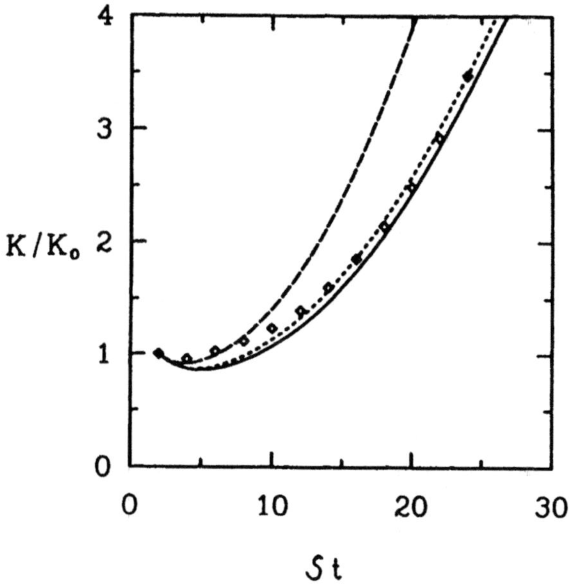

Figure 5.32. Model predictions for K with Sarkar *et al.* 's [74] model for ϵ_d and Sarkar's [72] model for $\overline{p'd'}$. Symbols as in figure 31. Taken from Speziale *et al.* [85] by permission.

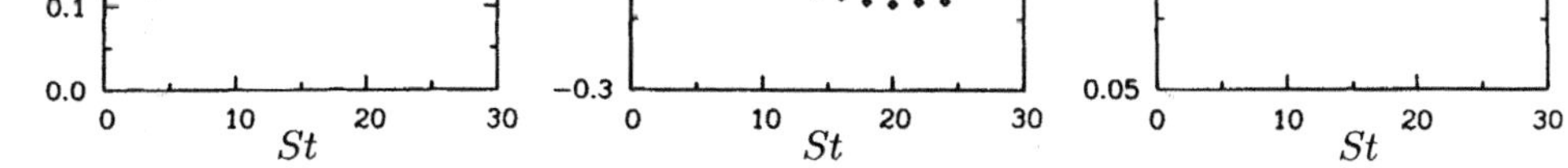

Figure 5.33. Model predictions for Reynolds stress anisotropies with explicit compressibility terms. Models and symbols as in figure 32. Taken from Speziale *et al.* [85] by permission.

allows for improved predictions of this complex flow. From grid coarsening studies it was concluded that still closer agreement with the experimental data must be possible with proper local grid refinement.

In the framework of an ESA turbulence modeling project, the main results of which are summarized by Lindblad *et al.* [56], the interaction of a $M = 5$ shock impinging at 10 degrees on a flat plate with constant wall temperature and interacting with the incoming turbulent boundary

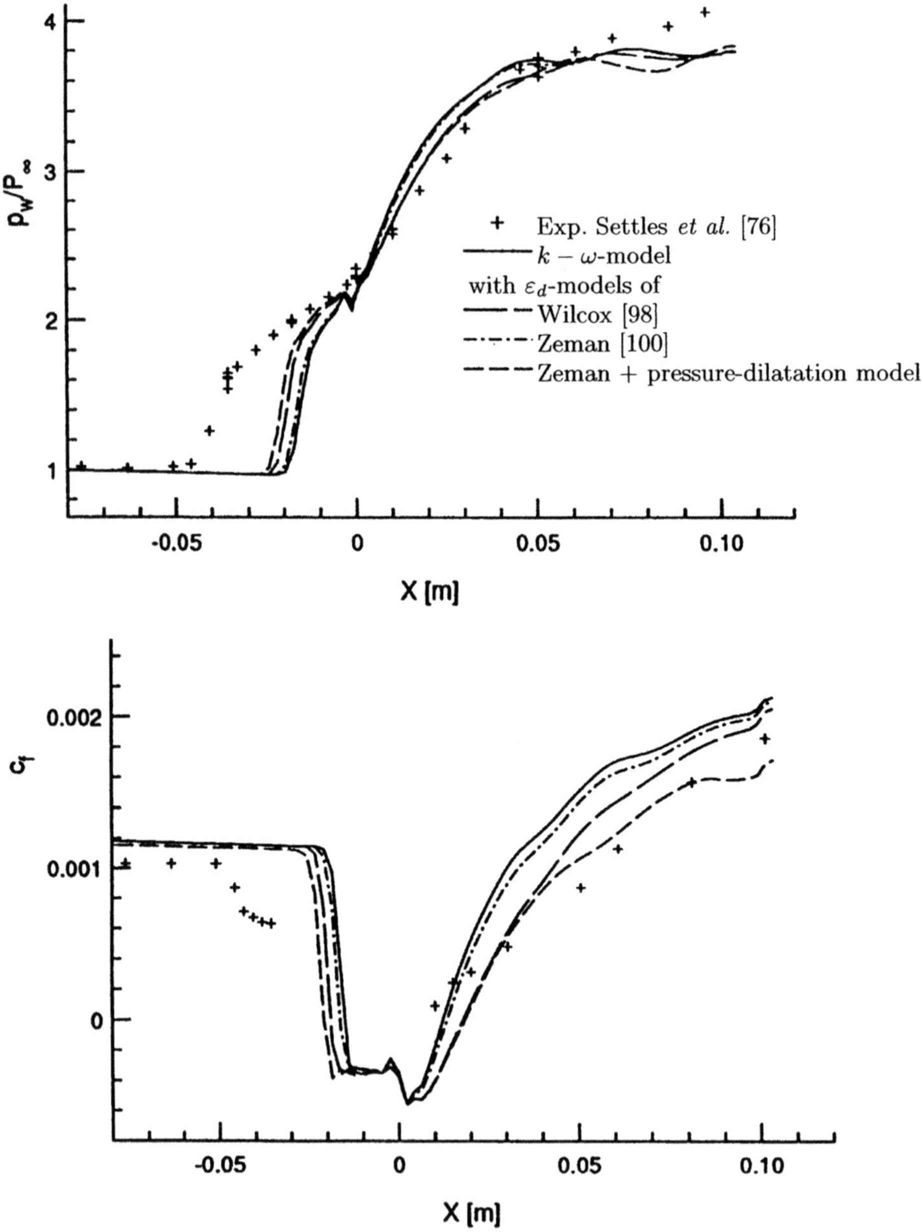

Figure 5.34. Effect of models for explicit dilatation terms on predictions of $M = 2.84$ flow over a 24° adiabatic compression ramp with the (K, ω) model. Wall pressure (top) and skin friction distributions (bottom) are compared with Settles *et al.* 's [76] experiment. Taken from Haidinger [39] and Haidinger and Friedrich [40] by permission.

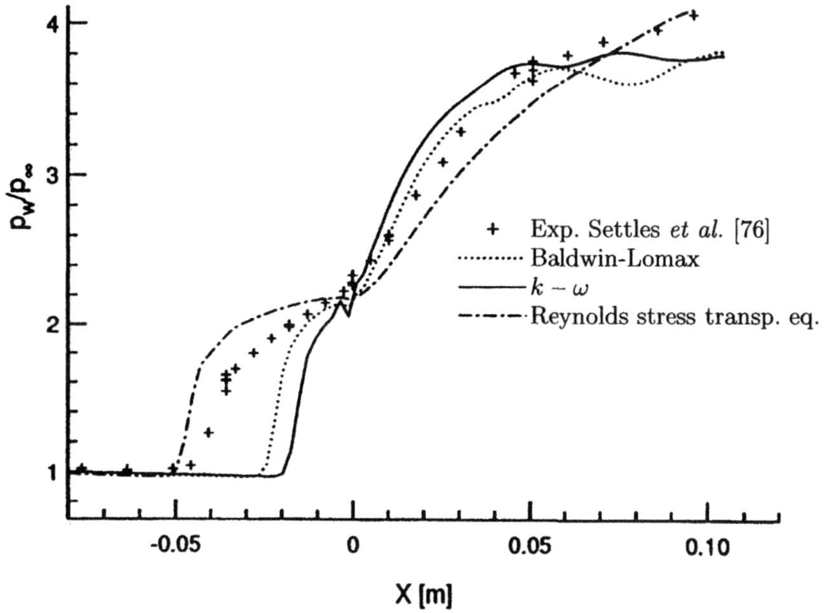

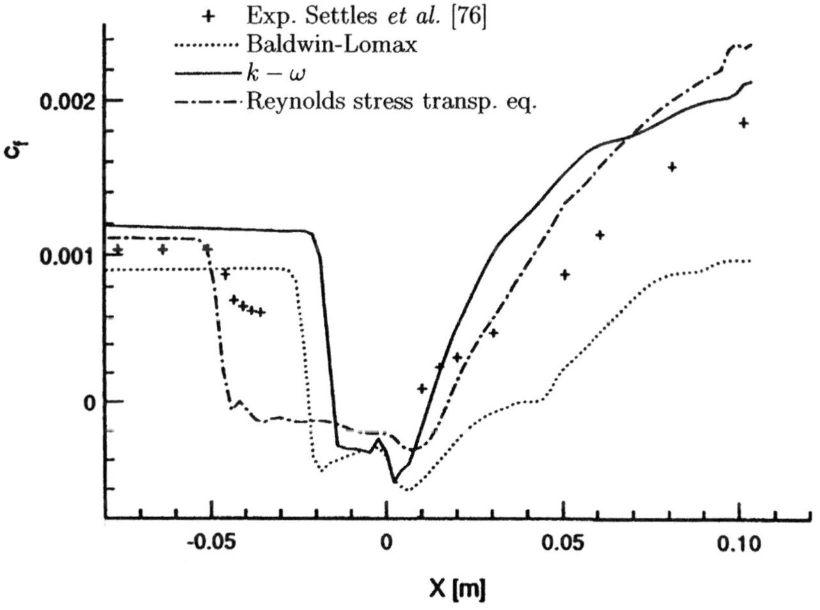

Figure 5.35. Predictions of ramp flow with the Baldwin-Lomax, the (K, ω) and the SSG Reynolds stress model. Wall pressure (top) and skin friction (bottom). Comparison with Settles *et al.* 's [76] experiment. Taken from Haidinger [39] and Haidinger and Friedrich [40] by permission.

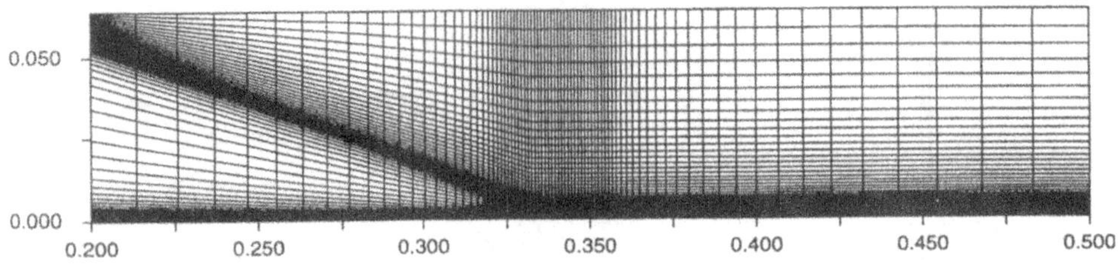

Figure 5.36. Computational grid with 80×80 cells for predictions of a Mach 5 shock impinging at 10 degrees on a flat plate with constant wall temperature ($T_w = 290K$).

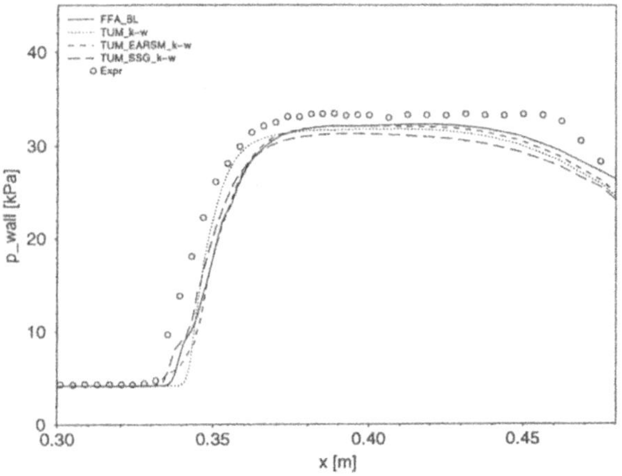

Figure 5.37. Predictions of wall pressure distribution for $M = 5$ impinging shock with four different turbulence models. ——— Baldwin-Lomax, - - - - - - (K,ω), – – – – EARSM of Wallin and Johansson [96], — — — SSG Reynolds stress model. Taken from Lindblad *et al.* [56] by permission.

layer ($Re = 4 \cdot 10^7$ $[1/m]$), was investigated with various turbulence models with the aim to study the performance of algebraic stress models. Figure 5.36 presents the adapted grid used in all computations. Figures 5.37 and 5.38 compare wall pressure and skin friction distributions obtained with the Baldwin-Lomax model (BL), the (K,ω)-model, the explicit algebraic Reynolds stress model of Wallin and Johansson [96] (EARSM_$K-\omega$) and the Reynolds stress transport model (SSG_$K-\omega$) used by Haidinger and Friedrich [40]. It is found that the explicit algebraic stress model and the transport model (SSG_$K-\omega$) produce similarly satisfactory results. The 'points' of pressure rise and separation, respectively reattachment are well

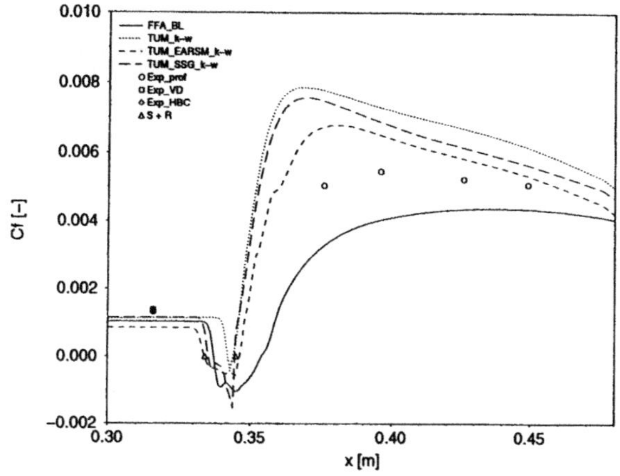

Figure 5.38. Predictions of skin friction distribution for $M = 5$ impinging shock with four different turbulence models, see fig. 37. Meaning of symbols: △ S+R, separation and reattachment points; ○ Exp_prof, cf computed using measured velocity profile; □ Exp_VD, Van Driest formula, ◇ Exp_HBC, Huang/Bradshaw/Coakley formula. Taken from Lindblad *et al.* [56] by permission.

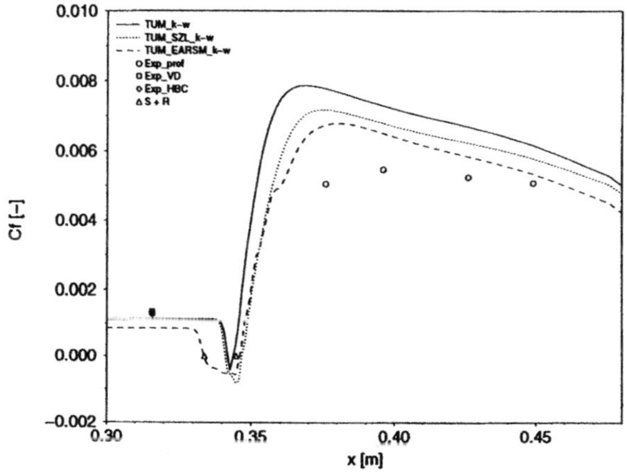

Figure 5.39. Predictions of skin friction distribution for $M = 5$ impinging shock. Comparison of two different explicit algebraic stress models. SZL_$K - \omega$, explicit algebraic stress model of Shih, Zhu and Lumley [78]. For remaining legend see figures 37,38. Taken from Lindblad *et al.* [56] by permission.

predicted. Figure 5.39 compares the behaviour of two algebraic stress models with that of the (K, ω)-model in predicting the skin friction. The second explicit algebraic stress model (SZL_K, ω) is that of Shih, Zhu and Lumley [78]. It was modified to account for low Reynolds number effects near the

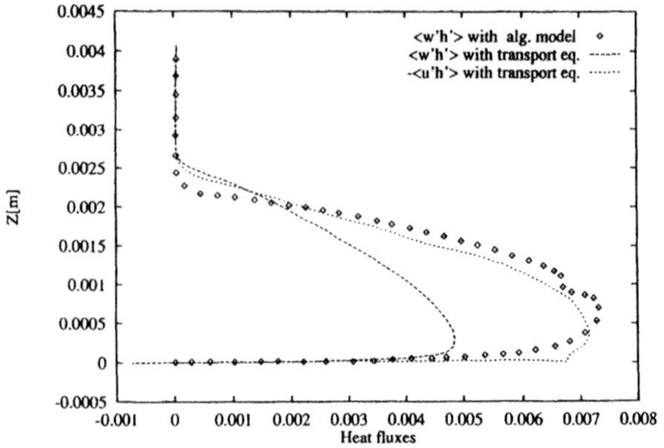

Figure 5.40. Turbulent heat fluxes in a ZPG $M = 3$ boundary layer along an adiabatic wall. Comparison between algebraic model and modelled transport equations for the heat flux. Taken from Le Ribault and Friedrich [53].

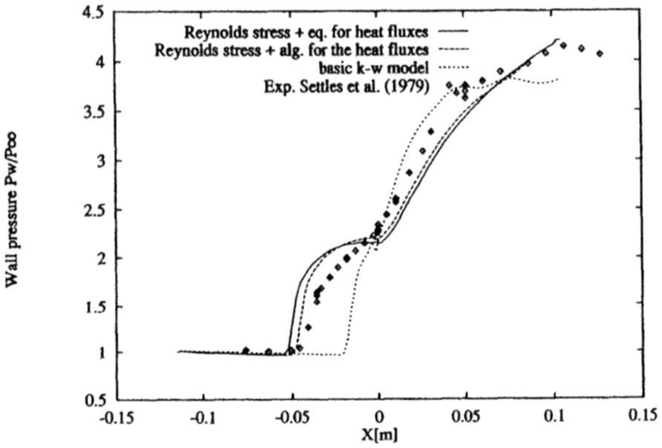

Figure 5.41. Wall pressure distribution for 24° ramp flow at $M = 2.84$. Comparison between algebraic heat flux model and modelled transport equations. Taken from Le Ribault and Friedrich [53].

wall. This model fails to predict the separation 'point' and in that sense resembles the (K, ω)-model.

Applying modelled transport equations for the turbulent heat fluxes, $\overline{\rho u_i'' h''}$, instead of a gradient hypothesis, leads to remarkable differences in the heat flux profiles, as shown in figure 5.40 for a zero pressure-gradient $M = 3$ adiabatic wall boundary layer. The Reynolds numbers based on momentum thickness range from 7000 to 9200. While the algebraic model predicts negligibly small streamwise flux, the transport equation model provides

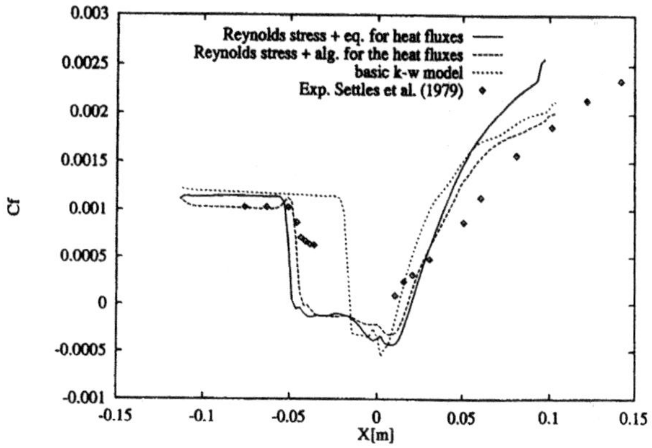

Figure 5.42. Skin friction distribution for 24° ramp flow at $M = 2.84$. Comparison between algebraic heat flux model and modelled transport equations. Taken from Le Ribault and Friedrich [53].

a peak value of the flux magnitude which is roughly 70% of the vertical flux maximum. The wall normal heat flux maximum computed with the algebraic model exceeds that obtained from the transport equation by a factor of 1.5 which is unacceptable. Finally, we show the effect of heat flux modeling on the wall pressure and skin friction distributions of the 24° supersonic ramp flow in figures 5.41 and 5.42, since measured Stanton number distributions were not available. The use of transport equations for the streamwise and wall normal turbulent heat fluxes moves the 'points' of pressure rise and separation further upstream, *i.e.* further away from experimental data. Starting from the experience that more sophisticated turbulence models should (and generally do) provide improved predictions we conclude that extensive grid refinement studies are necessary to clarify this issue.

Acknowledgment

The author is very grateful to T. J. Hüttl for his assistance in preparing this article. He also thanks all colleagues for providing figures to be included in this article. The permission of reproduction by the Cambridge University Press is finally gratefully acknowledged.

References

1. Abid, R., Morrison, J. H., Gatski, T. B., Speziale, C. G. 1996. Prediction of aerodynamic flows with a new explicit algebraic stress model. *AIAA J.*, **34**, No. 12,

2632-2635.
2. Abid, R., Rumsey, C., Gatski, T. B. 1995. Prediction of non-equilibrium turbulent flows with explicit algebraic stress models. *AIAA J.*, **33**, No. 11, 2026-2031.
3. Adams, N. A. 1998. Direct numerical simulation of turbulent compression ramp flow. *Theoret. Comput. Fluid Dynamics* **12**, 109-129.
4. Adams, N. A. 1999. *Direct numerical simulation of shock-turbulence interaction in boundary layers.* Inaugural dissertation. ETH Zürich.
5. Aupoix, B., Blaisdell, G. A., Reynolds, W. C., Zeman, O. 1990. Modelling the turbulent kinetic energy equation for compressible, homogeneous turbulence. Proceedings of the Summer Program, Center for Turbulence Research, Stanford.
6. Bardina, J. E., Huang, P. G., Coakley, T. J. 1997. Turbulence Modeling Validation, Testing, and Development. NASA Technical Memorandum 110446.
7. Batchelor, G.K. 1967. *An Introduction to Fluid Dynamics.* Cambridge University Press, U.K.
8. Batten, P., Loyau, H., Leschziner, M. A. (Eds.) 1997, *ERCOFTAC Workshop on Shock/Boundary-Layer Interaction*, March 25-26, 1997, UMIST, Manchester, UK. Proceedings available on the WEB site - http://sgp.me.umist.ac.uk/ercoftac97.
9. Bertoglio, J.P., Bataille, F., Marion, J.D. 1998. Two-point closures for weakly compressible turbulence. *Phys. Fluids*, under revision.
10. Bertolotti, F.P. 1997. The influence of rotational and vibrational energy relaxation on the instability of boundary layers in supersonic flows. Report ISRN DLR-FB-97-18, DLR Göttingen.
11. Blaisdell, G.A., Mansour, N.N., Reynolds, W.C. 1991. Numerical simulations of compressible homogeneous turbulence. Report No. TF-50, Dept. of Mechanical Engineering, Stanford University, Stanford, California.
12. Blaisdell, G.A., Mansour, N.N., Reynolds, W.C. 1993. Compressibility effects on the growth and structure of homogeneous turbulent shear flow. *J. Fluid Mech.* **256**, 443-485.
13. Blaisdell, G. A., Sarkar, S. 1993. Investigation of the pressure-strain correlation in compressible homogeneous turbulent shear flow. ASME FED-vol. 151, Transitional and Turbulent Compressible Flows, 133-138.
14. Bradshaw, P. 1977. Compressible turbulent shear layers. *Ann. Rev. Fluid Mech.* **9**, 33-54.
15. Cambon, C. 1982. Etude spectrale d'un champ turbulent incompressible, soumis à des effets couplés de deformation et de rotation, imposés extérieurement. Thèse d'Etat Es Sciences, Universitxié Claude Bernard de Lyon.
16. Cambon, C. 1990. Single and double point modelling of homogeneous turbulence. CTR Annual Research Briefs, Stanford University/NASA Ames.
17. Cambon, C., Coleman, G.N., Mansour, N.N. 1993. Rapid distortion analysis and direct simulation of compressible homogeneous turbulence at finite Mach number. *J. Fluid Mech.* **257**, 641-665.
18. Cebeci, T., Smith, A. M. O. 1974. Analysis of turbulent boundary layers. Academic Press, New York, London.
19. Coleman, G.N., Kim, J., Moser, R.D. 1995. A numerical study of turbulent supersonic isothermal - wall channel flow. *J. Fluid Mech.* **305**, 159-183.
20. Craya, A. 1958. Contribution à l'analyse de la turbulence associée à des vitesses moyennes. *P.S.T. Ministère de l'Air*, No. 345.
21. Dervieux, A., Braza, M., Dussauge, J.-P. (Eds.) 1998. Turbulence Modelling for Compressible Flow Arizing in Aeronautics. To appear in *Notes on Numerical Fluid Mechanics*, Vieweg Verlag, Braunschweig/Wiesbaden.
22. Durbin, P.A. 1991. Turbulence closure modeling near rigid boundaries. *Theoret. Comp. Fluid Dyn.* **3**, 1-13.
23. Durbin, P.A., Zeman, O. 1992. Rapid distortion theory for homogeneous compressed turbulence with application to modelling. *J. Fluid Mech.* **242**, 349-370.
24. Dussauge, J. P., Fernholz, H., Finley, P. J., Smith, R. W., Smits, A. J., Spina, E.

F. 1996. Turbulent boundary layers in subsonic and supersonic flow. AGARDOGRAPH 335, AGARD-AG-335.

25. Baz, A.M. El, Launder, B.E. 1993. Second-moment modelling of compressible mixing layers. In *Engineering Turbulence Modelling and Experiments 2* (Rodi,W. and Martelli, F., editors). Elsevier, Amsterdam, 63-72.
26. Fauchet, G. 1998. *Modélisation en deux points de la turbulence isotrope compressible et validation à l'aide de simulations numériques.* Thèse de doctorat. Univ. Claude Bernard-Lyon I, France.
27. Fauchet, G., Shao, L., Wunenberger, R., Bertoglio, J.P. 1997. An improved two-point closure for weakly compressible turbulence and comparisons with large-eddy simulation. Proc. 11th Symp. Turbulent Shear Flows, Grenoble, France, Sept. 8-10, 1997, Vol. 3, 32-13 to 32-18.
28. Fernholz, H.H., Finley, P. J. 1980. A critical commentary on mean flow data for two-dimensional compressible turbulent boundary layers. AGARD-AG-253.
29. Freund, J.B., Moin, P., Lele, S.K. 1997. Compressibility effects in a turbulent annular mixing layer. Report No. TF-72, Dept. Mech. Engg. Stanford Univ., Stanford Calif. 94305.
30. Friedrich, R., Bertolotti, F.P. 1997. Compressibility effects due to turbulent fluctuations. *Appl. Sci. Res.* **57**, 165-194.
31. Fu, S., Launder, B.E. and Tselepidakis, D.P. 1987. Accommodating the effects of high strain rates in modeling the pressure-strain correlation. UMIST Technical Report TFD/87/5.
32. Gatski, T. B. 1996. Turbulent flows: model equations and solution methodology. In *Handbook of Computational Fluid Mechanics* (Peyret, R., editor), Academic Press, London. Chapter 6, 339-416.
33. Gatski, T.B. 1997. Modelling compressibility effects on turbulence. In *New Tools in Turbulence Modelling*, (Métais, O. and Ferziger, J., editors), Les Editions de Physique, France and Springer, Berlin, New York.
34. Gatski, T. B., Speziale, C. G. 1993. On explicit algebraic stress models for complex turbulent flows. *J. Fluid Mech.*, **254**, 59-78.
35. Gaviglio, J. 1987. Reynolds analogies and experimental study of heat transfer in the supersonic boundary layer. *Int J. Heat Mass Transfer*, **30**, 911-926.
36. Guo, Y., Adams, N. A. 1994. Numerical investigation of supersonic turbulent boundary layers with high wall temperature. Proceedings of the Summer Program, Center for Turbulence Research, Stanford.
37. Haase, W. 1997. Case studies of applications of turbulence models in aerospace. In *AGARD FDP Special Course on Turbulence in Compressible Flows*, AGARD-R-819, 6-1 to 6-30.
38. Haase, W., Chabut, E., Elsholz, E., Leschziner, M.A., Müller, U.R. (Eds.) 1997. ECARP-European Computational Aerodynamics Research Project: Validation of CFD Codes and Assessment of Turbulence Models. *Notes on Numerical Fluid Mechanics (NNFM)*, Vol. 58, Vieweg Verlag Braunschweig/Wiesbaden.
39. Haidinger, F. A. 1993. *Numerische Untersuchung turbulenter Stoß-/Grenzschicht-Wechselwirkungen.* Ph.D. thesis, Technische Universität München.
40. Haidinger, F. A., Friedrich, R. 1995. Numerical simulation of strong shock/turbulent boundary layer interactions using a Reynolds stress model. *Zeitschrift für Flugwissenschaften und Weltraumforschung*, **19**, 10-18.
41. Herring, J.R. 1974. Approach of axisymmetric turbulence to isotropy. *Phys. Fluids* **17**, 859.
42. Huang, P.G., Coakley, T. J. 1993. Modeling hypersonic boundary layer flows with second-moment closure. In *Near-Wall Turbulent Flows* (So, R. M. C., Speziale, C. G. and Launder, B. E., editors), Elsevier, Amsterdam, 199-208.
43. Huang, P.G., Coleman, G.N. 1994. Van Driest transformation and compressible wall-bounded flows. *AIAA J.* **32**, 2110-2113.
44. Huang, P.G., Coleman, G.N., Bradshaw, P. 1995. Compressible turbulent channel

flows - DNS results and modelling. *J. Fluid Mech.* **305**, 185-218.
45. Jacquin, L., Cambon, C., Blin, E. 1993. Turbulence amplification by a shock wave and rapid distortion theory. *Phys. Fluids A* **10**, 2539-2550.
46. Jones, W. P., Launder, B. E. 1972. The prediction of laminarization with a two-equation model of turbulence. *Int. J. Heat Mass Transfer*, **15**, 301-314.
47. Kavasznay, L.S.G. 1953. Turbulence in supersonic flows. *J. Aeronaut. Sci.* **20**, 657-682.
48. Kim, J., Moin, P., Moser, R. 1987. Turbulence statistics in fully developed channel flow at low Reynolds number. *J. Fluid Mech.* **177**, 133-166.
49. Knight, D. D. 1997. Numerical simulation of compressible turbulent flows using the Reynolds-averaged Navier-Stokes equations. In *AGARD FDP Special Course on Turbulence in Compressible Flows*, AGARD-R-819, 5-1 to 5-52.
50. Launder, B. E. 1987. An introduction to single-point closure methodology. VKI Course, Brussels, Belgium.
51. Launder, B. E., Reece, G. J., Rodi, W. 1975. Progress in the development of a Reynolds stress turbulence closure. *J. Fluid Mech.* **68**, 537-566.
52. Launder, B.E., Sharma, B.I. 1974. Application of energy-dissipation model of turbulence to the calculation of flow near a spinning disc. *Letters in Heat and Mass Transfer*, **1**, 131-138.
53. Le Ribault, C., Friedrich, R. 1997. Investigation of transport equations for turbulent heat fluxes in compressible flows. *Int. J. Heat Mass Transfer*, **40**, 2721-2738.
54. Lele, S.K. 1994. Compressibility effects on turbulence.*Ann. Rev. Fluid Mech.*, **26**, 211 -254.
55. Lesieur, M. 1997. *Turbulence in Fluids.* Third revised and enlarged edition. Kluwer Academic Publishers, Dordrecht.
56. Lindblad, I. A. A., Wallin, S., Johansson, A. V., Friedrich, R., Lechner, R., Krogmann, P., Schülein, E., Courty, J.-C., Ravachol, M., Giordano, D. 1998. A prediction method for high speed turbulent separated flows with experimental verification. AIAA Paper 98-2547.
57. Lumley, J.L. 1978. Computational modeling of turbulent flows. *Advances in Applied Mechanics*, **18**, 123-173.
58. Mahesh, K., Lee, S., Lele, S.K., Moin, P. 1995. The interaction of an isotropic field of acoustic waves with a shock wave. *J. Fluid Mech.* **300**, 383-407.
59. Mansour, N. N., Kim, J., Moin, P. 1988. Reynolds-stress and dissipation-rate budgets in a turbulent channel flow. *J. Fluid Mech.* **194**, 15-44.
60. Marvin, J.G., Huang, P.G. 1996. Turbulence Modeling-Progress and Future Outlook. Keynote Lecture, 15th Int. Conf. on Numerical Methods in Fluid Dynamics, Monterey, Calif. Also: NASA Technical Memorandum 110414.
61. Menter, F. R. 1993. Zonal two-equation $k-\omega$ turbulence models for aerodynamic flows. AIAA Paper 93-2906.
62. Menter, F. R., Rumsey, L. C. 1994. Assessment of two-equation turbulence models for transonic flows. AIAA Paper 94-2343.
63. Morkovin, M.V. 1962. Effects of compressibility on turbulent flows. In *Mécanique de la Turbulence* (Favre, A., editor), Gordon and Breach, New York, 367 -380.
64. Pope, S. B. 1975. A more general effective viscosity hypothesis. *J. Fluid Mech.* **72**, 331-340.
65. Ristorcelli, J.R. 1995. A pseudo-sound constitutive relationship for the dilatational covariances in compressible turbulence: an analytical theory. ICASE Rep. 95-22. NASA Langley Research Center, Hampton, VA.
66. Ristorcelli, J. R. 1997. A pseudo-sound constitutive relationship for the dilatational covariances in compressible turbulence. *J. Fluid Mech.* **347**, 37-70.
67. Ristorcelli, J. R. 1998. Some results relevant to statistical closures for compressible turbulence. ICASE Report No. 98-1, NASA Langley Research Center, Hampton, VA.
68. Ristorcelli, J.R., Blaisdell, G.A. 1996. Consistent inital conditions for the DNS

of compressible turbulence. ICASE Report No. 96-49, NASA Langley Research Center, Hampton, VA 23681-0001.

69. Rubesin, M. W. 1976. A one-equation model of turbulence of use with the compressible Navier-Stokes equations. NASA Tech. Mem. X-73, 128.
70. Rubesin, M.W. 1990. Extra compressibility terms for Favre-averaged two-equation models of inhomogeneous turbulent flows. NASA CR-177556
71. Sandham, N.D., Reynolds, W.C. 1989. A numerical investigation of the compressible mixing layer. Rep. TF-45. Thermosciences Division, Mechanical Engineering Department, Stanford University, Stanford, California.
72. Sarkar, S. 1992. The pressure-dilatation correlation in compressible flows. *Phys. Fluids A* **4**, 2674-2682.
73. Sarkar, S. 1995. The stabilizing effect of compressibility in turbulent shear flow. *J. Fluid Mech.* **282**, 163-186.
74. Sarkar, S., Erlebacher, G., Hussaini, M. Y. 1991. Direct simulation of compressible turbulence in a shear flow. *Theor. Comput. Fluid Dyn.* **2**, 291-305.
75. Sarkar, S., Erlebacher, G., Hussaini, M. Y., Kreiss, H. O. 1991b. The analysis and modelling of dilatational terms in compressible turbulence. *J. Fluid Mech.* **227**, 473-493.
76. Settles, G.S., Fitzpatrick, T.J., Bogdonoff, S.M. 1979. Detailed study of attached and separated compression corner flow fields in high Reynolds number supersonic flow. *AIAA J.* **17**, 579-585.
77. Shih, T.-H. 1996. Constitutive relations and realizability of single-point turbulence closures. In *Turbulence and Transition Modelling* (Hallbäck, M., Henningson, D. S., Johansson, A. V. and Alfredsson, P. H., editors). ERCOFTAC Series, Kluwer Academic Publishers.
78. Shih, T.H., Zhu, J., Lumley, J.L. 1992. A realizable Reynolds stress algebraic equation model. NASA TM 105993.
79. Simone, A. 1995. *Etude théorique et simulation numérique de la turbulence compressible en présence de cisaillement ou de variation de volume à grande èchelle.* Thése. Ecole doctorale: Mècanique-Energètique. Ecole Centrale de Lyon.
80. Simone, A., Coleman, G.N., Cambon, C. 1997. The effect of compressibility on turbulent shear flow: A rapid distortion theory and direct numerical simulation study. *J. Fluid Mech.* **330**, 307 -338.
81. Smits, A.J., Dussauge, J.-P. 1996. Turbulent Shear Layers in Supersonic Flow. American Institute of Physics Press.
82. Sommer, T. P., So, R. M. C., Zhang, H. S. 1993. Near-wall variable - Prandtl-number turbulence model for compressible flows. *AIAA J.* **31**, 27-35.
83. Speziale, C. G. 1991. Analytical methods for the development of Reynolds stress closures in turbulence. *Ann. Rev. Fluid Mech.*, Vol. 23, 107-157.
84. Speziale, C. G. 1996. Modeling of turbulent transport equations. In *Simulation and Modeling of Turbulent Flows* (Gatski, T. B., Hussaini, M. Y. and Lumley, J. L., editors), ICASE/LaRC Series in Computational Science and Engineering, Oxford Univ. Press, 185-242.
85. Speziale, C.G., Abid, R., Mansour, N.N. 1995. Evaluation of Reynolds stress turbulence closures in compressible homogeneous shear flow. ZAMP, Special Issue, ed. by J. Casey & M.J. Crochet, S717-S736.
86. Speziale, C. G., Gatski, T. B., Sarkar, S. 1992. On testing models for the pressure-strain correlation of turbulence using direct simulations. *Phys. Fluids A* **4**, 2887-2899.
87. Speziale, C. G., Sarkar, S. 1991. Second-order closure models for supersonic turbulent flows. AIAA Paper 91-0217.
88. Speziale, C.G., Sarkar, S., Gatski, T.B., 1991. Modeling the pressure-strain correlation of turbulence: A invariant dynamical systems approach. *J. Fluid Mech.* **227**, 245-272.
89. Tennekes, H., Lumley, J.L. 1972. A first course in turbulence. The MIT Press.

Cambridge, Massachusetts and London, England.
90. Thompson, P.A. 1988. Compressible-Fluid Dynamics. McGraw-Hill, New York.
91. Uhlmann, M. 1997. *Etude de modèles de fermeture au second ordre et contribution à la résolution numérique des écoulements turbulents compressibles.* Thèse de doctorat. Ecole Centrale de Lyon.
92. Van Driest, E.R. 1951. Turbulent boundary layer in compressible fluids. *J. Aero. Sci.* **18**, 145-160 and 216.
93. Viala, S. 1995. *Effets de la compressibilité et d'un gradient de pression négatif sur la couche limite turbulente, Tome 1,2.* Thèse de doctorat. Ecole Nationale Supérieure de l'Aéronautique et de l'Espace, Toulouse.
94. Vreman, B. 1995. *Direct and large-eddy simulation of the compressible turbulent mixing layer.* Thesis, University of Twente, The Netherlands.
95. Vreman, A.W., Sandham, N.D., Luo, K.H. 1996. Compressible mixing layer growth rate and turbulence characteristics. *J. Fluid Mech.* **320**, 235-258.
96. Wallin, S., Johansson, A. V. 1998. A new explicit algebraic Reynolds stress model for incompressible and compressible turbulent flows. Submitted for publication.
97. Wilcox, D. C. 1988. Reassessment of the scale-determining equation for advanced turbulence models. *AIAA J.* **26**, 1299-1310.
98. Wilcox, D.C. 1991. Progress in hypersonic turbulence modelling. AIAA Paper 91-1785.
99. Wilcox, D. C. 1993. *Turbulence Modeling for CFD.* DCW Industries Inc., 5354 Palm Drive, La Cañada, Calif.
100. Zeman, O. 1990. Dilatation dissipation; The concept and application in modeling compressible mixing layers. *Phys. Fluids A* **2**, 178-188.
101. Zeman, O. 1991. On the decay of compressible isotropic turbulence. *Phys. Fluids A* **3**, 951-955.
102. Zeman, O. 1993. New model for super/hypersonic turbulent boundary layers. AIAA Paper 93-0897.
103. Zeman, O., Coleman, G. N. 1993. Compressible turbulence subjected to shear and rapid compression. *Turbulent Shear Flows 8* (Durst *et al.* editors), Springer-Verlag, 283-296.

Chapter 6

LARGE-EDDY SIMULATIONS OF INCOMPRESSIBLE AND COMPRESSIBLE TURBULENCE

O. MÉTAIS, M. LESIEUR AND P. COMTE
Laboratoire des Écoulements Géophysiques et Industriels
BP 53, 38041 Grenoble Cédex 9, France

6.1. Introduction

Direct-numerical simulations of turbulence (DNS) consist in solving explicitly all the scales of motion, from the largest l_I to the Kolmogorov dissipative scale l_D. It is well known from the statistical theory of turbulence that l_I/l_D scales like $R_l^{3/4}$, where R_l is the large-scale Reynolds number $u'l_I/\nu$ based upon the rms velocity fluctuation u'. Therefore, the total number of degrees of freedom necessary to represent the whole span of scales of a three-dimensional turbulent flow is of the order of $R_l^{9/4}$ in three dimensions. In the presence of obstacles, around a wing or a fuselage for instance, and if one wants to simulate three-dimensionally all motions ranging from the viscous thickness $\delta_v = \nu/v_* \approx 10^{-6}$ m up to 10 m, it would be necessary to put 10^{21} modes on the computer. Right now, the calculations done to the expense of not excessive computing times on the biggest machines take about 2. 10^7 grid points, which is very far from the above estimation. Even with the unprecedented improvement of scientific computers, it may take several tenths of years (if it becomes ever possible) before DNS permit to simulate situations at Reynolds numbers comparable to those encountered in natural conditions.

Large-Eddy Simulations (LES) techniques consist in trying to simulate deterministically only the large scales of the flow, which, from an engineering point of view, are responsible for an important part of turbulent transfers of momentum or heat for example. This is the large-eddy simulation

(LES) point of view, where the small scales are filtered out, but influence statistically the large-scale motions. As will be seen, LES are extremely useful in particular to understand the dynamics of coherent vortices and structures in turbulence.

6.2. Large-Eddy Simulation (LES) Formalism

To begin with, let us consider a simulation of Navier-Stokes equations carried out in physical space, using finite-difference or finite-volume methods. We assume first for sake of simplification that the spatial discretization is orthogonal and cubic, Δx being the grid mesh. One considers a spatial filter G of width Δx, which filters out the *subgrid-scales* of wavelength $< \Delta x$. The filtered field is defined as

$$\bar{u}_i(\vec{x},t) = \int u_i(\vec{y},t)G(\vec{x}-\vec{y})d\vec{y} = \int u_i(\vec{x}-\vec{y},t)G(\vec{y})d\vec{y} \quad , \tag{6.1}$$

and the subgridscale field is the departure of the actual flow with respect to the filtered field:

$$u_i = \bar{u}_i + u_i' \quad . \tag{6.2}$$

Because of the commutative properties of the filter and the derivatives, the continuity equation $\partial \bar{u}_j / \partial x_j = 0$ for the filtered field still holds. We consider first the incompressible Navier-Stokes equations. Compressible flows will be subsequently treated. Density variations are here only considered in the framework of the Boussinesq approximation. Furthermore, for generality, we work in a Cartesian frame of reference $\vec{x} = (x, y, z)$ rotating with a constant angular velocity Ω about the x_3 axis. When applying the filter G to the equations, it is then obtained:

$$\begin{aligned}\frac{\partial \bar{u}_i}{\partial t} + \frac{\partial}{\partial x_j}(\bar{u}_i\bar{u}_j) &= -\frac{1}{\rho_0}\frac{\partial \bar{p}}{\partial x_i} + 2\epsilon_{ij3}\Omega\bar{u}_j - g_i\delta_{i3}\frac{\bar{\rho}}{\rho_0} \\ &\quad + \frac{\partial}{\partial x_j}\left\{\nu\left(\frac{\partial \bar{u}_i}{\partial x_j} + \frac{\partial \bar{u}_j}{\partial x_i}\right) + T_{ij}\right\} \qquad (6.3)\\ \frac{\partial \bar{\rho}}{\partial t} + \frac{\partial}{\partial x_j}(\bar{\rho}\bar{u}_j) &= \frac{\partial}{\partial x_j}\left\{\kappa\frac{\partial \bar{\rho}}{\partial x_j} + R_{ij}\right\} \qquad (6.4)\end{aligned}$$

ρ is the density, and $\bar{p}$ the filtered pressure. T_{ij} and R_{ij} are the subgrid-scale tensors:

$$\begin{aligned}T_{ij} &= \bar{u}_i\bar{u}_j - \overline{u_i u_j} \\ &= -\left(\overline{u_i' u_j'} + \overline{\bar{u}_i u_j'} + \overline{u_i'\bar{u}_j} + \overline{\bar{u}_i\bar{u}_j} - \bar{u}_i\bar{u}_j\right) \qquad (6.5)\\ R_{ij} &= \bar{\rho}\bar{u}_j - \overline{\rho u_j} \\ &= -\left(\overline{\rho' u_j'} + \overline{\bar{\rho}u_j'} + \overline{\rho'\bar{u}_j} + \overline{\bar{\rho}\bar{u}_j} - \bar{\rho}\bar{u}_j\right) \quad , \qquad (6.6)\end{aligned}$$

which have to be modelled. Most subgrid-scale models make eddy-viscosity and eddy-diffusivity assumptions (Boussinesq's hypothesis) in order to model the subgrid-scale tensors:

$$T_{ij} = 2\nu_t \bar{S}_{ij} + \frac{1}{3} T_{ll} \delta_{ij} \quad ; R_{ij} = \kappa_t \frac{\partial \bar{\rho}}{\partial x_j} \tag{6.7}$$

where

$$\bar{S}_{ij} = \frac{1}{2}\left(\frac{\partial \bar{u}_i}{\partial x_j} + \frac{\partial \bar{u}_j}{\partial x_i}\right) \tag{6.8}$$

is the deformation tensor of the filtered field. The generalized LES Boussinesq equations then write

$$\begin{aligned} \frac{\partial \bar{u}_i}{\partial t} + \frac{\partial}{\partial x_j}(\bar{u}_i \bar{u}_j) &= -\frac{1}{\rho_0}\frac{\partial \bar{P}}{\partial x_i} + 2\epsilon_{ij3}\Omega \bar{u}_j - g_i \delta_{i3} \frac{\bar{\rho}}{\rho_0} \\ &\quad + \frac{\partial}{\partial x_j}\left\{(\nu + \nu_t)\left(\frac{\partial \bar{u}_i}{\partial x_j} + \frac{\partial \bar{u}_j}{\partial x_i}\right)\right\} \qquad (6.9) \\ \frac{\partial \bar{\rho}}{\partial t} + \frac{\partial}{\partial x_j}(\bar{\rho}\bar{u}_j) &= \frac{\partial}{\partial x_j}\left\{(\kappa + \kappa_t)\frac{\partial \bar{\rho}}{\partial x_j}\right\} \quad , \qquad (6.10) \end{aligned}$$

where $\bar{P} = \bar{p} - (1/3)\rho_0 T_{ll}$ is a modified pressure.

The question is now to determine the eddy-viscosity $\nu_t(\vec{x}, t)$ and eddy-diffusivity $\kappa_t(\vec{x}, t)$. Most of the existing models except the dynamic model and the spectral dynamic model discussed below assume a proportionality between ν_t and κ_t with a constant Prandtl number

$$Pr_t = \nu_t / \kappa_t = \text{ cte } . \tag{6.11}$$

6.2.1. LES AND UNPREDICTABILITY GROWTH

From a mathematical viewpoint, the LES problem is not very well posed. Indeed, let us consider the time evolution of the fluid as the motion of a point in a sort of phase space of extremely large dimension (*e.g.* $\sim 10^{21}$ around a wing, as seen above). At some initial instant, the flow computed with LES will differ from the actual flow, due to the uncertainty contained in the subgridscales. This initial difference between the actual and the computed flow will grow, due to nonlinear effects, as in a dynamical system having a chaotic behaviour. Therefore, the two points will separate in phase space, and, as time goes on, the LES will depart from reality. However, as will be seen below, LES permit to predict the statistical characteristics of turbulence, as well as the dynamics of coherent vortices and structures.

Note that chaos in dynamical systems with a low number of degrees of freedom is generally characterized by a positive Lyapounov exponent, with exponential growth of the distance between two points initially very close in phase space. In isotropic turbulence, one introduces for predictability studies the error spectrum $E_\Delta(k,t)$, characterizing the spatial-frequency distribution associated to the energy of the difference between two random fields $\vec{u}_1$ and $\vec{u}_2$ with same statistical properties:

$$\frac{1}{4} < [\vec{u}_1^2(\vec{x},t) - \vec{u}_2^2(\vec{x},t)] >= \int_0^{+\infty} E_\Delta(k,t)\, dk \quad , \tag{6.12}$$

the energy spectrum $E(k,t)$ being such that

$$\frac{1}{2} < \vec{u}_1^2 >= \frac{1}{2} < \vec{u}_2^2 >= \int_0^{+\infty} E(k,t)\, dk \quad . \tag{6.13}$$

The error rate

$$r(t) = \frac{\int_0^{+\infty} E_\Delta(k,t)dk}{\int_0^{+\infty} E(k,t)dk} \tag{6.14}$$

is zero when the two fields are completely correlated, and one when they are totally uncorrelated. In predictability studies, one takes generally an initial state such that complete unpredictability ($E(k) = E_\Delta(k)$) holds above $k_E(0)$, while $E_\Delta(k)$ is 0 for $k < k_E(0)$. Two-point closures of the EDQNM type (see [70] for details) show (in three or two dimensions) an inverse cascade of error, where the wavenumber $k_E(t)$ characterizing the error front decreases (see [87]). Thus, the error rate can be approximated by

$$r(t) \approx \frac{\int_{k_E(t)}^{\infty} E(k,t)dk}{\int_0^{+\infty} E(k,t)dk} \quad .$$

We assume that the turbulence is forced by external forces, so that the kinetic energy arising at the denominator of Eq. (6.14) is fixed. In three-dimensional turbulence, and if a $k^{-5/3}$ spectrum is assumed for $k > k_E$, the error rate will be proportional to $k_E^{-2/3}$. In fact, closures (see [87][70]) show that k_E^{-1} follows a Richardson's law ($k_E^{-1} \propto t^{3/2}$), so that the error rate grows linearly with time. This is in fact a slow increase compared with the exponential growth of chaotic dynamical systems, and quite encouraging concerning the potentialities of large-eddy simulations for three-dimensional turbulent flows.

6.3. Smagorinsky's Model

The most widely used eddy-viscosity model was proposed by Smagorinsky [110]. The latter, a meteorologist, was simulating a two-layer quasi-geostrophic model, in order to represent large (synoptic) scale atmospheric motions. He introduced an eddy-viscosity which was supposed to model three-dimensional turbulence that has approximately 3D Kolmogorov $k^{-5/3}$ cascade in the subgrid scales. Smagorinsky' model still proves very popular for several applications, since the pioneering work of Deardorff [29] for the channel flow.

In Smagorinsky's model, a sort of mixing-length assumption is made, in which the eddy-viscosity is assumed to be proportional to the subgrid-scale characteristic length scale Δx, and to a characteristic turbulent velocity based on the second invariant of the filtered-field deformation tensor. The model is

$$\nu_t = (C_S \Delta x)^2 |\bar{S}| \ , \tag{6.15}$$

where the local strain rate is defined by $|\bar{S}| = \sqrt{2\bar{S}_{ij}\bar{S}_{ij}}$. If one assumes that $k_C = \pi/\Delta x$, the cutoff wavenumber in Fourier space, lies within a $k^{-5/3}$ Kolmogorov cascade $E(k) = C_K \epsilon^{2/3} k^{-5/3}$ (C_K is the Kolmogorov constant), one can analytically determine the constant C_S. It is then found (see *e.g.* [77]):

$$C_S \approx \frac{1}{\pi} \left(\frac{3C_K}{2} \right)^{-3/4} \ . \tag{6.16}$$

The newly developed subgrid-scale models try to tackle the inherent weaknesses attached with the eddy-viscosity and eddy-diffusivity assumption given by (6.7) and with subgrid-scale models of Smagorinsky's type:

- These assume a one-to-one correlation between the subgrid-scale stress and the large-scale strain rate tensors. This is highly questionable and has never been verified experimentally or numerically (see *e.g.* [21], [78]).

- Equation (6.16) yields $C_S \approx 0.18$ for a Kolmogorov constant of 1.4. In fact, most workers prefer $C_S = 0.1$ (which represents a reduction by nearly a factor of 4 of the eddy-viscosity), a value for which Smagorinsky's model behaves reasonably well for free-shear flows and for channel flow [92]. This clearly indicates that C_S is not a universal constant and that assuming that $k_C = \pi/\Delta x$ lies within a $k^{-5/3}$ Kolmogorov cascade is too much a constraint, since even the smallest resolved scales can be far from being isotropic.

Some of these weaknesses are addressed by the various subgrid-scale models which have been developed since Smagorinsky's original model and which are presented in the following sections. The details of the various models can be found in [71], [69].

6.4. Spectral Eddy-viscosity and Eddy-diffusivity Models

We now work in Fourier space, in the context of three-dimensional isotropic turbulence. The filter consists in a sharp cut-off filter clipping all the modes larger than k_C, where $k_C = \pi/\Delta x$ is the cut-off wavenumber. The concept of k-dependent eddy-viscosity was first introduced by Kraichnan [54] for three-dimensional isotropic turbulence. The spectral eddy-diffusivity was introduced in [20] for a passive scalar.

Let us consider the spectral evolution equations for the supergrid-scale velocity, $\bar{E}(k,t)$, and scalar, $\bar{E}_T(k,t)$ spectra. $T_{>k_C}(k,t)$ and $T^T_{>k_C}(k,t)$ designated the kinetic energy and the scalar variance transfers across the cutoff k_C, corresponding to triadic interactions such that $k < k_C, p$ and (or) $q > k_C$ (see [70], for details). One poses

$$\begin{aligned} \nu_t(k|k_c) &= -\frac{T_{>k_C}(k,t)}{2k^2\bar{E}(k,t)} \\ \kappa_t(k|k_c) &= -\frac{T^T_{>k_C}(k,t)}{2k^2\bar{E}(k,t)} \quad , \end{aligned} \tag{6.17}$$

in such a way that the supergrid spectra $\bar{E}(k,t)$ and $\bar{E}_T(k,t)$ in the resolved scales ($k \leq k_C$) satisfy:

$$[\frac{\partial}{\partial t} + 2(\nu + \nu_t(k|k_c))k^2]\ \bar{E}(k,t) = T_{<k_C}(k,t) \tag{6.18}$$

$$[\frac{\partial}{\partial t} + 2(\kappa + \kappa_t(k|k_c))k^2]\ \bar{E}_T(k,t) = T^T_{<k_C}(k,t) \quad , \tag{6.19}$$

where $T_{<k_C}(k,t)$ and $T^T_{<k_C}(k,t)$ are the spectral transfers corresponding to resolved triads such that $k, p, q \leq k_C$. Using the Eddy-Damped Quasi-Normal Markovian (E.D.Q.N.M.) theory, one of the statistical closures of isotropic turbulence (see [70]), and assuming that k_C lies within a Kolmogorov cascade, we find that the eddy-viscosity may be written

$$\nu_t(k|k_C) = 0.441\ C_K{}^{-3/2}\nu_t^*(\frac{k}{k_C})\ \left[\frac{E(k_C)}{k_C}\right]^{1/2} \quad , \tag{6.20}$$

where $E(k_C)$ is the kinetic-energy spectrum at the cutoff k_C, and $\nu_t^*(k/k_C)$ is a non-dimensional eddy-viscosity, constant and equal to 1 for $k/k_C <\approx 0.3$, and rising for higher k up to $k/k_C = 1$ ("plateau-peak" or cusp behaviour, [54]). The function $0.441\ C_K{}^{-3/2}\nu_t^*(\frac{k}{k_C})$ is represented on Fig. 6.1, together with the normalized spectral eddy-diffusivity for the passive scalar and the corresponding turbulent Prandtl number. Both eddy coefficients have a plateau-peak behaviour and the spectral turbulent Prandtl number $\nu_t(k|k_C)/\kappa_t(k|k_C)$ is approximately constant and equal to 0.6. It is clear that the plateau part corresponds to the usual eddy-coefficients assumption when one goes back to physical space, so that the "peak" (Kraichnan called it "cusp") part goes beyond the scale-separation assumption inherent to the classical eddy-viscosity and diffusivity concepts. The peak is mostly due to semi-local interactions across k_C: near the cutoff wavenumber, the main non-linear interactions between the resolved and unresolved scales involve the smallest eddies of the former and the largest eddies of the latter (such that $p << k \sim q \sim k_C$). Note that it contains also possible backscatter contributions (which are however very small if k_C lies in a Kolmogorov or a Corrsin-Oboukhov cascade) coming from subgrid modes larger than k_C.

The plateau-peak behaviour of ν_t^* can approximately be expresssed with the following analytical expression:

$$\nu_t^* \left(\frac{k}{k_C}\right) = 1. + 34.5 e^{-3.03(k_C/k)} \quad . \tag{6.21}$$

For LES in Fourier space, the spectral eddy-viscosity (6.20) and eddy-diffusivity is plugged into the Navier-Stokes equation for the Fourier transforms $\hat{u}_i(\vec{k},t)$ and $\hat{T}(\vec{k},)$ of respectively the velocity field and the scalar field:

$$[\frac{\partial}{\partial t} + \nu k^2 + \nu_t(k|k_C)k^2]\hat{u}_i(\vec{k},t) = -ik_m P_{ij}(\vec{k}) \int_{\vec{p}+\vec{q}=\vec{k}}^{p,q\leq k_C} \hat{u}_j(\vec{p},t)\hat{u}_m(\vec{q},t)d\vec{p} \quad , \tag{6.22}$$

$$[\frac{\partial}{\partial t} + \kappa k^2 + \kappa_t(k|k_c)k^2]\hat{T}(\vec{k},t) = -ik_j \int_{\vec{p}+\vec{q}=\vec{k}}^{p,q\leq k_C} \hat{u}_j(\vec{p},t)\hat{T}(\vec{q},t)d\vec{p} \quad , \tag{6.23}$$

where $P_{ij}(\vec{k})$ is the projector onto the plane perpendicular to $\vec{k}$.

We present now spectral LES of decaying isotropic turbulence (see [88] for details) with passive scalar based upon the spectral-cusp eddy viscosity. These LES are used to compute the spectral eddy-viscosity and diffusivity, following a method employed by [32] for a direct numerical simulation: one defines a fictitious cutoff wave number $k_c' = k_c/2$, across which the kinetic energy transfer T and temperature transfer T^T are evaluated. Since we deal with a large-eddy simulation, the latter correspond to triadic interactions such that $k < k_c'$, p and (or) $q > k_c'$ and $p,q < k_c$: they are termed $T_{>k_c'}^{<k_c}(k,t)$ and $T_{k_c'}^{T\ <k_c}(k,t)$. $T_{>k_c'}^{<k_c}$ satisfies:

$$T_{>k_c'}^{<k_c}(k,t) = T_{>k_c'}(k,t) - T_{>k_c}(k,t) \tag{6.24}$$

where $T_{>k_c'}$ and $T_{>k_c}$ are the total kinetic energy transfers across k_c' and k_c. It is important to note that Eq. (6.24) is the exact equivalent of Germano's identity [43] in spectral space (see eq. 6.41 below). A similar relation holds for $T_{k_c'}^{T\ <k_c}$. Once divided by $-2k^2\ E(k,t)$ and $-2k^2\ E_T(k,t)$ (E_T temperature variance spectrum), they give the spectral eddy-viscosity and diffusivity. Figure 6.2 shows these functions normalized by $[E(k_c')/k_c']^{1/2}$.

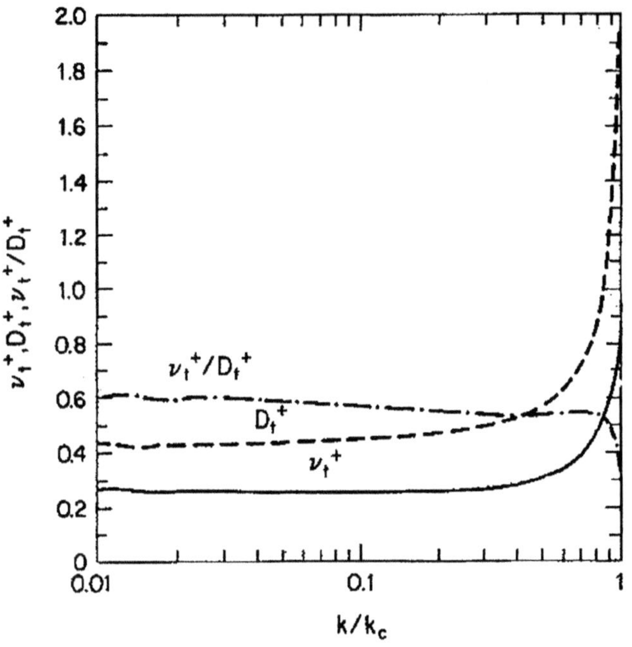

Figure 6.1. EDQNM spectral eddy coefficients in a 3D Kolmogorov cascade

It is clear that the plateau-peak behaviour does exist for the eddy viscosity, but is questionable for the eddy diffusivity. It was found in [88] that the anomaly disappears when the temperature is no more passive and coupled with the velocity within the frame of Boussinesq approximation (stable stratification). It is possible that the same holds for compressible turbulence, which would legitimate the use of the plateau-peak eddy diffusivity in this case.

6.4.1. TEMPORAL MIXING LAYER

We give here an application of Kraichnan's model (k_C in the Kolmogorov range) in a temporal mixing layer. The flow is initiated by a hyperbolic-tangent velocity profile, to which is superposed a small 3D white-noise random perturbation. Figure 6.3 shows the low-pressure field, with evidence for helical pairing, where vortex filaments oscillate out-of-phase in the spanwise direction, and reconnect, yielding a vortex-lattice structure. This was previously found in the DNS of [24] with the same initial conditions. On the other hand, and if the perturbation is quasi two-dimensional, the mixing layer evolves into a set of big quasi two-dimensional Kelvin-Helmholtz vortices which both undergo pairing and stretch intense longitudinal hairpin vortices in the stagnation regions between them. An illustration of these

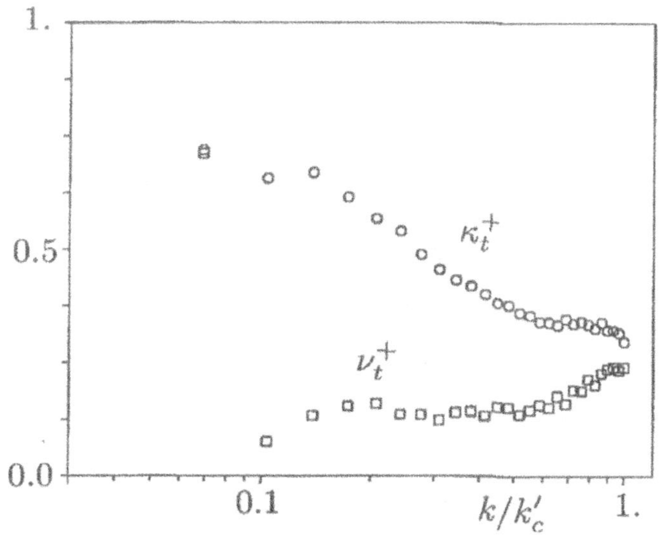

Figure 6.2. LES of 3D isotropic decaying turbulence, resolution 128^3; resolved eddy-viscosity and diffusivity calculated trough a double filtering (from [88])

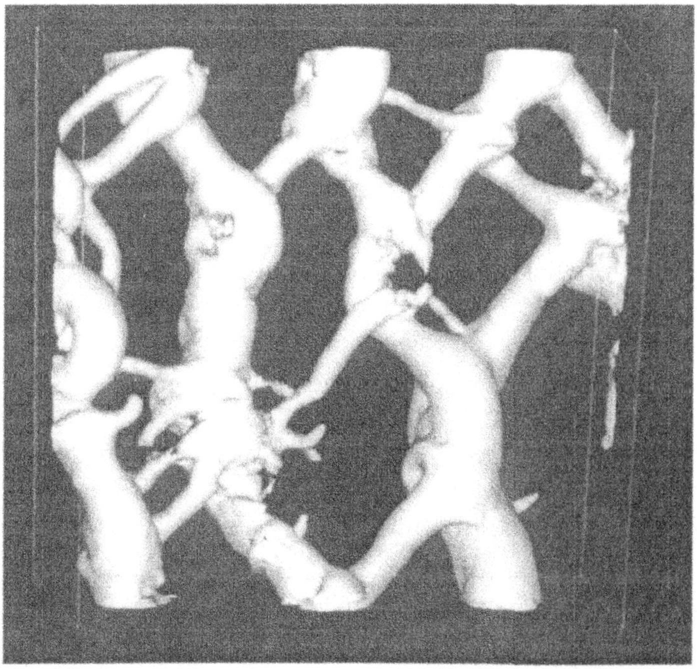

Figure 6.3. low-pressure field obtained in a spectral-cusp LES of a temporal mixing layer in the helical-pairing case (courtesy P. Comte)

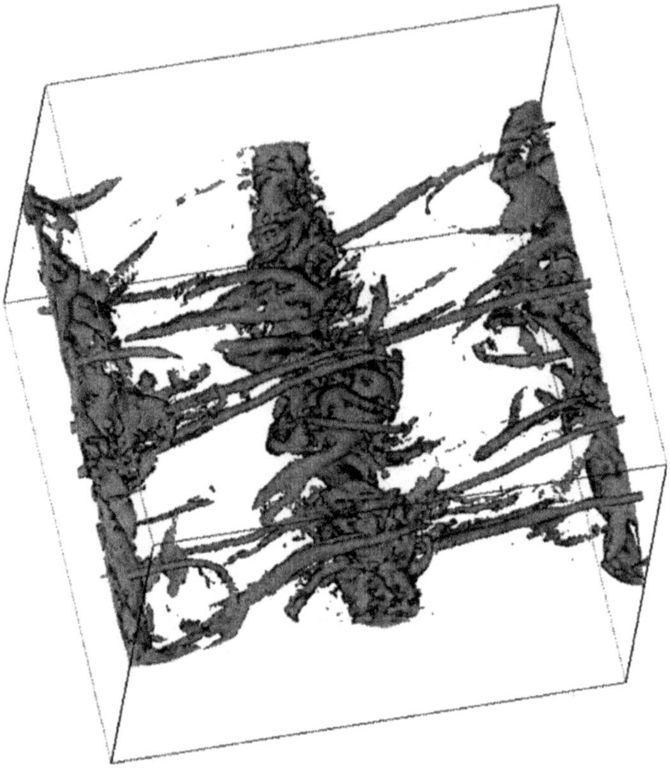

Figure 6.4. vorticity-modulus field obtained in the LES of a temporal mixing layer forced quasi two-dimensionally (courtesy G. Silvestrini)

vortices, whose circulation is of the order of that of the basic KH vortices, is shown on Figure 6.4. This stretching of longitudinal vortices was observed experimentally for a long time (see *e.g.* [12]). It is remarkable that the present LES, done at a quite low resolution (96^3 modes), are able to capture the longitudinal vortices, which are at quite small scales. The DNS of [24] at an equivalent resolution are unable to find organized intense longitudinal vortices, because their molecular Reynolds number is too low. It is only at a much higher resolution that DNS can capture these vortices[105].

6.4.2. SPECTRAL DYNAMIC MODEL

One of the nice feature of the spectral model given by equation (6.20) is that it can be easily extended to the case where k_C is not situated within a $k^{-5/3}$ Kolmogorov cascade. The plateau-peak model may indeed be adapted to kinetic-energy spectra $\propto k^{-m}$ for $k > k_C$, when the exponent m is not necessarily equal to 5/3. The spectral eddy viscosity is now

$$\nu_t(k|k_C) = 0.31 C_K{}^{-3/2} \sqrt{3-m}\frac{5-m}{m+1}\nu_t^*(\frac{k}{k_C})\left[\frac{E(k_C)}{k_C}\right]^{1/2} , \tag{6.25}$$

for $m \leq 3$. This expression is exact for $k << k_C$ within the EDQNM theory, as shown in [88]. We retain the peak shape through $\nu_t^*(k/k_C)$ in order to be consistent with the Kolmogorov spectrum expression of the eddy viscosity. For $m > 3$, the scaling is no more valid, and the eddy viscosity will be set equal to zero. In the spectral-dynamic model, the exponent m is determined through the LES with the aid of least-squares fits of the kinetic-energy spectrum close to the cutoff. On may also check that the turbulent Prandtl number is given by:

$$P_r^t = 0.18(5-m) \tag{6.26}$$

(see [70]).

6.4.3. INCOMPRESSIBLE PLANE CHANNEL

We show now how the spectral dynamic model may be applied to an incompressible turbulent Poiseuille flow between two infinite parallel flat plates. The channel has a width $2h$, and we define the macroscopic Reynolds number by $R_e = 2hU_m/\nu$, where U_m is the bulk velocity. We assume periodicity in the streamwise and spanwise directions. Calculations are carried out at constant U_m. They are initiated by a parabolic laminar profile perturbed by a small three-dimensional random noise, and pursued up to complete statistical stationarity. When turbulence has developed, we define a microscopic Reynolds number $h^+ = v_* h/\nu$, based on the friction velocity. We use a numerical code combining pseudo-spectral methods in the streamwise and spanwise directions, and compact finite-difference schemes of sixth order in the transverse direction (see [58]), with grid refinement close to the walls. This is a very precise code of accuracy comparable to a spectral method at equivalent resolution as shown by DNS presented on Figure 6.5 at $h^+ = 162$. It is compared with a DNS carried out in Ref. [56] using spectral methods at $h^+ = 150$. These DNS, which use very precise numerical methods, turn out to be in very good agreement with the experiments. We see on Figure 6.5-a that the logarithmic range begins at $y^+ = 30$. We show on Figure 6.5-b the r.m.s. velocity-fluctuations profile in terms of y. It is clear that there is a strong production of u' at the wall, with a peak at $y^+ = 12$. This corresponds in fact to the low- and high-speed streaks. Figure 6.5-e shows the Reynolds stresses, whose peak is higher (at the bottom of the logarithmic layer), which is the signature of ejections of vorticity from the wall. The same peak is observed for the pressure fluctuations (Figure 6.5-d), which is certainly due to low pressure associated to high vorticity at the tip of the ejected hairpin (see below). Finally Figure 6.5-f shows the r.m.s. vorticity fluctuations, a quantity very difficult to measure precisely experimentally. It indicates that the maximum vorticity produced is spanwise and at the wall, in fact under the high-speed streaks (see below). The vorticity perpendicular to the wall is about 40% higher than the longitudinal vorticity in the region $5 < y^+ < 30$, which indicates only a weak longitudinal vorticity stretching by the ambient shear. We next present two LES using the spectral-dynamic model, at $R_e = 6666$ ($h^+ = 204$, case A) and $R_e = 14000$ ($h^+ = 389$, case B). They are respectively subcritical and supercritical with respect to the linear-stability analysis of the Poiseuille profile. In the two simulations there is a grid refinement close to the wall, in order to simulate accurately the viscous sublayer. The kinetic-energy spectrum allowing to determine the eddy-viscosity is calculated at each time step by averaging in planes parallels to the walls, and is thus a function of (y, t). In fact, the original formula for the spectral-eddy viscosity considered a three-dimensional spectrum. It is possible, in the isotropic case and when spectra decrease as a power law, to relate the two-dimensional to the three-dimensional spectrum. LES of the channel seem to be insensitive to the particular spectrum chosen.

Figure 6.6 shows for case A the exponent m arising in the energy spectrum at the cutoff, as a function of the distance to the wall y^+. Regions where $m > 3$ correspond to a zero eddy viscosity and hence a direct-numerical simulation. This is the case in particular close to the wall, up to $y^+ \approx 12$ where we know that longitudinal velocity fluctuations are very intense, due to the low- and high-speed streaks. Therefore, and since the first point is very close to the wall ($y^+ = 1$), our LES has the interesting property of becoming a DNS in the vicinity of the wall, which enables us to capture events which occur in this region. Figure 6.7 shows the mean velocity profile in case A, compared with the LES of Piomelli [103] using the dynamic model of [43]. The latter is known to agree very well with experiments at these low Reynolds numbers. Our simulation with the spectral dynamic model coincides, with the right value for the Karman constant. On the other hand, the LES carried out with the classical spectral-cusp model with $m = 5/3$ gives an error of 20% for the Karman constant. Figure 6.8 shows for case A the rms velocity fluctuations, compared with the dynamic-model predictions of [103]. The agreement is still very good, with a correct prediction of the longitudinal velocity fluctuations peak. Concerning the supercritical case, the LES of case B are in very good agreement with a DNS at $h^+ = 395$ carried out by [3], both for the mean velocity and the rms velocity components. The latter are shown on Figure 6.9. Notice that the LES allows to reduce the computational cost by a factor of the order of 100, which is huge.

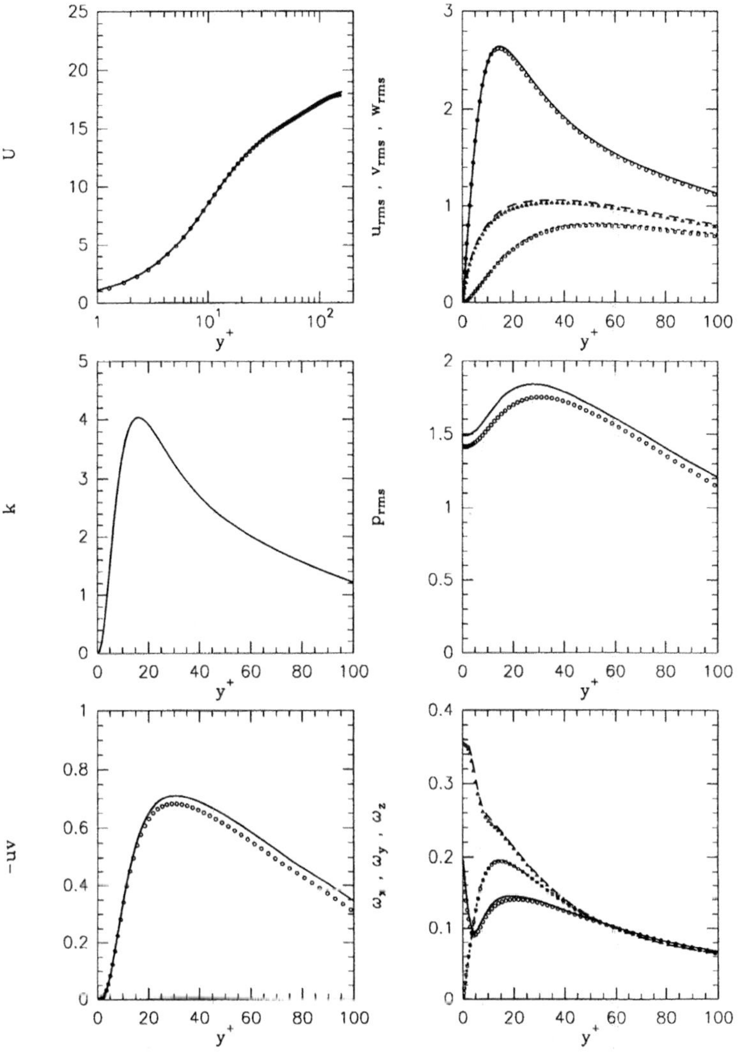

Figure 6.5. statistical data obtained in DNS of a turbulent channel flow by Lamballais (straight line) and Kuroda (symbols); from left to right and top to bottom, a) mean velocity, b) r.m.s. velocity fluctuations (respectively from top to bottom, longitudinal, spanwise, vertical), c) kinetic energy, d) r.m.s. pressure fluctuation, e) Reynolds stresses, f) r.m.s. vorticity (from top to bottom, spanwise, vertical, longitudinal)

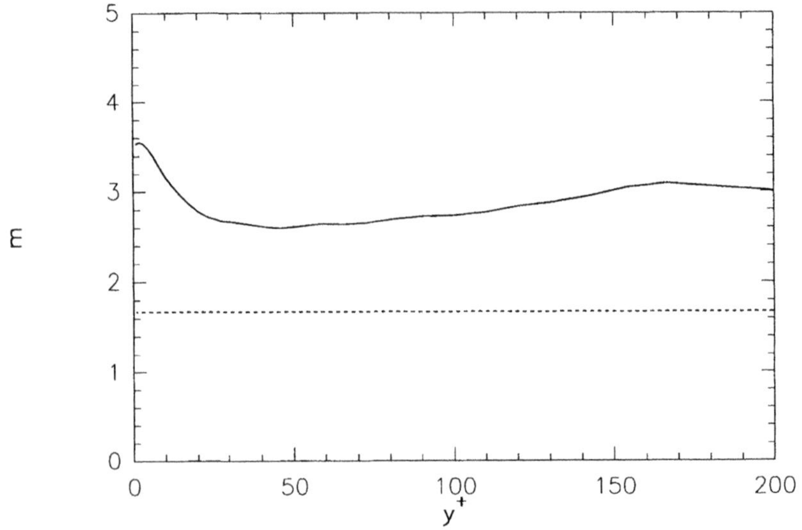

Figure 6.6. Spectral-dynamic LES of the channel flow (case A), exponent $m(y^+)$ of the kinetic-energy spectrum at the cutoff

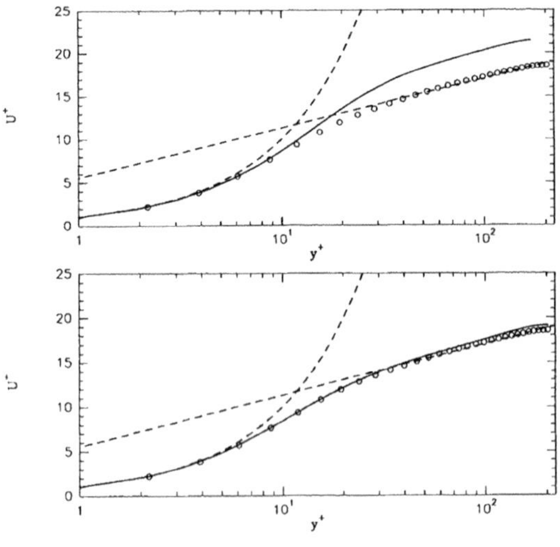

Figure 6.7. Mean velocity profiles in wall units. Lines : present simulation ($Re = 6666$) ; symbols : Piomelli [103] ($Re = 6500$). Case A (top) : $m = 5/3$ ($h^+ = 181$) ; Case B (bottom) : dynamic evaluation of $m(y,t)$ ($h^+ = 204$)

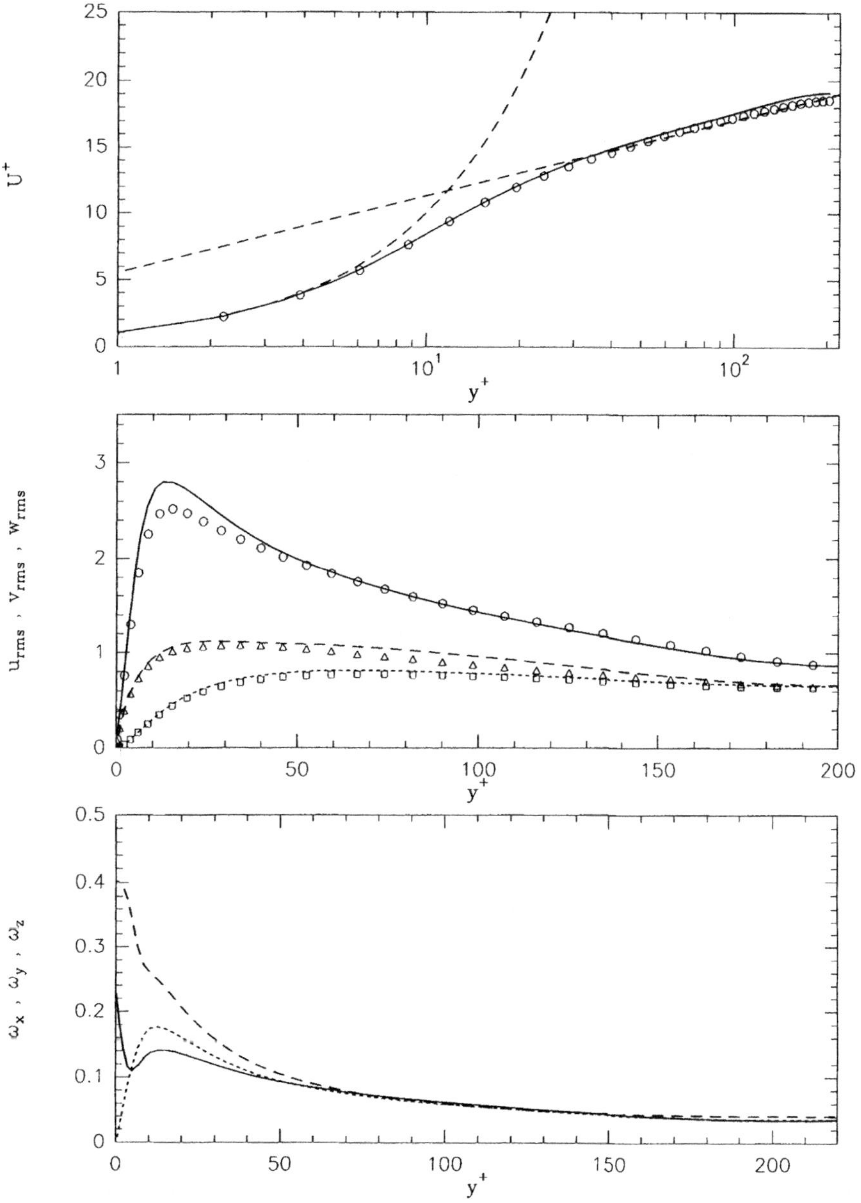

Figure 6.8. LES with the spectral dynamic model, same as Figure 6.7, but for the rms velocity fluctuations, from top to bottom longitudinal, spanwise and transverse

(a)

(b)

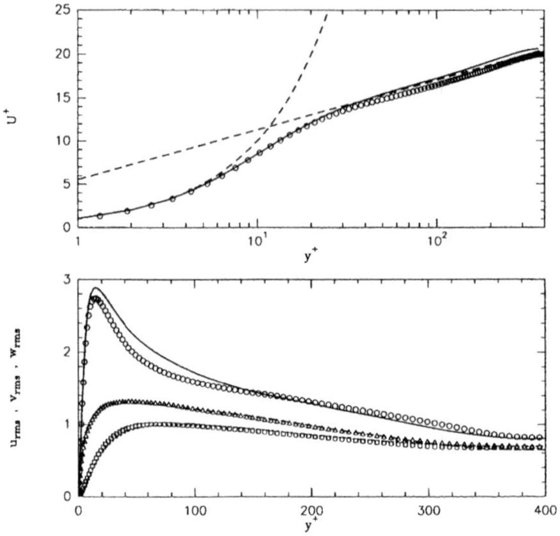

Figure 6.9. Turbulent channel flow, comparisons of the spectral-dynamic model (solid lines, $h^+ = 389$) with the DNS of Kim ([3], symbols, $h^+ = 395$); a) mean velocity, b) rms velocity components

We present finally on Figure 6.10 a map of the vorticity modulus at the same threshold for cases A and B. The flow goes from left to right. It is also clear that the LES do reproduce features expected from turbulence at higher Reynolds number, and display much more vortical activity in the small scales that the DNS.

6.5. Return to Physical Space

6.5.1. STRUCTURE-FUNCTION MODELS

This category of models, originally due to Métais and Lesieur [88], consists in using the spectral eddy-viscosity in physical space, while taking into account the intermittency of turbulence. If the peak behaviour is discarded, the eddy-viscosity given by equation (6.20) can be determined locally in physical space, through the local second-order velocity structure function. Let us consider the spectral eddy viscosity (still scaling on $\sqrt{E(k_C)/k_C}$) with no cusp, and adjust the constant as proposed by [68], by balancing in the inertial range the subgridscale flux with the kinetic energy flux ϵ in the energy spectrum evolution equation. This yields

$$\nu_t(k_C) = \frac{2}{3}\, C_K^{-3/2} \left[\frac{E(k_C)}{k_C}\right]^{1/2} \quad . \tag{6.27}$$

The problem with such an eddy-viscosity (if the energy spectrum may be computed) is that it is uniform in space when used in physical space. Ob-

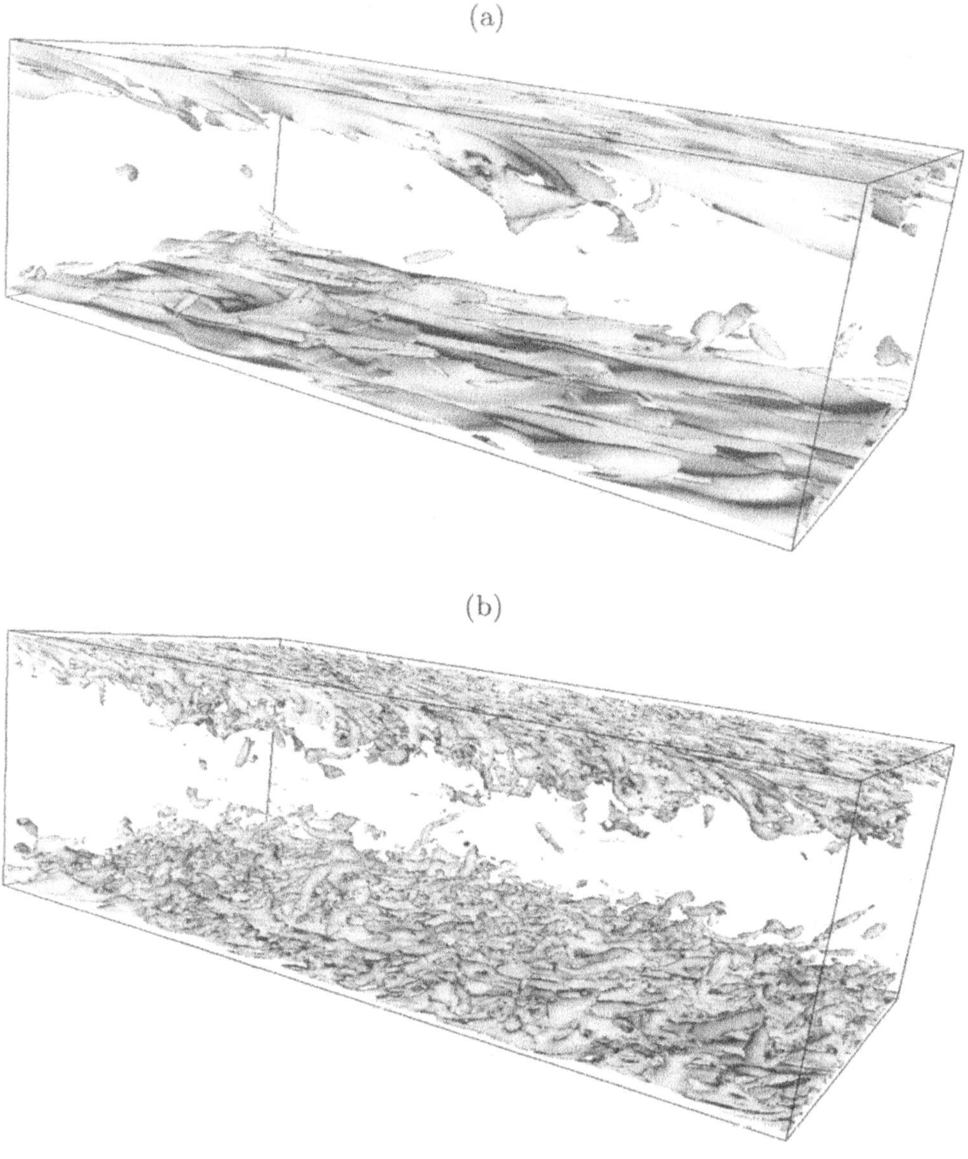

Figure 6.10. Turbulent plane channel, vorticity modulus; a) DNS ($h^+ = 165$), b) LES using the spectral-dynamic model ($h^+ = 389$), from [58]

viously, the eddy viscosity should take into account the intermittency of turbulence: there is no need for any subgridscale modelling in regions of space where the flow is laminar or transitional. On the other hand, it is essential to dissipate in the subgridscales the local bursts of turbulence if they become too intense. Considering also that turbulence in the small scales may not be too far from isotropy, it was proposed by [88] to come back to the classical formulation in the physical space, where the eddy viscosity is determined with the aid of (6.27). $E(k_C, \vec{x})$ is now a local kinetic energy spectrum, calculated in terms of the local second-order velocity structure function of the filtered field

$$F_2(\vec{x}, \Delta x) = \left\langle \|\vec{\bar{u}}(\vec{x}, t) - \vec{\bar{u}}(\vec{x} + \vec{r}, t)\|^2 \right\rangle_{\|\vec{r}\|=\Delta x} \tag{6.28}$$

as if the turbulence is three-dimensionally isotropic, with Batchelor's formula

$$F_2(\vec{x}, \Delta x) = 4 \int_0^{k_C} E(k) \left(1 - \frac{\sin(k\Delta x)}{k\Delta x}\right) dk \quad . \tag{6.29}$$

In fact, the original formula involves a k integral from 0 to ∞. But the filtered field has no energy at modes larger than k_C, which explains Eq. (6.29). This yields for a Kolmogorov spectrum

$$\nu_t^{SF}(\vec{x}, \Delta x) = 0.105\; C_K^{-3/2}\; \Delta x\; [F_2(\vec{x}, \Delta x)]^{1/2} \quad . \tag{6.30}$$

F_2 is calculated with a local statistical average of square velocity differences between $\vec{x}$ and the six closest points surrounding $\vec{x}$ on the computational grid. In some cases, the average may be taken over four points parallel to a given plane; in a channel, for instance, the plane is parallel to the boundaries. Notice also that if the computational grid is not regular (but still orthogonal), interpolations of (6.30) have been proposed by [71]. Let $\Delta c = (\Delta x_1 \Delta x_2 \Delta x_3)^{1/3}$ be a (geometric) mean mesh in the three spatial directions. Remembering Kolmogorov's (1941) law in physical space, which states that the second-order velocity structure-function scales like $(\epsilon r)^{2/3}$, one can in Eq. (6.30) replace Δx by Δc, with (in the six-point formulation)

$$F_2(\vec{x}, \Delta c) = \frac{1}{6} \sum_{i=1}^{3} F_2^{(i)} \left(\frac{\Delta c}{\Delta x_i}\right)^{2/3} \quad ,$$

with

$$F_2^{(i)} = [\|\vec{u}(\vec{x}) - \vec{u}(\vec{x} + \Delta x_i \vec{e}_i)\|^2 + \|\vec{u}(\vec{x}) - \vec{u}(\vec{x} - \Delta x_i \vec{e}_i)\|^2] \quad ,$$

where $\vec{e}_i$ is the unit vector in direction x_i.

One can also look at the relation of Smagorinsky's and the structure-function models when the differences in the structure-function are replaced (within a first-order approximation!) by spatial derivatives. It is found for the six-point formulation, in the limit of $\Delta x \to 0$:

$$\nu_t^{SF} \approx 0.777\; (C_S \Delta x)^2 \sqrt{2\bar{S}_{ij}\bar{S}_{ij} + \bar{\omega}_i\bar{\omega}_i} \quad ,$$

where $\vec{\bar{\omega}}$ is the vorticity of the filtered field, whereas C_S is Smagorinsky's constant defined by Eq. (6.16).

6.5.2. SELECTIVE AND FILTERED STRUCTURE-FUNCTION MODELS

As with Smagorinsky's model, however, the Structure-Function (SF) model is too dissipative for transition in a boundary layer for instance, and does not behave well in a channel: in fact, the spectrum $E_{\vec{x}}(k_C)$ is sensitive to the low-frequency oscillations. To overcome the difficulty with transition, two improved versions of the SF model have been developed: the selective structure-function model (SSF), and the filtered structure-function model (FSF). The dynamic model is another way of adapting the eddy-viscosity to the local conditions of the flow (see next section).

The selective structure-function (SSF) model was developed by David [28] (see also [22], [71] for details). The idea is to switch off the eddy-viscosity when the flow is not three-dimensional enough. The three-dimensionalization criterion is the following: one measures the angle between the vorticity at a given grid point and the average vorticity at the six closest neighbouring points (or the four closest points in the four-point formulation). If this angle exceeds 20^0, the most probable value according to simulations of isotropic turbulence at a resolution of $32^3 \sim 64^3$, the eddy-viscosity is turned on. Otherwise, there is only molecular dissipation which acts. In the filtered structure-function (FSF, see [35], [22], [71]), the filtered field $\bar{u}_i$ is submitted to a high-pass filter $\widetilde{(.)}$ in order to get rid of low-frequency oscillations which affect the local kinetic-energy spectrum $E_{\vec{x}}(k_C)$ in the SF model. The high-pass filter is a Laplacian discretized by second-order centered finite differences and iterated three times. It is found

$$\nu_t^{FSF}(\vec{x}, \Delta x) = 0.0014\ C_K^{-3/2}\ \Delta x\ [\tilde{F}_2(\vec{x}, \Delta x)]^{1/2} \quad , \qquad (6.31)$$

where $\tilde{F}_2(\vec{x}, \Delta x)$ is the second-order structure function of the high-pass filtered field $\tilde{\bar{u}}_i$.

Coming back to the temporally-growing mixing layer, figure 6.11 shows a comparison between Smagorinsky's model and the spectral-cusp model given by eq. (6.20) and the various structure function models (original, selective and filtered versions). The case of a temporally-growing incompressible mixing layer with a three-dimensional initial perturbation is here considered (helical pairing case; 64^3 Fourier modes). Notice the strong similarity between the results obtained with the spectral model, the filtered structure-function model and the selective structure-function model. This confirms that both modifications of the original structure-function model go in the right direction. Note also that the SF model seems to be slightly less dissipative that Smagorinsky's model.

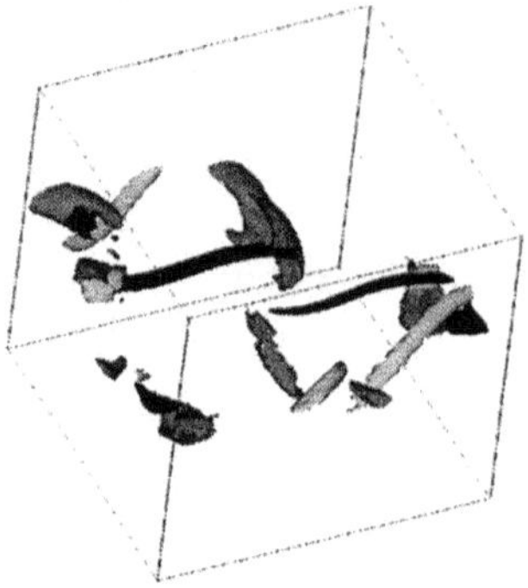

SM: $\max|\omega_x| = 2.92\ \omega_i$

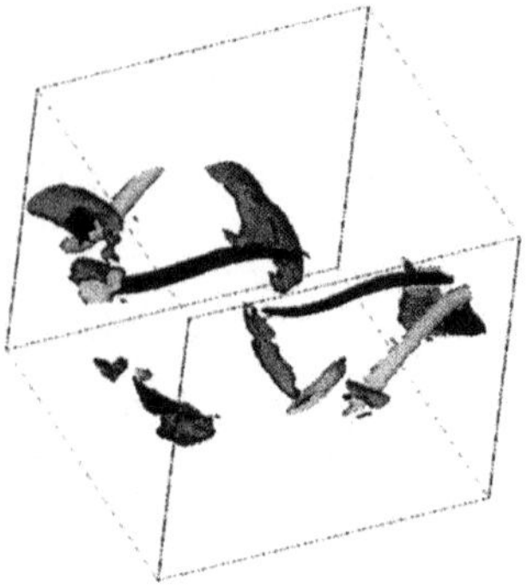

SF: $\max|\omega_x| = 2.86\ \omega_i$

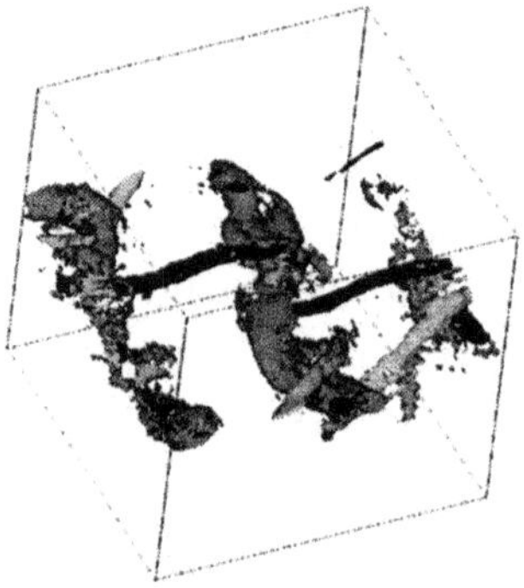

SC: $\max|\omega_x| = 4.75\ \omega_i$

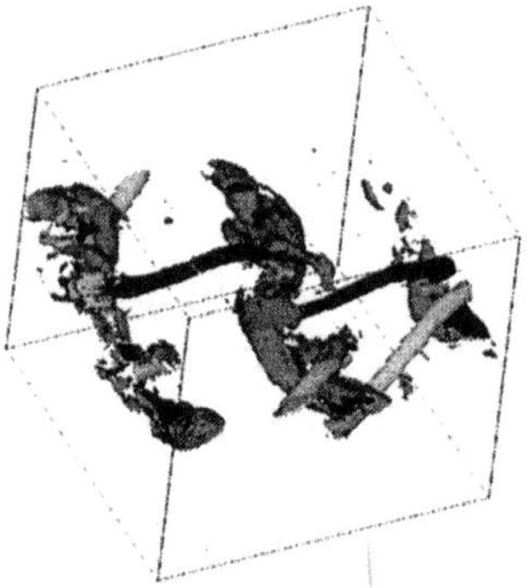

FSF: $\max|\omega_x| = 4.83\ \omega_i$

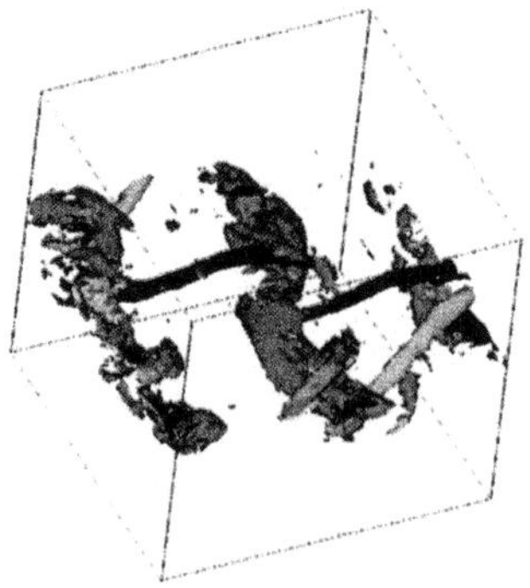

SSF: $\max|\omega_x| = 5.42\ \omega_i$

Figure 6.11. temporal mixing layer in LES with different SGS models, visualized by isosurfaces $\omega_x = \omega_i$ (black), $\omega_x = -\omega_i$ (light grey) and $\omega_z = \omega_i = 2U/\delta_i$ (dark grey).

Spatially Growing Mixing Layer

The temporal approximation is only a crude approximation of a mixing layer spatially developing, where one works in a frame traveling with the average velocity between the two layers. We present now LES using the FSF model of a spatial mixing layer, initiated upstream by a hyperbolic-tangent velocity profile superposed on the average flow, plus a weak random forcing regenerated at each time step.

We here compare a DNS at low Reynolds number ($Re = 100$) with a LES using the filtered structure-function subgrid scale model of a spatially developing incompressible mixing layer between two streams of velocity U_1 and U_2 ($U_1 > U_2$). Further details can be found in Silvestrini *et al.* [109]. The inflow consists of an hyperbolic-tangent velocity profile

$$\bar{u}(y) = \frac{U_1 + U_2}{2} + \frac{U_1 - U_2}{2} \tanh \frac{2y}{\delta_i} \quad . \tag{6.32}$$

where δ_i is the vorticity thickness. Two small-amplitude random perturbations of Gaussian p.d.f. are superimposed onto this profile: the first is three-dimensional (*i.e.* function of y, z and t); its kinetic energy is denoted $\varepsilon_{3D}U^2$. The second (of energy $\varepsilon_{2D}U^2$) depends only on y and t. The ratio $\varepsilon_{2D}/\varepsilon_{3D}$ is here set to 10 corresponding to a quasi-twodimensional perturbations. The DNS and the LES are henceforth referred to as DNSQ2D and FSFQ2D respectively. It is important to notice that the low Reynolds number DNS requires more grid points than the LES which will turn out to be much more turbulent. Precise numerical methods are compulsory when

TABLE 6.1. recapitulative table of the simulations presented here (L_x, L_y and L_z denote the domain's dimensions in the streamwise, transverse and spanwise directions, respectively. The corresponding numbers of collocation points N_x, N_y and N_z are such that the meshes are cubic of side $\Delta \approx 0.29\ \delta_i$).

Run	ε_{2D}	ε_{3D}	Re	$L_x/\delta_i,\ L_y/\delta_i,\ L_z/\delta_i$	N_x, N_y, N_z
DNSQ2D	10^{-3}	10^{-4}	100	140, 28, 14	480, 96, 48
FSFQ2D	10^{-4}	10^{-5}	-	112, 28, 14	384, 96, 48
FSF3D	0	10^{-4}	-	112, 28, 28	384, 96, 96

one wants to precisely describe the three-dimensional vortex dynamics. We solve the complete Navier-Stokes and LES equations for an incompressible fluid in a parallelepipedic domain. Sixth-order compact finite-differences [67] are used in the longitudinal direction x, along with pseudo-spectral methods on yz planes. Periodicity is assumed in the spanwise direction z. Sine/cosine expansions are used in the transverse direction y, enforcing free-slip boundary conditions. Non-reflective outflow boundary conditions are approximated by a multi-dimensional extension of Orlansky's discretization scheme, with limiters on the phase velocity (see [45], for a detailed description of the numerical code).

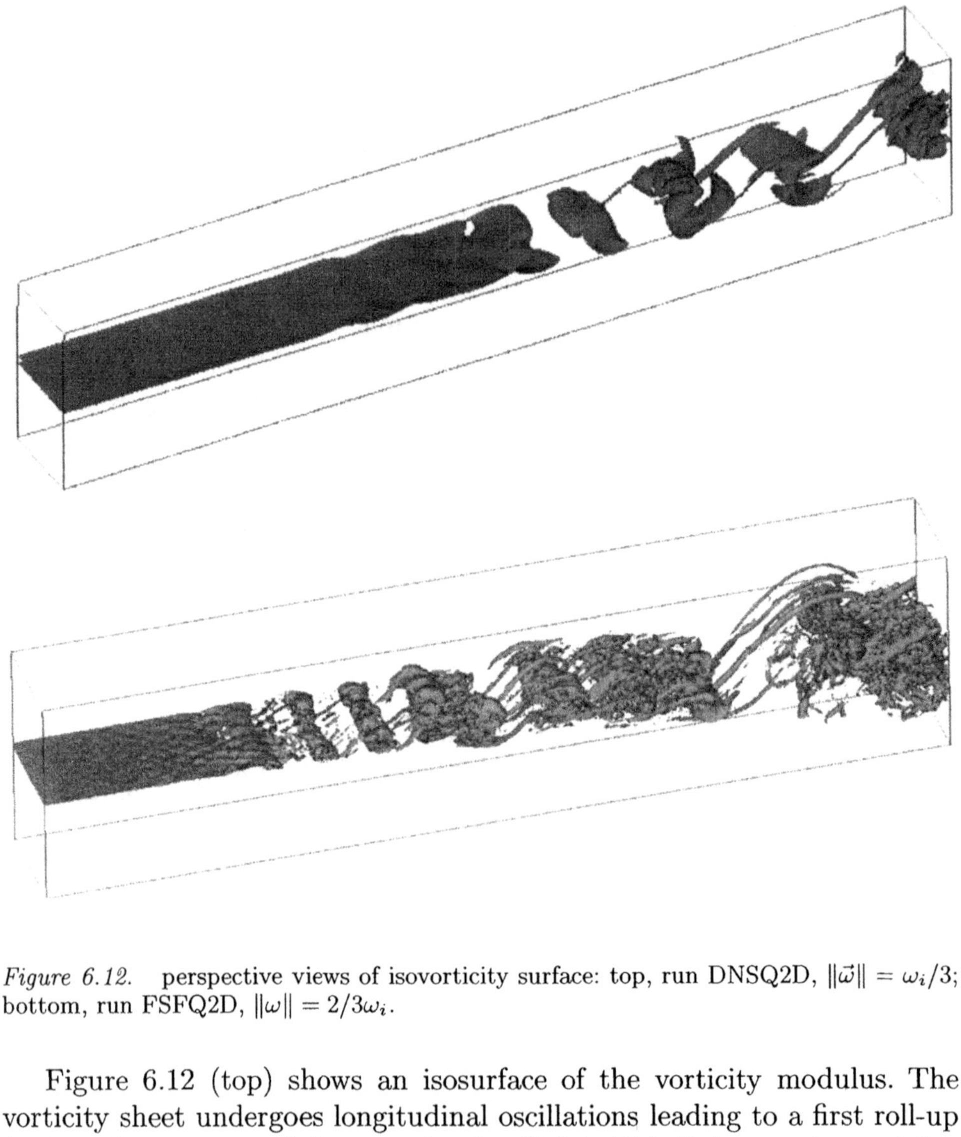

Figure 6.12. perspective views of isovorticity surface: top, run DNSQ2D, $\|\vec{\omega}\| = \omega_i/3$; bottom, run FSFQ2D, $\|\omega\| = 2/3\omega_i$.

Figure 6.12 (top) shows an isosurface of the vorticity modulus. The vorticity sheet undergoes longitudinal oscillations leading to a first roll-up further downstream. Subsequently, the Kelvin-Helmholtz vortices exhibit successive pairings. An important feature consists in thin intense longitudinal hairpin vortices which are stretched between the Kelvin-Helmholtz rollers as in Bernal and Roshko's [12] experiment. The vorticity magnitude within the longitudinnal vortices peaks at $2\omega_i$, where $\omega_i = 2U/\delta_i$ is the maximal vorticity magnitude introduced at the inlet. Run LESQ2D (figure 6.12, bottom) is visibly much more turbulent than DNSQ2D: maximal vor-

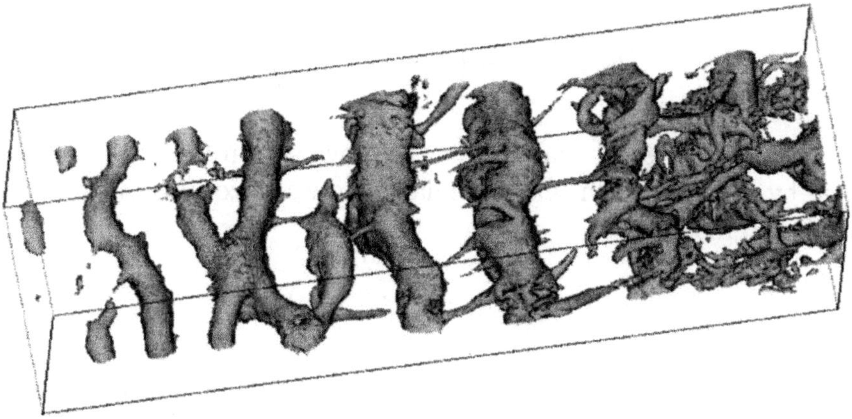

Figure 6.13. Same as Figure 6.12 (bottom), but with a three-dimensional upstream white-noise forcing (run FSF3D), low-pressure field

ticity magnitude is $\approx 4\ \omega_i$ for the whole run. Roll-up and pairing events occur much faster than in the DNS. Notice the complexity of the dynamics with a cluster of three fundamental Kelvin-Helmholtz vortices undergoing a first pairing and, at its downstream end, a billow made of 4 fundamental KH vortices whose second pairing is in progress. Experimentally observed trends such as the doubling of the spanwise spacing of the longitudinal vortices at every pairing also seem to be correctly reproduced. An interesting feature found is that longitudinal vortices of same sign may come close together and merge, contributing thus to the global self-similarity of the mixing layer. When the forcing is a three-dimensional random white noise (run FSF3D), helical pairing occurs upstream, as indicated by the low-pressure maps of Figure 6.13. But none of these simulations has reached self-similarity, since the kinetic-energy spectra in the downstream region are steeper than $k^{-5/3}$, and rms velocity fluctuations have a departure of about 20% with respect to the experiments. Thus calculations in longer domains are necessary, in order in particular to know in the helical-pairing case whether quasi two-dimensionality might not be restored further downstream.

6.5.3. GENERALIZED HYPERVISCOSITIES

One of the drawback of the structure-function model given by eq. (6.30) is the absence of a cusp near k_C. However, E.D.Q.N.M. data show that the exponential form given in equation (6.21) can be correctly approximated by a power law of the type:

$$\nu_t^*\left(\frac{k}{k_C}\right)=\left(1.+\nu_{tn}^*\left(\frac{k}{k_C}\right)^{2n}\right) \quad . \tag{6.33}$$

with $2n \approx 3.7$. Lesieur and Métais [71] have shown that ν_{tn}^* can be determined by considering the energy balance between explicit and subgrid-scale transfers. This yields:

$$\nu_{tn}^* = 0.512\left(\frac{3n}{2}+1\right) \quad . \tag{6.34}$$

In fact, the E.D.Q.N.M. value of $2n = 3.7$ is not so far from the exponent $2n = 4$ which would be obtained with a Laplacian operator iterated twice. Therefore, Lesieur and Métais [71] proposed a physical-space turbulent dissipative operator based upon the structure-function model and taking into account the "cusp" behaviour:

$$2\frac{\partial}{\partial x_j}\left[\nu_t^{SF}\bar{S}_{ij}\right]+\nu_t^{(2)}\left(\frac{\Delta x}{\pi}\right)^4\left(\frac{\partial^2}{\partial x_j^2}\right)^3\bar{u}_i \quad , \tag{6.35}$$

where $\bar{S}_{ij}$ is the deformation tensor of the field $\bar{u}_i$. $\left(\partial^2/\partial x_j^2\right)^3$ designates the Laplacian operator iterated three-times. ν_t^{SF} is given by the r.h.s. of equation (6.30) multiplied by $0.441/(2/3)$, $\nu_t^{(2)} = \nu_t^{SF} \times \nu_{t2}^*$, and ν_{t2}^* is given by equation (6.34) with $n = 2$. The expression (6.35) is interesting in the sense that it provides an eddy dissipation combining the structure function model with a hyperviscosity $(\nabla^2)^3\bar{u}_i$. The latter represents in physical space the action of the cusp in Kraichnan's spectral eddy viscosity.

6.5.4. HYPER-VISCOSITY

The model given by equation (6.35) bears some resemblance to hyperviscosity models which are widely used in the study of geophysical flows because of their simplicity. Indeed, the hyperviscosity consists in replacing the molecular dissipative operator $\nu\nabla^2$ by $(-1)^{\alpha-1}\nu_\alpha(\nabla^2)^\alpha$, where α is a positive integer. As opposed to equation (6.35), ν_α is here a constant (positive) coefficient which has to be adjusted. This has been widely used in two-dimensional isotropic turbulence (see [10]), with $\alpha = 2$ or $\alpha = 8$, as a way to shift the dissipation to the neighbourhood of k_c. This allows for a reduction of the number of scales strongly affected by viscous effects, and has rendered possible in the case of two-dimensional turbulence to demonstrate the existence of coherent vortices.

In three-dimensional turbulence, it was used by Bartello *et al.* [9] to study the influence of a solid-body rotation, with surprisingly good results.

6.5.5. SCALE-SIMILARITY AND MIXED MODELS

The lack of correlation between the subgrid-scale stress and the large-scale strain rate tensors has led Bardina *et al.* [8] to propose an alternative subgrid-scale model called the scale similarity model. This is based upon a double filtering approach and on the idea that the important interactions between the resolved and unresolved scales involve the smallest eddies of the former and the largest eddies of the latter. They suggest to evaluate the subgrid tensor as:

$$T_{ij} = \bar{\bar{u}}_i \bar{\bar{u}}_j - \overline{\bar{u}_i \bar{u}_j} \quad . \tag{6.36}$$

The analysis of DNS and experimental data [8], [78] have shown that the modelled subgrid-scale stress deduced from (6.36) exhibits a good correlation with the real (measured) stress. However, when implemented in LES calculations, the model hardly dissipates any energy. It is therefore necessary to combine it with an eddy-viscosity type model such as Smagorinsky's model to produce the "mixed" model. In the line of Bardina *et al.* model, new formulations have been proposed to correct this lack of dissipation. Liu *et al.* [78] have proposed the following model:

$$T_{ij} = C_L \left(\tilde{\bar{u}}_i \tilde{\bar{u}}_j - \widetilde{\bar{u}_i \bar{u}_j} \right) \quad , \tag{6.37}$$

where C_L is a dimensionless coefficient. The operator $\tilde{.}$ consists in a second filter of different width [78]. This concept of double filtering can be taken one step further, leading to the dynamic models presented in the next section.

6.5.6. DYNAMIC MODELS

We have already noted for Kraichnan's spectral eddy viscosity that the parameters defining it could be computed from a LES with a cutoff k_C, by defining a fictitious cutoff $k'_C = k_C/2$, and explicitly calculating the transfers across k'_C (see Refs [72][88]). This is the underlying philosophy of the dynamic model in physical space [43]. The method relies on a LES using a "base" subgrid-scale model such as Smagorinsky's model[1], with a grid mesh Δx. The computed fields $\bar{f}$ are filtered by a "test filter" $\tilde{.}$ of larger width $\alpha \Delta x$ (for instance $\alpha = 2$), to yield the field $\tilde{\bar{f}}$. If one applies the double filter to the Navier-Stokes equation (with constant density),

[1]But it may be used with other subgrid models.

the subgrid-scale tensor of the field $\tilde{\bar{u}}$ is readily obtained from with the replacement of the filter "bar" by the double filter "bar-tilde", that is:

$$\mathcal{T}_{ij} = \tilde{\bar{u}}_i \tilde{\bar{u}}_j - \widetilde{\overline{u_i u_j}} \quad . \tag{6.38}$$

We consider now the following resolved turbulent stress corresponding to the test-filter applied to the field $\bar{u}$:

$$\mathcal{L}_{ij} = \tilde{\bar{u}}_i \tilde{\bar{u}}_j - \widetilde{\bar{u}_i \bar{u}_j} \quad . \tag{6.39}$$

Finally we apply the filter "tilde" to Eq. (6.6), to yield

$$\widetilde{T_{ij}} = \widetilde{\bar{u}_i \bar{u}_j} - \widetilde{\overline{u_i u_j}} \quad . \tag{6.40}$$

Adding Eqs (6.39) and (6.40), using (6.38)

$$\mathcal{L}_{ij} = \mathcal{T}_{ij} - \widetilde{T_{ij}} \quad , \tag{6.41}$$

called Germano's identity. In this expression, $\mathcal{T}_{ij}$ and $\widetilde{T_{ij}}$ have to be modelled, while $\mathcal{L}_{ij}$ can be explicitly calculated by applying the test filter to the base LES results. Using Smagorinsky's model, we have

$$\widetilde{T_{ij}} - \frac{1}{3}\widetilde{T_{ll}}\ \delta_{ij} = 2\widetilde{\mathcal{A}_{ij}C} \quad , \tag{6.42}$$

whith $C = C_S^2$ and

$$\mathcal{A}_{ij} = (\Delta x)^2\ |\bar{S}|\bar{S}_{ij} \quad .$$

Still using Smagorinsky, we have

$$\mathcal{T}_{ij} - \frac{1}{3}\mathcal{T}_{ll}\ \delta_{ij} = 2\mathcal{B}_{ij}C \ , \tag{6.43}$$

whith

$$\mathcal{B}_{ij} = \alpha^2(\Delta x)^2\ |\tilde{\bar{S}}|\ \tilde{\bar{S}}_{ij} \quad .$$

$|\tilde{\bar{S}}|$ and $\tilde{\bar{S}}_{ij}$ are the quantities analogous to $|\bar{S}|$ and $\bar{S}_{ij}$ built with the doubly-filtered field $\tilde{\bar{u}}$. Substracting Eq. (6.42) from Eq. (6.43) yields with the aid of Eq. (6.41)

$$\mathcal{L}_{ij} - \frac{1}{3}\mathcal{L}_{ll}\ \delta_{ij} = 2\mathcal{B}_{ij}C - 2\widetilde{\mathcal{A}_{ij}C} \quad .$$

In order to obtain C, many people remove it from the filtering as if it were constant, leading to

$$\mathcal{L}_{ij} - \frac{1}{3}\mathcal{L}_{ll}\ \delta_{ij} = 2CM_{ij} \quad , \tag{6.44}$$

with

$$M_{ij} = \mathcal{B}_{ij} - \widetilde{\mathcal{A}_{ij}} \quad .$$

Now, all the terms of Eq. (6.44) can be determined with the aid of $\bar{u}$. There are however five independent equations for only one variable C, and the problem is overdetermined.

Two alternatives have been proposed to deal with this undeterminacy. A first solution (Ref. [43]) is to contract Eq. (6.44) by $\bar{S}_{ij}$ to obtain

$$C = \frac{1}{2}\frac{\mathcal{L}_{ij}\bar{S}_{ij}}{M_{ij}\bar{S}_{ij}} \quad , \tag{6.45}$$

since, due to incompressibility, $\bar{S}_{ij}$ is traceless. This permits in principle to "dynamically" determine the "constant" C as a function of space and time, to be used in the LES of the base field $\bar{u}$. In tests using channel flow data obtained from direct numerical simulations, it was however shown in [43] that the denominator in Eq. (6.45) could locally vanish or become sufficiently small to yield computational instabilities. To get rid of this problem, Lilly [80] chose to determine the value of C which "best satisfies" the system Eq. (6.44) by minimizing the error using a least squares approach. It yields

$$C = \frac{1}{2}\frac{\mathcal{L}_{ij}M_{ij}}{M_{ij}^2} \quad . \tag{6.46}$$

This removes the undeterminacy of Eq. (6.44).

The analysis of DNS data revealed, however, that the C field predicted by the models (6.45) or (6.46) varies strongly in space and contains a significant fraction of negative values, with a variance which may be ten times higher than the square mean. So, the removal of C from the filtering operation is not really justified and the model exhibits some mathematical inconsistencies. The possibility of negative C is an advantage of the model since it allows a sort of backscatter in physical space, but very large negative values of the eddy viscosity is a destabilizing process in a numerical simulation, yielding a non-physical growth of the resolved scale energy. The cure which is often adopted to avoid excessively large values of C consists in averaging the numerators and denominators of (6.45) and (6.46) over space and/or time, thereby losing some of the conceptual advantages of the "dynamic" local formulation. Averaging over direction of flow homogeneity has been a popular choice, and good results have been obtained in [43] and [103], who took averages in planes parallel to the walls in their channel flow simulation. Remark that the same thing has been done, with success, when averaging the dynamic spectral eddy viscosity in the channel-flow LES presented above. It can be shown that the dynamic model gives a zero subgrid-scale stress at the wall, where L_{ij} vanishes, which is a great advantage with respect to the original Smagorinsky model; it gives also the proper asymptotic behavior near the wall. Notice again that the use of Smagorinsky's model as a base for the dynamic procedure is not compulsory, and any of the models described in the present paper can be a candidate. As an example, Ref. [36] have applied the dynamic procedure to the structure-function model (see below) applied to a compressible boundary layer above a long cylinder.

6.5.7. ANISOTROPIC SUBGRID-SCALE MODELS

As stressed above, the subgrid-scale tensor and buoyancy flux given by (6.7) are assumed to be strictly proportional to the grid-scale strain rate tensor and buoyancy flux, respectively. Abbà *et al.* [1] have recently proposed an anisotropic formulation of (6.7) using eddy-viscosity and eddy-diffusivity tensors instead of scalar ones :

$$T_{ij} - \frac{1}{3}T_{ll}\ \delta_{ij} = 2\sum_{r,s} \nu^t_{ijrs}\bar{S}_{rs} - \frac{2}{3}\delta_{ij}\sum_{l,r,s,} \nu^t_{llrs}\bar{S}_{rs} \quad . \tag{6.47}$$

This formulation allows for a better description of the small-scale anisotropy. This model in conjunction with the dynamic procedure previously described has been used successfully in Large-Eddy Simulations of turbulent natural convection [1].

6.6. Vortex Control in a Round Jet

Our goal here is to demonstrate the ability of the LES to properly reproduce the coherent vortex dynamics in the transitional region of the jet. We also show the possibility of controlling the jet behaviour by manipulating the inflow conditions. The detailed results are presented in Urbin (1997) [114], Urbin and Métais [115] and Urbin *et al.* [116]. Due to the diversity of their coherent structures, axisymmetric jets constitute a prototype of free shear flows of vital importance from both a fundamental as well as a more applied point of view. Indeed, a better understanding of the jet vortex structures should make possible the active control of the jet (spreading rate, mixing enhancement...) for engineering applications (see *e.g.* Zaman *et al.* [120]). In the last five years, the progress in the experimental methods for detection and identification has made possible a detailed investigation of the complex three-dimensional coherent vortices imbedded within this flow. For instance, the influence of the entrainment of the secondary streamwise vortices has been studied by Liepmann and Gharib [76]. On the numerical side, several simulations of two-dimensionnal or temporally evolving jets have been performed. Very few have however investigated the three-dimensionnal spatial development of the round jet. We here show how LES can be used to perform a statistical and topological numerical study of the spatial growth of the round jet from the nozzle up to sixteen diameters downstream. The use of large-eddy simulations (LES) techniques allow us to reach high values of the Reynolds number: here, Re is 25000.

We have here used the *structure function model* in its selective version, which is well adapted for transitional flows and accepts non uniform grids. The LES filtered Navier-Stokes equations are solved using the TRIO-VF code. This is an industrial software developed for thermal-hydraulics applications at the Commissariat a l'Energie Atomique de Grenoble. It has been thoroughly validated in many LES of various flows (see *e.g.* Silveira *et al.* [108], for the backward facing step). It uses the finite volume element method on a structured mesh.

The experimental studies by Michalke and Hermann [90] have clearly pointed out the capital effect of the inflow momentum boundary layer thickness θ and of the ratio R/θ (R : jet radius) on the jet downstream development. It was shown that the detailed shape of the mean velocity profile strongly influences the nature of the coherent vortices appearing near the nozzle: either axisymmetric structures (vortex rings) or helical structure can indeed develop. Here, we did not simulate the flow inside the nozzle but we imposed a mean axial velocity profile in accordance with the experimental measurements:

$$W(r) = \frac{1}{2}W_o\left[1 - tanh\left(\frac{1}{4}\frac{R}{\theta}\left(\frac{r}{R} - \frac{R}{r}\right)\right)\right] \tag{6.48}$$

where W_o is the velocity on the axis. We have restrained ourself to a relatively small value of R/θ with $R/\theta = 10$ since a correct resolution of the shear zone at the border of the nozzle is crucial to correctly reproduce the initial development of the instabilities. We consider a computational domain starting at the nozzle and extending up to $16D$ downstream. The section perpendicular to the jet axis consists of a square $10D * 10D$, which has been shown to be sufficient to avoid jet confinement. The computational mesh is refined at the jet shear-layer (stretched mesh). The "natural" jet is forced upstream by the top-hat profile given by (6.48). to which is superposed a weak 3D white noise. The "excited" jet development is controlled with the aid of various deterministic inflow forcing (plus a white noise) designed to trigger specific types of three-dimensional coherent structures.

6.6.1. THE NATURAL JET

We have thoroughly validated our numerical approach by comparing the computed statistics with experimental results for the mean and for the r.m.s. fluctuating quantities. The frequency spectra have furthermore revealed the emergence of a predominant vortex-shedding Strouhal number, $Str_D = 0.35$ in good correspondance with the experimental value.

Temporal linear stability analysis performed on the inlet jet profile given by (6.48) (with $R/\theta = 10$) predicts a slightly higher amplification rate for the axisymmetric (varicose) mode than for the helical mode (see Michalke and Hermann [90]). For this reason, we have checked that the Kelvin-Helmholtz instability along the border of the jet yields, further downstream, vortex structures mainly consisting in axisymmetric toroidal shape. However, the simulations reveal that these structures are not always present and alternate with vortices of helical shape (see Urbin and Métais [115]).

The 3D visualization (figure 6.14) exhibits an original vortex arrangement subsequent to the varicose mode growth: the "alternate pairing". Such a structure was previously observed by Fouillet [40] and Comte *et al.* [23] in a direct simulation of a temporally evolving round jet at low Reynolds number ($Re = 2000$). The direction normal to the toroidal vortices symmetry plane, during their advection downstream, tends to differ from the jet axis. The inclination angle of two consecutive vortices appears to be of opposite sign eventually leading to a local pairing with an alternate arrangement. Note that vortex loop's inclination at the end of the potential core was experimentally observed by Petersen [102]. Experimental evidence of "alternate pairing" was recently shown by Broze and Hussain [17]. This alternate-pairing mode corresponds to the growth of a subharmonic perturbation (of wavelength double of the one corresponding to the rings) developing after the formation of the primary rings. It therefore presents strong analogies with the helical pairing mode observed in plane mixing layers (see above).

6.6.2. THE FORCED JET

Our natural jet simulations have therefore revealed three different types of vortical organization: the toroidal vortices (rings), the helical structure and the alternate-pairing. We now apply deterministic inflow perturbations to trigger one of these three particular flow organization and study the influence of the forcing on the statistics. Crow and Champagne [27] first noticed that the jet response is maximal with a preferred mode frequency corresponding to Str_D between 0.3 and 0.5. We then applied a periodic fluctuation associated with a frequency corresponding to $Str_D = 0.35$ in superposition to the white noise.

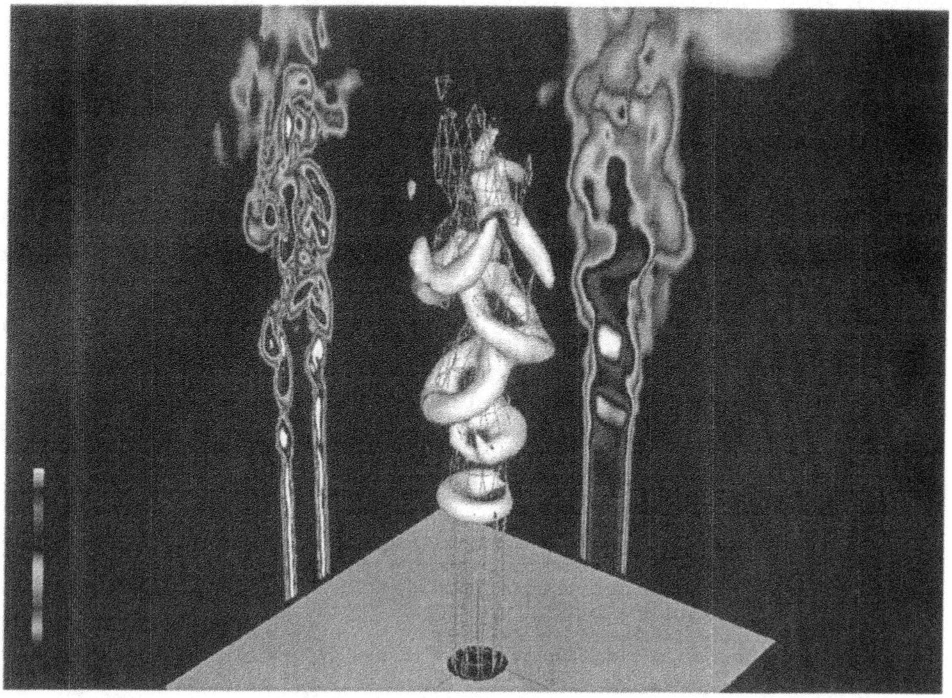

Figure 6.14. Natural jet: instantaneous visualization. Light gray: low pressure isosurface; wired isosurface of the axial velocity $W = W_o/2$; YZ cross-section (through the jet axis) of the vorticity modulus; XZ cross-section of the velocity modulus.

Varicose excitation

We first excitate the varicose mode by imposing a periodic perturbation (alternatively low-speed and high speed) to the axial velocity at the nozzle:

$$W(r) + \epsilon\ W_o\ sin\left(\frac{Str_D W_o}{D} t\right) \tag{6.49}$$

where $W(r)$ is given by equation (1) and $\epsilon = 1\%$.

Comparisons of the velocity fluctuations with experimental results, show that, as opposed to the unexcited jet, a strong and fast amplification of the instability appears.

The visualizations show that the varicose mode is now present at every instant at the beginning of the jet $Z/D < 6$. The vortex structures are more intense than in the natural case with well marked and organized pressure trough (Figure 6.15). The rings resulting from the varicose mode are linked together with longitudinal vortices. The maximum vorticity within these structures is about 40% of the vorticity of the associated rings. These have already been observed experimentally at moderate Reynolds number flow (see *e.g.* Lasheras, Lecuona and Rodriguez [62]; Monkewitz and Pfitzenmaier [94]; Liepmann and Gharib [76]). The present simulation indicates that they are also present at high Reynolds number. These longitudinal vortices are known to entrain and eject fluid outside, thus creating transverse side jets and "branches". The latter were studied numerically, in temporal simulations of Martin and Meiburg [83] and Abid and Brachet [2]. One possible explanation for the origin of these longitudinal vortices is an azimuthal oscillation of the vortex lines at the stagnation points between consecutive primary rings followed by a strong stretching mechanism by the latter (Lasheras, Lecuona and Rodriguez [62]). We have numerically checked that the number of longitudinal structures is directly linked with the most unstable azimuthal mode of the primary rings predicted by Widnall *et al.* [118].

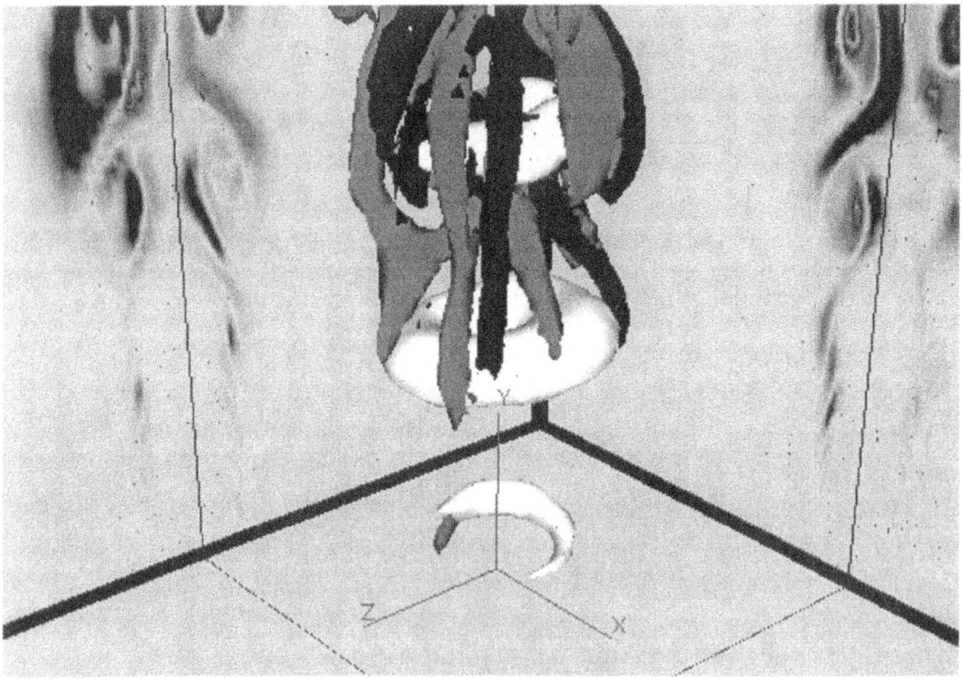

Figure 6.15. Jet with varicose mode excitation. Black and grey: positive and negative longitudinal vorticity isosurfaces corresponding to $\omega_z = \pm 1.2 W_o/D$. YZ and XZ cross-sections (through the jet axis) of the longitudinal vorticity component ($min = -4.W_o/D$; $max = +4.W_o/D$).

Helical excitation

The next excitation is designed to trigger the first helical mode by imposing the following inflow velocity profile:

$$W(r) + \epsilon\, W_o\, sin\left(\Theta - 2\pi \frac{Str_D W_o}{D} t\right) \frac{r}{D/2} \tag{6.50}$$

where Θ stands for the azimuthal angle.

The response of the jet indeed consists in the development of an helical coherent vortex structure (figure 6.16). This is in concordance with Kusek *et al.* [57] who experimentally observed the helical mode development with an appropriate inflow excitation. The signature of the helical excitation on the statistics consist in an increase of the potential core length as compared with the natural case, and a reduction of the spreading rate. For the present jet (no swirl), the velocity circulation on a circular contour of large radius, contained in a plane perpendicular to the jet axis and centered on the later remains zero. This implies that the longitudinal vorticity flux through the surface limited by this contour is also zero. The present excitation gives rise to an helix structure which rotates in the anti-clockwise direction, when moving away from the nozzle: it is therefore associated with a negative longitudinal vorticity component. This generation of negative longitudinal vorticity has necessarily to be compensated by regions of positive longitudinal vorticity. Indeed, in the vicinity of the nozzle, figure 6.16 shows the appearance of positive longitudinal vorticity on the helix border. However, further downstream both positive and negative longitudinal vortices do appear, but the former are more intense than the latter. At $Z = 4.5D$, the vorticity maximum within the positive vortices is $\approx 50\%$ of the vorticity (modulus) maximum within the helix, while it is only $\approx 25\%$ in the negative ones. Martin and Meiburg's [83] results, using vortex filament numerical techniques, display the same trend.

Alternate-pairing excitation

The excitation method is here based upon the same principle as previously described. The perturbation intensity is here $\epsilon = 5\%$. Its frequency is the same as before except that now half of the jet presents a speed excess, while a speed defect is imposed on the other half, and this alternatively. Note that this perturbation has a preferred direction, chosen along the Y axis. The resulting structures are analogous to figure 6.14 except that the alternatively inclined vortex rings now appear from the nozzle (see Figure 6.17). These inclined rings exhibit localized pairing and persist far downstream till $Z/D = 10$.

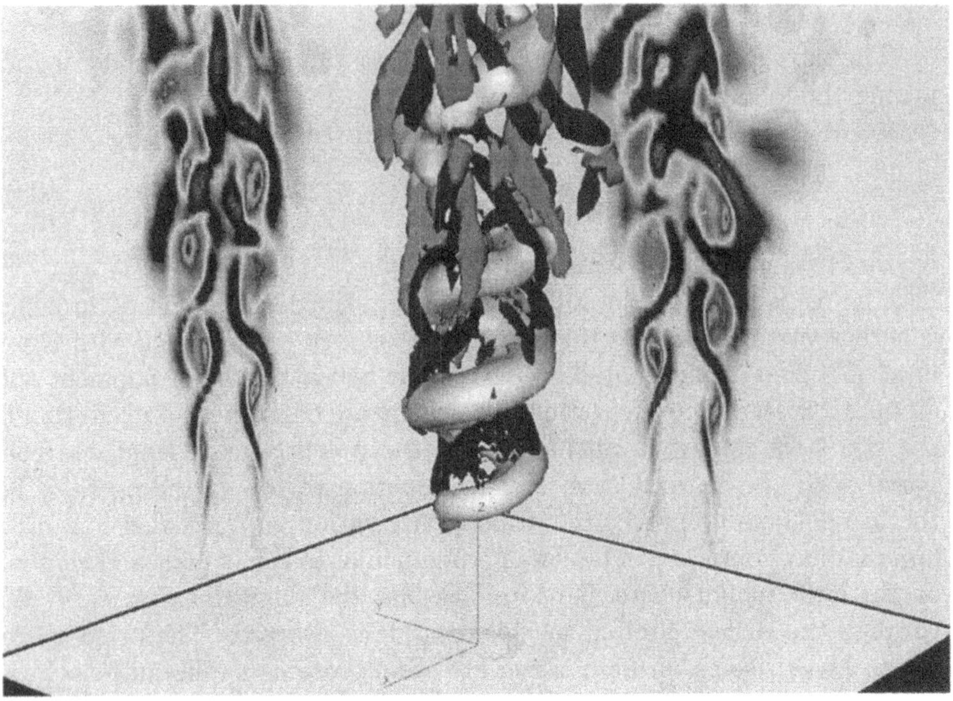

Figure 6.16. Jet with helical excitation: same vizualisation as figure 6.15.

One of the striking features is the very different spreading rates in the X and in the Y directions. In the XZ plane, the spreading rate is strongly increased as compared to the natural jet case: it reaches $50^o - 55^o$ at $Z/D = 6$. Conversely, it is close to zero (3° at $Z/D = 6$) in the YZ plane. Note that the present jet exhibits strong similarities with the "bifurcating" jet of Lee and Reynolds [65] (see also Parekh, Leonard and Reynolds [100]). They have experimentally showed, that a properly-combined axial and helical excitations can cause a turbulent round jet to split into two distinct jets. Such a bifurcation is indeed observed here (see figure 6.17). The streamlines originally concentrated close to the nozzle tend to clearly separate for $Z/D > 4$. Furthermore, the alternatively inclined vortex-rings seem to separate and move away from the jet centerline to form a Y-shaped pattern. More recently, the experiment performed by Longmire and Duong [81] has displayed similar vortex topology by using a specially designed nozzle made of two half nozzles. One of the important technological application of this peculiar excitation resides in the ability to polarize the jet in a preferential direction.

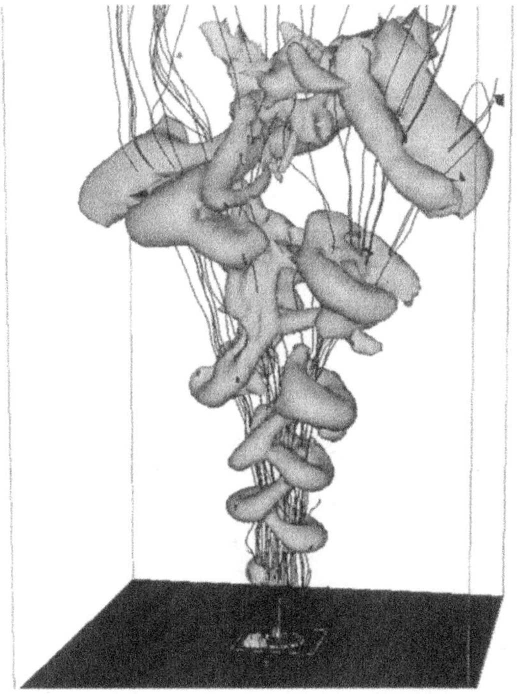

Figure 6.17. Bifurcation of the jet with alternate-pairing excitation. Instantaneous vizualisation of streamlines emerging from the nozzle. Low pressure isosurface in grey ($P = 25\%P_{min}$).

6.7. Rotating Flows

6.7.1. ROTATING CHANNEL FLOW

Our purpose here is to show the ability of the LES to accurately reproduce the detailed vorticity dynamics and flow statistics even in the presence of external forces like solid-body rotation.

Due to their numerous applications in turbo-machinery, the turbulent flow in a rotating channel with the rotation vector orientated along the spanwise direction has been the subject of extensive studies. Experimentally, it is not easy to cover a wide range of rotation regimes. Conversely, the introduction of the Coriolis force is rather straightforward in numerical codes simulating the three-dimensional Navier-Stokes equations. This may explain why more numerous numerical studies based either on Direct Numerical Simulations (DNS, see *e.g.* [55]) or Large-Eddy Simulations (LES, see *e.g.* [50, 91, 112, 104]) have been devoted to this topics than experimental ones (see *e.g.* [48, 95]). These previous works have been mainly devoted to weak rotation regimes and the analysis have principally concentrated

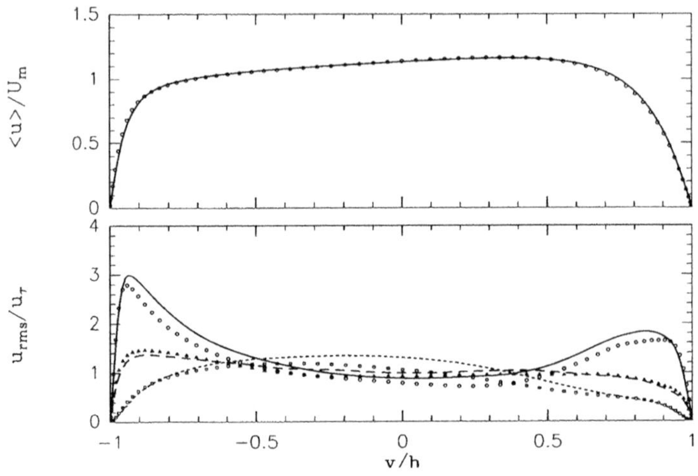

Figure 6.18. Mean velocity and turbulence intensities ($Re = 5700$, $Ro_g = 21$). —, ○, $\sqrt{\langle \overline{u}'^2 \rangle}/u_\tau$; ...,□, $\sqrt{\langle \overline{v}'^2 \rangle}/u_\tau$; - - -, △, $\sqrt{\langle \overline{w}'^2 \rangle}/u_\tau$. Lines : present LES ; Symbols : LES of Piomelli & Liu ([104]).

upon statistics directly issued from the velocity field, or upon the large-scale flow organization. Let $\langle \vec{u} \rangle$ be the mean velocity, $\langle \vec{u} \rangle = [\langle u \rangle (y), 0, 0]$ (x is the longitudinal direction and y the direction perpendicular to the wall). $\vec{\Omega}$ is oriented along the spanwise direction z, and may be positive or negative. For the channel flow, the vorticity vector associated with the mean velocity profile $\langle \vec{\omega} \rangle = (0, 0, -d \langle u \rangle /dy))$ is parallel to $\vec{\Omega}$ near one wall and antiparallel near the opposite wall: we refer to the two particular walls as cyclonic and anticyclonic. Various other terms are currently used. The names suction and pressure sides originate from the pressure gradient due to the Coriolis force, and the terms trailing and leading sides are borrowed from turbo-machinery. The previous studies have clearly shown that, due to the action of moderate rotation, the flow becomes very asymmetric with respect to the channel center, with a turbulent activity much reduced on the cyclonic side as compared with the anticyclonic side. The Rossby number is defined as $Ro_g = 3\ U_m/(2\Omega\ h)$ where U_m is the bulk velocity.

We first show a LES of a rotating channel flow based upon the spectral-dynamic model with the same characteristic parameters than Piomelli & Liu ([104]) in their LES using a localized version of the dynamic model : $Ro_g = 21$ and $Re = 5700$. Figure 6.18 shows that our mean velocity profile is in excellent agreement with Piomelli & Liu's predictions.

We next investigate rotation regimes for which anticyclonic destabilization is achieved: this corresponds to rotation rates such that $Ro_g > 1$. For these rotation rates, the combined effects of shear and solid-body rotation give rise to longitudinal (x-independent) instability called the shear-Coriolis instability. The computations are here performed with $Ro_g = \infty, 18, 6, 2$. We also study Reynolds number effects by comparing spectral-dynamic model based LES at $Re = 14000$ with DNS at $Re = 5000$.

6.7.2. SPATIAL ORGANIZATION

We first examine the three-dimensional flow structure. Figure 6.19 clearly shows that the vortex organization strongly differs in the rotating and in the non-rotating cases. We observe the following trends:

- The turbulent activity is gradually reduced near the cyclonic wall as the rotation rate is increased. For $Ro_g = 2$, the very flat isosurfaces indicate an almost complete flow relaminarization. This will be confirmed by the statistics.
- On the anticyclonic region, the flow presents a strong turbulence activity. We have checked, in the DNS computation, the existence of large scale longitudinal roll cells as already observed in the laboratory experiments of Johnston *et al.* ([48]) and in the numerical simulations of Kristoffersen and Andersson ([55]) and Piomelli and Liu ([104]). The roll cells are no longer present for $Ro_g = 2$.
- The vortical structures are more and more organized as rotation is increased, and their inclination with respect to the wall is reduced. This is clearly demonstrated by considering the statistics of the inclination angle of the vorticity vector (see ([59]).

It is important to note that the LES are capable to reproduce all the characteristic features of the flow organization. The subgrid scale model is indeed able to capture cyclonic relaminarization with inactive turbulent motions as well as detailed turbulent flow organization on the anticyclonic side (figure 6.19).

6.7.3. STATISTICS

Mean velocity profiles obtained for each couple (Re, Ro_g) are compared on figure 6.20. In the rotating case, the most important characteristic of the mean velocity profiles is a linear region of constant shear $d\langle u\rangle/dy$ and of slope 2Ω. This particular value corresponds to an anticyclonic region of nearly-zero mean spanwise absolute vorticity: $-d\langle u\rangle/dy + 2\Omega \approx 0$. In the channel flow, this characteristic mean velocity profile has already been no-

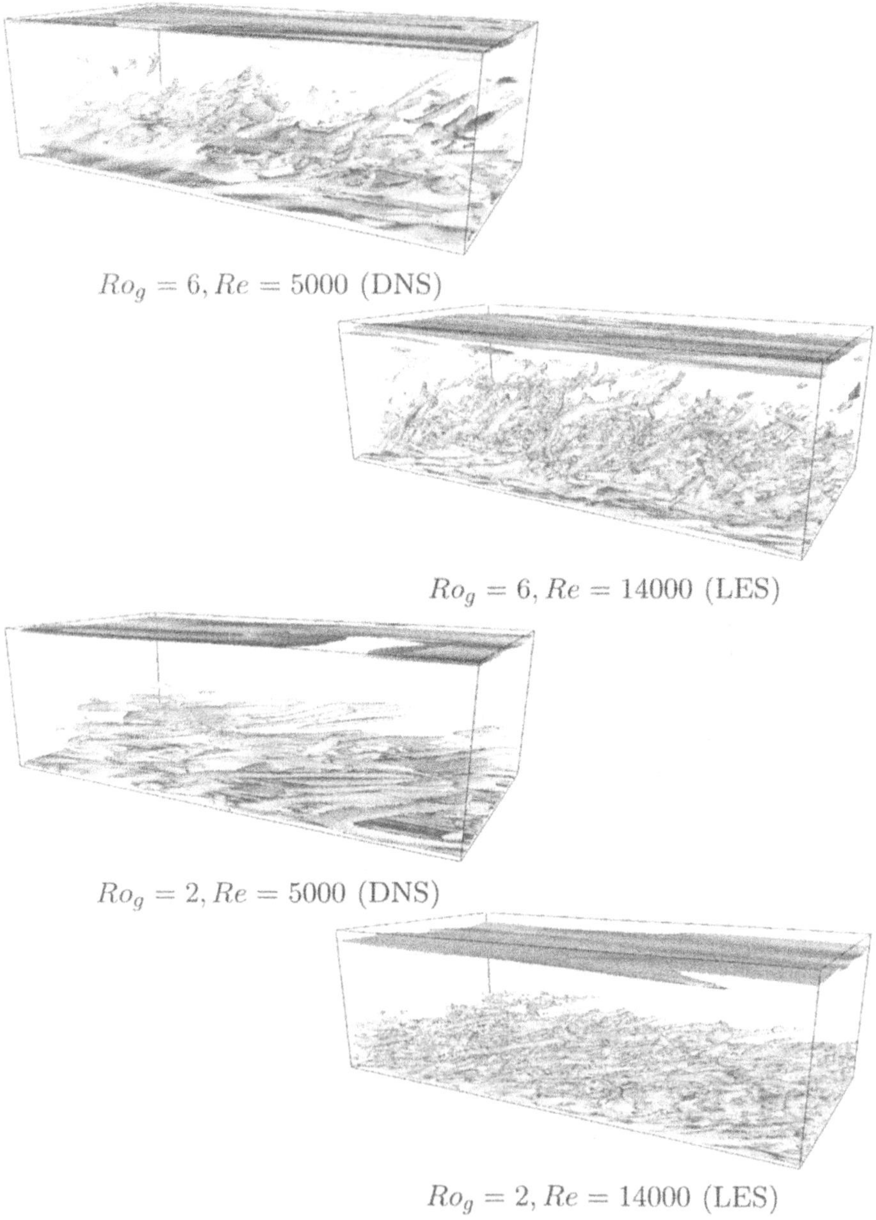

Figure 6.19. Isosurfaces of vorticity modulus $\omega = 3\, U_m/h$ for $Ro_g = 6$ or $\omega = 2.25\, U_m/h$ for $Ro_g = 2$ (for the DNS results, only a quarter of the computational domain is presented)

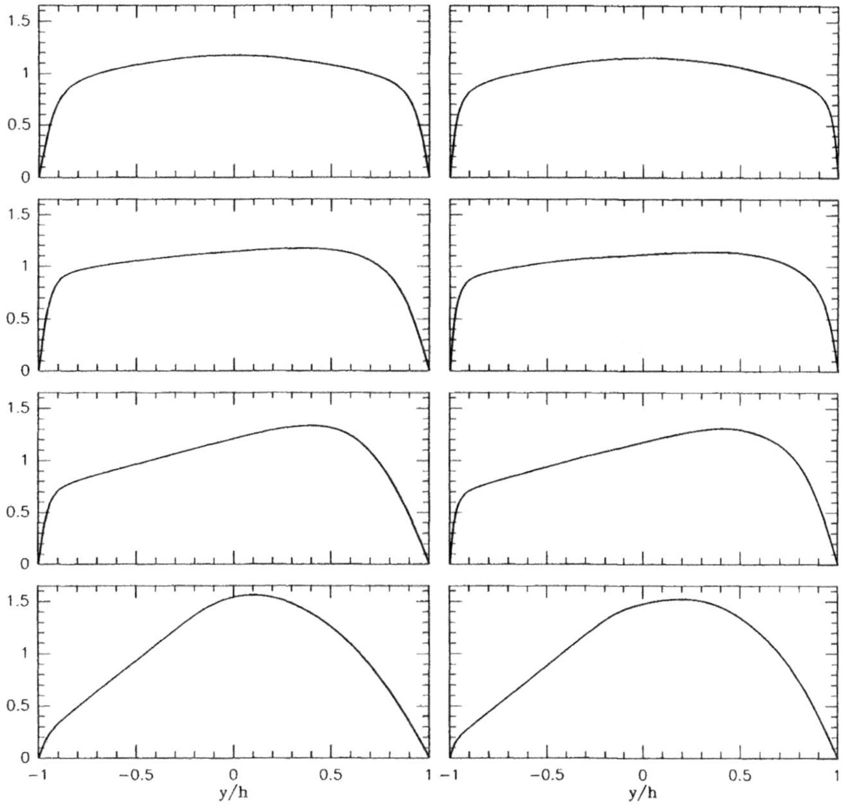

Figure 6.20. Mean velocity profiles : —, $\langle u \rangle / U_m$. From top to bottom, $Ro_g = \infty, 18, 6, 2$; left, DNS at $Re = 5000$; right, LES at $Re = 14000$

ticed both experimentally ([48]) and numerically ([50, 112, 55]). Moreover, the development of weak mean spanwise absolute vorticity regions is a common feature of various turbulent rotating shear flows, like mixing-layer or wake (see [89]) and turbulent plane Couette flow ([11]). The reasons of this general tendency are not totally well understood yet. We have checked that this constant-shear region corresponds to the location of the inclined vortex structures previously mentioned and to an intense longitudinal stretching of absolute vorticity.

We have already observed (see [89]) in the unstable anticyclonic ranges

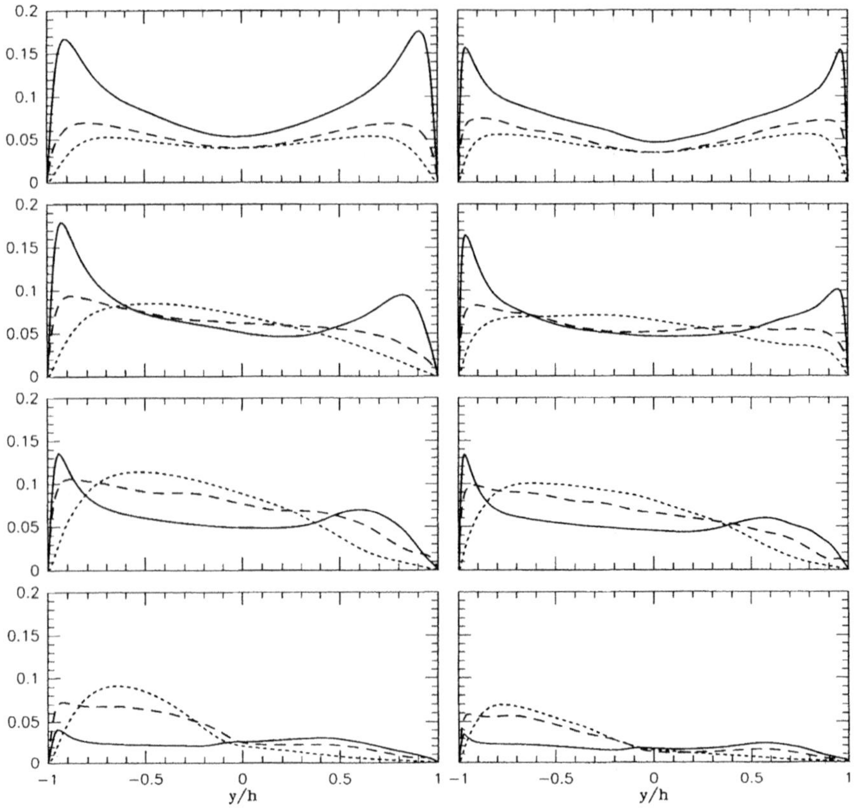

Figure 6.21. R.m.s. of the fluctuating velocity components : —, $\sqrt{\langle u'^2 \rangle}/U_m$; ..., $\sqrt{\langle v'^2 \rangle}/U_m$; - - -, $\sqrt{\langle w'^2 \rangle}/U_m$. From top to bottom, $Ro_g = \infty, 18, 6, 2$; left, DNS at $Re = 5000$; right, LES at $Re = 14000$.

of rotating free-shear flows, an initial linear growth of shear-Coriolis instability, followed by a strongly nonlinear longitudinal stretching of absolute vortex filaments, yielding the formation of very elongated alternate hairpins of weak absolute vorticity. Thus, spanwise absolute vorticity is nearly zero in the region of these hairpins. The same mechanisms are clearly at hand on the anticyclonic side of a rotating channel.

The r.m.s. velocity-component profiles are presented on figure 6.21 both for the DNS ($Re = 5000$) and the LES ($Re = 14000$) cases. Note that identical trends are observed at low and high Reynolds numbers. In the non-rotating case, the classical near-wall behaviour is recovered with a clear dominance of the longitudinal component $\sqrt{\langle u'^2 \rangle}$ over the spanwise component $\sqrt{\langle w'^2 \rangle}$ and the normal component $\sqrt{\langle v'^2 \rangle}$. This indicates the presence of the high- and low-speed streaks close to the wall. The turbulent quantities exhibit a strong asymmetric behaviour with respect to the channel center due to the rotation. When the rotation is increased, we observe on the cyclonic channel side a monotonic decrease of the r.m.s. velocity components with $\sqrt{\langle u'^2 \rangle} > \sqrt{\langle w'^2 \rangle} > \sqrt{\langle v'^2 \rangle}$. On the anticyclonic side at $Ro_g = 18$, rotation enhances the maximum fluctuations level of the three velocity components (with respect to the non-rotating case). $\sqrt{\langle u'^2 \rangle}$ undergoes a monotonic decrease for higher rotation rates ($Ro_g = 6$ and $Ro_g = 2$), and it is peak value is 4 times smaller than the corresponding non-rotating value at $Ro_g = 2$. It indicates the disappearance of streaks. Conversely, the normal and spanwise components of the fluctuating velocity reach maximum values at $Ro_g = 6$. Furthermore, and due to rotation, the flows undergoes anisotropy inversion in the fluctuating velocity components. For the three rotation rates corresponding to $Ro_g = 18$, $Ro_g = 6$ and $Ro_g = 2$, the normal component $\sqrt{\langle v'^2 \rangle}$ dominates the two other components in the flow region corresponding to $Ro(y) \approx -1$. Both close to the wall and at the upper end of the constant shear region, the spanwise component $\sqrt{\langle w'^2 \rangle}$ is the largest component. This seems to indicate that, in the $Ro(y) \approx -1$ range, the flow mainly consists of cells elongated in the direction normal to the wall, and composed of alternatively upward and downward motions with flow recirculating along the spanwise direction close to the wall and at the upper end of the linear region.

Figure 6.22 displays the r.m.s vorticity-component profile for the runs previously described. In the non-rotating case, longitudinal vorticity is low in the streak region, and the essential vorticity fluctuations are spanwise at the wall. Far from the walls, in the major part of the channel, the fluctuating vorticity components are equipartitioned. In the rotating case, and as stressed above, the constant-shear region coincides with a range where the longitudinal fluctuating vorticity component dominates now the other two components. This dominance is more and more pronounced as the rotation is increased, and, for $Ro_g = 2$, it extends up to the anticyclonic wall of the channel. This behaviour is similar for low and high Reynolds numbers. On the cyclonic side, the r.m.s. vorticity components are all monotonously reduced by rotation, indicating that all the scales of motion tend to relaminarize when rotation intensifies.

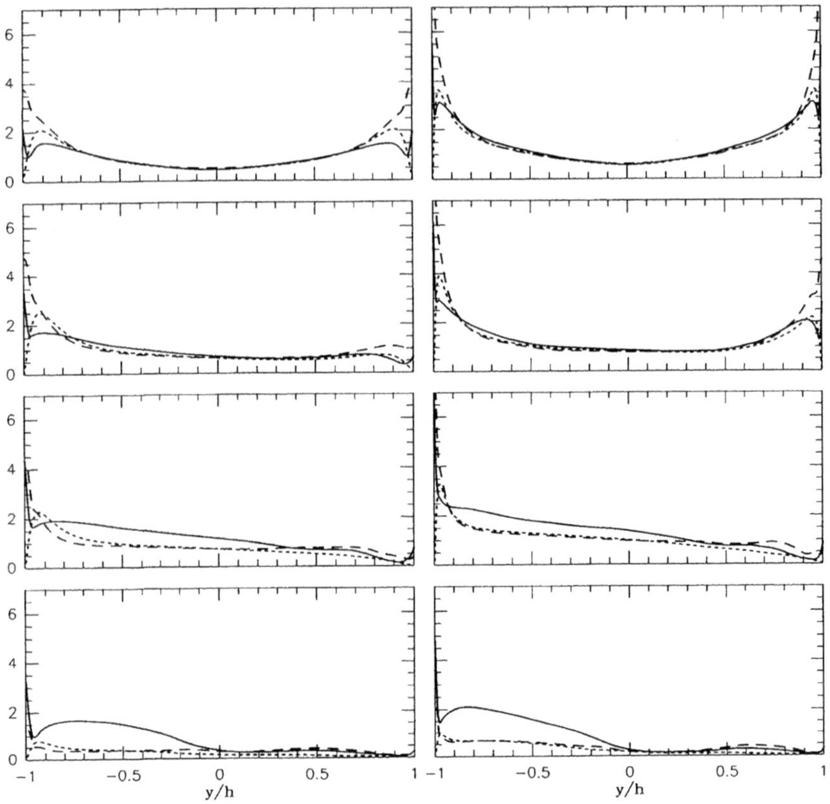

Figure 6.22. R.m.s. of the fluctuating vorticity components : —, $\sqrt{\langle \omega_x'^2 \rangle}/(U_m/h)$; ..., $\sqrt{\langle \omega_y'^2 \rangle}/(U_m/h)$; - - -, $\sqrt{\langle \omega_z'^2 \rangle}/(U_m/h)$. From top to bottom, $Ro_g = \infty, 18, 6, 2$; left, DNS at $Re = 5000$; right, LES at $Re = 14000$.

6.7.4. FLOWS OF GEOPHYSICAL INTEREST

The generalized hyperviscosity model previously presented has been applied successfully to the LES of flows submitted to the combined effects of vertical stable stratification and solid-body rotation. In particular we were able to study in the case of a baroclinic jet submitted to a rapid rotation the nature of the coherent vortices and in particular the asymmetry between cyclonic and anticyclonic eddies (see [41] for details). This demonstrates the importance of the cusp-like behaviour and the feasibility of the subgrid-

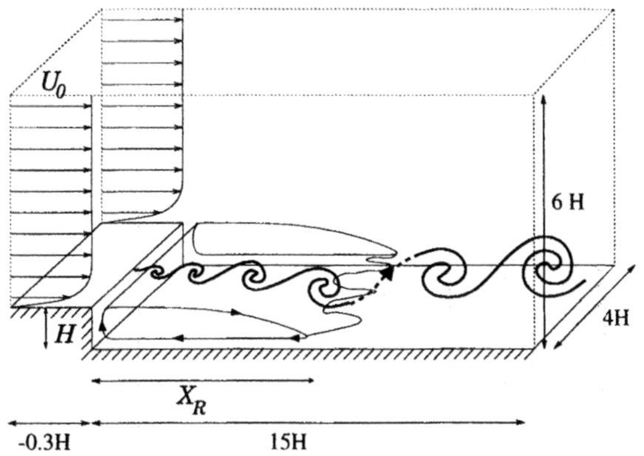

Figure 6.23. Schematic of the backward-facing step configuration (courtesy F. Delcayre).

scale model given by (6.35) for the LES of geophysical flows with quasi-twodimensional regions and sharp frontal regions.

6.7.5. SEPARATED FLOWS: THE BACKWARD FACING STEP

We now present a Large Eddy Simulation based upon the selective structure function model of a turbulent flow over backward-facing step. The detailed computations can be found in [30] and [31]. As for the jet results previously presented, we have used the TRIO code.

The flow configuration is the same as for the DNS of Le *el al.* [64], which closely corresponds to the experiment performed by Jovic & Driver [49]. The calculations were performed with an inlet mean velocity profile obtained from Spalart's [111] boundary layer DNS at Re_θ=670 (Re_{δ^*}=1000) where θ and δ^* are the momentum and displacement thicknesses. For this particular profile, the boundary layer thickness is $\delta \approx 6.1\delta^* = 1.2H$. The step-height Reynolds number corresponding to this profile is: $Re_H = \frac{U_0 H}{\nu} \approx 5100$. The inlet velocity profile is imposed at 0.3H (Fig. 6.23) upstream of the step with a three-dimensional random white noise of amplitude $\approx$ 1.25% superimposed on the shear zone. Periodicity and slip conditions are respectively imposed in the spanwise direction and on the upper boundary. The outlet condition is a convective condition based on Orlanski [98] formulation.

The computational domain extent is 15H downstream of the step, the spanwise direction size is 4H and in the vertical direction an expansion ratio of 1.2 is chosen which corresponds to a domain height of 6H.

The total resolution of the computational domain is $97 \times 34 \times 46$. The grid is uniform in the spanwise direction and stretched in the direction normal to the walls to resolve the boundary layer. Since the grid is structured, the stretching of the upstream boundary layer also resolves the shear layer which develops downstream of the step (Fig. 6.23). Thus, the spanwise grid spacing is constant and gives a $\Delta z^+ = 30$. The minimum streamwise resolution is $\Delta x^+_{min} = 11$ at the step and the maximum streamwise resolution (at the exit boundary) $\Delta x^+_{max} = 70$. In the vertical direction, the minimum resolution is $\Delta y^+_{min} = 3.75$ and the maximum resolution $\Delta y^+_{max} = 110$.

6.7.6. STATISTICS

The reattachment length X_R is overpredicted by the present LES. It is found $X_R = 7.2H$, which is far from the experimental measurement of Jovic & Driver [49] ($X_R = 6.15H$). This is certainly attributable to the fact that the inflow boundary layer upstream of the step is not explicitely simulated (see [64]). Indeed, the absence of the turbulent structures associated with the inflow boundary layer induces a delay in the transition of the shear layer and an increase of the reattachment length. It was however showed by Westphal *et al.* [117] ,using the renormalized coordinate $X = \frac{x-X_R}{X_R}$, that the reattachment process is quite independent of the inflow conditions. This is confirmed by the comparison between the experiment of Jovic & Driver [49], the DNS of Le *et al.* [64] and the present LES is presented at the reattachment point in Fig. 6.24. Here, the statistical quantities are averaged over the spanwise direction, using the periodicity condition, and time. The agreement is found to be good.

Next, the renormalized coordinate $X = \frac{x-X_R}{X_R}$ is used to compare the statistical results obtained in the calculation with the results obtained by Le *et al.*[64] in their DNS. Mean velocity profiles obtained in the present calculation in the reattachment region almost collapse with the DNS. Both profiles still exhibit an inflexional point at the exit boundary ($X = 0.66$). This indicates that the turbulent boundary layer in not completely developed (Fig. 6.25). The topological study of the next section will show the persistance of shear layer structures downstream of the reattachment which could explain this behaviour. Longitudinal turbulent intensities compare also well with DNS especially for $y/H < 1$. The slight underestimation of the longitudinal turbulent intensities for $y/H > 1$ in the LES could be attributable to the lack of coherent structures in the inlet boundary layer flow. At the end of the domain, we can notice the development of a peak in the near wall region. This peak is a proof of the redevelopment of the boundary layer. Nevertheless, canonical turbulent boundary layer profiles are not yet recovered.

6.7.7. COHERENT STRUCTURES

A usual way to characterize large scale coherent vortices consists in considering vorticity or pressure isosurfaces, as shown on Fig. 6.26. Another way is to use the so-called Q-criterion proposed by Hunt *et al.* [47]. This method

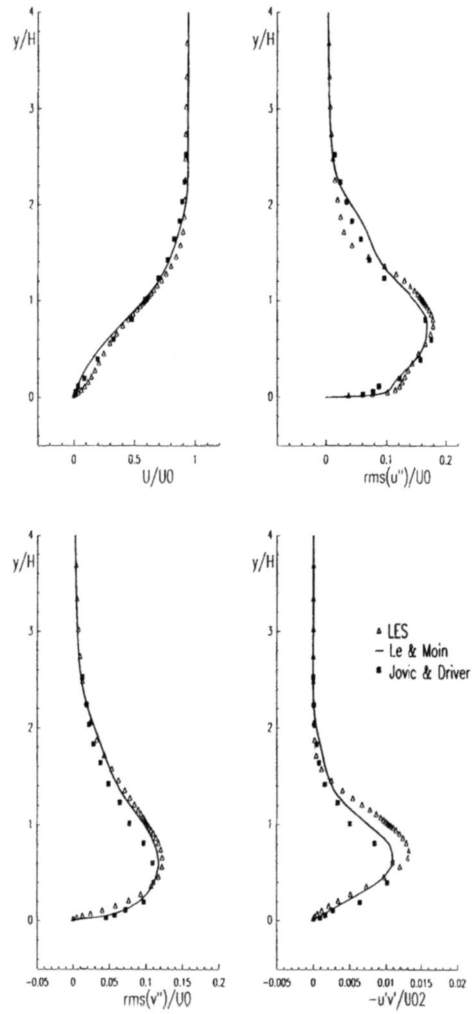

Figure 6.24. Statistical quantities at the reattachment length (courtesy F. Delcayre).

is particularly attractive in the present case since it consists in isolating the regions where the strain rate is lower than the vorticity magnitude. Hunt *et al.*[47] define a criterion based on the second invariant of the velocity gradient Q with $Q = (\Omega_{ij}\Omega_{ij} - S_{ij}S_{ij})/2$ where Ω_{ij} is the antisymmetrical part of $\partial u_i/\partial x_j$ and S_{ij} the symmetrical part. $Q > 0$ will define zones where rotation is predominant (vortex cores). Fig. 6.26 shows that the large-scales are approximately represented by pressure isosurfaces upstream of the reattachment, and more realistically downstream of the reattachment. The small scale field is accurately represented by the high-vorticity isosurfaces conditioned with $Q > 0$. As compared with non-conditioned high-vorticity

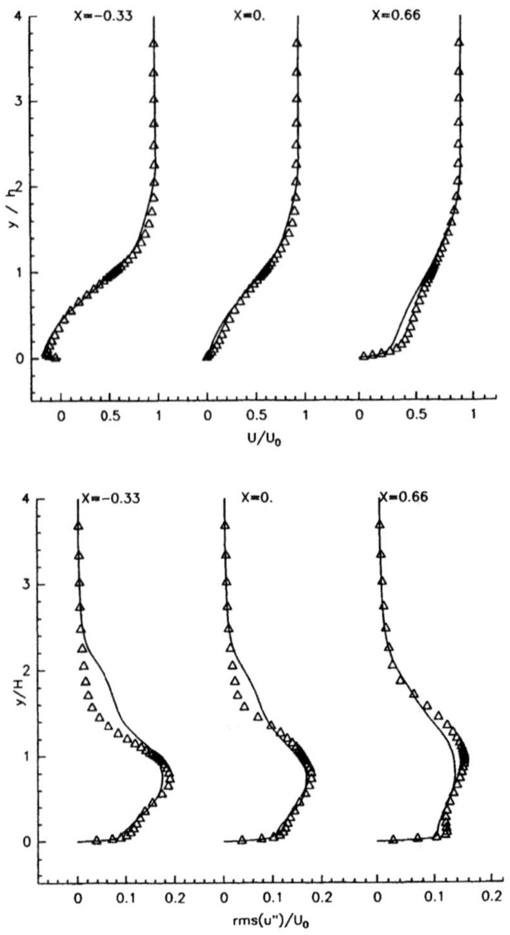

Figure 6.25. *Mean streamwise velocity and turbulent streamwise intensity profiles: continuous line DNS by Le et al.[64] (courtesy F. Delcayre).*

surfaces, the vortex sheets just behind the step and near the bottom wall are filtered out by the application of the Q-criterion.

Flow visualizations of the Q-conditioned vorticity modulus field and coloured by the ω_x value show the general tendency of the flow to reorient a spanwise vorticity field into streamwise vorticity. Figure 6.27 clearly indicates how Kelvin-Helmholtz billows shed downstream of the step oscillate in the spanwise direction and are subject to helical type of pairing (see above). Just after this pairing the three-dimensionalisation of the flow increases: the oscillation of spanwise vortices initiated by a staggered oscillation of the Kelvin-Helmholtz vortices is amplified, leading to the formation of large Λ-shaped vortices. Further downstream, these vortices impinge the wall and are stretched into big arch-like vortices as they are convected down-

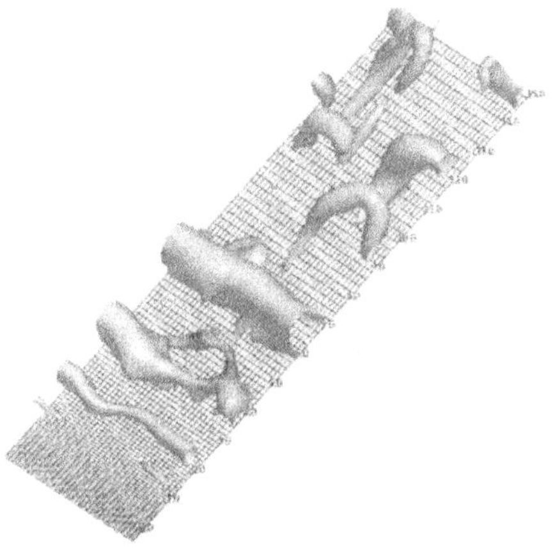

Pressure isosurface ($P = -0.08\rho_0 U_0^2 = P_{min} + 0.75(P_{max} - Pmin)$)

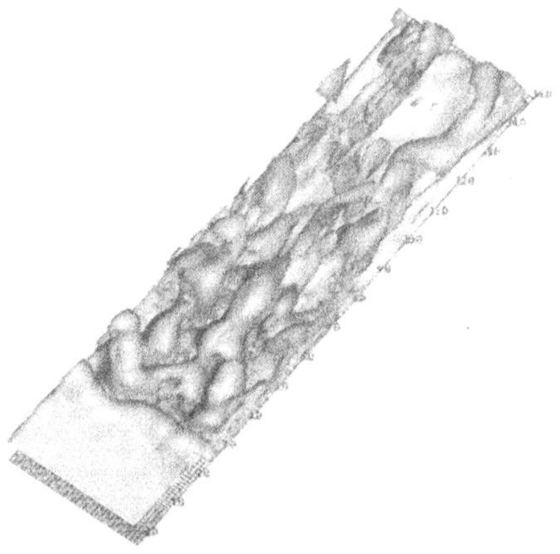

Vorticity modulus isosurface ($\|\vec{\omega}\| = 1.5 \frac{U_0}{H}$)

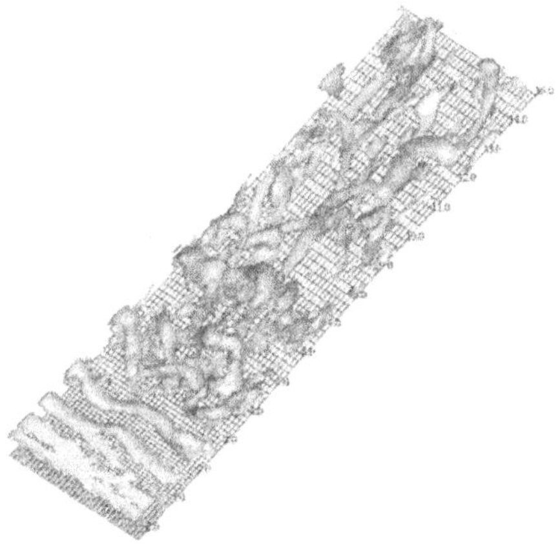

Vorticity modulus isosurface ($\|\vec{\omega}\| = 1.5 \frac{U_0}{H}$) *with* $Q > 0$.

Figure 6.26. Visualization of the backward-facing step flow structures by three different means (courtesy F. Delcayre).

stream of the reattachment. Note that the persistance of these Λ-shaped structures downstream of the reattachment is very likely to be a key factor in the slow relaxation of the mean streamwise profiles towards canonical turbulent boundary layer profiles. Bandyopadhyay [6], among others, has also experimentally highlighted the presence of large scale structures 39H downstream of the step and has shown their quasi-periodic occurrence.

Using pressure fluctuations spectra (Fig. 6.28), three dominant frequencies were found in the different flow regions. These spectra show a peak

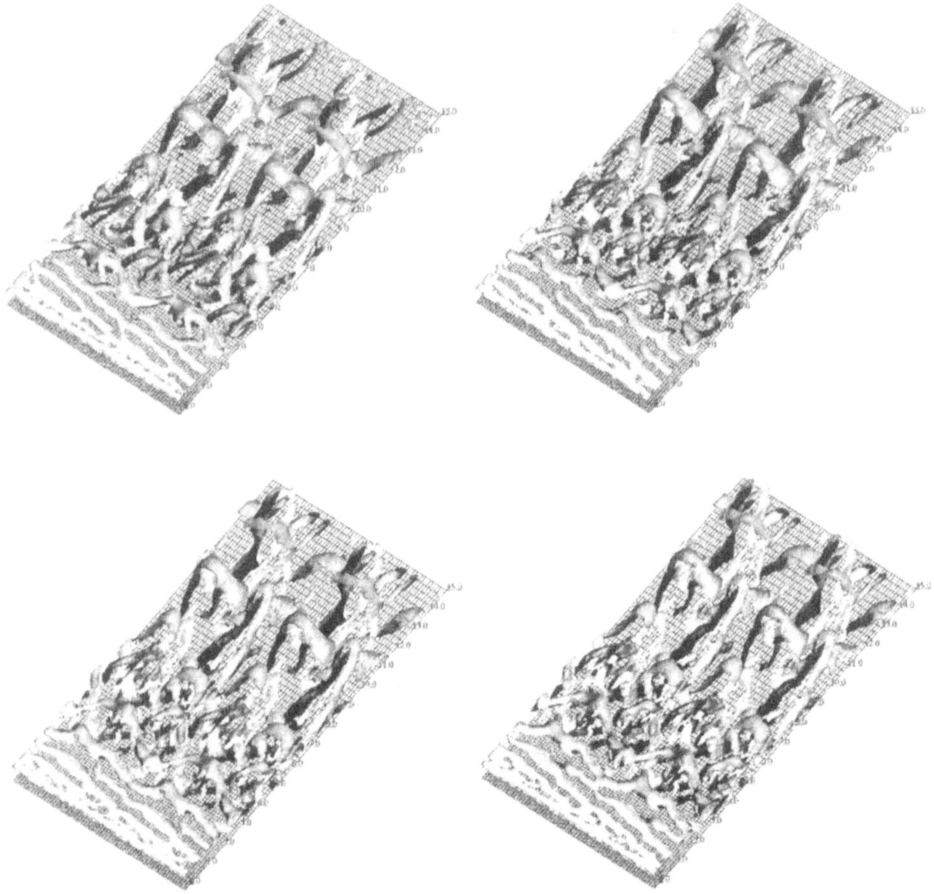

Figure 6.27. *Visualizations of the vorticity modulus ($\|\vec{\omega}\|$=1.5 $\frac{U_0}{H}$) with Q-criterion coloured by ω_x (black) $-1.5\frac{U_0}{H} < \omega_x < +1.5\frac{U_0}{H}$ (white) - Domain duplicated - The frames correspond to a separation of $\frac{H}{U_0}$, from left toright and top to bottom (courtesy F. Delcayre).*

($St_1 \approx 0.23$) in the first region ($x < 2H$) of the flow. It is the frequency corresponding with the Kelvin-Helmholtz shedding. Then, the frequency of the pairing appears ($St_2 \approx 0.12$). In the reattachment region another frequency emerges ($St_3 \approx 0.07$) which corresponds to a low-frequency flapping of the recirculation bubble whose mechanism is not yet well understood. Downstream of the reattachment, the three frequency peaks are associated with a similar amount of energy. These different Strouhal numbers associated with the different unsteady phenomena are in good agreement with those previously found by other authors (see [5]).

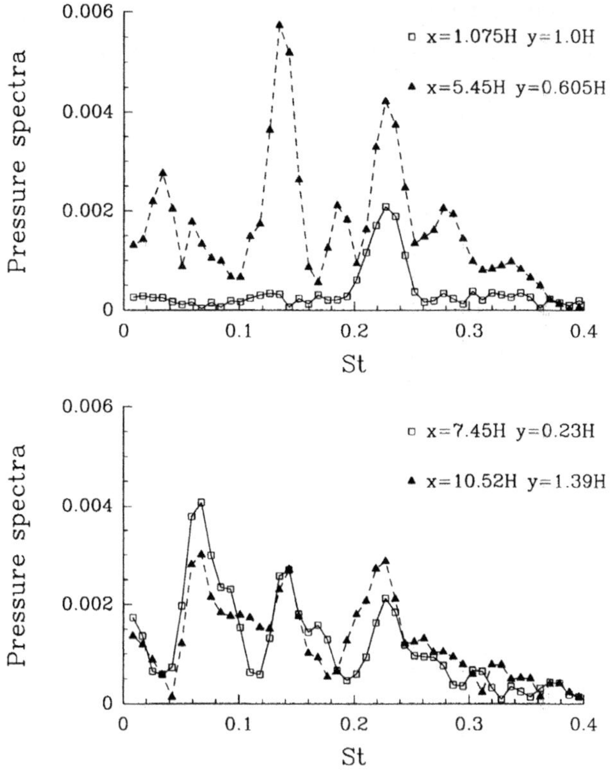

Figure 6.28. Spectra of spanwise averaged pressure fluctuations at different positions

6.8. Compressible LES Formalism

In Cartesian co-ordinates, the compressible Navier-Stokes equations can be cast in the so-called fast-conservation form

$$\frac{\partial U}{\partial t} + \frac{\partial F_1}{\partial x_1} + \frac{\partial F_2}{\partial x_2} + \frac{\partial F_3}{\partial x_3} = 0 \quad , \tag{6.51}$$

with

$$U = {}^T(\rho, \rho u_1, \rho u_2, \rho u_3, \rho e) \quad , \tag{6.52}$$

ρe being the total energy defined by, for an *ideal* gas,

$$\rho e = \rho \, C_v \, T + \tfrac{1}{2}\rho(u_1^2 + u_2^2 + u_3^2) \quad . \tag{6.53}$$

The fluxes F_i read, $\forall i \in \{1, 2, 3\}$,

$$F_i = \begin{pmatrix} \rho u_i \\ \rho u_i u_1 - \sigma_{i1} \\ \rho u_i u_2 - \sigma_{i2} \\ \rho u_i u_3 - \sigma_{i3} \\ \rho e u_i - u_j \sigma_{ij} - k \dfrac{\partial T}{\partial x_i} \end{pmatrix} \quad , \tag{6.54}$$

$k = \rho C_p \kappa$ being the thermal conductivity (and κ the thermal diffusivity).

The components σ_{ij} of the stress tensor are given by the Newton law

$$\sigma_{ij} = -p\ \delta_{ij} + \mu S_{ij} \quad , \tag{6.55}$$

in which

$$S_{ij} = \left[\frac{\partial u_j}{\partial x_i} + \frac{\partial u_i}{\partial x_j} - \frac{2}{3}(\boldsymbol{\nabla}.\boldsymbol{u})\delta_{ij}\right] \tag{6.56}$$

denotes the deviatoric part of the strain-rate tensor. Bulk viscosity is neglected (Stokes hypothesis), as commonly accepted except in extreme thermodynamical situations. This yields

$$F_i = \begin{pmatrix} \rho u_i \\ \rho u_i u_1 + p\ \delta_{i1} - \mu S_{i1} \\ \rho u_i u_2 + p\ \delta_{i2} - \mu S_{i2} \\ \rho u_i u_3 + p\ \delta_{i3} - \mu S_{i3} \\ (\rho e + p)u_i - \mu u_j S_{ij} - k\dfrac{\partial T}{\partial x_i} \end{pmatrix} \quad . \tag{6.57}$$

The Sutherland empirical law

$$\begin{aligned} \mu(T) = \ & \mu(273.15)\left(\frac{T}{273.15}\right)^{1/2} \frac{1 + S/273.15}{1 + S/T} \\ & \text{with } \mu(273.15) = 1.711\ 10^{-5} Pl \\ & \text{and } S = 110.4K \quad , \end{aligned} \tag{6.58}$$

and its extension to temperatures lower than 120 K :

$$\mu(T) = \mu(120)\ T/120 \quad \forall\, T < 120 \quad , \tag{6.59}$$

are prescribed for molecular viscosity. Conductivity $k(T)$ is obtained assuming the molecular Prandtl number $Pr = C_p\mu(T)/k(T)$ constant and equal to 0.7, as in air at ambiant temperature. The equation of state

$$p = R\ \rho\ T \tag{6.60}$$

closes the system, with $R = C_p - C_v = \frac{\mathcal{R}}{\mathcal{M}} = 287.06\ Jkg^{-1}K^{-1}$ for air.

6.8.1. COMPRESSIBLE FILTERING PROCEDURE

As in the incompressible regime, and whatever the numerical method used, the discretization of the above equations introduces a cut-off scale Δx which is by hypothesis larger than the Kolmogorov scale. We still account for this by a low-pass filter of width Δx, characterized by the convolution in space with a function $G_{\Delta x}(\vec{x})$. The operator $\overline{}$ commutes with the space and time derivatives in the case of uniform cubic meshes of side Δx. Convolution of the above equations therefore yields

$$\frac{\partial \overline{U}}{\partial t} + \frac{\partial \overline{F}_1}{\partial x_1} + \frac{\partial \overline{F}_2}{\partial x_2} + \frac{\partial \overline{F}_3}{\partial x_3} = 0 \quad , \tag{6.61}$$

with

$$\overline{\rho e} = \overline{\rho \; c_v \; T} + \overline{\tfrac{1}{2}\rho(u_1^2 + u_2^2 + u_3^2)} \tag{6.62}$$

and

$$\overline{p} = \overline{\rho \; R \; T} \tag{6.63}$$

At this level, it is convenient to introduce the density-weighted (or Favre [39]) filter $\tilde{}$ defined, for a given variable ϕ, by

$$\tilde{\phi} = \frac{\overline{\rho\phi}}{\overline{\rho}} \quad . \tag{6.64}$$

We then have

$$\overline{U} = {}^{T}(\overline{\rho}, \overline{\rho}\tilde{u}_1, \overline{\rho}\tilde{u}_2, \overline{\rho}\tilde{u}_3, \overline{\rho}\tilde{e}) \quad , \tag{6.65}$$

and the *resolved total energy*

$$\overline{\rho e} = \overline{\rho}\tilde{e} = \overline{\rho} \; C_v \; \tilde{T} + \overline{\tfrac{1}{2}\rho(u_1^2 + u_2^2 + u_3^2)} \quad . \tag{6.66}$$

The resolved fluxes $\overline{F_i}$ read

$$\overline{F_i} = \begin{pmatrix} \overline{\rho}\tilde{u}_i & & \\ \overline{\rho u_i u_1} + \overline{p}\; \delta_{i1} - \overline{\mu S_{i1}} & & \\ \overline{\rho u_i u_2} + \overline{p}\; \delta_{i2} - \overline{\mu S_{i2}} & & \\ \overline{\rho u_i u_3} + \overline{p}\; \delta_{i3} - \overline{\mu S_{i3}} & & \\ \overline{(\rho e + p) u_i} & - \overline{\mu S_{ij} u_j} & - \overline{k \frac{\partial T}{\partial x_i}} \end{pmatrix} \quad , \tag{6.67}$$

with the filtered equation of state

$$\overline{p} = \overline{\rho} R \tilde{T} \quad . \tag{6.68}$$

6.8.2. THE SIMPLEST POSSIBLE CLOSURE

The usual subgrid-stress tensor $\overline{\overline{\mathcal{T}}}$ of components

$$\mathcal{T}_{ij} = -\overline{\rho u_i u_j} + \overline{\rho}\tilde{u}_i\tilde{u}_j \tag{6.69}$$

is introduced and split into its isotropic and deviatoric parts, the latter being noted $\overline{\overline{\tau}}$:

$$\mathcal{T}_{ij} = \underbrace{\mathcal{T}_{ij} - \frac{1}{3}\mathcal{T}_{ll}\delta_{ij}}_{\tau_{ij}} + \frac{1}{3}\mathcal{T}_{ll}\delta_{ij} \quad . \tag{6.70}$$

Equations (6.67) and (6.66) then read

$$\overline{F_i} = \begin{pmatrix} \overline{\rho}\widetilde{u_i} \\ \overline{\rho}\widetilde{u_i}\widetilde{u_1} + (\overline{p} - \frac{1}{3}\mathcal{T}_{ll})\ \delta_{i1} - \tau_{i1} - \overline{\mu S_{i1}} \\ \overline{\rho}\widetilde{u_i}\widetilde{u_2} + (\overline{p} - \frac{1}{3}\mathcal{T}_{ll})\ \delta_{i2} - \tau_{i2} - \overline{\mu S_{i2}} \\ \overline{\rho}\widetilde{u_i}\widetilde{u_3} + (\overline{p} - \frac{1}{3}\mathcal{T}_{ll})\ \delta_{i3} - \tau_{i3} - \overline{\mu S_{i3}} \\ \overline{(\rho e + p)u_i} \qquad - \overline{\mu S_{ij} u_j} - \overline{k\frac{\partial T}{\partial x_i}} \end{pmatrix} \quad . \tag{6.71}$$

and

$$\overline{\rho}\widetilde{e} = \overline{\rho}\ C_v \widetilde{T} + \frac{1}{2}\overline{\rho}\ (\widetilde{u_1}^2 + \widetilde{u_2}^2 + \widetilde{u_3}^2) - \frac{1}{2}\mathcal{T}_{ll} \tag{6.72}$$

Let

$$M_{\text{sgs}}^2 = \mathcal{T}_{ll}/\overline{\rho c^2} \tag{6.73}$$

be the subgrid Mach number, which satisfies $\mathcal{T}_{ll} = \gamma M_{\text{sgs}}^2 \overline{p}$.

There are two options for the treatment of the uncomputable term $\mathcal{T}_{ll}$:

- simply neglect it, arguing as in [38] that M_{sgs} can be expected to be small when M_∞ is small.
- model it, as proposed by Yoshizawa [119], in a way which is consistent with the model chosen for $\overline{\overline{\tau}}$ (see *e.g.* [93]). Note that this was the initial choice of Erlebacher *et al.*[37].

We will here choose the first option, as in [97], bringing another argument: the incompressible LES formalism (see above), often introduces the *macro-pressure*

$$\varpi = \overline{p} - \frac{1}{3}\mathcal{T}_{ll} \quad . \tag{6.74}$$

It thus seems a good idea to re-write equation (6.72) as

$$\overline{\rho}\widetilde{e} = \overline{\rho}\ C_v \left(\widetilde{T} - \frac{1}{2C_v\overline{\rho}}\mathcal{T}_{ll}\right) + \frac{1}{2}\overline{\rho}\ (\widetilde{u_1}^2 + \widetilde{u_2}^2 + \widetilde{u_3}^2) \tag{6.75}$$

and introduce a *macro-temperature*

$$\vartheta = \widetilde{T} - \frac{1}{2C_v\overline{\rho}}\mathcal{T}_{ll} \quad , \tag{6.76}$$

computable out of $\overline{U}$ thanks to equation (6.75). The filtered equation of state (6.68) then reads

$$\begin{aligned} \varpi &= \overline{\rho}R\vartheta + \left(\frac{R}{2C_v} - \frac{1}{3}\right)\mathcal{T}_{ll} \\ &= \overline{\rho}R\vartheta + \frac{3\gamma - 5}{6}\ \mathcal{T}_{ll} \quad . \end{aligned} \tag{6.77}$$

Thus, for monoatomic gases like argon or helium (for which $\gamma \approx 5/3$), the contribution of $\mathcal{T}_{ll}$ to equation (6.77) is quite negligible whatever the Mach number, which makes ϖ computable in all cases. It is extremely tempting to generalize this to air (for which $\gamma \approx 1.4$) by assuming

$$\varpi \simeq \overline{\rho} R \vartheta \quad . \tag{6.78}$$

In other words, the first option amounts to assume $[(3\gamma - 5)/6]\ \gamma M_{\text{sgs}}^2 \ll 1$ in the equation of state only, which sounds slightly less stringent than assuming $\gamma M_{\text{sgs}}^2 \ll 1$ everywhere.

Considering from now on ϖ computable, it is sensible to involve it in the definition of a subgrid heat-flux vector, noted $\boldsymbol{Q}$, of components

$$Q_i = -\overline{(\rho e + p) u_i} + (\overline{\rho}\widetilde{e} + \varpi)\widetilde{u}_i \quad . \tag{6.79}$$

Provided acceptable models are proposed for $\overline{\overline{\mathcal{T}}}$ and $\boldsymbol{Q}$, the *resolved fluxes* already look more tractable:

$$\overline{F_i} = \begin{pmatrix} \overline{\rho}\widetilde{u}_i \\ \overline{\rho}\widetilde{u}_i\widetilde{u}_1 + \varpi\ \delta_{i1} - \tau_{i1} - \overline{\mu S_{i1}} \\ \overline{\rho}\widetilde{u}_i\widetilde{u}_2 + \varpi\ \delta_{i2} - \tau_{i2} - \overline{\mu S_{i2}} \\ \overline{\rho}\widetilde{u}_i\widetilde{u}_3 + \varpi\ \delta_{i3} - \tau_{i3} - \overline{\mu S_{i3}} \\ (\overline{\rho}\widetilde{e} + \varpi)\widetilde{u}_i \quad - Q_i - \overline{\mu S_{ij} u_j} - \overline{k \dfrac{\partial T}{\partial x_i}} \end{pmatrix} \quad . \tag{6.80}$$

The remaining non-computable terms are viscous terms, which can be considered of less importance when the Reynolds number is sufficiently large. We therefore simply replace (6.80) by

$$\overline{F_i} \simeq \begin{pmatrix} \overline{\rho}\widetilde{u}_i \\ \overline{\rho}\widetilde{u}_i\widetilde{u}_1 + \varpi\ \delta_{i1} - \tau_{i1} - \mu\widetilde{S_{i1}} \\ \overline{\rho}\widetilde{u}_i\widetilde{u}_2 + \varpi\ \delta_{i2} - \tau_{i2} - \mu\widetilde{S_{i2}} \\ \overline{\rho}\widetilde{u}_i\widetilde{u}_3 + \varpi\ \delta_{i3} - \tau_{i3} - \mu\widetilde{S_{i3}} \\ (\overline{\rho}\widetilde{e} + \varpi)\widetilde{u}_i \quad - Q_i - \mu\widetilde{S_{ij}}\widetilde{u_j} - k\dfrac{\partial \vartheta}{\partial x_i} \end{pmatrix} , \tag{6.81}$$

in which μ and k are linked to ϑ through the Sutherland relation (6.58) and the constant Prandtl number assumption $Pr = C_p\mu(\vartheta)/k(\vartheta) = 0.7$.

The system is finally closed with the aid of variable-density eddy-viscosity and diffusivity models, in the form

$$\tau_{ij} \simeq \overline{\rho}\nu_t\widetilde{S_{ij}} \tag{6.82}$$

$$Q_i \simeq \overline{\rho}\ \frac{\nu_t}{Pr_t}\frac{\partial \vartheta}{\partial x_i} \quad , \tag{6.83}$$

expressions for $\nu_t(\widetilde{\boldsymbol{u}})$ and Pr_t used in the following compressible simulations correspond to the incompressible *structure-function model* and its *filtered* and *selective* extensions, with a constant turbulent Prandtl number 0.6.

6.9. Compressible mixing layer

To demonstrate the compressibility effects, we first consider the temporally growing mixing layer, and we quote here excerpts of Ref. [70], with the permission of Kluwer academic publishers:

• Beginning of quotation:

In compressible mixing layers between two flows of parallel velocities U_1 and U_2 in unbounded domains, the relevant Mach numbers are the convective Mach numbers $M_c^{(1)}$ and $M_c^{(2)}$, built with the velocity difference of each layer with respect to U_c, the velocity of the large vortices [15], and respectively c_1 and c_2, the sound velocities in the two external flows. It can be shown, by assuming isentropy in the stagnation regions between the vortices, that

$$U_c = \frac{U_1 c_2 + U_2 c_1}{c_1 + c_2} \quad . \tag{6.84}$$

Then, within this assumption, the convective Mach numbers

$$M_c^{(1)} = \frac{U_1 - U_c}{c_1} \quad , \quad M_c^{(2)} = \frac{U_c - U_2}{c_2} \quad , \tag{6.85}$$

are both equal to

$$M_c = \frac{U_1 - U_2}{c_1 + c_2} = \frac{U}{\bar{c}} \quad , \tag{6.86}$$

where $2U$ is the velocity difference, and $\bar{c}$ an average speed of sound between the two layers. Note that Eq. (6.86) writes also as

$$U_c = U_1 \frac{1 + (U_2/U_1)\sqrt{\rho_2/\rho_1}}{1 + \sqrt{\rho_2/\rho_1}} \quad .$$

This expression allows to recover the value $U_c = (U_1 + U_2)/2$ in the incompressible uniform-density case. It may also be useful in an incompressible mixing layer with density differences, since it takes into account density effects which are not of gravitational type.

Returning to compressible mixing layers, laboratory experiments of [99] show that this hypothesis (identity of the two convective Mach numbers) is valid up to $M_c \approx 0.6$. Experiments show also a dramatic decrease of the spreading rate of the mixing layer, with respect to the incompressible value (...) between $M_c \approx 0.5$ and $M_c \approx 1$. What we call M_c is now the highest of the two convective Mach numbers. Above, it saturates at about 40% of the incompressible case.

The inviscid linear-stability analysis of the compressible mixing layer in the temporal case was performed by [74][75] and [14]. The stability diagram found by the latter (for $\gamma = 1.4$) shows that the maximum-amplification rate is a decreasing function of the initial Mach number $M_c^{(i)} = U/\bar{c}$, with a drastic change in the slope at $M_c^{(i)} = 0.6$. Two-dimensional DNS of [96] show an inhibition of Kelvin-Helmholtz instability for $M_c^{(i)} > 0.6$: there is hardly any roll-up of the vortices, which remain extremely flat and merge "longitudinally", without turning around each other. On the contrary, for $M_c^{(i)} \leq 0.6$, the roll-up and pairing occur qualitatively in the same fashion as in the incompressible case, although they are delayed by factors corresponding exactly to the amplification rates predicted by [14] (...).

A three-dimensional linear-stability analysis of the compressible temporal mixing layer was carried out by [106]. It turned out that oblique waves are more amplified than 2D waves when $M_c^{(i)}$ exceeds 0.6. Another result shown with the aid of DNS by [40] is that the helical pairing found in the incompressible case (with a 3D random forcing) is inhibitted above $M_c^{(i)} = 0.6 \approx 0.7$. The vortex structure of the mixing layer is then made of staggered Λ vortices, as shown in Figure 6.29-a, where the basic flow in the upper layer goes from top to bottom[2]. The corresponding pressure is displayed on Figure 6.29-b. It indicates a longitudinal reconnexion of pressure into tubes following the legs of the Λ's. This is an example where low pressure ceases to follow the coherent vortices.

Spatially-growing DNS of compressible mixing layers were also performed by [40]. Helical pairings was observed when the compressibility is low (upstream $M_c = 0.3$), as in the incompressible simulations presented above. At upstream $M_c = 0.7$ on the contrary, a pattern of very elongated staggered Λ vortices is obtained (...).

The saturation in the spreading rates observed experimentally when M_c exceeds 0.6 might be due to two causes. The first one is the reflexion of Mach waves on the walls of the facility. The second is the inhibition of the Kelvin-Helmholtz instability at this cross-over convective Mach number 0.6. Such an inhibition may be physically explained as follows. We first consider an incompressible mixing-layer, where the vortex cores correspond to pressure troughs, while pressure highs are located in the stagnation regions. We assume now that compressibility is present, but is not too high so that the same type of pressure distribution is preserved. We suppose also that the fluid is a barotropic ideal gas, where p/ρ^γ is conserved with the motion.

[2]The same structure was also found in the DNS of ([107]) with a quasi 2D initial forcing.

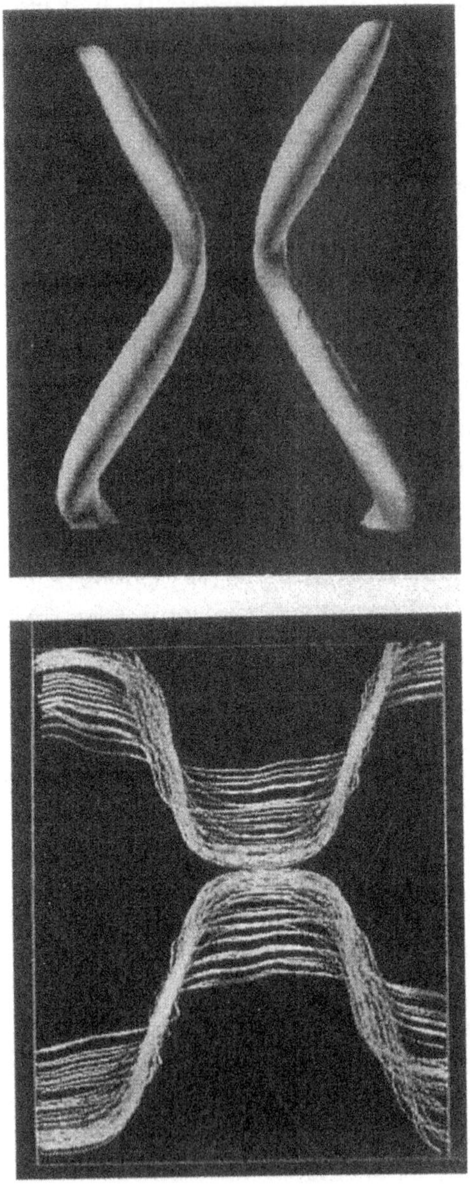

Figure 6.29. top view of vortex lines (a) and pressure (b) in the DNS of a compressible temporal mixing layer at convective Mach number 1 (from [40])

Therefore, fluid particles travelling from low to high pressures (...) will see their density increase when arriving at the stagnation points (which means convergence, that is, $D\rho/Dt > 0$ and $\vec{\nabla}.\vec{u} < 0$). Afterwards they will expand (..., $D\rho/Dt < 0$ and $\vec{\nabla}.\vec{u} > 0$). Let us now consider the vorticity equation

(...), which reduces, in this compressible two-dimensional case, to

$$\frac{D}{Dt}\frac{\omega}{\rho} = \frac{1}{\rho^3}(\vec{\nabla}\rho \times \vec{\nabla}p).\vec{z} + \frac{\nu}{\rho}\nabla^2\omega \quad .$$

[66] and [40] have verified in their numerical simulations that the baroclinic and the viscous terms are negligible, so that the vorticity dynamics reduces to the conservation of the "potential vorticity" ω/ρ . Thus, the convergence and divergence zones will be respectively a source and a sink of vorticity. This will work against Kelvin-Helmholtz instability, which tends to diminish the vorticity at the stagnation points, and increase it in the low-pressure regions (...)

- End of quotation

6.10. Compressible boundary layers on a flat plate

6.10.1. LES OF A SPATIALLY-DEVELOPING BOUNDARY LAYER AT MACH 0.5

The standard Structure Function model permits to go beyond transition in a temporal (periodic in the flow direction) compressible boundary layer upon an adiabatic wall at Mach 4.5 (see [34]). But it does not work for transition in a boundary layer at low Mach (or incompressible) where, like Smagorinsky, it is too dissipative and prevents TS waves to degenerate into turbulence . Conversely, in its filtered version (FSF model), it has been used for the simulation of a quasi-incompressible ($M_\infty = 0.5$) boundary layer developing spatially over an adiabiatic flat plate with a low level of upstream forcing. The numerical method is a Mac Cormack-type finite differences (Comte *et al.*, [25]). The numerical scheme is second order accurate in time and fourth order accurate in space. Periodicity is assumed in the spanwise direction. Non reflective boundary conditions (based on the Thompson characteristic method, Thompson, [113]) are prescribed at the outlet and the upper boundaries. With the minimal resolution of $650 \times 32 \times 20$ resolution points in the streamwise (x_1), transverse (x_2) and spanwise directions (x_3) respectively, covering a range of streamwise Reynolds numbers $Re_x \in [3.4 \times 10^5, 1.1 \times 10^6]$, transition has been obtained (see [35]) for 80 hours of time-processing on a CRAY2 (whereas DNS of the whole transition takes about ten times longer). The flow upstream is the superposition of the laminar profile at this Mach, a two-dimensional perturbation forcing the most-amplified Tollmien-Schlichting mode, and a three-dimensional white noise, such that :

$$U(0, x_2, x_3, t) = U_{\text{lam}}(x_2) + 5\times 10^{-3}\hat{U}(x_2) + 8\times 10^{-3}U_{\text{rand}}(x_2, x_3, t) \quad (6.87)$$

where $U_{\text{lam}}(x_2)$ is the laminar profile of the similarity equations, $\hat{U}(x_2)$ is the most amplified eigenmode of the two-dimensional Tollmien-Schlichting (TS) waves and $U_{\text{rand}}(x_2, x_3, t)$ is a randomly chosen three-dimensional white noise of variance U_∞^2. The upstream Reynolds number based on the displacement thickness $R_\delta = 1000$. It is seen how the TS wave generated

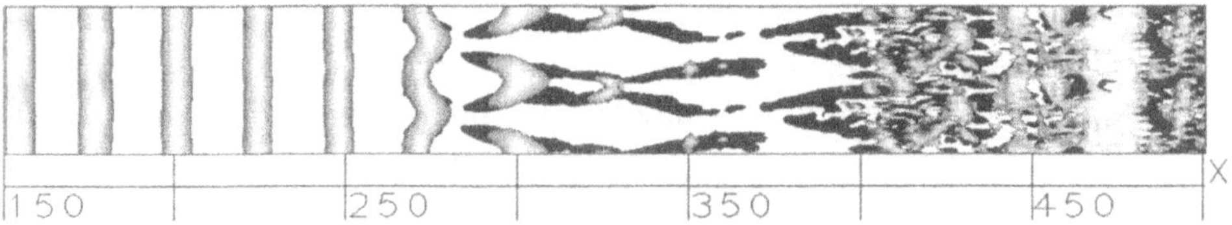

Figure 6.30. FSF structure-function based LES of a weakly-compressible spatially-developing boundary layer; isosurfaces of pressure ($p = 0.999 p_\infty$, grey) and longitudinal vorticity ($\omega_1 = \pm 0.1 U_\infty \delta_i$, dark) are shown

upstream propagates downstream. First, quasi two-dimensional billows of relatively low pressure and high vorticity form, and travel with the wave velocity. A top view of the low pressure and longitudinal vorticity in the transitional region is shown on Figure 6.30: just before the transition, TS waves give rise to straight lower pressure quasi two-dimensional rolls. During the transition, these rolls evolve into a staggered pattern which breaks down into turbulence. Meanwhile, the longitudinal velocity seems to develop a longitudinal mode close to the wall, as shown on Figure 6.31. The existence of this mode might be related to the low and high-speed streaks existing in the developed region. We show now on Figure 6.32 an enlarged view of a hairpin ejected away from the wall above a low-speed streak, just after transition. Such hairpins have a longitudinal vorticity which is low in front of the spanwise vorticities attained at the wall under the high-speed streaks, where most of the drag comes from. Another remark is that we could never find in these calculations coherent alternate longitudinal vortices at the wall. On the contrary, there are several hairpins ejected above one single low-speed streak.

Although it gives interesting qualitative information on the structure of turbulent boundary layers, the above LES does not have a sufficient resolution close to the wall (first point at $y^+ \approx 6$) to allow for good predictions of average quantities such as the friction coefficient at the wall or the shape factor. Here, we present new results with a finer resolution at the wall ($y^+ = 1$ or 2), a lower Mach number (0.3). It is known that transition in the boundary layer on a flat plate depends upon the type of perturbations exerted upstream on the flow (see [70]). In Klebannoff *et al.* [52] the boundary layer was forced upstream with a thin metal ribbon parallel to the wall

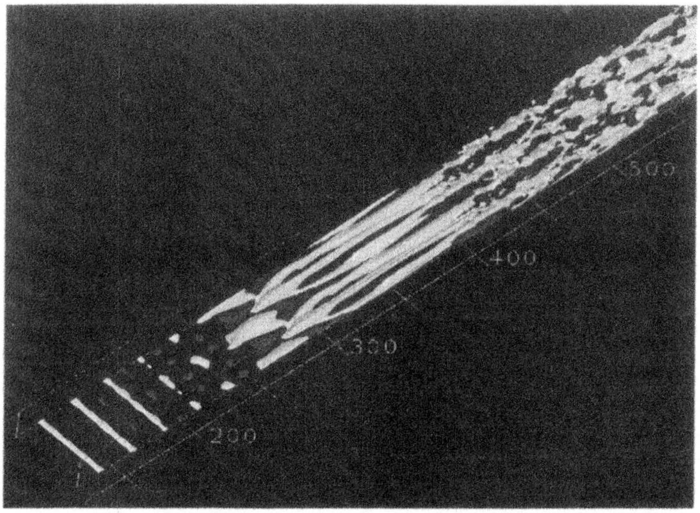

Figure 6.31. same calculation as Figure 6.30; isosurfaces of the longitudinal velocity fluctuations ($u_1' = 0.024U_\infty$, grey)

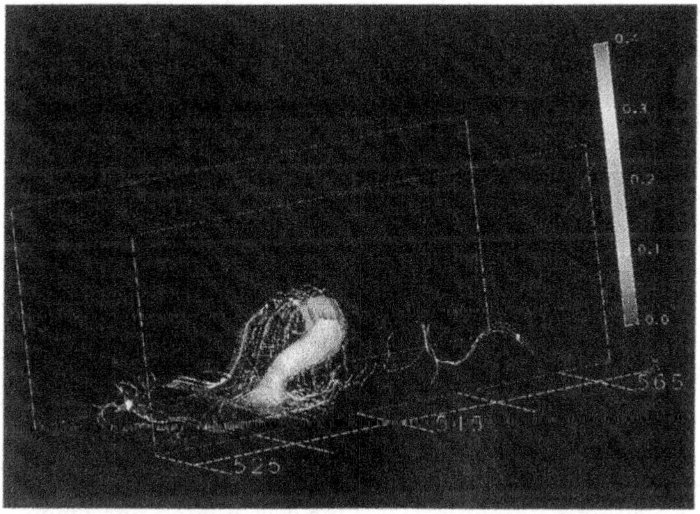

Figure 6.32. LES of the spatial boundary layer at Mach 0.5; vortex lines and low pressure characterizing a hairpin vortex ejected from the wall at the end of transition

and stretched in the spanwise direction, which vibrates two-dimensionally close to the wall. In this experiment, the 3D forcing was harmonic in the sense that the crests of the TS waves were oscillating in phase in the spanwise direction. This corresponds to what is generally referred to as the K-mode, where vortex filaments are aligned. On the other hand, if the par-

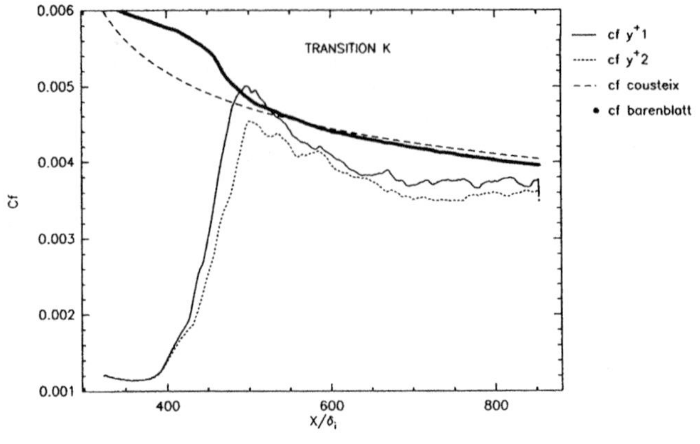

Figure 6.33. LES of a spatial boundary layer at Mach 0.3: friction coefficients against downstream distance, compared with theoretical predictions of Cousteix and Barenblatt (courtesy E. Briand).

turbation is subharmonic, the crests oscillate out of phase. This is called H-mode, from Herbert [46], and corresponds to a staggered organization of vortex filaments. For the temporal problem Herbert showed, with the aid of a secondary-instability analysis where a perturbation is superposed on a TS wave of finite amplitude, that the staggered mode was more amplified than the aligned mode. This should favour the emergence of H-mode during transition in natural situations. However, numerical simulations of the channel flow [51], [58] show instead the emergence of the aligned K-mode.

We now return to the LES of the spatially-developing boundary layer over a flat plate. It is here started with a different set of upstream conditions (harmonic K-mode or subharmonic H-mode) obtained with the aid of nonlinear PSE calculations[13] using a numerical code from the ONERA code[4]. To this upstream state (with still $R_\delta = 1000$), one superposes a weak 3D white-noise of amplitude 0.2 the amplitude of the PSE perturbation. The simulations involve up to five millions of grid points. Figure 6.33 shows for the K-case the downstream evolution of the friction coefficient at the wall, with comparison against the theoretical predictions of Van Driest discussed in [26] and Barenblatt[7]. One sees a good agreement of the LES with these predictions, with an improvement with a resolution of $y^+ = 1$. It is even better in the H-case. Figure 6.34 shows for the K-case (but results are very close in the H-case) the rms longitudinal velocity component u' at a downstream distance such that $R_\delta = 1670$, compared with Spalart's DNS[111] at Reynolds numbers of 1000 and 2000. Again, the agreement is good. We are now interested by the longitudinal[3] spectrum of u' at various distances from the wall, $E(k, y)$. We look for a self-similar evolution of the

[3]calculated through a Taylor hypothesis, as in the experiments

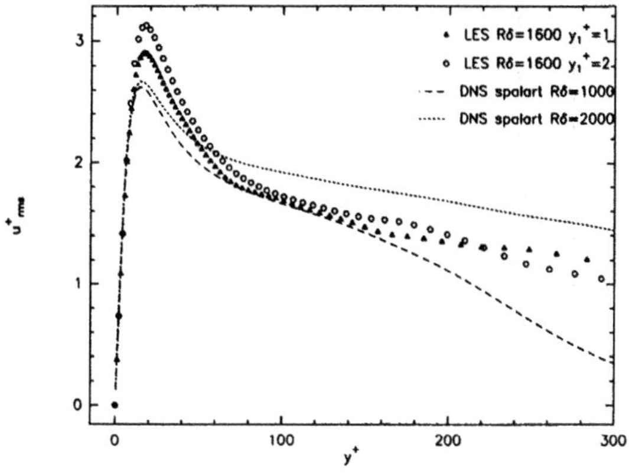

Figure 6.34. spatial boundary layer at Mach 0.3; rms velocity fluctuations compared with Spalart's DNS (courtesy E. Briand).

Karman-Howarth type

$$E(k,y) = v_*^2 y \; F(ky) \quad , \tag{6.88}$$

where F is a nondimensional function. In wall units, such a law corresponds to the following spectrum

$$E^+(k^+,y^+) = y^+ \; F(k^+y^+) \quad . \tag{6.89}$$

Figure 6.35 presents, for the Reynolds $R_\delta = 1670$, the function $E^+(k^+,y^+)/y^+$. The figure shows that the renormalization of Eq. (6.88) is quite good above $y^+ = 12$ at high wavenumbers ($k > k_I \approx y^{-1}$), where we have $F(ky) \sim (ky)^{-5/3}$. We have finally

$$E(k,y) \approx 0.13 \; v_*^2 y \; (ky)^{-5/3} \quad . \tag{6.90}$$

This is nothing more that a Kolmogorov-type law for 3D isotropic turbulence, if one assumes for the energy-dissipation rate $\epsilon \sim v_*^3 y^{-1}$. In fact, the $k^{-5/3}$ law is valid closer to the wall ($y^+ = 6$), but with a smaller constant. We have checked that the same scaling holds (with nearly the same constant, 0.12) at a different Reynolds $R_\delta = 1300$, just after the transition. However, the scaling does not work for the spanwise spectrum of u'. This is certainly due to an insufficient numerical resolution.

6.10.2. BOUNDARY LAYER UPON A GROOVE

The effect of a spanwise groove (whose dimensions are typically of the order of the boundary layer thickness) on the vortical structure of a turbulent boundary layer flow has recently regained interest in the field of turbulence

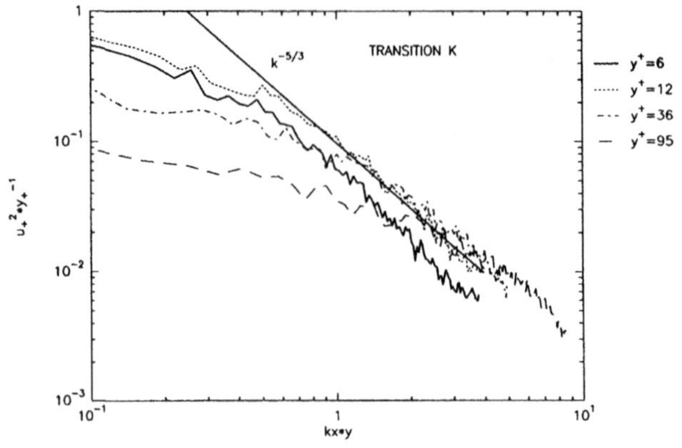

Figure 6.35. spatial boundary layer at Mach 0.3; longitudinal spectrum of u' at various distances of the wall, renormalized against the law of Eq. (6.89) (courtesy E. Briand).

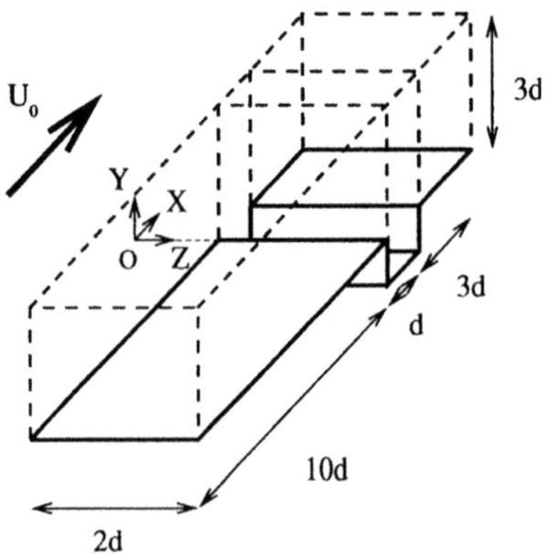

Figure 6.36. Sketch of the computational domain (courtesy Y. Dubief).

control (Choi & Fujisawa [18] and Pearson *et al.*[101]). The groove belongs to the category of passive devices able of manipulating skin friction in turbulent boundary layer flow. Depending on the dimensions of the cavity, the drag downstream of the groove can be increased or decreased.

In order to investigate the effects of a groove on the near-wall structure of turbulent boundary layer flows, Dubief and Comte [33] have recently performed a spatial numerical simulation of the flow over a flat plate with a spanwise square cavity embbeded in it. The goal here is to show the ability for the LES to handle geometrical singularities.

The width d of the groove is of the order of the boundary layer thickness, $d/\delta_0 = 1$. We here recall some of Dubief and Comte 's results. The simulation is carried out using the same compressible code as for the spatially-developing boundary layer previously presented. The computational domain is here decomposed into three blocks. The Mach number is 0.5. The

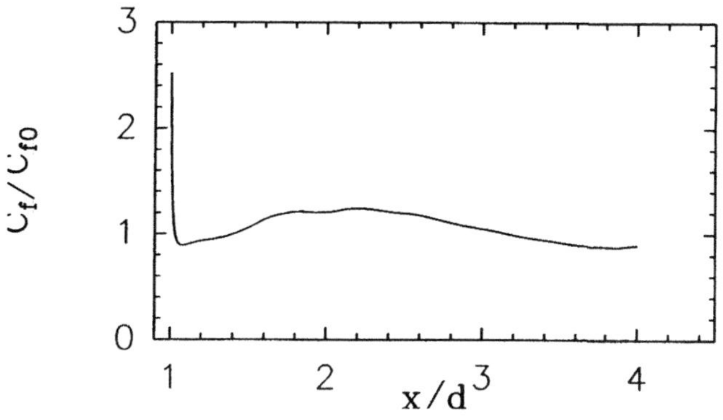

Figure 6.37. Longitudinal evolution of the skin friction coefficient normalised by its smooth wall value (courtesy Y. Dubief).

computational domain is sketched in figure 6.36. The large dimension of the upstream domain is required by the inlet condition. The coordinate system is located at the upstream edge of the groove. The resolution for the inlet, the groove and the downstream flat plate blocks are respectively $101 \times 51 \times 40$, $41 \times 101 \times 40$ and $121 \times 51 \times 40$. The minimal grid spacing at the wall in the vertical direction corresponds to $\Delta y^+ = 1$. The streamwise grid spacing goes from $\Delta x^+ = 3.2$ near the groove edges to 20 at the outlet. The spanwise resolution is $\Delta z^+ = 16$.

The Reynolds number of the flow is 5100, similar to the intermediate simulation of Spalart [111] at $R_\theta = 670$. The inflow is generated using the method of Lund *et al.* [82]. This method is based on the similarity properties of canonical turbulent boundary layers. At each time step, the mean and fluctuating velocities, temperatures and pressures are extracted from a plane, called the recycling plane and rescaled at the appropriate inlet scaling. The statistics are found in good agreement with Spalart's data.

Figure 6.37 shows the distribution of the skin friction coefficient C_f normalised by its smooth wall value on the downstream flat plate. Immediatly downstream of the groove the skin friction coefficient experiences a sharp rise, followed by a small undershoot. The skin friction eventually relaxes towards its smooth wall value in an oscillatory manner. This behavior is consistent with previous experimental results of Pearson *et al.* [101] with a smaller ratio d/δ_0. The local drag reduction observed in our simulation is smaller than that obtained by these authors, probably because of a larger width of the groove in our case. The high magnitude of the skin friction at the edge is caused by the impingment of the internal shear layer of the cavity. A similar overshoot is found at the upstream edge (not shown here), due to a local favorable pressure gradient which accelerates the near wall fluid approching the groove. The undershoot is still an unsolved issue.

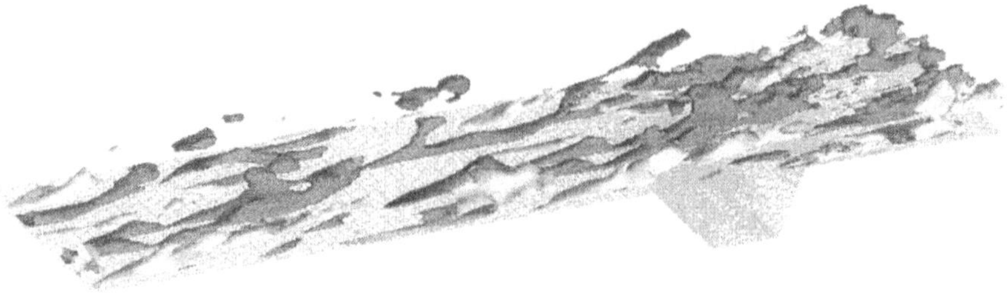

Figure 6.38. Isosurfaces of streamwise velocity fluctuations. Black $u' = -0.17$, white $u' = 0.17$ (courtesy Y. Dubief).

Spanwise correlations of u, v and w (not shown here) indicate a slight change in the streak spanwise wavelength. In the buffer layer, while negative u correlation peaks at $z^+ = 50$ in the upstream boundary layer, giving the right streak spacing $\lambda_z^+ = 100$, the spanwise wavelength is reduced downstream of the groove ($\lambda_z^+ = 70$). Figure 6.38 shows an instantaneous visualisation of u fluctation isosurfaces. The vertical extent of low-speed streaks is increased as they pass over the groove. The vorticity field is plotted using isosurfaces of the norm of the vorticity, conditioned by positive $Q = (\Omega_{ij}\Omega_{ij} - S_{ij}S_{ij})/2$. The structures downstream of the groove are smaller and less elongated in the streamwise direction (figure 6.39). It was checked that the statistics show a return towards a more isotropic state downstream of the groove. Figure 6.40 is a 3D view of the cavity where the upstream vertical wall is removed. Vortices aligned in the y direction can be isolated in the upstream part of the cavity. The curvature of the core of these vortices corresponds to the local curvature of the recirculating flow. The flow inside the groove is highly unsteady and there is obviously a high level of communication between the recirculating vortex and the turbulent boundary layer.

6.11. Conclusion

Direct numerical simulations of turbulence, which calculate all the scales of motion (from the large energetic scales to the small dissipative scales) explicitly can treat only small values of the Reynolds number. Therefore, they cannot reproduce the turbulence encountered in the atmosphere, the ocean, or most industrial devices. Attention is focussed on the large scales of motion since these are responsible for a significant part of heat and

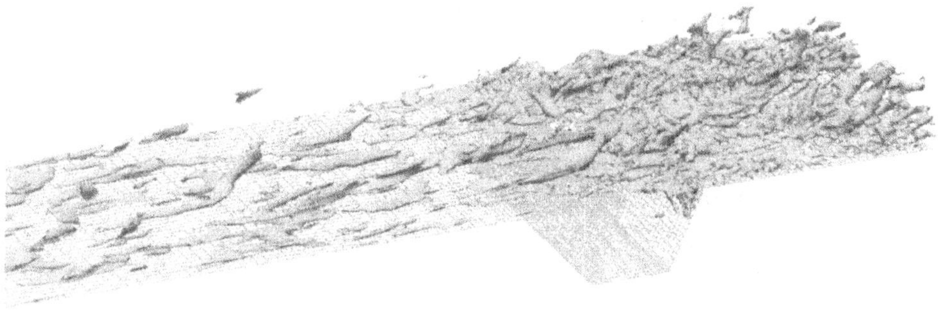

Figure 6.39. Isosurfaces of the norm the vorticity filtred by positive Q. $\omega = 0.3\omega_i$ (courtesy Y. Dubief).

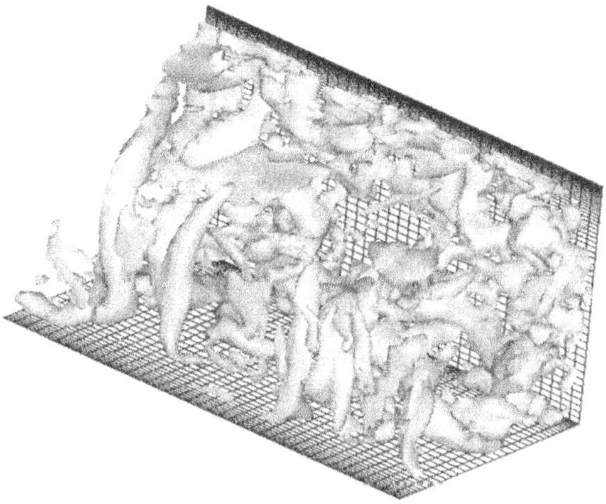

Figure 6.40. Isosurfaces of the vorticity fluctuations inside the cavity conditioned by $Q > 0$ (courtesy Y. Dubief).

mass transfer. The large turbulent scales can be simulated numerically by Large-Eddy Simulation (LES). LES requires correct representation of energy exchanges with the small scales that are not explicitly simulated: we have here presented the various "subgrid-scale" models developed since Smagorinsky's model in 1963.

Development of LES techniques is blooming at present. This is attributable to new ideas in subgrid-scale modelling and to tremendous progress in scientific computing. Once confined to very simple flow configurations such as isotropic turbulence or periodic flows, the field is evolving to include spatially growing shear flows, separated flows, pipe flows, riblet walls, and bluff bodies, among others. Classical turbulence modelling, based on one-point closures and a statistical approach allow computation of mean quantities. In many cases, it is necessary to have access to the fluctuating part of the turbulent fields such as the pollutant concentration or temperature: LES is then compulsory. Together with DNS, LES is indeed able to perform deterministic predictions (of flows containing coherent vortices, for instance) and to provide statistical information. The last is very important for assessing and improving one-point closure models, in particular for turbulent flows submitted to external forces (stratification, rotation, ...) or compressibility effects.

The complexity of problems tackled by LES is continuously increasing, and this has nowadays a decisive impact on industrial modelling and flow control. It is clearly demonstrated here that LES allow for a good knowledge of the vorticity topology, and that this knowledge is compulsory to design efficient turbulent flow control.

Acknowledgments

The results presented have greatly benefitted from the contributions of E. Briand, F. Delcayre, Y. Dubief, E. Lamballais, G. Silvestrini, and G. Urbin. We are indebted to P. Begou for the computational support. Most of the computating time used for the 3D calculations has been freely allocated by CNRS and CEA. Marcel Lesieur is supported by IUF.

References

1. Abbà, A., Cercignani, C., Valdettaro, L., and Zanini, P. (1997) LES of turbulent thermal convection, in J.P. Chollet, P.R. Voke and L. Kleiser (eds.), *Direct and Large Eddy Simulation II*, Kluwer Academic Publishers, pp. 147–156.
2. Abid, M. and Brachet, M.E. (1992) Mécanisme de génération des jets latéraux dans les jets axisymetriques forcés, *C.R.Acad.Sci.Paris.*
3. R. A. Antonia, M. Teitel, J. Kim, and L.W.B. Browne, Low-Reynolds-number effects in a fully developed turbulent channel flow. *J. Fluid Mech.* 236 (1992) 579–605.
4. Airiau, C., PhD, Toulouse University (1994).
5. Arnal, M., and FRIEDRICH, R., 1993 Large-eddy simulation of a turbulent flow with separation. In *Turbulent Shear Flows 8* (ed. F. Durst, R. Friedrich, B.E. Launder *et al.*, p. 169, Springer.
6. Bandyopadhyay, P.R. (1991) Instabilities and large structures in reattaching boundary layers. *AIAA Journal*, **29** , 1149–1155.
7. Barenblatt, G., *J. Fluid Mech.*, 248 (1993), 513-529.
8. Bardina, J., Ferziger, J.H., Reynolds, W.C. (1980) Improved subgrid model for large-eddy simulation, *AIAA paper N° 80-1357*.
9. Bartello, P., Métais, O., and Lesieur, M. (1994) Coherent structures in rotating three-dimensional turbulence. *J. Fluid Mech.* **273**, 1–29.

10. Basdevant, C., Sadourny R. (1983) Modélisation des échelles virtuelles dans la simulation numérique des écoulements bidimensionnels. *J. Mec. Theor. et Appl.*, Numéro Spécial, 243–269.
11. Bech, K.H. and Andersson, H.I. (1997) Turbulent plane Couette flow subject to strong system rotation, *J. Fluid Mech.* **347**, 289–314.
12. Bernal, L.P., and Roshko, A. (1986) Streamwise vortex structure in plane mixing layer, *J. Fluid Mech.* **170**, 499–525.
13. Bertolotti, P. & Herbert, T., *Theoret. and Comp. Fluids Dynamics*, **3** (1991) 117-124.
14. Blumen, W., 1970, Shear-layer instability of an inviscid compressible fluid, *J. Fluid Mech.*, **40**, pp 769-781.
15. Bogdanoff, D.W., 1983, Compressibility effects in turbulent shear layers, *AIAA J.*, **21**, pp 926-927.
16. Brown, G., and Roshko, A. (1974) On density effects and large structure in turbulent mixing layers, *J. Fluid Mech.* **64**, 775–816.
17. Broze, G. and Hussain, F. (1996) Transition to chaos in a forced jet: intermittency, tangent bifurcations and hysteresis. *J.Fluid Mech*, **311**, pp. 37–71.
18. Choi, K.S. and Fujisawa, N., 1993, Possibility of Drag Reduction using a *d*-type Roughness, *Appl. Sci. Res.*, **50**, pp. 315–324.
19. Chollet, J.P. and Lesieur, M., Parameterization of small scales of three-dimensional isotropic turbulence utilizing spectral closures. *J. Atmos. Sci.* 38 (1981) 2747-2757.
20. Chollet, J.P. and Lesieur, M., Modélisation sous maille des flux de quantité de mouvement et de chaleur en turbulence tridimensionnelle isotrope. *La Météorologie* 29-30 (1982) 183-191.
21. Clark, R.A., Ferziger, J.H., and Reynolds, W.C. (1979) Evaluation of subgrid-scale models using an accurately simulated turbulent flow, *J. Fluid Mech.* **91**, 1–16.
22. Comte, P., Ducros, F., Silvestrini, J., David, E., Lamballais, E., Métais, O. and Lesieur, M. (1994) Simulation des grandes échelles d'écoulements transitionnels, *AGARD Conference Proceedings 551*, pp. 14.1–14.12.
23. Comte, P., Fouillet, Y. and Lesieur, M., (1992) Simulation numérique des zones de mélange compressibles. *Revue scientifique et technique de la defense*, 3ème trimestre, 43–63.
24. P. Comte, M. Lesieur, and E. Lamballais, Large and small-scale stirring of vorticity and a passive scalar in a 3D temporal mixing layer. *Phys. Fluids A* 4 (1992) 2761-2778.
25. Comte, P., Silvestrini, J.H. and Lamballais, E., 1995, A Straightforward 3D Multi-block Unsteady Navier-Stokes Solver for a Direct and Large-Eddy Simulations of Transitional and Turbulent Compressible Flows, in *77th. AGARD Fluid Dynamic Panel Symposium "Progress and Challenges in CFD Methods and Algorithms"*, Seville, Spain, pp. 2–5.
26. Cousteix, J., *Turbulence et couche limite*, CEPADUES (1989).
27. Crow, S.C. and Champagne, F.H. (1971) Orderly structure in jet turbulence. *J.Fluid Mech*, **48**, pp. 547–591.
28. David, E. (1993) *Modélisation des Ecoulements Compressibles et Hypersoniques: une Approche Instationnaire*. PhD thesis. National Polytechnic Institute, Grenoble.
29. Deardorff, J.W. (1970) A numerical study of three-dimensional turbulent channel flow at large Reynolds number, *J. Fluid Mech.* **41**, 453–80.
30. Delcayre, F., and Lesieur, M. (1997) Topological features in the reattachment region of a backward-facing step. In *Advances in DNS/LES*, C. Liu and Z. Liu, eds., Greyden Press, Columbus, pp. 425–432.
31. Delcayre, F. (1997) Topology of coherent vortices in the reattachment region of a backward-facing step. Eleventh Symposium on *Turbulent Shear Flows*, Grenoble, France, 8–10 September, pp. 26-24/26-29.

Delcayre, F., (1997) Topological features in the reattachment region of a backward-facing step. In *Advances in DNS/LES*, C. Liu and Z. Liu, eds., Greyden Press,

Columbus, pp. 425–432.
32. Domaradzki, J.A., Metcalfe, R.W., Rogallo R.S. & J.J. Riley 1987 Analysis of subgrid-scale eddy viscosity with the use of results from direct numerical simulations. *Phys. Rev. Lett.* **58**, pp 547-550.
33. Dubief, Y. and P. Comte, 1997, Large-Eddy simulation of a boundary layer flow passing over a groove, in *Turbulent Shear Flows 11*, Grenoble, France, pp. 1-1/1-6.
34. Ducros, F., Comte, P., and Lesieur, M., 1995, Direct and large-eddy simulations of a supersonic boundary layer. *Selected Proceedings of Turbulent Shear Flows 9*, pp. 283–300, Springer.
35. Ducros, F., Comte, P., and Lesieur, M. (1996) Large-eddy simulation of transition to turbulence in a boundary layer developing spatially over a flat plate, *J. Fluid Mech.* **326**, 1–36.
36. El-Hady, N.M., and Zang T.A. (1995) Large-eddy simulation of nonlinear evolution and breakdown to turbulence in high-speed boundary layers, *Theoret. Comput. Fluid Dynamics* **7**, 217–240.
37. Erlebacher G., Hussaini M.Y., Speziale C.G., Zang T.A., ICASE Report, 87-20, 1987.
38. Erlebacher G., Hussaini M.Y., Speziale C.G., Zang T.A., J. Fluid Mech., 238, 1992, 155.
39. Favre A., J. de Mécanique, **4**, 1965, 361.
40. Fouillet, Y. (1992) *Contribution à l'étude par expérimentation numérique des écoulements cisaillés libres. Effets de compressibilité.* PhD thesis. National Polytechnic Institute, Grenoble.
41. Garnier, E., Métais, O., and Lesieur, M. (1998) Synoptic and frontal-cyclone scale instabilities in baroclinic jet flows, *J. Atmos. Sci.*, **55**, N° 8, pp. 1316–1335.
42. Garnier, E., Métais, O., and Lesieur, M. (1996) Instabilités primaire et secondaire dans un jet barocline, *C.R. Acad. Sci. Paris* Série II b **323**, 161–168.
43. Germano, M., (1992) Turbulence, the filtering approach, *J. Fluid Mech.* **238**, 325–336.
44. Germano, M., Piomelli, U., Moin, P., and Cabot, W. (1991) A dynamic subgrid-scale eddy-viscosity model, *Phys. Fluids A.* **3**(7), 1760–1765.
45. Gonze, M.A. (1993) *Simulation Numérique des Sillages en Transition à la Turbulence.* PhD thesis. National Polytechnic Institute, Grenoble.
46. Herbert, T., 1988, Secondary instability of boundary layers, *Ann. Rev. Fluid Mech.*, **20** pp. 487–526.
47. Hunt, J.C.R., Wray, A.A. and Moin, P. (1998) Eddies, stream, and convergence zones in turbulent flows. Center for Turbulence Research Rep., **CTR-S88**, p. 193.
48. Johnston, J.P., Halleen, R.M., and Lezius, D.K. (1972) Effects of spanwise rotation on the structure of two-dimensional fully developed turbulent channel flow, *J. Fluid Mech.*, **56**, 533–557.
49. Jovic, S. and Driver, M. (1994) Backward-facing step measurement at low Reynolds number $Re_h = 5000$, Ames Research Center, NASA Technical Memorandum, 108807.
50. Kim, J. 1983 The effect of rotation on turbulence structure. In *Proc. 4th Symp. on Turbulent Shear Flows, Karlsruhe*, pp. 6.14-6.19.
51. Kim, J., and Moser, R.D., 1989, On the secondary instability in plane Poiseuille flow, *Phys. Fluids A*, **1**, pp. 775–777.
52. Klebanoff, P.S., Tidstrom, K.D. and Sargent, L.M., 1962, The three-dimensional naure of boundary layer instability, *J. Fluid Mech.*, **12**, pp. 1–34.
53. Kolmogorov, A.N. (1941) The local structure of turbulence in incompressible viscous fluid for very large Reynolds numbers. *Dokl. Akad. Nauk SSSR* **30**, 301–305.
54. Kraichnan, R.H. (1976) Eddy viscosity in two and three dimensions. *J. Atmos. Sci.* **33**, 1521–1536.
55. Kristoffersen, R. and Andersson, H.I. 1993 Direct simulations of low-Reynolds-number turbulent flow in a rotating channel. *J. Fluid Mech.* **256**, 163-197.

56. Kuroda, A., 1990, Direct-numerical simulation of Couette-Poiseuille flows, PhD thesis, University of Tokyo, 1990.
57. Kusek, Corke and Reisenthel, Seeding of helical modes in the initial region of an axisymmetric jet (1990), *Experiments in Fluids*, **10**, pp. 116–124.
58. Lamballais, E. (1996) *Simulations numériques de la turbulence dans un canal plan tournant*, PhD thesis. National Polytechnic Institute, Grenoble.
59. Lamballais, E., Lesieur, M., Métais, O. (1996) Effects of spanwise rotation on the vorticity stretching in transitional and turbulent channel flow, *Int. J. Heat and Fluid Flow* **17** (3), 324–332.
60. Lamballais, E., Lesieur, M., and Métais, O. (1996) Influence of a solid-body rotation upon coherent vortices in a channel, *C. R. Acad. Sci.* Série II b **323**, 95–101.
61. Lamballais, E., Métais, O., and Lesieur, M. (1996) Influence of a spanwise rotation upon the coherent-structure dynamics in a turbulent channel flow, in J.P. Chollet, P.R. Voke and L. Kleiser (eds.), *Direct and Large Eddy Simulation II*, Kluwer Academic Publishers, pp. 225–236.
62. Lasheras, J.C., Lecuona A. and Rodriguez, P. (1991) Three dimensionnal structure of the vorticity field in the near region of laminar co-flowing forced jets. In *The Global Geometry of Turbulence*, edited by J.Jimenez (Plenum Press, New-York).
63. Le, H. and P. Moin, 1994, Direct Simulation of Turbulent Flow Over a Backward-Facing Step, NASA 58.
64. Le, H., Moin, P. and Kim, J. (1997) Direct numerical simulation of turbulent flow over a backward-facing step, *J. Fluid Mech.*, **330**, 349–374.
65. Lee, M., Reynolds, W.C. (1985) Bifurcating and blooming jets at high Reynolds number.*Fifth Symp. on Turbulent Shear Flows*, Ithaca, New York 1.7-1.12.
66. Lele, S.K., 1989, Direct-numerical simulation of compressible free-shear flows, A.I.A.A. Paper, n^0 89-0374.
67. Lele, S.K. (1992) Compact finite difference schemes with spectral-like resolution. *J. Comput. Phys.* **103**, 16–42.
68. D.C. Leslie and G.L. Quarini, The application of turbulence theory to the formulation of subgrid modelling procedures. *J. Fluid Mech.* 91 (1979) 65-91.
69. Lesieur, M. (1997) Recent approaches in large-eddy simulations of turbulence, in O. Métais and J. Ferziger (eds), *New Tools in Turbulence Modelling*, Les Éditions de Physique, Springer-Verlag, pp. 1–28.
70. Lesieur, M. (1997) *Turbulence in Fluids, Third Revised and Enlarged Edition*, Kluwer Academic Publishers, Dordrecht.
71. Lesieur, M., and Métais, O. (1996) New trends in large-eddy simulations of turbulence", *Annu. Rev. Fluid Mech.* **28**, 45–82.
72. M. Lesieur and R. Rogallo, Large-eddy simulation of passive-scalar diffusion in isotropic turbulence. *Phys. Fluids A* 1 (1989) 718-722.
73. Lesieur, M., Yanase, S., and Métais, O. (1991) Stabilizing and destabilizing effects of a solid-body rotation on quasi-two-dimensional shear layers, *Phys. Fluids A* **3**, 403–407.
74. Lessen , M., Fox, J.A. and Zien, H.M., 1965, On the inviscid stability of the laminar mixing of two parallel streams of a compressible fluid, J. Fluid Mech., **23**, pp 355-367.
75. Lessen, M., Fox, J.A. and Zien, H.M., 1966, Stability of the laminar mixing of two parallel streams with respect to supersonic disturbances, J. Fluid Mech., **25**, pp 737-742.
76. Liepmann, D. and Gharib, M. (1992) The role of streamwise vorticity in the near-field entrainment of round jets. *J.Fluid Mech*, **245**, pp. 643–668.
77. Lilly, D.K. (1987) J.R. Herring and J.C. McWilliams (eds), *Lecture Notes on Turbulence*, World Scientific, pp. 171–218.
78. Liu, S., Meneveau, C., and Katz, J. (1994) On the properties of similarity subgrid-scale models as deduced from measurements in turbulent jet, *J. Fluid Mech.*

275,83–119.
79. Lilly, D.K. (1992) A proposed modification of the Germano subgrid-scale closure method, *Phys. Fluids A.* **4** (3), 633–635.
80. Lilly D.K., in Computational Fluids Dynamics, Les Houches; session *LIX*, 1993 Lesieur, Comte, Zin-Justin eds., Elsevier Sciences Publishers, B.V., 1996, p. 353.
81. Longmire, E.K. and Duong, L.H. (1996) Bifurcating jets generated with stepped and sawtooth nozzles. *Phys. Fluids*, **8** (4), pp. 978–992.
82. Lund, T.S., Wu, X. and Squires, K. D., 1996, On the Generation of Turbulent Inflow Conditions for Boundary Layer Simulations, *Ann. Res. Briefs*, Stanford, pp 287–295.
83. Martin, J.E. and Meiburg, E. (1991) Numerical investigation of three-dimensionally evolving jets subject to axisymmetric and azimuthal perturbations. *J.Fluid Mech*, **230**, pp. 271–318.
84. Mason, P. J. (1994) Large-eddy simulation: a critical review of the technique, *Q.J.R. Meteorol. Soc.* **120**, 1–26.
85. McWilliams, J.C. (1985) A uniformly valid model spanning the regimes of geostrophic and isotropic, stratified turbulence: Balanced turbulence. *J. Atmos. Sci.* **42**, 1773–1774.
86. Métais, O., Flores, C., Yanase, S., Riley, J.J. and Lesieur, M. (1995) Rotating free shear flows Part 2 : Numerical simulations, *J. Fluid Mech.* **293**, 41–80.
87. Métais, O. and Lesieur, M. (1986) Statistical predictability of decaying turbulence, *J. Atmos. Sci.* **43**, 857–870.
88. Métais, O., and Lesieur M. (1992) Spectral large-eddy simulations of isotropic and stably-stratified turbulence. *J. Fluid Mech* **239**, 157–194.
89. Métais, O. Flores, C., Yanase, S., Riley, J.J. and Lesieur, M. (1995) Rotating free shear flows Part 2 : Numerical simulations, *J. Fluid Mech* **293**, 41–80.
90. Michalke, A. and Hermann, G. (1982) On the inviscid instability of a circular jet with external flow. *J.Fluid Mech*, **114**, pp. 343–359.
91. Miyake, Y. and Kajishima, T. (1986) Numerical simulation of the effects of Coriolis force on the structure of turbulence, *Bull. JSME*, **29**, N° 256, 3347–3351.
92. Moin, P., and Kim, J. (1982) Numerical investigation of turbulent channel flow, *J. Fluid Mech.* **118**, 341–377.
93. Moin P., Squires K., Cabot W., Lee S., Phys. Fluids A 3, 11, 1991, 2746.
94. Monkewitz, P.A. and Pfizenmaier, E. (1991) Mixing by 'side jets' in strongly forced and self-excited round jets. *Phys. Fluids A*, **3** (5), pp. 1356–1361.
95. Nakabayashi, K. and Kitoh, O. (1996) Low Reynolds number fully developed two-dimensional turbulent channel flow with system rotation. *J. Fluid Mech.*, **315**, 1–29.
96. Normand, X., 1990, Transition à la turbulence dans les écoulements cisaillés libres et pariétaux, PhD thesis. National Polytechnic Institute, Grenoble.
97. Normand X., Lesieur M., Theor. and Comp. Fluid Dyn., 3, 1992, 231.
98. Orlanski, I. (1976) A simple boundary condition for unbounded hyperbolic flows, *J. Comp. Phys.*, **21**, 251–269.
99. Papamoschou, D. and Roshko, A., 1988, The compressible turbulent shear layer: an experimental study, J. Fluid Mech., **197**, pp 453-477.
100. Parekh, D.E., Leonard, A. and Reynolds, W.C. (1988) Bifurcating jets at high Reynolds number. Air Force Office of Scientific Research contractor Rep n°AF-F49620-84-K-0005 and n°AF-F49620-86-K-0020.
101. Pearson, B.R., Elavarasan, R. Antonia, R.A., 1997, The Response of a Turbulent Boundary Layer to a Square Groove, *J. Fluids Eng.*, To be published.
102. Petersen, R.A. (1978) Influence of wave dispersion on vortex pairing in a jet. *J.Fluid Mech*, **89**, pp. 469–495.
103. Piomelli, U. (1993) High Reynolds number calculations using the dynamic subgrid-scale stress model, *Phys. Fluids A* **5** (6), 1484–1490.
104. Piomelli, U. and Liu, J. (1995) Large-eddy simulation of rotating channel flows

using a localized dynamic model, *Phys. Fluids A*, **7** (4), 839–848.
105. Rogers, M. & Moser, R., 1994 *Phys. Fluids A*, **6**, 903.
106. Sandham, N.D. and Reynolds, W.C., 1989, 'The compressible mixing layer: linear theory and direct simulation, AIAA Paper 89-0371.
107. Sandham, N.D. and Reynolds, W.C., 1991, Three-dimensional simulations of large eddies in the compressible mixing layer, J. Fluid Mech., **224**, pp 133-158.
108. Silveira-Neto, A., Grand, D. Métais, O. Lesieur, M. (1993) A numerical investigation of the coherent structures of turbulence behind a backward-facing step. *J.Fluid Mech*, **256**, pp. 1–55.
109. Silvestrini, J., Comte, P., and Lesieur, M. (1995) DNS and LES of spatial incompressible mixing layer. *Proc. of the 10th Symposium on Turbulent Shear Flows.*
110. Smagorinsky, J. (1963) General circulation experiments with the primitive equations, *Mon. Weath. Rev.* **91**, 3, 99–164.
111. Spalart, P.R. (1988) Direct Simulation of a Turbulent Boundary Layer up to $R_\theta =$ 1410, *J. Fluid Mech.*, **187**, 61–98.
112. Tafti, D. K. and Vanka, S. P. (1991) A numerical study of the effects of spanwise rotation on turbulent channel flow, *Phys. Fluids A*, **3** (4), 642–656.
113. Thompson, K.W., 1987, Time Dependent Boundary Conditions for Hyperbolic Systems, *J. Comp. Phys.*, **68**, pp. 506–517.
114. Urbin, G., (1998) Etude numérique par simulation des grandes échelles de la transition à la turbulence dans les jets. PhD thesis. National Polytechnic Institute, Grenoble.
115. Urbin, G. and Métais, O. (1997) Large-eddy simulation of three-dimensional spatially-developing round jets, in *Direct and Large-Eddy Simulation II*, J.P. Chollet, L. Kleiser and P.R. Voke eds., Kluwer Academic Publishers, pp. 35–46.
116. Urbin, G., Brun, C. and Métais, O. (1997) Large-eddy simulations of three-dimensional spatially evolving roud jets, 11th symposium on Turbulent Shear Flows, Grenoble, September 8-11, pp. 25-23/25-28.
117. Westphal, R. V., Johnston, J. P. and Eaton, J. K. (1984) Experimental study of flow reattachment in a single-sided sudden expansion, Stanford University, MD 41.
118. Widnall, S.E., Bliss, D.B. and Tsai, C. (1974) The instability of short waves on vortex ring. *J.Fluid Mech*, **66** (1), pp. 35–47.
119. Yoshizawa Y., Phys. fluids, **29**, 1986, 2152.
120. Zaman, K.B.M.Q., Reeder, M.F. and Samimy, M. (1994) Control of an axisymmetric jet using vortex generators. *Phys.Fluids*, **6** (2), pp. 778–794.
121. Zang, Y., Street, R.L., and Koseff J.R. (1993) A dynamic mixed subgrid scale model and its application to turbulent recirculating flows. *Phys. Fluids A* **5** (12), 3186–3196.

Chapter 7

DIRECT NUMERICAL SIMULATIONS OF COMPRESSIBLE TURBULENT FLOWS: FUNDAMENTALS AND APPLICATIONS

SANJIVA K. LELE

Department of Aeronautics and Astronautics and Department of Mechanical Engineering Stanford University, Stanford, CA 94305-4035

7.1. Introduction

The equations governing a time-dependent compressible flow are first discussed to bring out the key features associated with compressibility and the limiting behavior at low speed. Concepts of linear eigenmodes representing vortical, entropic and acoustic disturbances are introduced via analysis of disturbances in a compressible parallel shear flow. Linear and nonlinear coupling of such modes is stressed.

Fundamentals of numerical methods used in direct numerical simulations of compressible turbulent flows including high-bandwidth space- and time- discretization techniques, initial conditions, and boundary conditions are given. Attention is drawn to the numerical artifacts which must be controlled to achieve successful simulations of physical flow phenomena. Applications of these techniques in simulating compressible homogeneous turbulence, turbulent free-shear layers and shock-turbulence interaction are discussed. Recent use of direct simulation in aeroacoustic studies is also highlighted. Emphasis is placed on the physical phenomena revealed by numerical simulations and open issues requiring further research are identified.

7.2. Physical nature of compressible turbulent flows

Turbulent flows involve fluctuations in the flow variables over a broad range of space and time scales. The large scales of such a flow are related to the geometrical constraints, forcing, and boundary conditions. These may be quasi-organized and quasi-deterministic, and at high Reynolds number break down into small-scale motions via nonlinear processes. Measurements taken at different positions in a turbulent flow typically exhibit decorrelation with finite space- and time-scales. The large-scale motions are often regarded as controlling the transport of heat, chemical species and momentum across the flow. Unlike most wave motions, the fluid motions in a turbulent flow are dissipative. Turbulent eddies dissipate their energy in a time scale of the order of the eddy turnover time, *i.e.* the time scale for a fluid parcel to tumble from one side of an eddy to the other. Again the rate of energy dissipation is set by the energy-containing scales, but the dissipative process which converts the mechanical energy into heat occurs at the small scales. The scale ratio between the large and small scales is dependent on Reynolds number and typically the dissipative scales are also well described by the continuum hypothesis as they are many times larger than the molecular mean free-path (*e.g.* Tennekes and Lumley [158]).

In a compressible flow the volume (or density) of a fluid-element does not remain constant. This volume change may arise from a variety of causes. For example the volume of a fluid-element containing a mixture of a fuel and oxidizer may change due to the heat released from combustion of the fuel. Alternatively, the volume of a fluid-element may change due to varying pressure, such as during the compression stroke in an internal combustion engine or in the flow of exhaust gases in a rocket nozzle. It is not necessary to treat all processes which change the volume of a fluid-element by the fully-compressible governing equations. For instance many aspects of low-speed combustion are well described by the so-called low Mach number equations. This distinction is also expected to apply to some processes occuring in a turbulent flow. It is possible that the mean or averaged properties of a flow require a compressible description but the fluctuations are approximated by a reduced set of equations.

If the volume of a fluid-element changes appreciably in response to a change in pressure this change must be regarded as an intrinsically compressible change. An acoustic (sound) wave is intrinsically compressible. The acceleration of a gas flow through the 'sonic state' to a supersonic flow in a converging-diverging nozzle is intrinsically compressible. But the turbulent fluctuations in the supersonic boundary layer within a propulsion nozzle could still be essentially incompressible. Evidently, parameters besides the mean flow Mach number are needed to make such distinctions. The equations governing the instantaneous motions contain this information.

7.3. Governing equations

For simplicity consider the compressible Navier Stokes equations for an ideal, calorically perfect gas. Dimensional variables are denoted by superscript $\star$. At position $x_i^\star$ and time $t^\star$ the density is $\rho^\star$, the velocity is $u_i^\star$, the pressure is $p^\star$, the temperature is $T^\star$, the enthalpy per unit mass is $h^\star$ and the entropy per unit mass is $s^\star$. Mass conservation requires that

$$\frac{\partial \rho^\star}{\partial t^\star} + \frac{\partial \rho^\star u_i^\star}{\partial x_i^\star} = 0. \tag{7.1}$$

Momentum conservation requires that

$$\frac{\partial \rho^\star u_i^\star}{\partial t^\star} + \frac{\partial \rho^\star u_i^\star u_j^\star}{\partial x_j^\star} = \rho^\star \frac{D u_i^\star}{D t^\star} = -\frac{\partial p^\star}{\partial x_i^\star} + \frac{\partial \tau_{ij}^\star}{\partial x_j^\star}. \tag{7.2}$$

Energy conservation requires that

$$\frac{\partial \rho^\star h^\star}{\partial t^\star} + \frac{\partial \rho^\star h^\star u_i^\star}{\partial x_i^\star} = \rho^\star \frac{D h^\star}{D t^\star} = \frac{D p^\star}{D t^\star} + \Phi^\star - \frac{\partial q_i^\star}{\partial x_i^\star} + Q^\star, \tag{7.3}$$

where $Q^\star$ the rate of 'external' heat addition per unit volume (*e.g.* due to combustion), $D/Dt^\star = \partial/\partial t^\star + u_i^\star \partial/\partial x_i^\star$, is the material derivative following the flow $u_i^\star$. The viscous stress $\tau_{ij}^\star$, the molecular heat flux $q_i^\star$, and the rate of viscous dissipation of energy $\Phi^\star$ are given by:

$$\tau_{ij}^\star = 2\mu^\star \{S_{ij}^\star - \frac{1}{3}\delta_{ij} S_{kk}^\star\}, \qquad q_i^\star = -k^\star \frac{\partial T^\star}{\partial x_i^\star}, \qquad \Phi^\star = \tau_{ij}^\star S_{ij}^\star,$$

where the strain rate $S_{ij}^\star = \frac{1}{2}\{\frac{\partial u_i^\star}{\partial x_j^\star} + \frac{\partial u_j^\star}{\partial x_i^\star}\}$, and Stokes hypothesis with zero bulk viscosity is used to obtain the viscous stresses. Fluid properties such as the viscosity $\mu^\star$ and thermal conductivity $k^\star$ are regarded as functions of temperature $T^\star$. The ideal gas equation of state is

$$p^\star = \rho^\star \Re^\star T^\star / W \tag{7.4}$$

where $\Re^\star$ is the universal gas constant and W is the molecular weight of the gas and for calorically perfect gas the specific heats are

$$C_p^\star = \frac{\Re^\star \gamma}{(\gamma - 1)W}, \qquad C_v^\star = \frac{\Re^\star}{(\gamma - 1)W}.$$

Since the internal energy per unit mass, $e^\star = h^\star - p^\star/\rho^\star$, it follows that

$$\frac{\partial \rho^\star e^\star}{\partial t^\star} + \frac{\partial \rho^\star e^\star u_i^\star}{\partial x_i^\star} = \rho^\star \frac{De^\star}{Dt^\star} = -p^\star \frac{\partial u_i^\star}{\partial x_i^\star} + \Phi^\star - \frac{\partial q_i^\star}{\partial x_i^\star} + Q^\star \tag{7.5}$$

Similarly the total energy per unit mass, $e_t^\star = e^\star + u_i^\star u_i^\star/2$, satisfies

$$\frac{\partial \rho^\star e_t^\star}{\partial t^\star} + \frac{\partial \rho^\star e_t^\star u_i^\star}{\partial x_i^\star} = \rho^\star \frac{De_t^\star}{Dt^\star} = -\frac{\partial p^\star u_i^\star}{\partial x_i^\star} + \frac{\partial \tau_{ji}^\star u_i^\star}{\partial x_j^\star} - \frac{\partial q_i^\star}{\partial x_i^\star} + Q^\star \tag{7.6}$$

The energy equation can also be expressed using the pressure, $p^\star$, as the dependent variable. The equation of state (7.4) requires that

$$\frac{1}{p^\star}\frac{Dp^\star}{Dt^\star} = \frac{1}{\rho^\star}\frac{D\rho^\star}{Dt^\star} + \frac{1}{T^\star}\frac{DT^\star}{Dt^\star}$$

Using mass balance (7.1) and energy balance (7.5) in the above equation gives

$$\frac{Dp^\star}{Dt^\star} + \gamma p^\star \frac{\partial u_i^\star}{\partial x_i^\star} = (\gamma - 1)(\Phi^\star - \frac{\partial q_i^\star}{\partial x_i^\star} + Q^\star) \tag{7.7}$$

Still another form of the energy equation is found using the Gibbs relation for the entropy per unit mass, $s^\star$:

$$T^\star ds^\star = dh^\star - \frac{dp^\star}{\rho^\star},$$

Applying this to a 'fluid particle' moving with the velocity $u_i^\star$ gives the energy equation in entropy form

$$\rho^\star T^\star \frac{Ds^\star}{Dt^\star} = \Phi^\star - \frac{\partial q_i^\star}{\partial x_i^\star} + Q^\star \tag{7.8}$$

7.3.1. NON-DIMENSIONALIZATION

The governing equations can be rendered nondimensional in a number of ways. Nondimensional equations bring out the key nondimensional parameters governing a problem. When one or more of such nondimensional parameters take on extreme values the governing equations can be simplified. The simplified set of equations expresses a specific balance of physical effects. Often an overall flow phenomena requires a different physical balance in different regions of the flow. A single set of simplified equations is then inappropriate to describe the entire flow. This is the so-called singular perturbation problem. Two particular choices for making the equations nondimensional will be described. First the incompressible scaling and later the acoustic scaling are recalled. Problems of low-speed aeroacoustics involve singular perturbation (Crow [38], Crighton [34]) and are discussed in section 7.7. The familiar inviscid-viscous singular perturbation of boundary layer flows is not discussed. Within the incompressible scaling some further choices are possible. The choice leading to the so-called low Mach number equations for low-speed variable-density flow is described. This low Mach number limit is contrasted with the more familiar limit yielding the conventional constant density incompressible flow.

Incompressible Scaling

Choosing the length scale L, velocity scale V, time scale $\tau = L/V$, density scale ρ_r, pressure scale $\rho_r V^2$, temperature scale T_r, internal energy, total energy and enthalpy scale $C_{pr}T_r$ and entropy scale C_{pr}, yields equations with the following non-dimensional parameters: Reynolds number, $R = \rho_r VL/\mu_r$, Prandtl number, $Pr = \mu_r C_{pr}/k_r$, and Mach number $M = V/C_r$ (C_r is the speed of sound at reference conditions). This choice of scales also produces following relations among the non-dimensional dependent variables: (equation of state) $p = \rho T/(\gamma M^2)$; (internal energy definition) $e = T/\gamma$; (enthalpy definition) $h = T = e + (\gamma - 1)M^2 p/\rho$; (total energy definition) $e_t = e + (\gamma - 1)M^2 u_i u_i/2$, and $Q = Q^\star L/(\rho_r C_{pr} T_r V)$. The resulting equations are:

$$\frac{\partial \rho}{\partial t} + \frac{\partial \rho u_i}{\partial x_i} = 0 \tag{7.9}$$

$$\rho \frac{Du_i}{Dt} = -\frac{\partial p}{\partial x_i} + R^{-1}\frac{\partial \tau_{ij}}{\partial x_j} \tag{7.10}$$

$$\rho \frac{DT}{Dt} = (\gamma - 1)M^2 \frac{Dp}{Dt} + \frac{(\gamma - 1)M^2}{R}\Phi - \frac{1}{RPr}\frac{\partial q_i}{\partial x_i} + Q \tag{7.11}$$

where

$$D/Dt = \partial/\partial t + u_i \partial/\partial x_i, \quad \tau_{ij} = 2\mu^\star/\mu_r\{S_{ij} - \frac{1}{3}\delta_{ij}S_{kk}\},$$

$$q_i = -\frac{k^\star}{k_r}\frac{\partial T}{\partial x_i}, \quad \Phi = \tau_{ij}S_{ij},$$

with $S_{ij} = \frac{1}{2}\{\frac{\partial u_i}{\partial x_j} + \frac{\partial u_j}{\partial x_i}\}$. Other equivalent forms of energy balance are:

$$\rho \frac{De}{Dt} = -(\gamma - 1)M^2 p \frac{\partial u_i}{\partial x_i} + \frac{(\gamma - 1)M^2}{R}\Phi - \frac{1}{RPr}\frac{\partial q_i}{\partial x_i} + Q \tag{7.12}$$

$$\rho\frac{De_t}{Dt} = -(\gamma-1)M^2\frac{\partial pu_i}{\partial x_i} + \frac{(\gamma-1)M^2}{R}\frac{\partial \tau_{ji}u_i}{\partial x_j} - \frac{1}{RPr}\frac{\partial q_i}{\partial x_i} + Q, \quad (7.13)$$

$$\frac{Dp}{Dt} + \gamma p\frac{\partial u_i}{\partial x_i} = \frac{(\gamma-1)\Phi}{R} - \frac{1}{PrRM^2}\frac{\partial q_i}{\partial x_i} + \frac{Q}{M^2} \quad (7.14)$$

$$\rho T\frac{Ds}{Dt} = \frac{(\gamma-1)M^2}{R}\Phi - \frac{1}{PrR}\frac{\partial q_i}{\partial x_i} + Q \quad (7.15)$$

The so-called low Mach number equations, commonly used in low-speed combustion, are readily obtained by decomposing pressure as $p = 1/(\gamma M^2) + \widetilde{p}$, where $\widetilde{p}$ contains the pressure departure from its reference state and is $\mathcal{O}(1)$ as $M \to 0$. Substituting this in (7.14) yields:

$$\frac{1}{M^2}\{\frac{\partial u_i}{\partial x_i} + \frac{1}{RPr}\frac{\partial q_i}{\partial x_i} - Q\} = -\frac{D\widetilde{p}}{Dt} - \gamma\widetilde{p}\frac{\partial u_i}{\partial x_i} + \frac{(\gamma-1)\Phi}{R},$$

where the largest terms have been isolated on the left hand side. It may be concluded that in a low Mach number flow the volume of a fluid element changes appreciably only if there is significant heat-transfer or significant external heat addition [1]. Thus in a low Mach number flow $\frac{\partial u_i}{\partial x_i} \approx Q - \frac{1}{RPr}\frac{\partial q_i}{\partial x_i}$. This heat-transfer or heat-addition dominated low Mach number limit is valid only on the convective time scale, *i.e.* $t^\star V/L \sim \mathcal{O}(1)$. This simplified equation set which 'filters out' the fast-acoustic waves of time scale C_r/L is obtained by letting: $\rho = \widetilde{\rho}^{(0)} + M^2\widetilde{\rho}^{(1)} + \ldots$, $T = \widetilde{T}^{(0)} + M^2\widetilde{T}^{(1)} + \ldots$, $u_i = \widetilde{u}_i^{(0)} + M^2\widetilde{u}_i^{(1)} + \ldots$, $p = 1/(\gamma M^2) + \widetilde{p}^{(1)} + M^2\widetilde{p}^{(2)} + \ldots$. At leading order the equation of state reduces to $\widetilde{\rho}^{(0)}\widetilde{T}^{(0)} = \gamma$ while the equation set (7.9)-(7.15) gives:

$$\frac{\partial\widetilde{\rho}^{(0)}}{\partial t} + \frac{\partial\widetilde{\rho}^{(0)}\widetilde{u}_i^{(0)}}{\partial x_i} = 0 \quad (7.16)$$

$$\widetilde{\rho}^{(0)}\frac{D_0\widetilde{u}_i^{(0)}}{Dt} = -\frac{\partial\widetilde{p}^{(1)}}{\partial x_i} + R^{-1}\frac{\partial\tau_{ij}^{(0)}}{\partial x_j} \quad (7.17)$$

$$\widetilde{\rho}^{(0)}\frac{D_0\widetilde{T}^{(0)}}{Dt} + \frac{1}{RPr}\frac{\partial\widetilde{q}_i^{(0)}}{\partial x_i} - Q = 0 \quad (7.18)$$

where $D_0/Dt = \partial/\partial t + \widetilde{u}_i^{(0)}\partial/\partial x_i$, and other symbols with superscript '(0)' are similarly defined. Auxiliary forms of the energy equation give

[1] Volume changes also occur due to molecular diffusion of different chemical species. This effect is neglected for simplicity in present analysis.

$$\frac{\partial \widetilde{u}_i^{(0)}}{\partial x_i} + \frac{1}{PrR}\frac{\partial \widetilde{q}_i^{(0)}}{\partial x_i} - Q = 0 \qquad (7.19)$$

$$\widetilde{\rho}^{(0)}\widetilde{T}^{(0)}\frac{D_0\widetilde{s}^{(0)}}{Dt} = -\frac{1}{PrR}\frac{\partial \widetilde{q}_i^{(0)}}{\partial x_i} + Q \qquad (7.20)$$

These low Mach number equations were obtained under the assumption of a constant reference pressure at leading order in M. Evidently, entropy fluctuations associated with heat-transfer or heat addition are dominant and the fluid motion is not incompressible even at low Mach number. The change in volume of a fluid element (dilatation) is prescribed by the energy balance (7.19) and occurs 'instantaneously'. The actual acoustic time delay for volume changes is $\mathcal{O}(L/C_r)$ and $\mathcal{O}(1/M)$ faster than the convective time scale. The derivation which allows for time evolution only on the convective time scale suppresses the fast acoustic waves. This is permissible as long as the heat addition is slow.

A different low Mach number limit is appropriate in a low speed flow when significant heat addition is absent. The departure from strictly incompressible flow are then at leading order isentropic, rather than the situation of dominant entropy fluctuations revealed in (7.20). Both of these limits were studied in recent studies of low-Mach number flows [169], and [10]. The nearly-isentropic low-Mach number limit can be found in compressible flow books (*e.g.* [81], [156]) and obtained by recognizing that in absence of strong heating, *i.e.*

$$\hat{Q} \equiv \frac{Q}{M^2} = \frac{(\gamma-1)Q^\star\tau}{\rho_r V^2} \ll 1,\ \rho = 1 + M^2\hat{\rho},\ p = \frac{1}{\gamma M^2} + \hat{p},$$

and

$$T = 1 + M^2\hat{T} \text{ as } M \to 0.$$

Defining $\hat{q}_i = -k^\star/k_r \frac{\partial \hat{T}}{\partial x_i}$ transforms (7.9)-(7.15) to:

$$\frac{\partial u_i}{\partial x_i} + M^2\{\frac{\partial \hat{\rho}}{\partial t} + \frac{\partial \hat{\rho} u_i}{\partial x_i}\} = 0 \qquad (7.21)$$

$$\frac{Du_i}{Dt} + \frac{\partial \hat{p}}{\partial x_i} - R^{-1}\frac{\partial \tau_{ij}}{\partial x_j} + M^2\{\hat{\rho}\frac{Du_i}{Dt}\} = 0 \qquad (7.22)$$

$$\frac{D\hat{T}}{Dt} - (\gamma-1)\frac{D\hat{p}}{Dt} - \frac{(\gamma-1)}{R}\Phi + \frac{1}{RPr}\frac{\partial \hat{q}_i}{\partial x_i} - \hat{Q} + M^2\{\hat{\rho}\frac{D\hat{T}}{Dt}\} = 0 \quad (7.23)$$

$$\gamma\frac{\partial u_i}{\partial x_i} - \frac{(\gamma-1)\Phi}{R} + \frac{1}{PrR}\frac{\partial \hat{q}_i}{\partial x_i} - \hat{Q} + M^2\{\frac{D\hat{p}}{Dt} + \gamma\hat{p}\frac{\partial u_i}{\partial x_i}\} = 0 \quad (7.24)$$

$$\frac{Ds}{Dt} - M^2\{\frac{(\gamma-1)}{R}\Phi - \frac{1}{PrR}\frac{\partial \hat{q}_i}{\partial x_i} + \hat{Q} - (\hat{\rho} + \hat{T} + M^2\hat{\rho}\hat{T})\frac{Ds}{Dt}\} = 0 \quad (7.25)$$

This rescaled set of equations reduces to the uniform-density incompressible equations at leading order in M. A systematic expansion of the dependent variables in powers of M can be used to obtain the departure from the incompressible limit. Equation (7.25) brings out the nearly-isentropic nature of this limit; if the Reynolds number is large and heating-effects are minor the thermodynamic fluctuations are, at leading order, isentropic. The fluid dilatation associated with such nearly-isentropic fluctuations is obtained from (7.24). Specifically, setting $u_i = v_i^{(0)} + M^2 v_i^{(1)} + \ldots$ and $\hat{p} = \hat{p}^{(0)} + M^2\hat{p}^{(1)} + \ldots$ gives: (at high R)

$$\frac{\partial v_i^{(0)}}{\partial x_i} = 0, \text{ and } \frac{\partial v_i^{(1)}}{\partial x_i} = -\frac{1}{\gamma}\frac{D_0\hat{p}^{(0)}}{Dt},$$

where $D_0/Dt = \partial/\partial t + v_i^{(0)}\partial/\partial x_i$. The latter equation gives fluid dilatation (correction) in terms of the rate of change of pressure along a material trajectory in the (leading order) incompressible flow. Ristorcelli [130] uses this 'diagnostic equation' for constructing models for dilatational covariances.

Acoustic Scaling

A different choice of scales is appropriate for acoustic waves. Denoting the length scale L, velocity scale C_r, time scale $\tau = L/C_r$, density scale ρ_r, pressure scale $\rho_r C_r^2$, where C_r is a reference speed of sound, temperature scale $\hat{T}_r \equiv C_r^2/C_{pr}$, internal energy, enthalpy and total energy scaled with $\hat{T}_r C_{pr}$ and entropy scale C_{pr} yields the following relations $h = T/(\gamma-1)$, $e = T/[\gamma(\gamma-1)]$, $e_t = e + u_i u_i/2$ and $p = \rho T/\gamma$. Define $\widetilde{Q} = Q^\star\tau/(\rho_r C_{pr} T_r)$ and the Reynolds number based on a velocity scale V (to be chosen later) as $R = \rho_r V L/\mu_r$. The resulting equations are:

$$\frac{\partial \rho}{\partial t} + \frac{\partial \rho u_i}{\partial x_i} = 0 \quad (7.26)$$

$$\rho\frac{Du_i}{Dt} = -\frac{\partial p}{\partial x_i} + \frac{M}{R}\frac{\partial \tau_{ij}}{\partial x_j} \quad (7.27)$$

$$\rho \frac{Dh}{Dt} = \frac{Dp}{Dt} + \frac{M}{R}\Phi - \frac{M}{RPr}\frac{\partial q_i}{\partial x_i} + \widetilde{Q} \tag{7.28}$$

where

$$D/Dt = \partial/\partial t + u_i \partial/\partial x_i, \quad \tau_{ij} = 2\mu^\star/\mu_r\{S_{ij} - 1/3\delta_{ij}S_{kk}\},$$

$$q_i = -\frac{k^\star}{k_r}\frac{\partial T}{\partial x_i}, \quad \Phi = \tau_{ij}S_{ij},$$

with $S_{ij} = 1/2\{\frac{\partial u_i}{\partial x_j} + \frac{\partial u_j}{\partial x_i}\}$. Other equivalent forms of energy balance are:

$$\rho \frac{De}{Dt} = -p\frac{\partial u_i}{\partial x_i} + \frac{M\Phi}{R} - \frac{M}{RPr}\frac{\partial q_i}{\partial x_i} + \widetilde{Q} \tag{7.29}$$

$$\rho \frac{De_t}{Dt} = -\frac{\partial p u_i}{\partial x_i} + \frac{M}{R}\frac{\partial \tau_{ji}u_i}{\partial x_j} - \frac{M}{RPr}\frac{\partial q_i}{\partial x_i} + \widetilde{Q}, \tag{7.30}$$

$$\frac{Dp}{Dt} + \gamma p \frac{\partial u_i}{\partial x_i} = \frac{M\Phi}{R} - \frac{M}{PrR}\frac{\partial q_i}{\partial x_i} + \widetilde{Q} \tag{7.31}$$

$$\rho T \frac{Ds}{Dt} = \frac{M\Phi}{R} - \frac{M}{PrR}\frac{\partial q_i}{\partial x_i} + \widetilde{Q} \tag{7.32}$$

This set of non-dimensional equations is often adopted as the starting point in developing numerical solutions to compressible flow problems. Numerical methods suitable for direct numerical simulation of compressible turbulent flows (to be outlined in section 7.4) are also based on this set of equations. However, before discussing the numerical issues it helps to consider simplified versions of this equation set. The simplified equations for various idealized steady compressible flows are well-known, *e.g.* the compressible potential flow equations and its simplification to small disturbance equations for subsonic, transonic and supersonic steady flows can be found in well-known text books ([93], [156], [137]). Treatment of steady compressible flows with vorticity can be found in [39] and [162]. Specific solutions for steady rotational compressible flows are less well-developed. This is, however, not a significant limitation for the present discussion. Turbulence inherently involves unsteady dynamics of vorticity and nonlinearity is one of it's hallmarks. This is why turbulence (incompressible or compressible) remains a major unsolved problem in nonlinear physics. Fortunately, flows in which the mean flow profiles are inviscidly unstable such as jets, wakes and mixing layers, some insights into the mechanics of large-scale eddies can be gained from the linearized theory of unsteady disturbances on a suitable unidirectional sheared flow. This analysis leads to a *modal decomposition* of the fluctuations in a compressible turbulent flow.

7.3.2. LINEARIZED EQUATIONS AND MODAL DECOMPOSITION

In the non-dimensional equations (7.26)-(7.8) (which used acoustic scaling) suppose that:

$$\rho = \overline{\rho} + \epsilon\rho', \; p = \overline{p} + \epsilon p', \; T = \overline{T} + \epsilon T', \; s = \overline{s} + \epsilon s', \; u_i = \overline{u}_i + \epsilon u_i',$$

where ϵ is a nondimensional measure of the disturbance amplitude about the reference flow state denoted by $\overline{\rho}$, $\overline{p}$, $\overline{T}$, $\overline{s}$, and $\overline{u}_i$. For simplicity also assume that the reference flow is steady and depends only on the coordinate transverse to the flow x_2, *i.e.* $\overline{\rho} = \overline{\rho}(x_2)$, $\overline{p} = constant = \overline{\rho}(x_2)\overline{T}(x_2)/\gamma$, $\overline{T} = \overline{T}(x_2)$, $\overline{s} = \overline{s}(x_2)$, and $\overline{u}_i = \{\overline{U}(x_2), 0, 0\}$. Note the Gibbs relation yields $\overline{s}(x_2) = -\log\overline{\rho}(x_2) - \frac{1}{\gamma}\log\gamma$. The equations governing the disturbances are (only the inviscid terms are shown for simplicity):

$$\frac{\overline{D}}{Dt}\rho' + \frac{d\overline{\rho}}{dx_2}u_2' + \overline{\rho}\frac{\partial u_i'}{\partial x_i} = -\epsilon R_\rho,$$

$$\overline{\rho}\{\frac{\overline{D}}{Dt}u_1' + \frac{d\overline{U}}{dx_2}u_2'\} + \frac{\partial p'}{\partial x_1} = -\epsilon R_1$$

$$\overline{\rho}\frac{\overline{D}}{Dt}u_2' + \frac{\partial p'}{\partial x_2} = -\epsilon R_2$$

$$\overline{\rho}\frac{\overline{D}}{Dt}u_3' + \frac{\partial p'}{\partial x_3} = -\epsilon R_3$$

$$\overline{\rho}\{\frac{\overline{D}}{Dt}h' + \frac{d\overline{h}}{dx_2}u_2'\} - \frac{\overline{D}p'}{Dt} = -\epsilon R_h,$$

where $\frac{\overline{D}}{Dt} = \frac{\partial}{\partial t} + \overline{U}\frac{\partial}{\partial x_1}$, is the material derivative following the reference flow. Terms denoted by R_ρ, R_1, R_2, R_3, and R_h represent nonlinear terms. They involve quadratic or cubic products of disturbance variables. Their precise form is easily obtained from the exact equations.

The auxiliary forms of the energy equation are

$$\frac{\overline{D}}{Dt}p' + \gamma\overline{p}\frac{\partial u_i'}{\partial x_i} = -\epsilon\{u_i'\frac{\partial p'}{\partial x_i} + \gamma p'\frac{\partial u_i'}{\partial x_i}\},$$

$$\frac{\overline{D}}{Dt}s' + \frac{d\overline{s}}{dx_2}u_2' = -\epsilon u_i'\frac{\partial s'}{\partial x_i} - \epsilon^2\frac{p'}{\overline{p}}u_i'\frac{\partial s'}{\partial x_i}.$$

The equivalence of these can be algebraically verified by noting the equation of state

$$p' = \frac{\gamma - 1}{\gamma}\{\rho'\overline{h} + \overline{\rho}h' + \epsilon(\rho'h' - \overline{\rho'h'})\},$$

or its equivalent forms

$$\frac{p'}{\overline{p}} = \frac{\rho'}{\overline{\rho}} + \frac{T'}{\overline{T}} + \epsilon\frac{(\rho'T' - \overline{\rho'T'})}{\overline{\rho}\overline{T}}; \; s' = \frac{p'}{\gamma\overline{p}} - \frac{\rho'}{\overline{\rho}} + \frac{\epsilon}{2}\{(\frac{\rho'}{\overline{\rho}})^2 - (\frac{p'}{\overline{p}})^2\} + \ldots$$

Dropping all terms explicitly involving ϵ (*i.e.* nonlinear terms) yields the *linearized disturbance equations.* The linearized equations are the basis of the modal decompostion discussed in the next section. Modal interactions in the form of nonlinear terms can then be regarded as 'sources' forcing the dynamics of individual modes. This is a generalized interpretation of the weak mode-interactions discussed by Chu and Kovasznay [23].

Modal decomposition

The linearized equations have the form:

$$\frac{\overline{D}}{Dt}\rho' + \frac{d\overline{\rho}}{dx_2}u_2' + \overline{\rho}\frac{\partial u_i'}{\partial x_i} = 0$$

$$\overline{\rho}\{\frac{\overline{D}}{Dt}u_1' + \frac{d\overline{U}}{dx_2}u_2'\} + \frac{\partial p'}{\partial x_1} = 0$$

$$\overline{\rho}\frac{\overline{D}}{Dt}u_2' + \frac{\partial p'}{\partial x_2} = 0$$

$$\overline{\rho}\frac{\overline{D}}{Dt}u_3' + \frac{\partial p'}{\partial x_3} = 0$$

$$\overline{\rho}\{\frac{\overline{D}}{Dt}h' + \frac{d\overline{h}}{dx_2}u_2'\} - \frac{\overline{D}p'}{Dt} - 0$$

$$\frac{\overline{D}}{Dt}p' + \gamma\overline{p}\frac{\partial u_i'}{\partial x_i} = 0$$

$$\frac{\overline{D}}{Dt}s' + \frac{d\overline{s}}{dx_2}u_2' = 0$$

$$p' = \frac{\gamma - 1}{\gamma}\{\rho'\overline{h} + \overline{\rho}h'\}$$

$$\frac{p'}{\overline{p}} = \frac{\rho'}{\overline{\rho}} + \frac{T'}{\overline{T}}; \; s' = \frac{p'}{\gamma\overline{p}} - \frac{\rho'}{\overline{\rho}}$$

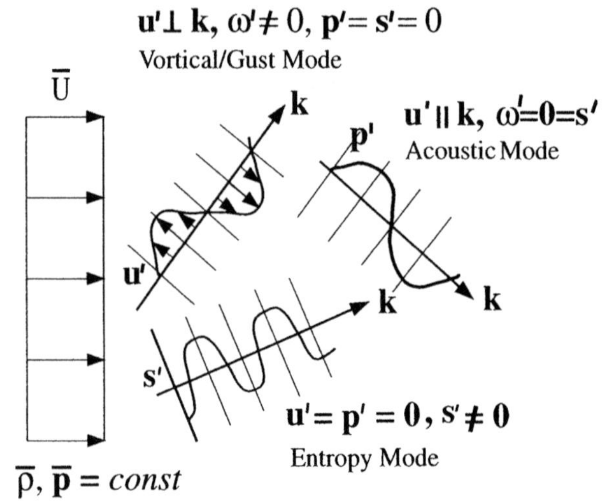

Figure 7.1. Modal decomposition of linearized disturbances in a uniform flow.

If the reference flow $\{\overline{U}(x_2), 0, 0\}$ is a uniform flow, *i.e.* $\overline{U}$ does not depend on x_2, the linearized equations can be decoupled into transverse and longitudinal modes. This can be observed by introducing the Helmholtz decomposition for the disturbance velocity in terms of its scalar potential ϕ, and vector potential $\vec{A}$, *i.e.* $u_i{}' = \nabla\phi + \nabla \times \vec{A}$. The vector potential is related to the vorticity $\nabla \times \nabla \times \vec{A} = -\nabla^2 \vec{A} = \nabla \times \vec{u'} = \vec{\omega'}$ and describes the transverse mode $\vec{u_\perp} = \nabla \times \vec{A}$, where the gauge condition $\nabla \circ \vec{A} = 0$ has been adopted. The scalar potential is related to the dilatation $\nabla^2\phi = \nabla \circ \vec{u'}$ and describes the longitudinal mode $\vec{u_\parallel} = \nabla\phi$. It is straightforward to show that the disturbance pressure p' is entirely associated with the longitudinal mode $\vec{u_\parallel}$ and obeys the stratified-medium wave equation

$$\frac{\overline{D}^2}{Dt^2}p' - \nabla \circ \{\overline{C}^2 \nabla p'\} = 0, \tag{7.33}$$

where $\overline{C}^2(x_2) = \gamma\overline{p}/\overline{\rho}$. The transverse or vortical mode is one-way coupled to the longitudinal mode due to baroclinic torques. However, if startification is absent, *i.e.* $\overline{C}^2 = constant$ (and $\overline{U} = constant$, as before) the transverse mode decouples completely. In this situation the linearized disturbance equations admit three decoupled modes ([81, p. 184], [80], [113]) : the vortical or transverse gust or shear wave mode $\vec{u}_\perp$ which corresponds to all of the vorticity, is solenoidal, has no pressure disturbance and convects frozen with the uniform reference flow; the acoustic or longitudinal mode $\vec{u_\parallel}$ has all of dilatation and no vorticity, accounts for all of the pressure disturbance and propagates as an acoustic wave with speed $\overline{C}$ relative to the reference flow; and the entropy mode corresponding to the nonuniform entropy disturbance in the medium, has no associated velocity or pressure, and convects frozen with the uniform reference flow. Both acoustic and entropic modes have associated disturbances in density and temperature. Hence pressure and entropy disturbances are more convenient variables for defining the modal decomposition. The schematic in Figure 7.1 depicts such a modal decomposition.

For a sheared base flow $\overline{U}(x_2)$ the equations given above can be reduced to a single third-order wave equation for the pressure disturbance:

$$Lp' = \frac{\overline{D}}{Dt}\{\frac{\overline{D}^2 p'}{Dt^2} - \nabla \circ (\overline{C}^2 \nabla p')\} + 2\overline{C}^2 \frac{d\overline{U}}{dx_2} \frac{\partial^2 p'}{\partial x_2 \partial x_1} = 0 \qquad (7.34)$$

In aeroacoustics literature this equation is often called the linearized Lilley's equation [57]. Substituting the normal mode form

$$p' = Real\{\hat{p}(x_2) e^{i(k_1 x_1 + k_3 x_3 - \omega t)}\}$$

yields a second order differential equation for the mode shape $\hat{p}(x_2)$

$$D[\frac{D\hat{p}}{M_r^2}] - [\frac{k^2 - k_1^2 M_r^2}{M_r^2}]\hat{p} = 0, \qquad (7.35)$$

where, $D \equiv \frac{d}{dx_2}$, $k^2 = k_1^2 + k_3^2$, and $M_r(x_2) \equiv \frac{\overline{U}(x_2) - \omega/k_1}{\overline{C}(x_2)}$ is the local Mach number of the base flow in a frame moving at the phase speed of the normal mode. Another equivalent form of this equation is

$$D^2\hat{p} - 2\{\frac{D\overline{U}}{\overline{U} - \omega/k_1} - \frac{D\overline{C}}{\overline{C}}\}D\hat{p} - (k^2 - k_1^2 M_r^2)\hat{p} = 0, \qquad (7.36)$$

which can be readily used to determine the asymptotic behavior of the pressure disturbances in the free-stream. In many studies the base flow and the disturbance equations are formulated in Howarth-transformed coordinates (see [135, p. 340]) which explains the apparent difference in the equations (*e.g.* [88] , [72]).

As discussed in section 7.3.1, a different choice of velocity and pressure scales are appropriate for flows at low Mach number. Rescaling the variables as $\hat{p} = M^2\hat{\tilde{p}}$, $\overline{U} = M\widetilde{U}$ and $\omega = M\tilde{\omega}$, transforms (7.35) with $M \to 0$ to

$$D[\frac{D\hat{\tilde{p}}}{\overline{\rho}(x_2)(\widetilde{U}(x_2) - \tilde{\omega}/k_1)^2}] - [\frac{k^2}{\overline{\rho}(x_2)(\widetilde{U}(x_2) - \tilde{\omega}/k_1)^2}]\hat{\tilde{p}} = 0, \qquad (7.37)$$

which is recognizable from incompressible hydrodynamic stability.

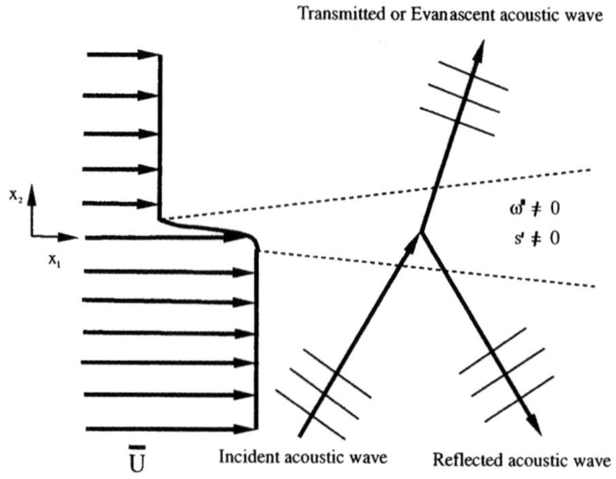

Figure 7.2. Linearized disturbances in a non-uniform flow.

An equation analogous to (7.34) can also be expressed in terms of the transverse velocity $u_2{}'$. In terms of the normal mode of the form

$$u_2{}' = Real\{\hat{v}(x_2)e^{i(k_1x_1+k_3x_3-\omega t)}\},$$

This equation is (*e.g.* Gropengiesser [61])

$$D\{\frac{(\overline{U}-\omega/k_1)D\hat{v}-D\overline{U}\hat{v}}{G}\}-\frac{(\overline{U}-\omega/k_1)}{\overline{C}^2}\hat{v}=0, \tag{7.38}$$

or equivalently,

$$D^2\hat{v}-\frac{DG}{G}D\hat{v}-\{\frac{D^2\overline{U}}{(\overline{U}-\omega/k_1)}-\frac{DG}{G}\frac{D\overline{U}}{(\overline{U}-\omega/k_1)}+(k^2-M_r^2k_1^2)\}\hat{v}=0, \tag{7.39}$$

where $G=\overline{C}^2(k^2-M_r^2k_1^2)$, and is the compressible analog of the Rayleigh equation of hydrodynamic stability.

Equations (7.35) and (7.38) apply to all types of linearized disturbances in a transversely sheared base flow regardless of whether they are predominantly acoustic, entropic or vortical in nature. It is possible to discriminate between these three basic types (or modes) of disturbances on the basis of their behavior in regions of uniform flow. In the non-uniform flow regions the modes defined in this way are not pure, *i.e.* the continuation of an acoustic 'mode' into a sheared region also involves a vortical disturbance and so on. This is schematically indicated in Figure 7.2. If the transverse domain is finite, a sequence of transverse (duct) modes results for the acoustic modes above their respective cut-on frequencies, with higher modes possessing progressively more nodes (zero-crossings) in their transverse mode shape. Vortical modes of a subsonic base flow are the inviscid instability eigenmodes, with discrete or continuous spectra. For supersonic base flows both vortical and mixed-vortical-acoustic instability modes are possible. The former have subsonic phase speed relative to the base flow at any point, while the latter have supersonic phase speed relative to that portion of the base flow where the mode is oscillatory (*e.g.* [98], [150], [151]). When the transverse domain is unbounded, no discrete acoustic modes are possible in a subsonic flow.

Plane acoustic waves of different orientations may be considered in the free-stream and their continuation into the non-uniform flow region gives a continuous spectrum of acoustic modes. Transmitted, reflected and evanascent acoustic wave components must be considered togather with the incident wave. Evidently an incident acoustic wave is associated with a vortical disturbance and an entropic disturbance in the non-uniform flow region. However, for a strictly parallel base flow the linearized disturbance equations are homogeneous in x_1 and x_3. As a result the wavenumbers k_1 and k_3 are constant and determined by the incident disturbance. For a plane incident acoustic wave of Figure 7.2 scattering into 'non-acoustic' disturbances is not possible in a parallel flow. However, if the base flow is non-parallel or if non-linearity is allowed such scattering becomes possible. Additionally, if the incident disturbance is not a plane wave, *e.g.* in a transient wave or a wave with spatially varying amplitude, non-acoustic response can be triggered. Goldstein [59] provides a generalization of the gust-mode of uniform flow to a general transversely sheared mean flow. A general classification of linearized unsteady disturbances into globally-defined modes is given by Goldstein [58] for arbitrary irrotational steady mean flows. His analysis highlights the linear-coupling between such modes in a non-uniform flow. Localized changes in geometry or boundary conditions also introduce coupling amongst the linear modes and one type of modal disturbance may scatter into another type of disturbance (see [57], [59] for many examples). In section 7.6.2 an example of linear mode coupling due to linearized shock-jump relations at a disturbed shock is discussed.

As stated earlier non-linearity introduces coupling between different modal disturbances of a base flow. In this way nonlinearity may modify the dynamics of a given mode, *e.g.* via coupling between instability modes of different frequencies. Nonlinearity also act as 'source' of an otherwise non-existent mode. The generation of aerodynamic noise due to nonlinear evolution of instability waves in a jet column is an example of this. Problems of aerodynamically generated sound are discussed further in section 7.7.

7.4. Numerical Methods

Several different types of numerical algorithms exist for solving the flow equations outlined in the previous section. However, for direct numerical simulations of turbulence it is paramount to choose algorithms which provide an accurate representation of the turbulent fluctuations over a wide range of temporal and spatial scales. Spectral methods ([21]) would be a natural choice but are difficult to extend to complex domains and cumbersome to use when non-periodic boundary conditions are required. However, the efficiency offered by spectral methods with periodic boundary conditions is unsurpassed and their use in homogeneous directions (*i.e.* allowing periodic boundary conditions) is strongly recommended. High order compact finite-difference schemes (Lele [91] provide good representation of a wide-band of spatial scales and have found applications in a broad range of turbulence simulations both compressible and incompressible.

7.4.1. BASIC DISCRETIZATION IN SPACE AND TIME

Consider the numerical evaluation of terms in the governing equations which involve spatial derivatives of the dependent variables. For example, the convective terms, the pressure gradient and the viscous terms have this form in the momentum equation. Consider first the discretization for the inviscid terms. The viscous terms and heat-conduction are treated slightly differently. It is notable that in (7.26)-(7.8) all inviscid terms involve a first order spatial derivatives. They can be numerically evaluated by applying the compact differentiation scheme for a first derivative. The basic idea is as follows.

The finite difference approximation f_i' to the first derivative $\frac{df}{dx}(x_i)$ at the node i on an equally spaced mesh with spacing h is evaluated by solving a tridiagonal system of the form (one equation for each interior node):

$$\alpha f_{i-1}' + f_i' + \alpha f_{i+1}' = b\frac{f_{i+2} - f_{i-2}}{4h} + a\frac{f_{i+1} - f_{i-1}}{2h}. \qquad (7.40)$$

The relations between the coefficients a,b and α are derived by matching the Taylor series coefficients of various order. With

$$a = \frac{2}{3}(\alpha + 2), \qquad b = \frac{1}{3}(4\alpha - 1), \qquad (7.41)$$

a family of fourth order schemes is obtained. It may be noted that as $\alpha \to 0$ this family merges into the well known fourth order central difference scheme. Similarly for $\alpha = \frac{1}{4}$ the classical Pade' scheme is recovered. Furthermore for $\alpha = \frac{1}{3}$ the leading-order truncation error coefficient vanishes and the scheme is formally sixth order accurate. Most of the simulations illustrated here used this sixth order scheme. More general schemes of

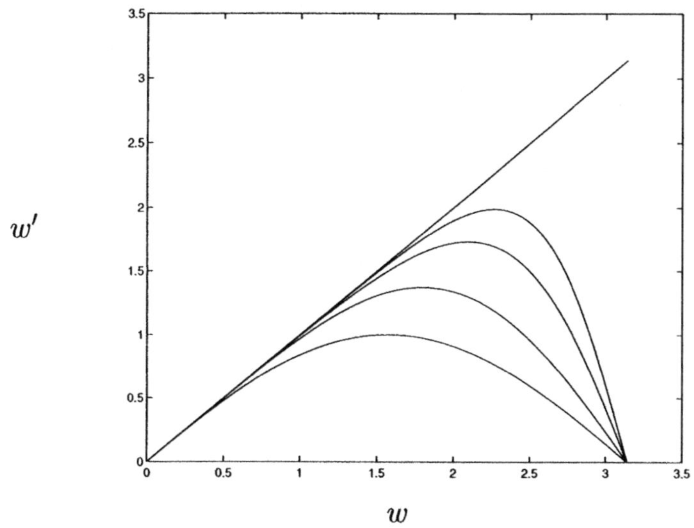

Figure 7.3. Modified wavenumber for different first derivative schemes. The plotted curves are in the sequence of second-order central difference, fourth-order central difference, fourth-order compact, sixth-order compact, exact differentiation.

this type were presented in [91] along with an analysis of the discretization errors. It is straightforward to show that a sinusoidal function over domain $0 \leq x \leq L$ with wavenumber $w = 2\pi k/L$, defined by $f(x) = \hat{f}_k \exp(iwx)$, where $i = \sqrt{-1}$, is differentiated by the scheme in equation (7.40) to yield $f'_n(x) = iw'\hat{f}_k \exp(iwx)$, where

$$w'(w) = \frac{a\sin(w) + \frac{b}{2}\sin(2w)}{1 + 2\alpha\cos(w)}.$$

Evidently, the departure of the modified wavenumber $w'(w)$ from w is a measure of how accurately a numerical scheme differentiates this sinusoidal wave. Varying the wavenumber w of the test function over the range $1 \leq k \leq N/2$ (*i.e.* $2\pi/L \leq w \leq \pi N/L$) which can be represented on a uniformly spaced mesh of N (independent) points with mesh spacing $h = L/N$, provides full information on how well this scheme treats the full range of representable wavenumbers. As depicted in Figure 7.3 the compact schemes provide a more accurate representation for a wide range of waves than the traditional central difference schemes. The straight line on the figure represents a spectrally accurate scheme. The improved representation of the shorter spatial scales by the compact scheme is evident in the figure. Further, the modified wavenumber is purely real-valued and thus the discretization of the inviscid terms, by itself, does not introduce any numerical dissipation. The modified wavenumber plot characterizates a numerical scheme over the full range of wavenumbers and hence is much more informative than the classical notion of trunction error. As noted in [91] differentiation schemes can be optimized on the basis of the modified wavenumber $w'(w)$ to achieve schemes with better broadband resolution than is achieved by minimizing the truncation error.

If the dependent variables are periodic then the system of relations (7.40) written for each node can be solved togather as a linear system of equations. The general non-periodic case requires additional relations appropriate for the near-boundary nodes. Generally, schemes with asymmetric stencils are needed for the boundary node. A third order scheme

$$f_1' + 2f_2' = \frac{1}{2h}(-5f_1 + 4f_2 + f_3), \tag{7.42}$$

at the boundary node and the fourth order compact scheme ($\alpha = \frac{1}{4}$ in equation 7.40) at the next to the boundary node have been used to complete the system of equations necessary to solve for the unknown derivative values at the node points. A mathematically defined mapping between a non-uniform physical mesh and a uniform computational mesh provides derivatives on a non-uniform mesh. The basic method outlined above can be used directly on the uniformly spaced computational mesh.

The viscous and diffusive terms are expanded to express them as a sum involving the Laplacian and other terms involving products of spatial gradients. For example, in (7.27)

$$\frac{\partial \tau_{ij}}{\partial x_j} = \mu \frac{\partial^2 u_i}{\partial x_j \partial x_j} + \frac{\mu}{3} \frac{\partial}{\partial x_i} \frac{\partial u_j}{\partial x_j} + 2 \frac{d\mu}{dT} \frac{\partial T}{\partial x_j} \{S_{ij} - \frac{1}{3}\delta_{ij} S_{kk}\}$$

The viscous Laplacian is then evaluated by directly applying the compact scheme to the second derivative. This is numerically superior than two successive applications of the numerical scheme for the first derivative. A similar remark also applies to the heat conduction term in (7.28)-(7.32).

The second derivatives are evaluated by solving a system similar to (7.40), viz.

$$\alpha f_{i-1}'' + f_i'' + \alpha f_{i+1}'' = b \frac{f_{i+2} - 2f_i + f_{i-2}}{4h^2} + a \frac{f_{i+1} - 2f_i + f_{i-1}}{h^2}, \tag{7.43}$$

where f_i'' represents the finite difference approximation to the second derivative at node i. With

$$a = \frac{4}{3}(1 - \alpha), \; b = \frac{1}{3}(-1 + 10\alpha), \tag{7.44}$$

a one parameter family of fourth order schemes is obtained. Again as $\alpha \to 0$ this family coincides with the well known fourth order central difference

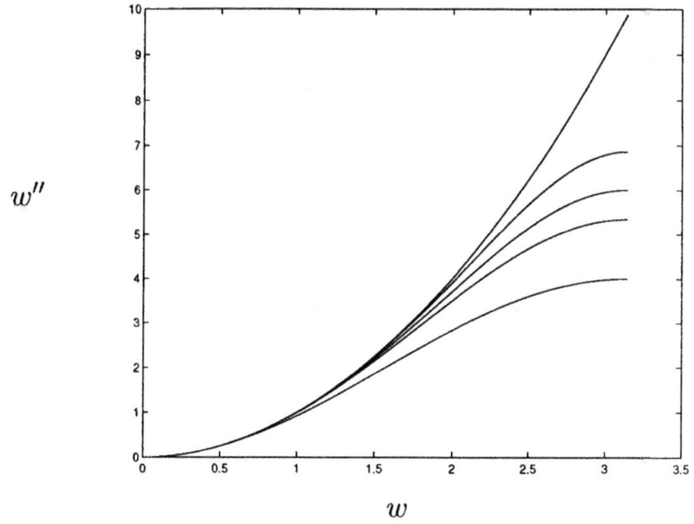

Figure 7.4. Modified wavenumber for different second derivative schemes. Curves are plotted in the sequence of second-order central difference, fourth-order central difference, fourth-order compact, sixth-order compact, exact differentiation.

scheme. For $\alpha = \frac{1}{10}$ the classical Pade' scheme is recovered. For $\alpha = \frac{2}{11}$ a sixth order tridiagonal scheme is obtained. This scheme with

$$\alpha = \frac{2}{11},\ a = \frac{12}{11},\ b = \frac{3}{11}$$

is used in the simulations illustrated here. It may also be noted that the schemes (7.43) provide an accurate evaluation of the second derivative over a wide range of length scales. The error associated with the second derivative evaluation for a variety of schemes is shown on Figure 7.4. The spectrally accurate evaluation is the parabola on the figure. The improvement of the present scheme in representing the shorter scales is again evident.

As with the first derivative scheme, the system of equations represented by (7.43), one for each interior node, needs to be appended with second derivative schemes appropriate for near boundary nodes. Relatively simple schemes can be adopted for this unless the viscous effects are dominant near the computational boundary (as they would be for a wall-bounded viscous flow). The following approximation is used for the boundary node

$$f_1'' + 11 f_2'' = \frac{1}{h^2}(13 f_1 - 27 f_2 + 15 f_3 - f_4). \tag{7.45}$$

and the interior scheme (7.43) with $\alpha = \frac{1}{10}$ is used at the next-to-boundary node. As noted in [91] boundary schemes such as (7.42) and (7.45) have trunction errors which are a factor of 10 lower than the corresponding explicit schemes on the same stencil.

Second derivatives can be evaluated on a nonuniformly spaced grid in the physical domain by employing an analytical mapping to a uniformly spaced computational domain. This process also requires the computation of the first derivative since for $f(x(s))$ it follows that

$$\frac{d^2 f}{dx^2} = \{\frac{d^2 f}{ds^2} - \frac{d^2 x}{ds^2}\frac{df}{dx}\}(\frac{dx}{ds})^{-2},$$

where s is the computational coordinate with a uniform grid spacing, and $x(s)$ is the analytically defined non-uniform physical coordinate. However the requisite first derivative values can be saved and reused when needed for other terms.

The spatial derivative schemes given in this section are easily combined with explicit time advancement schemes. A compact storage version of the third order Runge-Kutta scheme due to A. Wray [167] or the fourth-order Runge-Kutta scheme are popular choices. Explicit schemes place the usual CFL restriction on the time step for ensuring numerical stability. These stability limits for linear problems can obtained by standard methods, see [91]. Due to the improved wideband resolution of the compact schemes this CFL restriction is more severe than for a second order scheme. However, since DNS studies require time-accurate unsteady flows the time step must also be chosen to maintain sufficient temporal accuracy. Carpenter *et al.* [22] give a comprehensive assessment of the numerical stability properties of high-order boundary treatments for compact finite difference schemes, and provide stable boundary schemes of order higher than those discussed in this section. For example, fifth order boundary schemes which are stable in combination with the sixth-order compact scheme (7.41) with $\alpha = \frac{1}{3}$ are derived in their paper.

7.4.2. BOUNDARY CONDITIONS

Careful attention must be given to the physical and numerical boundary conditions before a reliable simulation of some chosen compressible flow can be achieved. Often the boundary conditions used constitute the weakest link in the simulation methodology. This is not because of a lack of appreciation for the critical importance of the boundary conditions, but simply because the boundary conditions used represent a *modeling compromise* often arrived at by limited numerical experimentation. Boundary conditions applied at a location where substantial unsteady, nonlinear, nonuniform flow exists represent an attempt to 'model' the influence of the flow in the exterior region on the flow being simulated. Lacking precise mathematical information about the flow in the exterior problem, it becomes essential to seek numerical procedures which mimic the physical effects of this exterior flow.

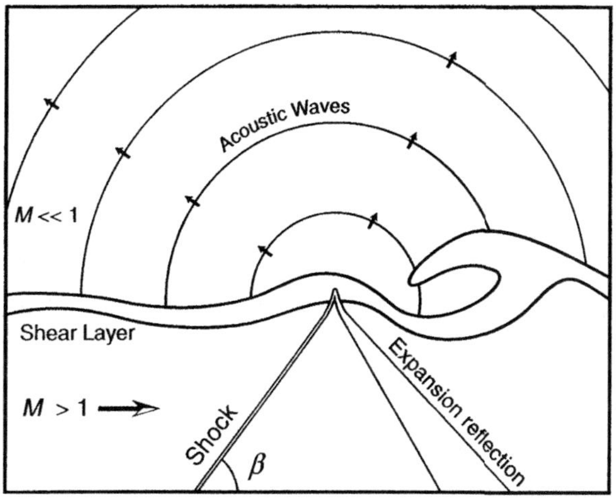

Figure 7.5. Schematic of shock-vortex interaction problem relevant to jet screech. Adapted from Manning and Lele [104].

To highlight the physical, mathematical and numerical difficulties consider the schematic in Figure 7.5 from Manning and Lele [104]. At the left-edge of the physical domain a mixing layer divides the incoming flow into supersonic and subsonic regions (zero co-flow in the present case). A similar division applies at the right-edge of the physical domain. The bottom-edge of the physical domain lies in an entirely supersonic flow while the top-edge is within a subsonic ambient region. The model problem studied by Manning and Lele focuses on the sound generated by the interaction of instability waves of a supersonic shear layer with standing shock-cells. It is highly desireable to eliminate any feedback between the upstream-travelling acoustic waves and the shear-layer instability waves. This translates into the following desired modeling attributes of the left-edge inflow boundary: prescribe the desired shear layer mean flow, prescribe the desired unsteady instability wave perturbation, minimize the reflection of upstream-travelling acoustic disturbances into other modes (acoustic and instability wave). Similarly the downstream right-edge should allow the unsteady flow approaching that boundary to quietly exit the domain. More specifically large amplitude vortical disturbances need to exit the domain without significant reflection into upstream propagating acoustic waves or other spurious numerical waves. The right-edge must also be sufficiently 'anechoic' to the incident acoustic disturbances and the same is required of the top-edge and bottom-edge boundaries as well. The bottom-edge boundary must also allow a 'controlled imposition' of an incident standing shock and 'absorb' the reflected wave resulting from the interaction of the incident standing wave and the shear layer.

This long list of challenging requirements for numerical/boundary conditions might seem too ambitious. In fact, numerical algorithms based strictly on a mathematical treatment of wave propagation in a *uniform* flow fail to achieve these target requirements. As shown by Colonius *et al.* [29] the non-uniform flow of a shear layer sets up significant reflection of the vortical disturbances into upstream-travelling acoustic disturbances when boundary conditions are based on linearized equations about a uniform flow. More flexibility is required to reduce the various undesired cross-couplings or reflection phenomena. One method to gain such flexibility is the notion of boundary zones, buffer regions/domains, sponge regions or absorbing layers which are appended to the physical domain being simulated. Within such boundary zones a variety of numerical algorithms can be applied to *simulate* the desired physical attributes of the boundary (*i.e.* the exterior domain which is not being simulated). Actual boundary conditions are applied only at the end of the boundary zones.

A variety of methods have been proposed as boundary conditions, and although some of these have a rigorous mathematical basis, they perform poorly when applied without boundary zones to non-uniform flow problems such as in Figure 7.5. Included in this class are characteristic based approaches [157], [120] which are based on local one-dimensional characteristic decomposition of the flow variables. They give good results when the waves incident on the boundary are nearly normal to it. When the mean flow near the computational boundary is nearly uniform, radiation boundary conditions which are based on the asymptotic solutions to the exterior flow problem [46], [54], [11] have been derived. Similar asymptotic boundary conditions based on the discrete approximations have been developed by Tam and Webb [153]. The so-called DtN boundary conditions [55], [56], which in their nonlocal form are exact for linear time-harmonic problems, are also based on modeling the external flow as a uniform medium. Absorbing boundary conditions [66], [79] are based on the idea of absorbing the incident disturbances at the boundary.

A variety of methods are also available for simulating boundary zones. Absorbing sponges or Exit zones [29] are based upon the notion of progressively attenuating the outgoing disturbance in the 'sponge' region before it encounters the computational boundary. Methods of attenuation include filtering on a non-uniform stretched mesh [29] , stretched mesh with upwinding or artificial dissipation [123], and the addition of 'cooling-terms' without grid stretching [71]. The recent development [68] of boundary conditions based on 'perfectly matched layer' (PML) is an elegant extension of the absorbing sponge idea. Another method [147], [145] called 'buffer-zone technique' is based on adding new terms to the equations in the buffer-zone so that the local flow is gradually made 'supersonic', *i.e.* all charateristics are downwind directed. Freund [51] presents a method based on smoothly modifying the governing equations within the boundary zone by changing the inviscid characteristics and addition of damping terms. Freund's method was used by Manning and Lele [104] for the problem in Figure 7.5.

Specifying a known unsteady inflow while maintaining non-reflection of outgoing disturbances has also received significant attention recently. Several methods for prescribing inflow disturbances are evaluated in the Proceedings of the second Computational Aeroacoustics (CAA) workshop on Benchmark Problems, Tam and Hardin editors, 1997, NASA Conference Publication 3352 (hereafter referred as CAA-2) . Splitting of flow variables into a part which is specified and other parts which are obtained with a numerical procedure is a common theme. The latter can be solved for using local one-dimensional characteristic decomposition [120], with non-reflecting treatment [54], absorbed via sponges [28] , absorbed via PML (Hu in CAA-2), or treated via a buffer-zone (Nark in CAA-2). For special problems for which analytic representation of the outgoing disturbances in terms of duct modes or other eigen-modes is available, these can also be directly exploited (Tam in CAA-2).

Some numerical aspects of boundary conditions are revisited in section 7.4.3, where specific attention is drawn to its numerical artifacts which must be controlled in a successful simulation.

Initial Conditions

Initial conditions suitable for the physical problem at hand are necessary even if only statistical information is being sought from a DNS study. A consistent specification of all field variables is needed. For example, a compressible axisymmetric vortex has significant density and pressure variations with distance from its center. If these variations are not consistent with the velocity distribution in the vortex, strong transient acoustic waves are generated as the simulation is started. For simple flows a choice of initial conditions consistent with the governing equations and boundary conditions can often be made (*e.g.* by making appropriate use of a suitable image system) but this is more difficult for a complex flow. The solution to a related but approximate model can be used to construct the initial conditions. For example, the solution to a related incompressible flow, or the solution to a suitable potential flow/linearized flow can be used in this way. Computational fields obtained from a related flow also serve well. For example, if it desired to simulate the interaction of a compressible vortex-pair with an airfoil, it helps to first simulate just this vortex-pair by itself and obtain a consistent field comprising of this dipole by letting the transient waves to propagate out of the desired region. This simulation data can be subsequently used in the more complex flow involving vortex-dipole airfoil interaction.

For a compressible turbulent flow, one method of specifying initial conditions is to use a modal superposition (based on the decomposition discussed in section 7.3.2) with the amplitudes scaled according to the desired spectra and phases chosen randomly. Blaisdell *et al.* [12] studied the evolution of compressible isotropic turbulence and compressible homogeneous sheared turbulence using this method for specifying the initial conditions. Ristorcelli and Blaisdell [131] have emphasized that a low Mach number expansion (section 7.3.1) can be exploited in the construction of consistent initial conditions, and shown its effectiveness in reducing the undesired transients.

Inflow and Outflow of Turbulence

Simulations of turbulent flows in open domains require that a realistic turbulent flow be provided at the inlet to the computational domain. If the inlet turbulence is not realistic a significant part of the flow domain may be occupied by the adjustment processes which attempts to bring the specified inflow data towards a more realistic turbulent state. Lee *et al.* [84] synthesized the inflow data using Fourier modes scaled to obtain desired spectral properties but with random phases. Taylor's hypothesis was used to transform the streamwise wavenumber into frequency. Since the inflow lacked the phase correlations appropriate for large-scale turbulent eddies a significant adjustment length was required. A similar inefficiency was faced by [83] in their simulations of a backward-facing step flow.

This adjustment region can be reduced if the inflow data is specified using slices from a precomputed realization of the turbulent flow expected at the inflow boundary. For simple inflow states, such as homogeneous isotropic turbulence, this data can be provided from a temporal simulation. This method was demonstrated by Mahesh *et al.* [102]. Na and Moin [114] used Spalart's boundary layer DNS [143] to generate inflow data. Both studies introduced a small amplitude jitter into the precomputed DNS slices to ensure that inflow data is not time-periodic. Lund *et al.* [97] have used scaling approximations for spatial growth [143] to generate inflow data. This scheme has been used to specify turbulent inflow data in a planar diffuser flow [74] , and in the flow over a trailing-edge of an airfoil [166]. Slices from a temporal DNS with amplitude jitter have been used by Freund *et al.* [53] in simulations of a supersonic jet.

Shock Waves

Compact schemes described earlier yield significant Gibbs oscillations around discontinuities if applied directly. If the internal structure of the physical discontinuity is well-resolved then, as expected, no spurious oscillations appear. However, as discussed in detail by Mahesh *et al.* [102] the approach of resolving the internal structure is feasible only for relatively weak shocks. For stronger shock waves it is necessary that differentiation across the shock front be avoided. Lee *et al.* [87] and Mahesh *et al.* [103] used a combination of high-order (typically sixth-order) Essentially Non-Oscillatory schemes (ENO) developed by Harten and Osher [65], Shu and Osher [139] and Shu [138] in a prescribed region around the shock along with the standard compact scheme of section 7.4.1 elsewhere. In these applications it was necessary to ensure that the ENO-region was wide enough to contain the distorted shock-front and the grid spacing near the shock was small enough to accurately resolve the shock motion. With grid refinement near the shock the spurious entropy-fluctuations generated by the numerical scheme (see [105] for illustrative examples) were less than 0.02% while the level of physical fluctuations incident on the shock was typically 2-3%. These errors were regarded as small enough to not interfere with the physical phenomena being studied. There is, however, a need for further improvements in such hybrid-schemes.

Variants of compact schemes have been proposed to deal with shock waves and other unresolved phenomena without compromising their wide-band resolution in smooth regions of the flow. Adams and Shariff [1] introduce upwinded compact schemes with upwinding applied to the inviscid-fluxes using a characteristics-based flux-splitting. In regions containing the shocks they couple the upwinded compact scheme to an ENO scheme. Non-linear compact schemes have been proposed by Cockburn and Shu [26], Deng and Maekawa [40] and by others as well. The development of high accuracy schemes capable of simultaneously simulating shock-dynamics and the vortex-dynamics of turbulence remains an active area of research.

Density Interfaces

Flows with significant density (entropy) non-uniformity may evolve regions of steep density (entropy) gradients. The large-scale stirring associated with energy-containing scales of turbulence moves fluid parcels with different initial entropy into close proximity and local straining further amplifies these gradients. Molecular diffusion processes (heat-conduction, species-diffusion) act to limit the sharpness of the gradient. In DNS the sharpness of these gradients must be adequately resolved otherwise spurious oscillations akin to Gibbs phenomena result. An order of magnitude estimate of the peak gradients in density and entropy can be obtained using the equations developed earlier. For illustration consider the so-called braid region between two large-scale eddies in a mixing layer, or the 'interface' between the hot near-wall fluid and the colder free-stream in a supersonic adiabatic-wall boundary layer [62]. Although some pressure disturbance would be invariable present at such interfaces, a useful estimate of the interface sharpness can be obtained by neglecting the pressure perturbation. With this approximation the linearized relations of section 7.3.2 give:

$$s' = -\frac{\rho'}{\bar{\rho}} = \frac{T'}{\bar{T}}, \tag{7.46}$$

which shows that entropy and density perturbations are negatively correlated, and entropy and temperature perturbations are positively correlated. This equation is quantitatively accurate for small-amplitude perturbations. We will use it for order of magnitude estimates.

When two fluid parcels of different temperature (entropy) are brought close togather an appreciable heat flow arises between them. A balance between the convective heat transport due to the straining motions and heat conduction determines the thickness, δ, of the interface between them. Evidently, $\delta \sim \sqrt{\kappa/S}$, where κ is the thermal diffusivity and S is the large-scale strain rate. Replacing S by $\Delta U/L$ gives

$$\delta/L \sim Pr^{-\frac{1}{2}} Re_L^{-\frac{1}{2}}. \tag{7.47}$$

This region of intense heat-exchange between two fluid parcels also induces volume changes (thermal expansion) in the flow. These can be estimated using the energy equation (7.31). In accordance with the assumptions made earlier the $\frac{Dp}{Dt}$ term, viscous dissipation and external heat addition are neglected in this energy equation. The outcome is

$$\gamma\bar{p}\frac{\partial u_i}{\partial x_i} = -\frac{M}{Pr R}\frac{\partial q_i}{\partial x_i}, \tag{7.48}$$

thus the fluid elements close to the interface on the hotter side shrink their volume while those on the cooler side thermally expand. This process prevents further sharpening of the density interface, and is readily observed in compressible flow DNS. Figure 7.6 is taken from [90] and depicts such a region in a mixing layer. The numerical consequences of such regions must be considered in DNS studies. Although the mass conservation equation (7.26) is devoid of any explicit diffusion terms, the molecular processes of heat and species diffusion are critical in determining the sharpness of density interfaces and hence the numerical resolution needed. Equation (7.47) provides an estimate of interface thickness. It should be stressed that such density/entropy interfaces arise when ever two fluid lumps of different entropy are brought togather. Sharp gradients of density are not confined to the edges of a turbulent region, or the near-wall region in a wall-bounded flow.

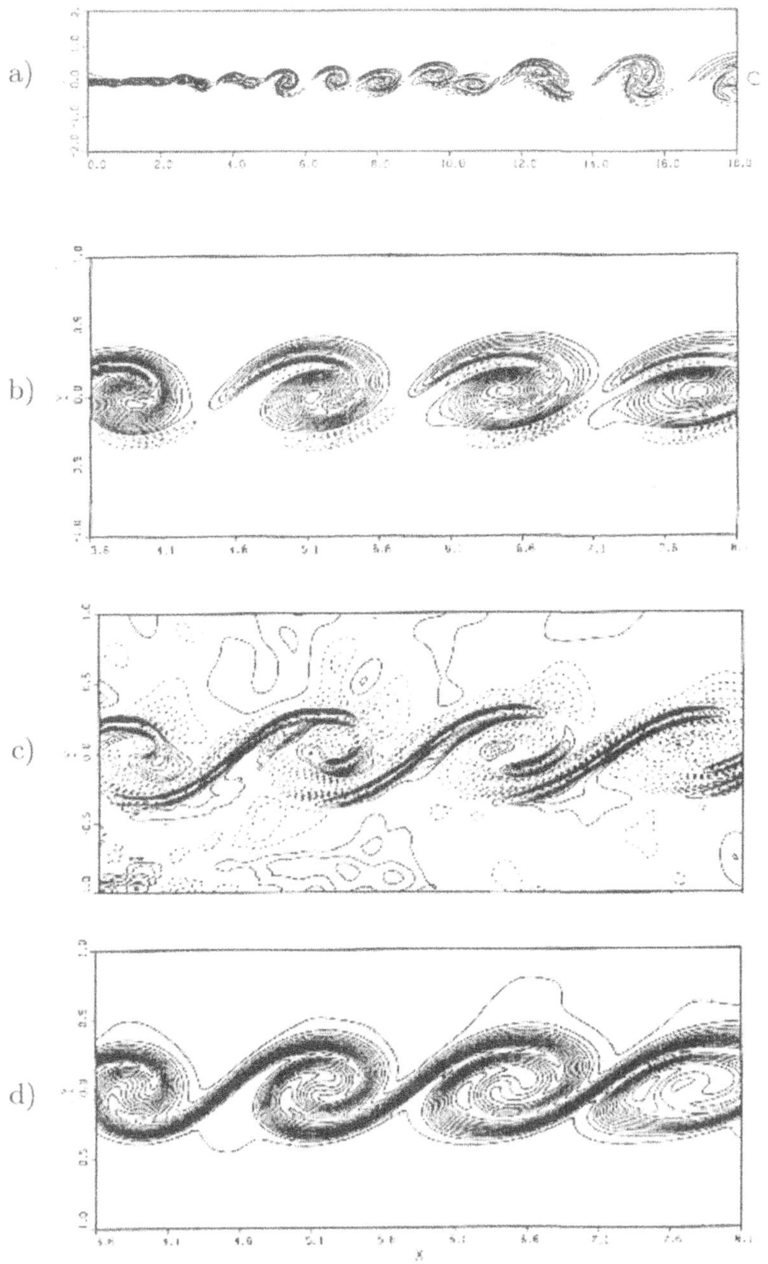

Figure 7.6. A two-dimensional spatially developing shear layer with $M_1 = 1.5$, $M_2 = 1.5$, $U_2/U_1 = 0.6$. Adapted from Lele [90]. a) Contours of vorticity from a simulation with two vortex pairings; b) Expanded view of vorticity contours from a simulation with single frequency forcing; c) Contours of temperature in a region same as figure 6b; d) Contours of dilatation in the same region.

A scaling relation for interface sharpness can also be obtained directed from the mass conservation equation (7.26). Taking a spatial gradient this equation yields:

$$\frac{D}{Dt}\frac{\partial \rho}{\partial x_i} + \frac{\partial u_j}{\partial x_i}\frac{\partial \rho}{\partial x_j} + \theta \frac{\partial \rho}{\partial x_i} + \rho \frac{\partial \theta}{\partial x_i} = 0, \tag{7.49}$$

where $\theta = \frac{\partial u_j}{\partial x_j}$. The second term in this equation represents the steepening of the density gradient due to straining/non-uniform advection. Evidently this effect can be balanced (on both sides of the interface) only by the last term, *i.e.* differential volume expansion. Taking the density change across the interface to be $\Delta\rho$, the large-scale strain rate to be S, dilatation scale to be Θ and interface thickness to be δ gives

$$S\Delta\rho/\delta \sim \overline{\rho}\Theta/\delta \quad or \quad S\Delta\rho/\overline{\rho} \sim \Theta, \tag{7.50}$$

where it was assumed that dilatation is also concentrated in a region of scale δ. To see that this estimate is consistent with (7.48), we note that $\frac{\partial q_i}{\partial x_i} \sim \Delta T/T_r(L/\delta)^2$ for a temperature difference ΔT across the interface, so it follows that $\frac{\partial u_i}{\partial x_i} \sim M\Delta T/T_r$, which is same as (7.50) in view of the acoustic scaling of (7.26)-(7.8).

Pressure perturbations were neglected in the estimates of interface sharpness derived here. This is *not* a low-Mach number assumption. In fact DNS data from supersonic boundary layers by Guarini [62] and channel flow [27], and experimental data from supersonic boundary layers (reviewed by Spina *et al.* [142]) also support this assumption. However, there are other more complex situations involving density interfaces, *e.g.* shock-shock interactions, which must be dealt with separately.

7.4.3. ARTIFACTS OF NUMERICAL DISCRETIZATION

The very act of discretization (unavoidable in a numerical simulation) brings with it many numerical artifacts. The degree to which these artifacts are displayed in a numerical simulation depends upon the numerical schemes and boundary conditions (procedures), and how well the physical phenomena occuring in the simulation are resolved. It is important to understand the origin of the numerical artifacts, so that a simulation can be designed which reduces the adverse impact of these artifacts.

First we consider those artifacts which are intrinsic to the basic numerical method without invoking any boundary conditions, or more precisely for problems posed with spatially periodic boundary conditions. It is well known that the discretized form of the acoustic wave equation cannot *exactly* represent the dispersion relation of the acoustic waves. This is true even for the more elementary advection equation in a single dependent variable. Numerical discretization (in space and in time) can be regarded as converting the original non-dispersive system into a dispersive discretized system, *e.g.* Vichnevetsky and Bowles [161, chapters 6, 7]. This discrete system exhibits wave phenomena of two kinds: a) long wavelength components which approach the solution of the original partial differential equation (PDE) as the grid is refined and, b) short wavelength spurious waves, also called parasites [159], which have no counterpart in the PDE. The discrete dispersion relation for the long wavelength component is an approximation of the exact dispersion relation and becomes more and more accurate as the mesh is refined. The numerical method used in a simulation (*e.g.* the choice of spatial and temporal discretization scheme) determines the discrete dispersion relation. This presents an opportunity to choose numerical methods whose discrete dispersion relation is closer to the exact dispersion relation for a given mesh (*i.e.* given spatial mesh size, *e.g.* Δx, and time step Δt). First- and second-order finite difference or finite element schemes perform poorly in this. Schemes with larger stencils are conventionally chosen to give higher order formal accuracy (*i.e.* lower truncation error) which improves the error for the best resolved wave component. An alternative better suited for problems with a wideband spectrum of wave components is to choose the coefficients of the scheme to represent more accurately the wave components over a wider range of wavenumbers. Such procedures lead to compact finite difference schemes (described in section 7.4.1), and to the DRP schemes introduced by Tam and Webb [152].

The analysis of dispersion errors for a time-harmonic problem (in frequency domain) gives the wavenumbers supported by the numerical scheme [161, chapter 7], and [31, for compact scheme]. For each given frequency two (or more, depending on the stencil) wavenumbers result. The long wave components, as discussed above, are an approximation to the exact dispersion relation, and have a group velocity which is a function of the frequency. The error in the group velocity is small for low frequencies (long wavelengths, which are computationally represented by many mesh points) and increases with frequency. The second (and other) wavenumber(s) correspond to spurious error waves which are supported by the numerical method. These have a short wavelength (*i.e.* involve oscillations with wavelengths only a few times the mesh spacing) and possess a group velocity with a wrong sign! If the physical disturbance travels from left to right, the spurious error wave travels from right to left.

It is important to know the types of spurious waves which are supported by a chosen numerical method, and processes which excite the spurious waves. In absence of other effects discussed below, the degree to which the wave phenomena are resolved on a given mesh controls the extent to which

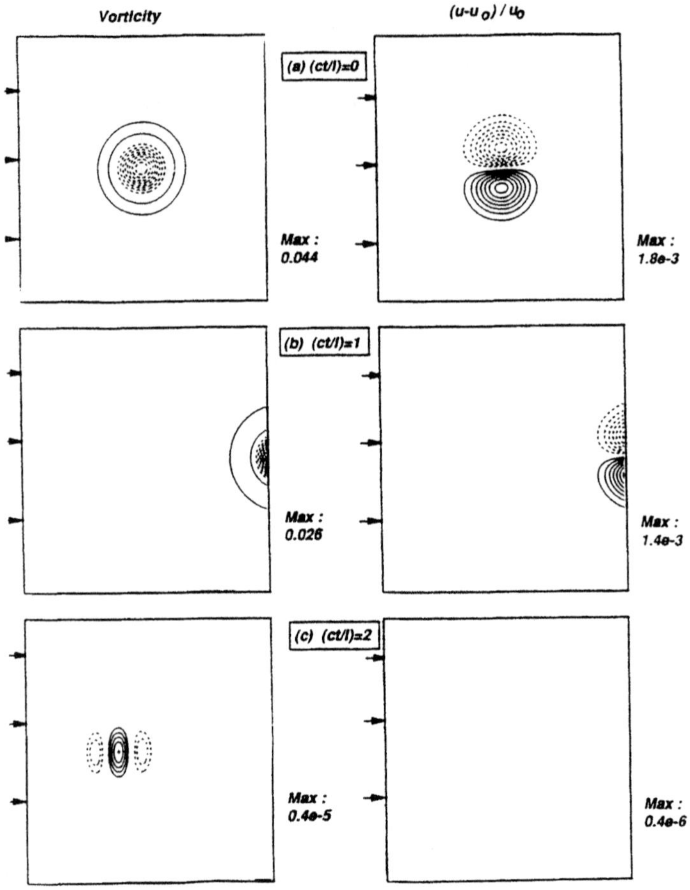

Figure 7.7. Example of spurious numerical feedback. Adapted from Poinsot and Lele [120]. The test shows that as a vortex passes out of the computational domain, it generates small amplitude short wavelength spurious waves which interact with the inflow boundary to create spurious moderate wavelength vortical disturbances.

the spurious waves are excited. For the same mesh, a steeper wave profile generates more spurious oscillations than a less steep wave profile. This is directly linked to the Fourier coefficients of the initial wave profile, which for a steeper wave are greater at higher wavenumbers. Similarly, a sudden temporal change in the boundary conditions, or inconsistent initial conditions also generate spurious waves. Spurious waves are excited by rapid grid stretching (fine to coarse or vice-versa), and by changes in the differencing scheme (*e.g.* across a multi-block interface, or at a computational boundary). It should be stressed that boundary conditions can also 'disguise' the spurious waves into the long wavelength physical waves.

This is readily demonstrated by solving a one-dimensional advection equation, with a compact initial disturbance. Suppose that the physical solution is a right going disturbance (which preserves its shape). Numerically this is roughly what is observed (with some dispersion errors) up to a time before the disturbance has reached the 'outflow' boundary. Upon reaching the outflow boundary a spurious left-going disturbance is excited (often at small amplitude). When the packet of spurious waves reaches the left boundary it is disguised into a right going long wave disturbance. For the advection equation this inter-conversion between the physical and spurious waves is relatively harmless and eventually the system returns to its undisturbed state. However, for systems with convective instabilities this inter-conversion provides an artificial feedback between the outflow and inflow, and if not sufficiently controlled sets up 'self-excited' oscillations in the system, and its original convectively unstable behavior is completely masked by the spurious numerical feedback [120], [29]. Figure 7.7 shows an example of spurious feedback from [120].

Spurious waves of another form are associated with the imposition of the inviscid no-penetration boundary condition at a hard wall. Tam and Dong [154] consider the oblique wave reflection problem at a hard wall. Their analysis shows that the numerical scheme now permits two types of spurious waves: a freely propagating spurious wave and a trapped spurious wave, trapped near the hard wall. An optimized boundary treatment using ghost points is presented which reduces these spurious waves. Other methods which also reduce the boundary induced spurious waves and do not require ghost points and allows for an implicit-time advancement have also been studied [28].

Different approaches for controlling the spurious waves have been explored. These range from upwinding or upwind biasing the numerical method, introducing artificial (solution dependent) viscosity, hyperviscosity or selective damping of the short wavelength components. The need for controlling the spurious waves are greatest in inviscid solutions where either significant nonlinear steepening of waves occur, or when rapid changes in boundary conditions occur in the domain (*e.g.* at sharp leading or trailing edges), or when rapid changes occur in the mesh.

7.5. DNS of Compressible Free-Shear Flows

Compressible free shear layers provide the clearest example of a turbulent flow where compressibility has a major effect on the turbulence. It is widely reported that the spreading rate of a compressible mixing layer decreases as the flow Mach number is increased while keeping other parameters fixed. Slow growth of high speed mixing regions also implies a slower rate of mixing of fuel with an oxidizer stream in hypervelocity scramjets, an application which has motivated much of recent research on high-speed mixing. It has been known since the pioneering study of Brown and Roshko [20], referred to hereafter as BR, that the suppression of the spreading rate is not related to changes in the mean density across the mixing region but a genuine compressibility effect. However, detailed data on the changes in the turbulence characteristics in high-speed mixing layers has not been available in past. Several experiments were undertaken in the 90's to gather some of this basic information and to explore means for enhancing the mixing rate. Dutton [45] and Dimotakis [42] give a recent survey of this experimental data. These recent experiments used non-intrusive optical techniques, although some studies have used hot-wire measurements. Much new information on the structure of the eddies (*e.g.* [24], [25]) and the changes in certain turbulence statistics with Mach number has been gathered. However, it has not been possible with experimental means alone to reveal the mechanism at work which brings about the suppressed growth. DNS has played a key role in studies of this fundamental issue. Virtually any flow property can be postprocessed from a suitable DNS database. This allows careful testing of theoretical hypotheses and modeling assumptions within the parametric range available in DNS. This section will serve to illustrate this powerful role of DNS.

7.5.1. FLOW DEFINITION

Consider the mixing layer formed between two free-streams flowing in x_1 direction which are brought into contact at $x_1 = 0$. The mixing layer width $\delta(x_1)$ grows with x_1. Suppose that the conditions in the high and low speed free-streams are denoted by U_1, ρ_1, T_1, C_1, and U_2, ρ_2, T_2, C_2, respectively. Here U is the mean velocity, ρ is the density, T is the static temperature and C is the speed of sound. Let $r = \frac{U_2}{U_1}$, and $s = \frac{\rho_2}{\rho_1}$. The static pressures in the two streams are assumed to be equal, $p_1 = p_2$. The shear across the mixing layer is, $\Delta U = U_1 - U_2$. When $\Delta U/C_1$ and $\Delta U/C_2$ are both much smaller compared to unity, the compressibility effect on the the mixing layer is negligible. Assuming that the turbulent velocity fluctuations scale with ΔU, it follows that the 'hydrodynamic' pressure fluctuations $p_i{}'$ scale as $\rho_r \Delta U^2$, where ρ_r is a representative density scale, and consequently $p_i{}'/p_1 \ll 1$. The low-speed mixing between the two streams thus occurs at nearly constant pressure (isobaric). Significant density fluctuations can still exist in this low-speed flow. To be specific, consider the low speed mixing layer between two gaseous streams. The equation of state suggests two independent ways in which significant density variations can arise: a) due to temperature difference, $\Delta T = T_1 - T_2$, between the two streams but with the same gas composition, and b) when $T_1 = T_2$ but the composition of the two streams is different. With the same operating gas in both streams it is difficult, in practice, to obtain large density contrast between the two streams without simultaneously increasing the Mach number, so variable density effects at low speeds have been explored using dissimilar gases in the two streams.

7.5.2. INCOMPRESSIBLE MIXING LAYER

It is well known that the spreading rate of the low speed uniform density ($\rho_1 = \rho_2$) turbulent mixing layer, $\frac{d\delta^*}{dx^*}$, correlates linearly with the velocity ratio $\lambda = \frac{U_1 - U_2}{U_1 + U_2}$. This dependence can be understood as follows: the mixing layer spreads due to the entrainment of free-stream fluid at a rate proportional to $\Delta U/2$ into the dominant eddies which convect at a speed $U_c = \frac{U_1 + U_2}{2}$. The entrainment is, however, asymmetric, and slightly more high-speed fluid is entrained [41]. When $\rho_1 \neq \rho_2$ two dominant changes occur. The convection speed U_c shifts towards the denser stream and assuming that the entrainment velocity continues to scale with ΔU, it follows that (for given U_1 and U_2) the spreading rate is enhanced when the heavier fluid is carried in the slower stream. This is borne out by the BR experiments and is also consistent with the rapid initial spreading of Helium jets in air [43]. It is worth noting that a factor of 50 change in the density ratio (at the same velocity ratio) in a mixing layer changes the growth rate by only a factor of 2. This observation stresses that the growth rate reduction observed in high-speed mixing layers is a compressibility effect (BR).

A formula for the convection speed U_c can be obtained to account for the density effects. In the moving frame the stagnation pressure of the two streams must balance at the stagnation point which exists between any pair of large eddies, *i.e.* $\rho_1(U_1 - U_c)^2 = \rho_2(U_c - U_2)^2$ or $U_c = \frac{U_1 + \sqrt{s}U_2}{1+\sqrt{s}}$. Since $U_1 - U_c \neq U_c - U_2$ the mixing layer entrains more volume of the lighter fluid than the heavier fluid. A broadening of the mean density profile toward the dense stream is also observed in the BR experiments but the mean velocity profile shows little change. This broadening is most dramatic when the heavy stream is also the slower stream. Brown and Roshko interpret this as follows: the density interface between the two streams is convoluted by the engulfment associated with large eddies, but continuity of pressure implies that velocities induced in the denser stream are smaller. This is suggested as the reason why in traversing away from a low speed dense stream the mean composition changes appreciably, well before the mean velocity shows any change.

7.5.3. CONVECTIVE/RELATIVE MACH NUMBER

In the previous section a conceptual picture of shear-layer entrainment process and growth was discussed. Energetic eddies convecting at speed U_c entrain fresh fluid from either free-stream into the mixing region. If the Mach numbers of the free-streams relative to the actively entraining eddies is not small, a compressibility effect is expected. Such reasoning motivates the use of a *convective Mach number* for each free-stream: $M_{c1} = (U_1 - U_c)/C_1$ and $M_{c2} = (U_c - U_2)/C_2$. Bogdanoff [13] and Papamoschou and Roshko [117] advocated the convective Mach number hypothesis. They also showed that the convection velocity U_c could be determined by applying an isentropic form of the Bernoulli equation to match the stagnation pressure of the two streams in the frame moving with the eddies. For the case when the specific heat ratio $\gamma_1 = \gamma_2$ this model gives the convection velocity of the large eddies as, $U_c = \frac{C_2U_1+C_1U_2}{C_1+C_2}$, and $M_{c1} = M_{c2} = \frac{\Delta U}{C_1+C_2}$. For dissimilar gases $\gamma_1 \neq \gamma_2$, the expression for U_c is more involved and $M_{c1} \neq M_{c2}$ and it is not clear whether a single compressibility parameter can be used. A compromise is the relative (or average) Mach number $\mathbf{M}_r = \frac{\Delta U}{C_1+C_2}$, which is also equal to M_c when $\gamma_1 = \gamma_2$. $\mathbf{M}_r$ is a parameter commonly adopted to highlight the effects of compressibility [45].

Experiments have supported the usefulness of the relative/convective Mach number as a parameter to describe the compressibility effects. Following [13] and [117] the intrinsic compressibility effect on the growth rate can be isolated by plotting the growth rate of the compressible mixing layer normalized by the growth rate of the corresponding incompressible mixing layer at the same velocity and density ratio, against the convective Mach number M_c (see data compiled in [92], or [45]). However, such attempts to collapse the growth rate data on a single curve also show significant 'scatter'. Different definitions of mixing layer thickness, different degree of self-similarity in different experiments, differences in the state of the boundary layer on the splitter-plate, differences in the degree of confinement, presence of acoustic duct modes are suggested as possible causes for the scatter. Using only the data based on the same thickness measure (*e.g.* vorticity thickness) appears to significantly reduce the scatter. Since only a few conditions have been studied in each laboratory set up, it is difficult to assign a degree of sensitivity to each of the listed causes of 'scatter'.

7.5.4. TURBULENCE AND EDDY STRUCTURES

Experiments have also reported some structural changes in turbulence as $\mathbf{M}_r$ is increased (see the review by Dutton [45] for details). Normalized shear-stress $-\overline{u'v'}/(U_1-U_2)^2$ and normalized transverse fluctuations $\overline{v'^2}/(U_1-U_2)^2$ are reported to decrease with $\mathbf{M_r}$ in all experiments, but there is disagreement on the behavior of the streamwise component $\overline{u'^2}/(U_1 - U_2)^2$. Some studies report that this quantity decreases while others report no significant change. Similarly whether the structure parameter $b_{12} = \overline{u'v'}/q^2$, where $q^2 = \overline{u'^2} + \overline{v'^2} + \overline{w'^2}$, changes with $\mathbf{M}_r$ is also in dispute.

Instantaneous images of the flow obtained with non-intrusive diagnostics (Clemens and Mungal 1992, 1995 and by other groups) show that as M_c increases the quasi-two-dimensional Brown-Roshko roller eddies which span across the flow are no longer discernable. Instead highly three-dimensional turbulent eddies are observed. This behavior is consistent with predictions of linear stability theory (based on equation 7.34 discussed in section 7.3.2) which shows that above M_c of 0.6 oblique three-dimensional modes are linearly most-unstable. Spatial correlations in such images have been quantified [45] and show that the streamwise correlation length of the imaged scalar increases with M_r. Whether the eddies dominating the momentum and species transport span the width of the shear layer at high M_c is not clear. Some differences between the two edges of the shear layer, in the three-dimensional activity and intermittancy, have also been reported by Bonnet *et al.* [14].

Obtaining images of the flow requires that the flow be suitably seeded. Different techniques have been utilized to 'visualize' different parts of the flow. Miller *et al.* [106] used techniques which primarily visualize the fluid originating in either the high or low speed streams, or only the 'mixed' fluid resulting from the mixing between the two streams. Such measurements allow a detailed probing of passive scaler mixing within the layer. Probability density function (pdf) of a conserved passive scalar within the mixing layer has been inferred from measurements. The pdf's show a 'tilted' behavior across the layer [76] which is consistent with the lack of quasi-two-dimensional eddy patterns in the high $\mathbf{M}_r$ flows [106]. Pairs of passive scalar images obtained within a short elapsed time have been analyzed to infer the convection velocity [115], [99], [119] and find it to be consistently different from the 'isentropic' formula noted in section 7.5.3; the measured velocity is close to either the high or the low speed free-stream velocity. Other techniques which rely on space-time covariance between two pressure transducers or hot-wires do not show a marked deviation from the isentropic formula. Instantaneous images of the velocity field within the shear layer have recently been acquired by Urban and Mungal [160] and should prove valuable in resolving the convection velocity issue.

7.5.5. PROPOSED EXPLANATIONS/MODELING

The suppression of linear instability growth rates with increasing M_c is well known. This is responsible for the stabilization of the two-dimensional

Kelvin-Helmholtz instability of a compressible vortex-sheet for $M_c \geq \sqrt{2}$ for $\rho_1 = \rho_2$ case [82]. The finite thickness inviscid case has been explored by many authors ([89], [61], [132], [124], [72]). For $M_c \geq 0.6$ (approx.) three-dimensional oblique disturbances become most unstable. For supersonic M_c, acoustic modes are destabilized and become the most unstable modes at hypersonic conditions [151], [72]. If a connection between the inviscid linear instability modes of a shear-layer and the 'large-scale' eddies of a turbulent mixing layer which are responsible for turbulence production, entrainment and its spreading, is accepted a correlation between the instability growth rate and shear-layer spreading rate is expected. Such a correlation has been noted by Monkewitz and Huerre [70] for incompressible shear layers, and by Ragab and Wu [124], Sandham and Reynolds [132], and Lu and Lele [96] for compressible shear layers.

A very different physical picture was proposed by Zeman [170] who assumed that as M_c increased above 0.3 significant additional dissipation of turbulence would result from eddy-shocklets. He proposed a model for this additional shocklet or dilatational dissipation based on the turbulence Mach number $M_t = q/\overline{C}$. An alternate model which also emphasized the role of dilatational dissipation was proposed by Sarkar, Erlebacher and Hussaini [133] based on their low Reynolds number DNS of isotropic compressible turbulence. Both models exhibited the growth rate suppression of supersonic shear layers and could be calibrated on such data. However, more recent studies (see section 7.5.6) show that direct (energetic) effects associated with compressibility are too small to explain the experimental observations.

Suppressed growth of turbulence was observed in DNS of compressible homogeneous sheared turbulence by Blaisdell, Mansour and Reynolds [12]. Analysis of turbulent kinetic energy balance showed that explicit compressibility terms (pressure-dilatation correlation and dilatational dissipation) were important. Changes in Reynolds stress anisotropy were also observed but their critical importance emerged only after a revisit to this problem by Sarkar [134]. Sarkar showed that principal contributor to the reduced turbulence growth rate was the reduced normalized shear-stress. As the gradient Mach number was increased the turbulent eddies became less efficient in extracting energy from the mean flow. Changes in the dissipative range were less important.

Papamoschou [116] and Papamoschou and Lele [118] have shown that the effectiveness of acoustic communication between two spatially separated positions diminishes as the convective Mach number is increased. An intuitive hypothesis is that acoustic communication is necessary across an active large-scale eddy within a turbulent flow for it to dynamically maintain its coherence/organization. Breidenthal [15] proposed a 'sonic eddy hypothesis' that only eddies with a Mach number difference of unity or less participate actively in the shear layer entrainment, and observed its consequences in suppressing the growth rate of a compressible mixing layer.

A physical explanation of the reduced growth which is based on vorticity dynamics in a compressible flow was offered by Lele [90] using 2-D simulations. The shear layer growth was linked to the large-scale reorganization of vorticity by inviscid instability. The effect of compressibility was revealed by comparing the importance of different processes causing vorticity redistribution. The inviscid vorticity equation for a two-dimensional flow may be written as:

$$\frac{D(\omega/\rho)}{Dt} = \frac{\nabla\rho \times \nabla p}{\rho^3}, \tag{7.51}$$

$$\frac{\partial \omega}{\partial t} + U_c \frac{\partial \omega}{\partial x} = -(u - U_c)\frac{\partial \omega}{\partial x} - v\frac{\partial \omega}{\partial y} - \omega \nabla \cdot \vec{u} + \frac{\nabla\rho \times \nabla p}{\rho^2}, \tag{7.52}$$

showing that the rate of change of vorticity observed in a frame moving at speed U_c (the intrinsic frame of a large-scale eddy) is due to three effects: 1) advection relative to this frame (first two terms on the r.h.s.), 2) change due to dilatation of fluid elements (third term), and 3) baroclinic change (fourth term). These individual terms were extracted from DNS data for mixing layers at different M_c and density ratio, s. The uniform density case $s = 1$ shows the physical effect of compressibility most clearly. At M_c of 0.4 the most important term causing the vorticity redistribution is the advection term. Advection may be seen as moving vorticity away from the braid region (the region containing stagnation points between two vortices) and bringing it to the vortex centers. This is precisely the incompressible instability mechanism causing the shear layer instability (Batchelor [9, p. 516]). The dilatation term, while not dominant at $M_c = 0.4$, provides clues on how the stabilizing effect associated with compressibility arises. Vortex roll up brings about a depression of pressure in the vortex core, and similarly elevates the pressure in the braid region between each pair of vortices. As equation (7.51) shows the potential vorticity ω/ρ is conserved in an inviscid two-dimensional flow in absense of baroclinic torques. The same effect is alternatively expressed by the third term on the r.h.s. of (7.52). The change in density counteracts the Kelvin-Helmholtz redistribution of vorticity and thus slows down its rate. As M_c increases the relative magnitude of density change (and dilatation) increase. At $M_c = 0.6$ the dilatation effect on vorticity transport is comparable to the advection term.

Simulations with $s = 1$ reveal that a 'quadropole' pattern of dilatation is associated with each vortex as it rolls up and convects in the mixing layer.

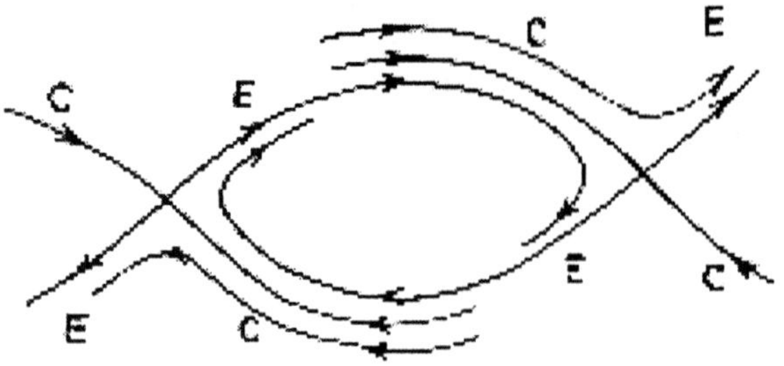

Figure 7.8. A schematic depicting the flow field around a vortex during its roll up in a frame convecting with the vortex. Adapated from Lele [90].

This pattern is simply understood via Figure 7.8 showing a schematic of the flow field around the vortices in a frame of reference moving with speed U_c, the speed of the vortices. It was observed in [90] that the 'stagnation points' between the vortices also move at the same speed as the vortices and thus are stationary in this frame. Fluid moving away from the stagnation points expands as it accelerates towards the vortex center. Past this point the flow compresses and decelerates towards the other stagnation point. This expansion and compression process may be seen in the characteristic quadrupole pattern of the dilatation field. At low M_c this dilatation field can also be predicted by the 'diagnostic equation' of Ristorcelli [130] discussed in section 7.3.1. As M_c is increased above 0.4 nonlinear distortions arise, compression regions become steeper and expansions become broader. Further increase in M_c produces 'shocklets' around the vortices.

In mixing layers with unequal free-stream density baroclinic effects become important. This is essentially an incompressible effect. As vortex roll up procceeds the density interface remains sharp at the 'braids' and the pressure maximum (at the stagnation point in the convected frame) produces regions with dynamically significant baroclinic torque. If both free-streams have the same stagnation enthalpy the density of the slower stream is lower. Assuming the faster stream to be the upper stream, the baroclinic torques tend to enhance the vorticity in the lower part of the braid and cause suppression of vorticity in the upper part of the braid. But if free-streams at equal Mach number but unequal velocity are chosen, the density of the faster stream is lower and a reversal of the situation just described takes place. The vortices evolving in both cases develop a layered vorticity distribution which is strongly affected by the baroclinic torques (see the vorticity contours in Figure 7.6). Such cases were studied by Lele [90] who also noted that the net circulation of the vortices remained relatively uninfluenced by this redistribution. Several recent studies have investigated the baroclinc effects in 3-D situations. Staquet [144] shows secondary roll-ups of the baroclinically generated vorticity in the braid regions of a stratified mixing layer. Cortesi *et al.* [32] study stratified mixing layers for fluids with different Prandtl numbers and show the competetion between the stabilizing effect of thermal diffusion and the enchanced entrainment due to baroclinic vorticity.

As just noted, vorticity dynamics gives useful insights into the large-scale unsteady flow behavior. However, these insights are yet to be incorporated in engineering prediction methods for turbulent flows.

7.5.6. INSIGHTS FROM RECENT DNS STUDIES

Two independent DNS studies Vreman, Sandham and Luo [163], and Freund *et al.* [52] have provided new insights into the mechanics of the unsteady compressible flow within a high-speed shear layer. Both studies focus on compressibility effects using temporally-evolving shear layers, the former uses a planar shear layer and the latter an annular configuration. Despite the geometrical differences the two flows behave in much the same way.

Both studies show that direct effects of compressibility on the energy balance of turbulence are *not* responsible for the spreading rate suppression. While the contribution of dilatational dissipation to TKE dissipation ϵ does significantly increase as M_c is increased, its level is too small (less than 2% of ϵ at $M_c = 1$ [52] to have caused the factor of 3-4 suppression in spreading rate. Both studies report observing eddy-shocklets in their supersonic M_c cases, but their contribution to kinetic energy dissipation is small. These findings support the view that the dominating compressibility effects are *structural* rather than *energetic* by Simone *et al.* [141]. Both DNS studies consider the balance equations for Reynolds stress components $R_{ij} \equiv \bar{\rho}\widetilde{u''_i u''_j}$ and show that pressure-strain correlations $(PS)_{ij} \equiv \overline{p'(\frac{\partial u''_i}{\partial x_j} + \frac{\partial u''_j}{\partial x_i})}$ are suppressed with increasing M_c. At the level of Reynolds stresses the picture which emerges is that $(PS)_{11}$, the 'redistribution' terms in R_{11} equation, is suppressed and so are $(PS)_{22}$, and $(PS)_{33}$. This suppresses the energy input into R_{22} and R_{33}. Since R_{22} controls the production of the shear-stress R_{12}, there is suppression of shear-stress (and spreading rate) and this suppressed shear-stress in turn suppresses the production of R_{11}. This cycle of linked processes was noted by Blaisdell *et al.* [12] in their DNS study of compressible homogeneous sheared turbulence, but its dominance in bringing about the structural changes in shear flow turbulence has emerged with recent DNS studies.

Suppression of the intercomponent redistribution was linked to the suppression of pressure fluctuations p'_{rms} (relative to the dynamic head). Vreman *et al.* explain this change by appeal to the sonic eddy hypothesis. A more direct approach is taken by Freund *et al.* who show that the transverse

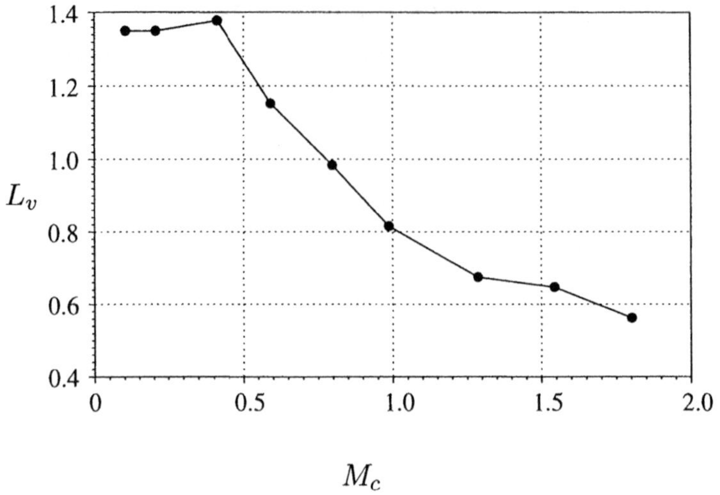

Figure 7.9. Suppression of transverse correlation length L_v with M_c. Adapted from [52].

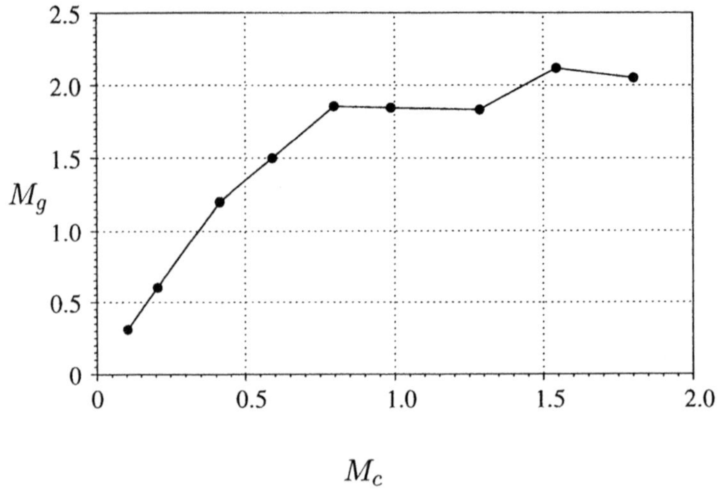

Figure 7.10. The saturation of gradient Mach number M_g with M_c. Adapted from [52].

correlation length scale, L_v, (defined as the distance at which the correlation in transverse velocity fluctuations falls to 0.1) decreased relative to the shear layer thickness δ_ω, as M_c is increased. This data is reproduced in Figure 7.9. The implication of this trend is that reduced L_v is responsible for the suppression of p'_{rms} and consequently for suppressing v'_{rms} and the shear-stress. They also showed that the reduced L_v was consistent with the sonic-eddy notion of acoustic communication limiting the eddy size; the ratio of acoustic time L_v/C and the mean shear time scale $(dU/dy)^{-1}$, which

(a)

(b)

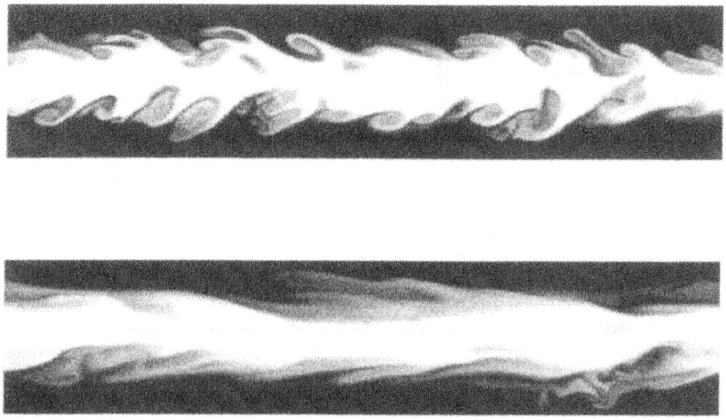

Figure 7.11. Grey scale contours of a passive scalar in low and high Mach number shear layers. Adapted from Freund *et al.* [52]. a) $M_c = 0.4$; b) $M_c = 1.0$.

is also the gradient Mach number based on L_v, was found to increase with M_c at low M_c, but saturate to a constant value at high M_c. This shown in Figure 7.10. Consequences of the reduced L_v on the structure of the vortical eddies, and on scalar-mixing were also studied by Freund *et al.* Figure 7.11 contrasts the side-views for $M_c = 0.4$ and $M_c = 1.0$ shear layers marked by a passive scalar (in grey scale). Dark regions denote the ambient fluid (outside the jet), and the whitest regions mark the pure jet fluid.

It would be natural to seek a turbulence model consistent with such a physical picture of turbulence in a compressible shear layer, but such a predictive model is currently not available. It should be noted that Simone *et al.* [141] show that DNS results on homogeneous shear flow can be predicted using linearized equations, but a 'turbulence-model' containing these effects is not available. It is also important to ask whether the DNS predictions on the structural changes to turbulence due to compressibility effects are borne out by laboratory experiments. A critical question concerns the size of the active eddies in a high M_c shear layer. Do these eddies become smaller than the shear layer thickness as predicted by DNS ? If this is verified, its consequence on scalar-mixing and chemical reactions within a high-speed mixing region and how they should be modelled would naturally follow. The puzzle of the convection speed, and the role of eddy-shocklets can also be decisively resolved once the physical picture suggested by DNS is confirmed by the experiments.

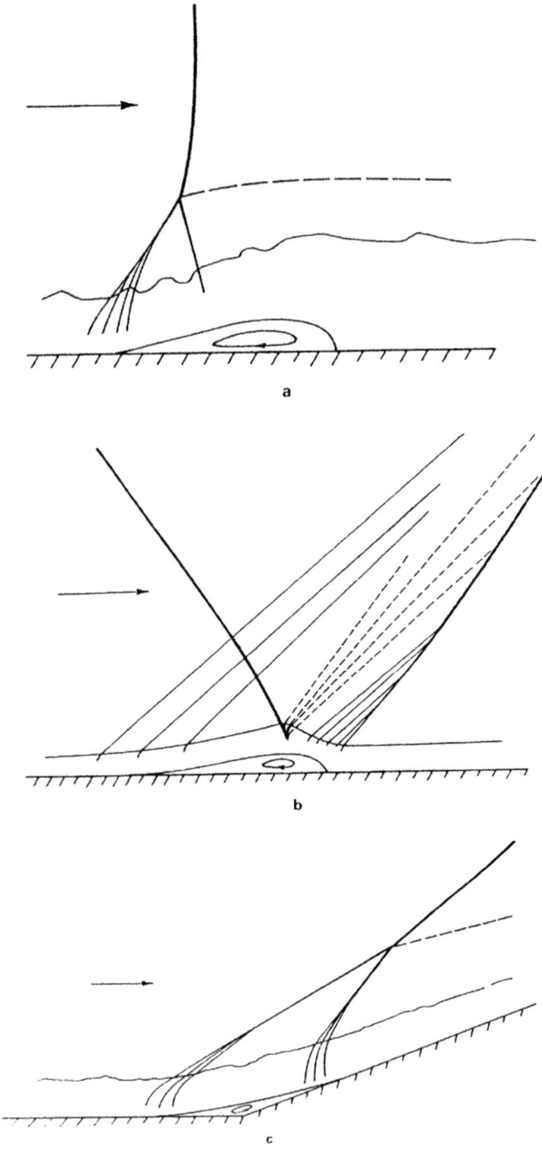

Figure 7.12. Typical interactions between a shock waves and boundary layers. Adapted from Adamson and Messiter [3]. a) Incident normal shock case; b) Incident oblique shock case; c) Compression ramp or corner.

7.6. DNS of shock-turbulence interaction

Interactions between shocks and turbulent shear flows are common in high-speed flows associated with propulsion systems. Supersonic inlets, supersonic diffusers and exhaust plumes contain complex systems of oblique and normal shock waves. Fuel injection in a hot supersonic flow (*e.g.* scramjet) also involves complex shock patterns. Shock waves cause rapid changes in flow; both the mean and turbulence properties are strongly modified. The pressure rise associated with the shock is transmitted to the upstream boundary layer via the subsonic flow region. The boundary layer thickens in response and affects the shock structure. If the pressure gradient is sufficiently adverse to cause separation the inviscid outer flow is strongly altered, producing a viscous-inviscid coupling. Figure 7.12 adapted from Adamson and Messiter [3] shows typical shock-boundary layer interactions in two-dimensional mean flows. Shock-induced separation is often unsteady and its effect persists over a long distance downstream. Calculation methods which are based on solving the averaged Navier Stokes equations have difficulties in accurately predicting these phenomena. Accurate prediction of heat-transfer to surfaces near complex shock-boundary layer interactions is also difficult (see Knight [78] for a recent assessment of RANS predictions). In this context, DNS of idealized shock-turbulence interactions provides an avenue to understand the fundamental interactions and help develop models for improving the prediction of more complex situations.

7.6.1. IDEALIZED SHOCK-TURBULENCE INTERACTION

Consider the idealized problem of a normal shock wave interacting with an unstream field of homogeneous turbulence depicted in Figure 7.13. The upstream turbulence can be specified independently of the shock-strength and in accord with the modal decomposition of section 7.3.2 may be a chosen blend of vortical, acoustic and entropic disturbances. Predicting the changes in the attributes of turbulence as it interacts with the shock, and the concomitant shock dynamics are the major goals of the study. Being the simplest configuration which isolates the primary effects of a generic shock-turbulence interaction, this idealized problem has received significant attention in theoretical, experimental and simulation studies. Major results of these studies are highlighted in the subsections to follow. It should be noted that more complex building block problems of shock-turbulence interaction can also be concieved. For example, the interaction depicted in Figure 7.5 may be considered as a shock turbulent-shear-layer interaction. More realistic mean flow effects can be introduced in parallel to the flow situations in Figure 7.12. In fact, the compression ramp flow is the subject of an on going study by Adams [2].

7.6.2. LINEARIZED ANALYSIS OF SHOCK-TURBULENCE INTERACTION

The interaction of turbulence with a shock involves three time scales (in order of decreasing magnitude): the eddy turnover time, $\tau_t = L/u'$ of an

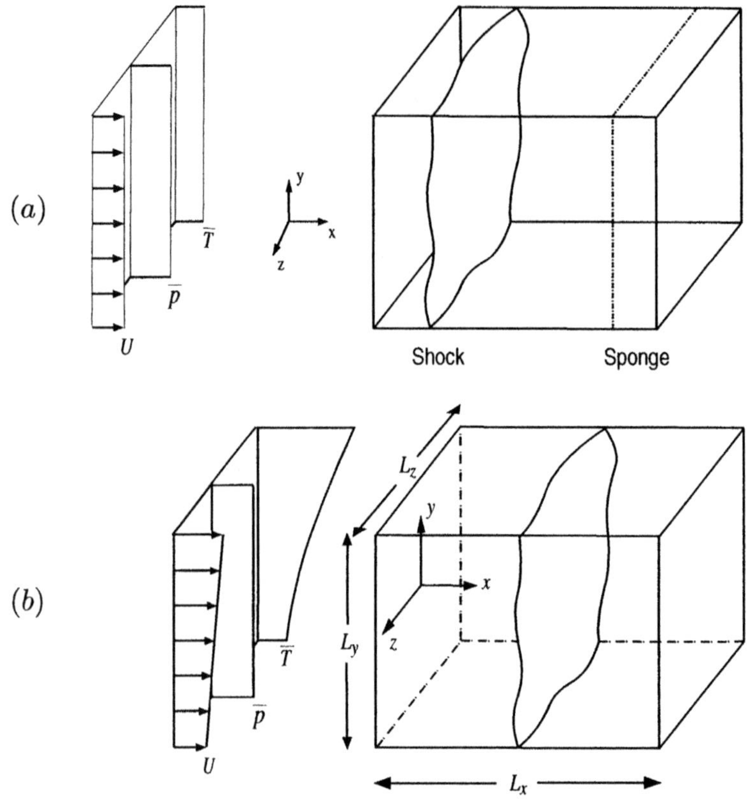

Figure 7.13. Two idealized problems of shock-turbulence interaction. a) Transversely homogeneous turbulence interacting with a normal shock, adapted from Mahesh *et al.* [103]; b) Transversely sheared turbulence interacting with a normal shock, adapted from Mahesh *et al.* [102].

eddy of length scale L and velocity scale u', the time scale for passage of the eddy of through the shock, $\tau_c = L/\overline{U}_n$, and the time scale of compression within the shock itself $\tau_d = \delta_s/(\overline{U}_1 - \overline{U}_2)$, where δ_s is the shock-thickness. Under most conditions $\tau_d << \tau_t$, *i.e.* the deformation within the shock is sufficiently rapid that the turbulence may be considered as 'frozen' during the interaction. The shock wave also moves unsteadily and distorts due to its interaction with the oncoming turbulence, and if the turbulence is weak this shock dynamics is given by a linearized theory. A rather complete picture of the linearized interaction of a shock with turbulence (represented as a superposition of linear modes of section 7.3.2) is available (Ribner [125], [126], [128], and Moore, [112]) and has been used to study specific physical effects in detail. Anyiwo and Bushnell [5] studied different mechanism of turbulence amplification across a shock, Ribner [127] applied the linear theory to study acoustic wave generation due to a vortex interacting with a

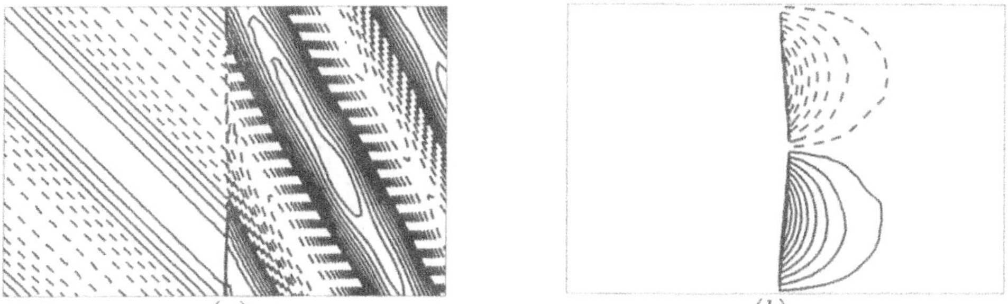

Figure 7.14. Interaction of vorticity-entropy wave with a normal shock. Adapted from Mahesh *et al.* [102]. a) and b) correspond to incidence angles of 45° and 90°, respectively.

normal shock, Ribner [128] studied the changes to turbulence spectra - an aspect further explored by Lee *et al.* [86], [87]. Mahesh *et al.* [101] applied this theory to study the interaction of a random field of acoustic waves (noise) with a shock, and Mahesh *et al.* [103] emphasize the importance of upstream entropy fluctuations in changes to turbulence across a shock.

The linearized theory predicts statistics of the turbulence downstream of the shock in terms of the upstream statistics, with the transfer functions based on the interaction of elementary waves (modes) with the shock. An interesting aspect of these wave interactions problems is the occurance of a 'critical angle'. Waves with incidence angles beyond the critical angle induce disturbances on the shock front whose surface wave speed is subsonic relative to transmitted side, hence they yields 'evanascent' transmitted acoustic waves which decay to zero in the near-field of the shock. This feature is highlighted in Figure 7.14 taken from Mahesh *et al.* [102] via contour plots of transverse velocity u_2' for the interaction of vorticity-entropy waves of different orientation with a $M_1 = 1.5$ shock. In case (a) the incident wave fronts are oriented at 45° to the shock which is within the propagating regime. The u_2 component of the incident and transmitted waves are readily apparent in the figure. However, case (b) corresponds to an incidence angle of 90° which yields evanascent waves. The downstream decay of these waves is notable and as expected no u_2 disturbance is evident upstream of the shock; this vorticity-entropy wave has wave fronts aligned with the x_2 axis so the disturbance velocity is aligned with x_1. Statistical quantities affected by the evanascent waves show a non-monotonic downstream evolution. Such behavior is intrinsic to the linear theory, non-linear effects are not required to explain it - a point overlooked in early DNS studies [86]. Another interesting aspect is that for normal shocks with $M_1 > 2$ (approx.) the theory predicts that the 'far-field' v'_{rms} exceeds u'_{rms}, while the reverse is true for $M_1 < 2$.

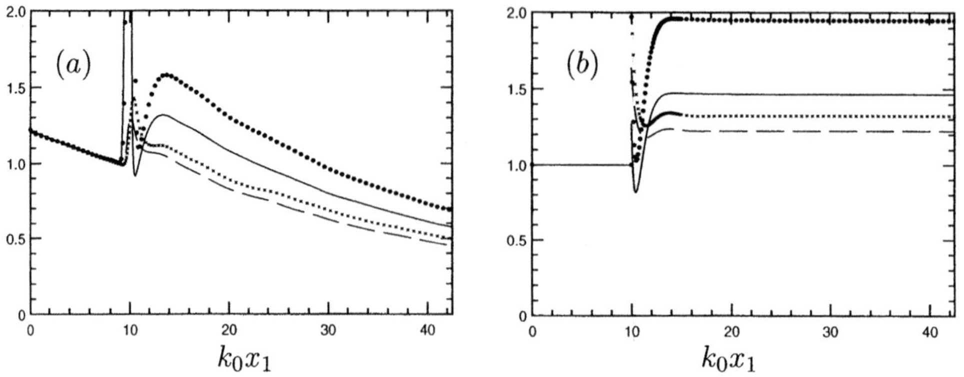

Figure 7.15. Evolution of velocity fluctuations in shock turbulence interaction. Incoming turbulence contains vorticity and entropy fluctuations and $M_1 = 1.29$ and $M_t = 0.14$ just before the interaction. Adapted from Mahesh *et al.* [103]. a) Results of DNS. All curves are normalized by their value immediately upstream of the shock. Solid line and solid circles show $\overline{u'^2}$ without and with upstream entropy fluctuations, and dashed line and crosses show $\overline{v'^2}$ without and with upstream entropy fluctuations; b) Prediction of inviscid linear analysis. Same line- and symbol-types as in Figure 7.15(a)

7.6.3. OBSERVATIONS FROM DNS

Before summarizing the key oberservations from DNS studies of shock turbulence interaction it is important to note the regimes accessible to such DNS studies. Early DNS studies were limited to weak shocks since they were based on resolving the viscous shock-structure. More recently numerical methods for shock-capturing have been combined with high-fidelity schemes (as outlined in section 7.4.2). This has allowed stronger shock cases to be simulated. However, in all cases the turbulence Reynolds number is relatively low. The necessity to accurately track the unsteady distortions of the shock requires grid clustering in the vicinity of the shock and limits the time step which can be used. Mahesh *et al.* [102] discuss these numerical issues. Moin and Mahesh [111] also give an estimate of the grid-requirements for shock-capturing DNS.

The most ubiquitous feature of shock-turbulence interaction is the amplification of turbulence across the shock. The incoming vortical fluctuations are subjected to an essentially one-dimensional compression not unlike that in the compression stroke of a piston engine, but much more rapid. The vorticity components in the plane of the shock are amplified by the compression while the shock-normal vorticity is unaffected. Shock curvature and incident entropy non-uniformities also cause baroclinic vorticity generation. These effects combine to amplify turbulence across a shock, the turbulence downstream of the shock is more vigorous and anisotropic. A significant increase in the dissipation rate of turbulence also results. However, more interestingly the turbulence just downstream of a shock is in a non-equilibrium state. Its subsequent evolution reveals the importance of nonlinear processes [103]. DNS results on turbulence amplification were compared with linear theory in [87] and [103]. Figure 7.15 depicts an example from Mahesh *et al.* [103] study. The agreement is quite remarkable. The low Reynolds number of the DNS, however, makes it difficult to compare amplification ratios etc. in a fully quantitative manner. Laboratory experiments (Barre *et al.* [7]) have also reported similar success of linear theory. In general, however, the extent of turbulence amplification in laboratory experiments depends on other details of the experiments. Jacquin *et al.* [73] observed almost no amplification of turbulence in a shock-containing jet. The linear theory, when extended to include the effects of upstream entropy and acoustic waves also shows sensitivity of turbulence amplification to the 'composition' (vortical- entropy-acoustic wave mixture) of the upstream state. Negative correlation between entropy fluctuations and streamwise velocity fluctuations (as expected in adiabatic wall boundary layers) enhances the turbulence amplification and positive correlation has a suppressing effect (Mahesh *et al.* [103]). Hannappel and Friedrich [63] considered the effect of 'acoustic' fluctuations on the turbulence amplification through a shock. Their numerical study showed that the turbulence amplification is lower when significant dilatational fluctuations are present. This effect has been explained by Mahesh *et al.* [101] using linear theory.

When the intensity of the upstream turbulence is high ($M_t^2 > 0.1(M_1^2 - 1)$ for isotropic vortical upstream state, according to [86]) the structure of the instantaneous shock is significantly altered. The shock is no longer a wrinkled pseudo-laminar shock; a complex structure with multiple shocks and 'holes' is formed. On a streamline passing through the 'holes' the fluid is compressed in a sequence of waves. Details of the dynamics in this strongly disturbed regime have not been studied. A model problem relavant to this regime has been studied by Kevlahan [77]. Shock perturbations due to several different two-dimensional non-uniform incident velocity fields were studied and shock-focusing phenomena including the formation of shock-shocks (discontinuity in shock strength) were analyzed.

Linear theory and DNS both show that most length scale attributes of turbulence reduce across a shock. The transverse correlation length based on $\overline{u_3'(x_1, x_2, x_3, t)u_3'(x_1, x_2 + r_2, x_3, t)}$, however, shows an increase. Lee *et al.* [87] discuss the length scale changes in detail and point out that some experimental observations are at odds with the expected length scale reduction. Recent experiments by Barre *et al.* [7] do report length scale reduction which is consistent with theory (also see [129] and authors' response). Recent studies by Agui [4] in a modified shock-tube with a porous end-wall also report data consistent with linear theory. This confirms that data processing used with earlier shock-tube experiments (with closed ends) is highly suspect. The mean flow speed behind a reflected shock is essentially zero when the end of the shock-tube is closed. The absence of any appreciable mean flow completely invalidates Taylor's hypothesis used in data processing.

Many of the trends from the idealized shock-turbulence interaction were also observed in Mahesh's study [102] of the interaction of a normal shock with a uniformly sheared flow. Homogeneous rapid distortion theory predicts a suppression of the $-\overline{u_1'u_2'}$ shear-stress, and a change in its sign for sufficiently strong shocks (*e.g.* Mahesh *et al.* [100]). Shear stress suppression was observed in the shear flow DNS. However, the Reynolds number of the simulation was low and the inhomogeneity effects made the statistical sampling problem more severe. Simulations of shock-wave boundary layer interaction or a compression corner flow [2] are subject of current studies. Such simulations are quite challenging and benchmark studies should be very useful.

7.7. DNS of Aerodynamically-Generated Sound

In recent years direct numerical simulations are being applied to study fundamental aeroacoustic problems. Before describing some illustrative applications it helps to note the special challenges of aeroacoustic phenomena (*e.g.* Crighton [37]). In most aeroacoustic problems there is a large disparity between the energy levels of the unsteady flow fluctuations and the sound. Even in those cases where the radiated noise is very loud, such as in the near-acoustic field of a supersonic jet (at about 10 jet-diameters away), the acoustic disturbance amplitudes are about three orders of magnitude smaller than the flow disturbance. For supersonic jets only about one percent of the mechanical power of the jet is radiated as noise (see Seiner [140]). In other cases the radiation efficiency of aeroacoustic source processes is considerable smaller.

When the characteristic flow speed (in the source region) is small compared to the speed of sound propagation in the medium, there is also a great disparity in the length scales between the unsteady flow and the sound. This is because the time scale or frequency of the unsteady flow (regarded here as a stationary sound source) and the sound must match. In low Mach number flows ($M \ll 1$) this gives an acoustic wavelength which is M^{-1} times longer than the flow length scale. A numerical simulation attempting to represent both the source process and the sound must resolve the dynamics within the source region and the sound field. Besides this overall disparity in the physical length scales, scattering problems (*e.g.* with sharp leading/trailing edges) and linear interaction problems, where one type of physical disturbance is scattered into another type of disturbance such as in a localized receptivity phenomena, often involve rapid variations of the acoustic field. These need to be carefully represented in the computation.

Since the human ear is sensitive to acoustic frequencies over a wide range (*e.g.* 20 Hz to 20,000 Hz) simulations dealing with community annoyance need to span this wide range. Often the sound-fields need to be propagated over a long range, and spectral distortions due to the cumulative non-linear effects as well as dissipation of acoustic energy need to be taken into account [35]. These studies are further complicated by the need to take into account the scattering of sound due to atmospheric irregularities and turbulence. In view of these physical challenges, it is critical that special attention be given in numerical studies to the possible numerical artifacts described in section 7.4.3.

7.7.1. DIRECT COMPUTATION OF SOUND GENERATION

Sound radiation from a co-rotating vortex pair is one of the simplest sound generation problems. It requires straightforward radiation boundary conditions making it attractive as an initial test case for numerical algorithms. In this flow the assumptions needed to rigorously justify the acoustic analogy theory of Lighthill [94] are also satisfactorily met (see Crow [38]). In fact the theory of vortex-sound by Mohring [110], and Kambe [75] is asymptotically valid at low Mach number for acoustically compact vortex flows. In the DNS study by Mitchell *et al.* [107] each vortex is represented by a smooth distribution of vorticity and the unsteady vortical flow and the radiated sound were calculated togather as a single calculation based on the compressible Navier Stokes equations. The vortex pair exhibited co-rotation about their common centroid for several cycles and this circulatory motion radiated sound.

In one of the cases studied the Mach number of co-rotation of the vortex pair M_r was 0.06. Since the co-rotating solution repeats itself after half a revolution the acoustic wavelength generated is 52.5 times the half-vortex-separation. The numerical simulations were designed to fit two acoustic wavelengths in the computational domain and maintained sufficient resolution near the central region to resolve the vortex dynamics. After several cycles the effects of vorticity diffusion became appreciable and the vortex pair underwent a rapid merger. The vortex cores merged and formed a nearly circular vortex within one period of co-rotation. This process is highlighted by contours of vorticity at different times in Figure 7.16, reproduced from [107].

The sound radiation from this flow was most intense during the sudden merger and decreased substantially after the merger as the vortex core be-

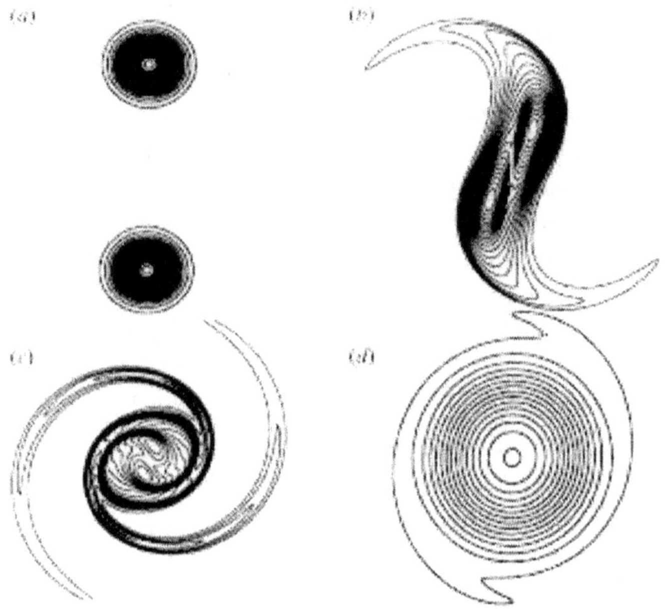

Figure 7.16. Evolution and merger of co-rotating vortices. Adapted from Mitchell *et al.* [107]. Iso-contours of vorticity are plotted. See original paper for sampling times.

came nearly axisymmetric. The acoustic fields obtained in the simulations were compared to the prediction based on the vortex-sound theory. A nearly perfect agreement was found (see Figure 7.17) which motivated extensive comparison to other acoustic theories. These also gave very good comparisons, once some technical issues associated with the two-dimensional version of Lighthill's theory were resolved. At higher co-rotation Mach number $M_r = 0.18$ significant compressibility effects were expected, and according to the point-vortex thoery of Yates [168] the vortex pair would act as a non-compact source. The simulations indeed showed that at higher M_r the predictions of the incompressible vortex-sound theory of compact sources showed a 65% difference from the DNS data. The incompressible Lighthill prediction also showed a similar difference. Accounting for the density variation via Powell's [121] form of the acoustic analogy gave good agreement.

The roll-up and pairing of vortices in a mixing layer provides the next step in building towards realistic shear flows. Not surprizingly the inflow/outflow boundary conditions become the most critical element of the simulation. Colonius *et al.* [30] simulated the spatial evolution of a two-dimensional mixing layer under prescribed inlet perturbation (specified via the linear stability eigenmodes for the fundamental and its three subharmonics).

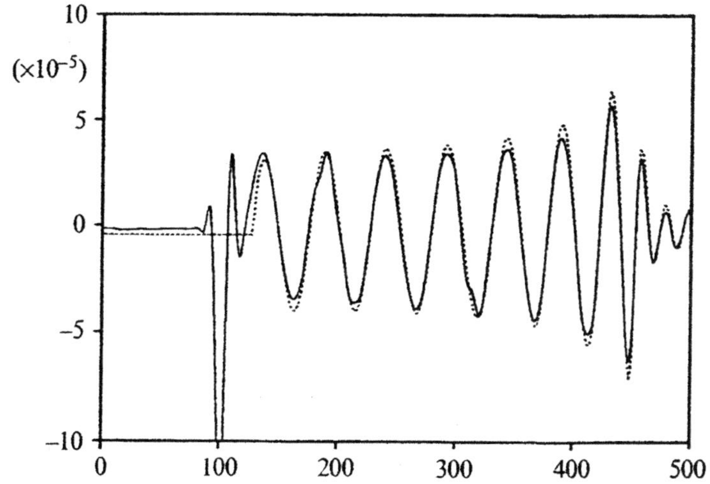

Figure 7.17. Comparison of sound radiated by co-rotating vortices with prediction using Mohring's vortex sound theory. Adapted from Mitchell *et al.* [107]. Data is taken at $\frac{r}{\lambda} = 2$, where λ is the acoustic wavelength.

The transverse domain size was chosen so that two acoustic wavelengths of the second subharmonic frequency would be captured in the domain. The simulation results clearly show the acoustic radiation emanating from the pairing process in the mixing layer. Dilatation was chosen to be the primary acoustic variable. In the radiation field it is simply related to other acoustic variables and was Fourier analyzed to obtain the acoustic fields for a range of frequencies. This highlighted some deficiencies of the boundary conditions. The exit zone outflow boundary conditions [29] worked well for the frequencies being forced – but low levels of acoustic fields with irregular radiation patterns were found for the *unforced* frequencies and were regarded spurious. A very slow drift in the mean pressure was also detected and choosing dilatation as the acoustic variable helped to reduce the contamination from this slow drift in time.

Depicting the mixing layer dynamics as growth, saturation and subsequent decay of vortical instability waves suggests that only the *supersonic* wave components are responsible for radiation. This description of the radiation process localizes the 'sources' to the region near the pairing location. Radiation from supersonic wave components is a critical element in the theory of Tam and Morris [148], and Crighton and Huerre [36]. Some modification to the source model used by Crighton and Huerre was necessary to achieve a reasonable comparison between the model predictions and DNS

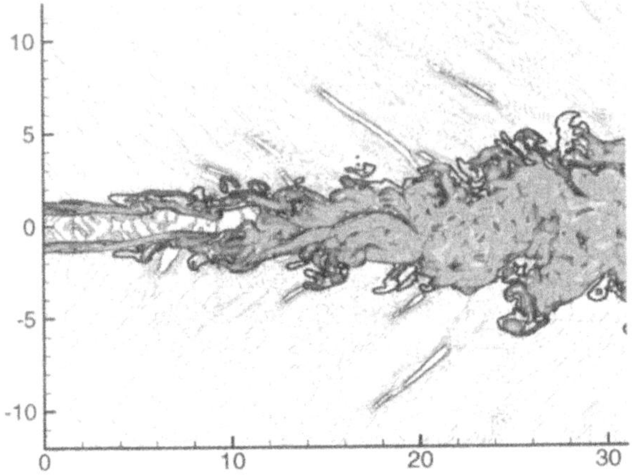

Figure 7.18. Instantaneous snapshot from the DNS of a $M_j = 1.92$ jet. Adapted from Freund *et al.* [53]. Thick contours show vorticity magnitude and thin contours show dilatation.

data. The simulations were also used to test the adequacy of acoustic theories of sound generation in a flow with extensive unsteady vortical flow. A theory originating with Lilley [95] and developed in full and refined by Goldstein [57], [60] was tested. The source terms in Lilley's equation (equation 7.34 with source terms due to nonlinearity) were provided from the DNS data. Lilley's equation was formulated in the wavenumber-frequency domain which naturally emphasized the supersonic wavenumbers which contribute to a radiating solution. The solution method was validated on test problems (such as scattering of the sound from a prescribed source by the non-uniform flow of the mixing layer). After some careful refinement of the post-processing excellent comparison with the DNS acoustic data were obtained. This is regarded as the first instance where detailed predictions from an acoustic analogy have been validated in a shear flow. The study also underscored the sensitivity of the predictions to a proper treatment of flow-acoustic interactions.

In recent years direct simulations have been extended to study sound radiation from jets. The work of Freund *et al.* [53] represents the current state of the art in direct computation of sound from a turbulent flow. Figures 7.18 and 7.19 (reproduced from this study) are snapshots of a supersonic $M_j = 1.92$ jet. Contours of vorticity magnitude mark the turbulent flow. Contours of dilatation in a selected range are overlayed to highlight the dominant acoustic radiation from the jet. It may be noted that the radiation is highly directional and resembles Mach wave radiation. The Reynolds number of the jet based on the jet diameter was 2000 and a mesh

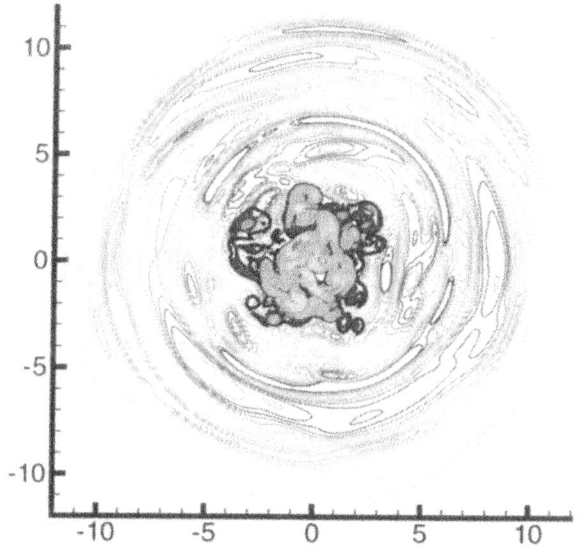

Figure 7.19. Side views of the $M_j = 1.92$ jet. Adapted from Freund *et al.* [53]. Thick contours show vorticity magnitude and thin contours show dilatation.

of $640 \times 270 \times 128$ grid points was used along axial, radial and azimuthal directions, respectively. The evolution of the mean flow and Reynolds stresses, two-point correlations and energy spectra were studied. The significance of nonlinearity in the near-acoustic field was noted. Detailed analysis of the noise generation in this jet, and its modeling are topics of continuing study. J. Freund (private communication, 1998) has also simulated a low Reynolds number $M_j = 0.9$ jet which duplicates the conditions of Stromberg *et al.* [146] experiment. Such databases will be invaluable in future studies of jet-noise.

It is also instructive to look at simpler jet flows and their noise radiation. Mitchell *et al.* [109] carried out studies of sound radiation from the pairing of axisymmetric vortices in low Reynolds number 'subsonic' jets at Mach numbers ($M_j = U_0/C_\infty$) of 0.4, 0.8 and 1.2. The initial momentum thickness of the jet shear layer was taken be 0.1 of the jet radius to make it easier to resolve the shear layer dynamics. The jet was forced at the inflow boundary with linear stability eigenmodes at the fundamental and its two subharmonic frequencies, and the computational domain chosen to contain two wavelengths of the lowest frequency sound. The acoustic data obtained from the simulations was analyzed into its frequency components.

The acoustic field obtained from the direct simulations was compared to a prediction based on data taken at a Kirchhoff surface. A cylindrical Kirch-

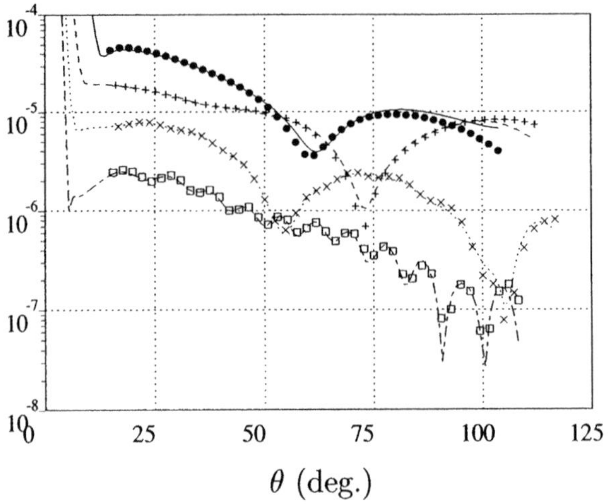

Figure 7.20. Comparison of the sound radiation due to vortex-pairing in a $M_j = 0.4$ jet obtained by DNS (lines) with a Kirchhoff-surface based prediction (symbols). Adapted from Mitchell *et al.* [109]; Solid line and solid circles for $0.5f_0$; dashed line and pluses for f_0; dotted line and crosses for $1.5f_0$; and chain dashed line and squares for $2f_0$, where f_0 is the most unstable frequency at the jet inlet. The vertical axis is the magnitude of dilatation at frequency ω scaled as $x|\hat{\Theta}|R_0/(RC_\infty)$, and the horizontal axis is the angle θ.

hoff surface was placed at a radial distance of 10 initial jet radii, and this was shown to be a nearly ideal location. As Figure 7.20 shows, the acoustic data from DNS matched nearly perfectly with the Kirchhoff surface based predictions. For all cases investigated with different frequencies and Mach numbers and different shear-layer thicknesses the comparison was excellent. This gave further confidence in the DNS data. In fact, a perfect agreement with the Kirchhoff method is not expected, as the Kirchhoff surface is, of necessity, an open surface. Estimates of the correction associated with an open Kirchhoff surface were analyzed by Freund *et al.* [50] using a high frequency approximation. They showed that if the ray from the source to the listener intersects the Kirchhoff surface within the portion where data is available the error associated with the unavailable data on the 'open portion' of the Kirchhoff surface is negligible. Asymptotic correction to account for the open surface were also derived and tested by Freund *et al.* [50].

The acoustic data at the pairing frequency showed that in the lowest Mach number case ($M_j = 0.4$) the radiation field resembled the field of an axisymmetric point quadrupole. Laboratory experiments by Bridges and Hussain [19] also observed such directivity in a low-speed forced jet. However at other frequencies and at higher Mach numbers the DNS data could not be interpreted using a simple localized source model. In fact, the radiation pattern became progressively more concentrated at small angles to the jet with increasing M_j. Complex interference effects were observed for the unforced frequencies, and in the simulations with thinner shear layers. Overall the intensity of the radiated acoustic field scaled well with Lighthill's U_j^8 scaling, and more detailed comparisons with Lighthill's theory were undertaken.

7.7.2. PREDICTIONS BASED ON ACOUSTIC ANALOGIES

The acoustic analogy framework pioneered by Lighthill [94] for predicting aerodynamic noise requires a prescription of the 'quadrupole' sources distributed in the unsteady flow volume and 'dipole' and 'monopole' sources distributed on immersed body surfaces. These methods begin with the Ffowcs Willams-Hawkings integral equation [49] or its reformulation (see Brentner and Farassat [18]). Quotation marks are placed around 'quadrupole' etc. to stress that these labels are intended primarily for bookkeeping. Such source-data is available from a DNS of the flow. Mitchell *et al.* [109] compute only the flow region some distance downstream of the nozzle lip, and ignore any direct effect of the nozzle lip on the flow or the radiated sound. This allows the use of a free-space Green's function in calculating the radiated noise. The computational domain was split into small volumes each of which had a small enough cell-Helmholtz number to justify a compact source approximation within its cell-volume. The overall Lighthill acoustic integral was thus the superposition of the compact quadrupole (source) radiation fields associated with each cell. Numerical experiments showed that a cell-Helmholtz number of 0.05 in the axial direction was sufficient. An additional issue was the finite axial domain size. The DNS data on the Lighthill acoustic sources showed that the source strength decayed quite slowly in the axial direction beyond the saturation location of each frequency. It was necessary to find a method to 'extrapolate' the data to larger distances; an abrupt termination would cause a dominating spurious radiation from the boundary. The decay of sources was fitted by functions whose envelope decayed slowly in the axial direction and the contribution of these 'external' sources to the radiation was analytically calculated. The predictions of Lighthill's theory obtained with these methods compared quite well with the DNS data. Figure 7.21 reproduces one such comparison. The agreement with DNS data was not as close as the Kirchhoff surface prediction, and to some extent depended on the method of extrapolating the source terms. This sensitivity increased at higher Mach numbers and higher frequencies, but the predictions based on Lighthill's theory tracked all major trends of the data quite successfully in amplitude levels and directivity shapes.

Mitchell *et al.* [108] conducted a computational study of Mach wave

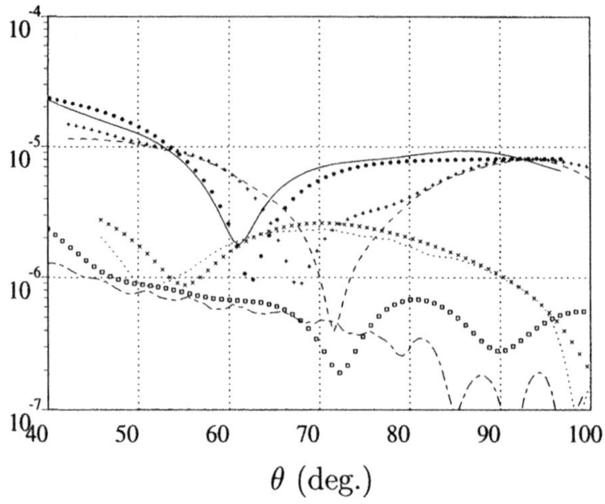

Figure 7.21. Comparison of the sound radiation due to vortex-pairing in a $M_j = 0.4$ jet obtained by DNS (lines) with Lighthill acoustic analogy based prediction (symbols). Adapted from Mitchell *et al.* [109]. Line- and symbols-types are same as in 7.20. The vertical axis is the magnitude of dilatation at frequency ω scaled as $x|\hat{\Theta}|R_0/(RC_\infty)$, and the horizontal axis is the angle θ.

radiation from supersonic jets. In their flow supersonically travelling instability waves undergo amplification and decay without having any noticeable impact on the mean flow of the jet. This is possible because at a low Reynolds number the laminar spreading of the jet is significant and limits the total amplification which instability waves can undergo. This rather artificial regime is unique in that a weakly-nonparallel linear theory for the disturbances is available; more realistic high Reynolds number jets are turbulent and require models which mimic the effects of turbulence. An intense field of Mach waves with wavefronts oriented at the Mach angle Ψ (measured from the negative jet axis) was observed in the DNS. Here the Mach angle $\sin\Psi = 1/M_i$, where $M_i = U_i/C_\infty$ is the Mach number at which the instability waves are propagating with respect to the ambient medium. Figure 7.22 shows a contour plot of dilatation corresponding to the radiation at the most unstable frequency. For this frequency $\Psi \approx 40\,$deg. The highly directional character of the radiation with its peak at an angle Θ (measured from the jet axis) of about 40 deg can be noted.

The acoustic data obtained from the DNS was compared to Kirchhoff surface based prediction which again showed a nearly perfect comparison. Further comparisons were made to predictions based on theory of Tam and Burton [149], and the prediction based on Lighthill's equation with the source terms provided from DNS. The Tam and Burton theory uses an asymptotic matching between the instability wave representation of supersonic vortical disturbances in the jet and an acoustic wave-field at large

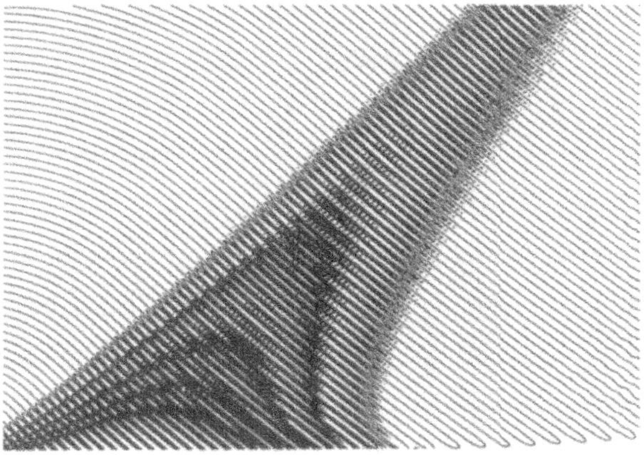

Figure 7.22. Mach wave radiation from supersonic instability waves in a $M_j = 2.0$ jet. Adapted from Mitchell *et al.* [108]. Contours of dilatation are plotted for the most unstable frequency.

distances. The theory requires that the mean flow changes are gradual so that their impact on the instability waves can be accounted via a non-parallel correction. This theory was applied to the Mach 2 laminar jet, and good comparison with the DNS data was obtained. As expected the viscous and non-parallel effects on disturbance evolution were quite important for the low Reynolds number simulation. The DNS data also allowed a test of the Lighthill's acoustic analogy in the context of Mach wave radiation. The quadrupole acoustic source terms for the Lighthill equation were taken from the DNS data, and predictions of the radiation field were made from them with due account of the source non-compactness. Good comparison with the DNS data was again achieved. These different methods for predicting the Mach wave radiation as a function of the angle θ measured from the jet axis are compared in Figure 7.23. This study confirmed that the highly directional radiation associated with Mach wave emission is consistent with the representation of acoustic sources in the Lighthill formulation as a non-compact axial distribution of acoustic quadrupoles. Further analysis using such source representations for a model problem, however, showed that a very accurate representation of the spatially extensive sources is needed in order to accurately preserve the interferences in the quiet region away from the angle of peak radiation. This is a deficiency of the solution method formulated in the $x - \omega$ domain and alternatives solution methods are desireable. Avital [6] has also considered radiation problems with distributed sources with special attention to Mach wave radiation and source-truncation issues.

The high-speed flow situations described so far require that the unsteady

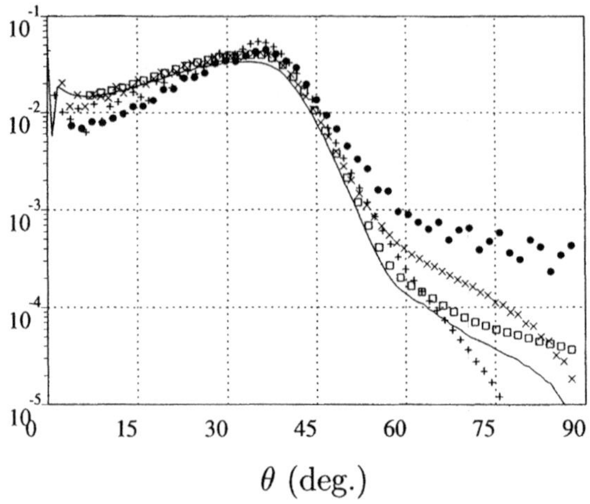

Figure 7.23. Comparison of predictions of Mach wave radiation using different methods (symbols) with DNS data (line). Adapted from Mitchell *et al.* [108]. Circles for Tam and Burton [149] theory; crosses Lighthill sources; pluses Lighthill sources based on linear instability theory; squares Kirchhoff surface method. The vertical axis is the magnitude of dilatation at frequency ω scaled as $x|\hat{\Theta}|R_0/(RC_\infty)$, and the horizontal axis is the angle θ.

flow be obtained by solving the compressible flow equations. There are, however, other circumstances where an incompressible or nearly-incompressible description of the unsteady flow would also be sufficient. Sound generated by a localized region of flow separation in a low-Mach number flow, *e.g.* the vortex shedding noise from a landing-gear strut or the noise due to a partly closed valve in a duct are examples of such cases. These problems can be described by hybrid methods which combine the computations of nonlinear unsteady incompressible flowfields in a limited region of space with an appropriate acoustic formulation. Preserving the correct multipole structure of the acoustic sources, and satisfying the physical boundary conditions on nearby solid-surfaces are additional issues which need to be carefully considered in such hybrid methods. In the area of rotorcraft noise, hybrid methods which combine unsteady flow solutions with a suitable formulation of an acoustic analogy are common, see [47], [16]. An arbitrary motion of bodies is included in the formulation and typically the unsteady flow around airfoil/blade geometries is computed using the compressible equations. However, as noted by Brentner [17] the details of numerical implementation of the acoustic integrals are quite important. Wang *et al.* [165] illustrate the application of Lighthill's acoustic analogy to compute vortex-shedding noise from a low speed flow over an airfoil. A method for treating the unsteady outflow in the context of the acoustic analogy is also given by them.

Wang *et al.* [164] used a hybrid formulation to study the noise radiated by the local breakdown of a boundary layer instability wave packet into a turbulent spot. The simulations introduce a localized region of small amplitude disturbances in a Blasius boundary layer. The disturbances amplified in a compact wavepacket, caused lift up of shear layers from the wall, and its subsequent three-dimensional breakdown into a compact region of turbulence. This incompressible flow simulation provided the acoustic source-field data. Image quadrupoles in the rigid plane were used, Powell [122], to eliminate the surface pressure integral. This left the 'sources' as the net unsteady viscous force on the surface due to the local turbulent spot in the boundary layer, and the unsteady quadrupoles in the disturbed flow region (and its image). The latter was found to generate high frequency sound with a characteristic frequency 5-7 times the basic (TS) instability wave frequency. The contribution of the net unsteady viscous force on the wall to sound radiation remained an open problem (see Dowling [44] for a survey of the controversy about the role of viscous shear stresses on a rigid infinite wall). The simulation showed that if the viscous stresses on the wall are interpreted as acoustic sources, in fact their contribution would overwhelm the quadrupole noise. However, given that the simulations used were incompressible a definitive conclusion could not be reached. It would seem that this is precisely the type of problem where compressible flow computations can help resolve a long standing debate. In fact, promising steps in this direction have already been taken. Shariff and Wang [136] reported results from their compressible flow simulations on a model problem chosen to shed light on the shear-stress controversy. They considered the unsteady flow and sound radiation associated with in-plane wall displacements prescribed over a compact region of an otherwise rigid wall. An unsteady Stokes boundary layer gets set up to satisfy the wall boundary conditions. The simulations yielded a clear dipole radiation pattern due to the net unsteady drag on the wall. This radiation field also agreed perfectly with acoustic analogy if the shear-stress on the wall was taken as a sound source. While the results of the model problem are convincing its connection to the original problem, where the unsteady shear-stress distribution is flow-induced, is not direct, and further work is needed.

Further work is also needed on hybrid methods for aeroacoustic prediction in low-speed flows. Hardin and Pope [64] propose a scheme of splitting the flow variables into (nearly) incompressible and acoustic parts. The nearly incompressible flow is, at leading-order in Mach number M, strictly incompressible. The pressure variations required to maintain a strict divergence-free velocity field in this incompressible flow are linked to an isentropic 'hydrodynamic' density perturbation. This near-field description is consistent with the nearly-isentropic low-Mach number limit discussed in section 7.3.1. To obtain the acoustic field Hardin and Pope [64] subtract the equations governing the 'incompressible' flow from the exact mass and momentum conservation equations. Pressure perturbations are isentropically linked to density perturbations to avoid the need to invoke the exact energy equation. The resulting nonlinear set is viewed as a set of equations appropriate for the acoustic field and is discretized on an acoustic mesh which is chosen with a suitably large mesh spacing so that only the expected large-scale acoustic field is represented. The nearly incompressible flow is advanced on a different (much finer) hydrodynamic mesh. This procedure is algebraically correct and posesses many physically attractive ingredients. There are, however, some subtle aspects to its implementation. These include the fine-to-coarse transfer of variables, suppression of fast-acoustic waves supported on the fine-mesh and the treatment of boundary conditions. Hardin and Pope present results for the classical problems of radiation from a pulsating or oscillating sphere. These are compared to a matched asymptotic solution which accounts for weak-nonlinearity. The tests shown compare very well with the theory and the authors proceed to study the acoustic radiation from a low Mach number cavity flow. Some basic issues pertaining to the method, however, require clarification. The disturbance variables interpreted as 'acoustic' variables by [64] can not be purely acoustic in a general flow. For example, the 'hydrodynamic' density fluctuation being time dependent must also have an accompanying 'hydrodynamic' dilatation and its associated velocity fluctuation which do not correspond to sound radiation. Since the equations used by [64] are 'exact' (within the isentropic assumption) the psuedo-sound associated disturbances are contained in their 'acoustic' variables.

The full method of Hardin and Pope solves nonlinear equations for the 'acoustic' variables. The model problems of sound generation by prescribed unsteady boundary motions are simpler if the only small amplitude motions are considered, *i.e.* a linear forced response is sought. When the technique of Hardin and Pope is analyzed for these simpler cases, it becomes apparent that the numerical technique is closely related to Ribner's theory of sound generation by unsteady fluid dilatations, *i.e.* to using a volume distribution of monopole sources and one is reminded of Crow's [38] criticism. The numerical solutions given for the oscillating sphere case (yielding a dipole radiation) are for a frequency $\omega a/C_0 = ka = 1.25$, where a is the radius of the sphere, ω the frequency (and k the associated wavenumber) and C_0 the ambient sound speed, which is far from being a compact radiator of sound. It is for the compact case ($ka \ll 1$) that the length scale disparity $\lambda/a = 2\pi/(ka)$ becomes extreme and the radiated sound levels drop due to the destructive interference. How well the numerical technique works for compact radiators (dipole, quadrupole, etc.) in a low Mach number flow is presently not clear.

Many issues remain open with regard to the development of other hybrid methods for acoustic prediction also. The use of large eddy simulation (LES) for acoustic predictions is just begining (see for example Bastin [8], Wang [166]), issues regarding how to process LES data for acoustic predictions, the noise impact of the subgrid model, and the general issue of the accuracy needed in the source representation need to be explored. Methods for efficiently dealing with spatially extensive source regions are important to LES applications as well. Ensuring that spurious sound is not generated as the flow exits the computational domain may also need further development. How to best treat the effects of flow-acoustic interaction in a hybrid prediction method, particularly when the flow is not a parallel flow, accounting for the effects of flow compressibility when treating non-compact source-fields, the inclusion of edge-scattering [48], [33], and accounting for nonlinear distortion, and viscous attenuation in long range propagation problems remain challenges for future work.

7.8. Concluding Remarks

These notes are meant to be read as an introduction to the fundamental physical and numerical issues associated with direct numerical simulation of compressible flows. Some illustrative applications of compressible flow DNS from recent literature were included to give a sense of the current state of the art. Specific references, which are cited throughout the paper, should be consulted for technical details. The topics of wall-bounded turbulent flows, turbulent flows with chemical reactions, and the technique of large-eddy simulation were not discussed. Each of these is an important topic in itself and it is hoped that other contributions in this collection of notes will help the reader gain an appreciation of the issues involved.

Acknowledgements

The author is grateful to the organizers of the ERCOFTAC Summer School for the invitation to present this material in lecture form at the second summer school on Transition and Turbulence Modeling at KTH in July 1998, and for their patience during the subsequent preparation of these notes. The author is grateful to Cambridge University Press, American Institute of Physics, American Institute of Aeronautics and Astronautics, Academic Press and Annual Reviews for their kind permission to reproduce various figures. Original sources of the figures are cited along with the figure caption.

References

1. Adams, N. A. and Shariff, K. 1996. A high-resolution hybrid compact-ENO scheme for shock-turbulence interaction problems. *J. Comput. Phys.* **127**, 27-51.
2. Adams, N. A. 1998. Direct numerical simulation of turbulent compression corner flow. *Theor. Comput. Fluid Dyn.* **12**, 109-129.
3. Adamson, T. C. and Messiter, A. F. 1980. Analysis of two-dimensional interactions between shock waves and boundary layers. *Ann. Rev. Fluid Mech.* **12**, 103-138.
4. Agui, J. H. 1998. Shock wave interactions with turbulence and vortices. Ph. D. Thesis, City Univ. of New York, New York.
5. Anyiwo, J.C. and Bushnell, D.M. 1982. Turbulence amplification in shock-wave boundary-layer interaction. *AIAA J.* **20**, 893-899.
6. Avital, E. 1998. A computational and analytical study of sound emitted by free shear flows. Ph. D. Thesis. Department of Engineering, Queen Mary and Westfield College, London.
7. Barre, S., Alem, D. and Bonnet, J. P. 1996. Experimental study of a normal shock/homogeneous turbulence interaction. *AIAA J.* **34**, 968-974.
8. Bastin, F. 1996. Jet noise using large eddy simulation. Annual Research Briefs, 1996, 115-132, Center for Turbulence Research, Stanford University.
9. Batchelor, G. K. 1967. *An Introduction to Fluid Dynamics.* Cambridge University Press.
10. Bayley, B. J., Levermore, C. D., Passot, T. 1992. Density variations in weakly compressible flows. *Phys. Fluids A* **4**, 945-954.
11. Bayliss, A. and Turkel, E. 1982. Far field boundary conditions for compressible flow. *J. Comput. Phys.* **48**, 182-199.
12. Blaisdell, G. A., Mansour, N. N., and Reynolds, W. C. 1993. Compressibility effects on the growth and structure of homogeneous turbulent shear flow. *J. Fluid Mech.* **256**, 443-485.
13. Bogdanoff, D. 1983. Compressibility effects in turbulent shear layers. *AIAA J.* **21**, 926-927.
14. Bonnet, J. P., Debisschop, J. R. and Chambres, O. 1993. Experimental studies of the turbulent structure of supersonic mixing layers. *AIAA* Paper 93-0217.
15. Breidenthal, R. 1990. The sonic eddy - a model for compressible turbulence. *AIAA* Paper 90-0495.
16. Brentner, K. S. and Farassat, F. 1992. Helicopter noise prediction: the current status and future direction. *J. Sound Vib.* **170**, 79-96.
17. Brentner, K. S. 1996. Numerical algorithms for acoustic integrals - the devil is in the details. AIAA Paper 96-1706.
18. Brentner, K. S. and Farassat, F. 1998. Analytical comparison of the acoustic analogy and Kirchhoff formulation for moving surfaces. *AIAA J.* **36**, 1379-1386.
19. Bridges, J. and Hussain, F. 1992. Direct evaluation of aeroacoustic theory in a jet. *J. Fluid Mech.* **240**, 469-501.
20. Brown, G. and Roshko, A. 1974. On density effects and large structure in turbulent mixing layers. *J. Fluid Mech.* **64**, 775-816.
21. Canuto, C., Hussaini, M. Y., Quarteroni, A. and Zang, T. A. 1988. *Spectral Methods in Fluid Dynamics.* Springer Verlag.
22. Carpenter, M. H., Gottlieb, D., Abarbanel, S. 1993. The stability of numerical boundary treatments for compact high-order finite-difference schemes. *J. Comput. Phys.* **108**, 272-295.
23. Chu, B. T. and Kovasznay, L. 1958. Nonlinear interactions in a viscous heat conducting compressible gas. *J. Fluid Mech.* **3**, 494-514.
24. Clemens, N. T. and Mungal, M. G. 1992. Two- and three-dimensional effects in the supersonic mixing layer. *AIAA J.* **30**, 973-981.
25. Clemens, N. T. and Mungal, M. G. 1995. Large-scale structure and entrainment in the supersonic mixing layer. *J. Fluid Mech.* **284**, 171-216.

26. Cockburn, B, and Shu, C. W. 1994. Nonlinearly stable compact schemes for shock calculations. *SIAM J. Numer. Anal.* **31**, 607-627.
27. Coleman, G. N., Kim, J. and Moser, R. D. 1995. A numerical study of turbulent supersonic isothermal-wall channel flow. *J. Fluid Mech.* **305**, 159-183.
28. Collis, S. S. and Lele, S. K. 1997. A computational investigation of receptivity in high-speed flow near a swept leading-edge. Report TF-71, Flow Physics and Computation Division, Department of Mech. Engrg., Stanford University.
29. Colonius, T., Lele, S. K. and Moin, P. 1993. Boundary conditions for direct computation of aerodynamic sound generation. *AIAA J.* **31**, 1574-1582.
30. Colonius, T., Lele, S. K. and Moin, P. 1997. Sound generation in a mixing layer. *J. Fluid Mech.* **330**, 375-409.
31. Colonius, T. 1996. Numerically nonreflecting boundary and interface conditions. AIAA Paper 96-1661.
32. Cortesi, A. B., Smith, B. L., Yadigaraglu, G. and Banerjee, S. 1999. Numerical investigation of the entrainment and mixing processes in a neutral and stably-stratified mixing layers. *Phys. Fluids*, **11**, 162-185.
33. Crighton, D. G. and Leppington, F. G. 1971. On the scattering of aerodynamic noise. *J. Fluid Mech.* **46**, 577-597.
34. Crighton, D. G. 1975. Basic principles of aerodynamic noise generation. *Prog. Aerospace Sci.* **16**, 31-96.
35. Crighton, D. G. and Bashforth, S. 1980. Nonlinear propagation of broadband jet noise. AIAA Paper 80-1039.
36. Crighton, D. G. and Huerre, P. 1990. Shear layer pressure fluctuations and superdirective acoustic sources. *J. Fluid Mech.* **220**, 355-368.
37. Crighton, D. G. 1993. Computational aeracoustics for low Mach number flows. in *Computational Aeroacoustics*, eds. Hardin, J. C. and Hussaini, M. Y., Springer Verlag, pp. 50-68.
38. Crow, S. C. 1970. Aerodynamic sound emission as a singular perturbation problem. *Studies in Appl. Math.* **49**, 21-44.
39. Courant, R. and Friedrichs, K. 1976. *Supersonic flow and shock waves.* Springer-Verlag.
40. Deng, X. and Maekawa. H. Compact high-order accurate nonlinear schemes. *J. Comput. Phys.* **130**, 77-91.
41. Dimotakis, P. E. 1986. Two-dimensional shear-layer entrainment. *AIAA J.* **24**, 1791-1796.
42. Dimotakis, P. A. 1991. Turbulent free shear layer mixing and combustion. in *High-speed Flight Propulsion Systems*, Prog. in Astronaut. Aeronaut., ed. S. N. B. Murthy, E. T. Curran, AIAA-Washington D. C., 265-340.
43. Djeridane, T., Amielh, M., Anselmet, F. and Fulachier, L. 1996. Velocity turbulence properties in the near-field region of axisymmetric variable density jets. *Phys. Fluids* **8**, 1614-1630.
44. Dowling, A. 1992. Flow noise of surfaces. in *Modern Methods in Analytical Acoustics*, chapter 16, 452-509.
45. Dutton, J. C. 1997. Compressible turbulent free-shear layers. in AGARD Report 819, Turbulence in Compressible Flows.
46. Engquist, B. and Majda, A. 1979. Radiation boundary conditions for acoustic and elastic wave calculations. *Comm. Pure and Appl. Math.* **32**, 313-357.
47. Farassat, F. 1993. The acoustic analogy as a tool of computational aeroacoustics. in *Computational Aeroacoustics*, eds. Hardin, J. C. and Hussaini, M. Y., Springer Verlag, 133-155.
48. Ffowcs Williams and Hall, L. H. 1970. Aerodynamic sound generation by turbulent flow in the vicinity of a scattering half plane. *J. Fluid Mech.* **40**, 657-670.
49. Ffowcs Williams, J. E. and Hawkings, D. L. 1969. Sound generation by turbulence and surfaces in arbitrary motion. *Phil. Trans. Royal Soc. A* **264**, 321-342.
50. Freund, J. B., Lele, S. K. and Moin, P. 1996. Calculation of the radiated sound

field using an open Kirchhoff surface. *AIAA J.* **34**, 909-916.
51. Freund, J. B. 1997. Proposed inflow/outflow boundary condition for direct computation of aerodynamic sound. *AIAA J.* **35**, 740-742.
52. Freund, J. B., Moin, P. and Lele, S. K. 1997. Compressibility effects in a turbulent annular mixing layer. Technical Report TF-72, Flow Physics and Computation Division, Mechanical Engineering, Stanford University.
53. Freund, J. B., Lele, S. K. and Moin, P. 1998. Direct simulation of a Mach 1.92 jet and its sound field. AIAA/CEAS Paper 98-2291.
54. Giles, M.B. 1990. Nonreflecting boundary conditions for Euler equation calculations. *AIAA J.* **28**, 2050-2058.
55. Givoli, D. 1991. Non-reflecting boundary conditions. *J. Comput. Phys.* **94**, 1-29.
56. Givoli, D. 1992. *Numerical methods for problems in infinite domains.* Elsevier, Amsterdam.
57. Goldstein, M. E. 1976. *Aeroacoustics.* McGraw Hill, New York.
58. Goldstein, M. E. 1978. Unsteady vortical and entropic distortions of potential flows round arbitrary obstacles. *J. Fluid Mech.* **89**, 305-329.
59. Goldstein, M. E. 1979. Scattering and distortion of the unsteady motion on transversely sheared mean flows. *J. Fluid Mech.* **91**, 601-632.
60. Goldstein, M. E. 1984. Aeroacoustics of turbulent shear flows. *Annual Rev. Fluid Mech.* **16**, 263-285.
61. Gropengeisser, H. 1970. Study of the stability of boundary layers in compressible fluids. NASA TT-F-12.
62. Guarini, S. 1998. A direct numerical simulation of turbulence in a compressible turbulent boundary layer. Ph. D. Thesis, Department of Aeronautics and Astronautics, Stanford University.
63. Hannappel, R. and Friedrich, R. 1994. DNS of a M=2 shock interacting with isotropic turbulence. In *Direct and Large-Eddy Simulation I*, eds. Voke, P.R., Kleiser, L. and Chollet, J. P., Kluwer Academic Publishers.
64. Hardin, J. C. and Pope, D. S. 1994. An acoustic/viscous splitting technique for computational aeroacoustics. *Theor. Comput. Fluid Dyn.* **6**, 323-340.
65. Harten, A. and Osher, S. 1987. Uniformly high-order accurate nonoscillatory schemes, I. *SIAM J. Num. Anal.* **24**, 279-309.
66. Higdon, R. L. 1986. Absorbing boundary conditions for difference approximations to the multidimensional wave equation. *Math. Computation* **47**, 629-651.
67. Hu, F. Q., Hussaini, M. Y. and Manthey, J. L. 1996. Low-dissipation and low-dispersion Runge-Kutta schemes for computational acoustics. *J. Comput. Phys.* **124**, 177-191.
68. Hu, F. Q. 1996. On absorbing boundary conditions for linearized Euler equations in a perfectly matched layer. *J. Comput. Phys.* **129**, 201-219.
69. Huang, P. G., Coleman, G. N. and Bradshaw, P. 1995. Compressible turbulent channel flows: DNS results and modeling. *J. Fluid Mech.* **305**, 185-218.
70. Monkewitz, P. A. and Huerre, P. 1982. Influence of the velocity ratio on the spatial stability of mixing layers. *Phys. Fluids* **25**, 1137.
71. Israeli, M. and Orszag, S. A. 1981 Approximation of radiation boundary condition. *J. Comput. Phys.*, **41**, 115-135.
72. Jackson, T. L. and Grosch, C. E. 1989. Inviscid spatial stability of a compressible mixing layer. *J. Fluid Mech.* **209**, 609-637.
73. Jacquin, L., Blin, E. and Geffroy, P. 1991. Experiments on free turbulence/shock wave interaction. *Eighth Symposium on Turbulent Shear Flows*, Munich, p. 1-2-1 to 1-2-6.
74. Kaltenbach, H. J., Fatica, M , Mittal, R, Lund, T. S. and Moin, P. 1999. Study of flow in a planar asymmetric diffuser using large-eddy simulation. to appear in *J. Fluid Mech.*
75. Kambe, T. 1986. Acoustic emissions by vortex motions. *J. Fluid Mech.* **173**, 643-666.

76. Karasso, P. S. and Mungal, M. G. 1996. Scalar mixing and reaction in plane liquid shear layers. *J. Fluid Mech.* **323**, 23-63.
77. Kevlahan, N. 1996. The propagation of weak shocks in non-uniform flows. *J. Fluid Mech.* **327**, 161-197.
78. Knight, D. D. 1997. Numerical simulations of compressible turbulent flows using the Reynolds-averaged Navier-Stokes equations. in AGARD Report 819, Turbulence in Compressible Flows.
79. Kosloff, R. and Kosloff, D. 1986. Absorbing boundaries for wave propagation problems. *J. Comput. Phys.* **63**, 1986, 363-376.
80. Kovásznay, L.S.G. 1953. Turbulence in supersonic flow. *J. Aero. Sci.* **20**, 657-682.
81. Lagerstrom, P. A. 1964. *Laminar Flow Theory.* Princeton Univ. Press. reprinted from *Theory of Laminar Flows*, F. K. Moore, Editor, Vol. IV of *High Speed Aerodynamics and Jet Propulsion.*
82. Landau, L. D. 1944. Stability of tangential discontinuities in compressible fluid. *Dokl. Akad. Nauk. SSSR* **44**, 139-141.
83. Le, H., Moin, P. and Kim, J. 1997. Direct numerical simulation of turbulent flow over a backward-facing step. *J. Fluid Mech.* **330**, 349-374.
84. Lee, S., Lele, S. K. and Moin, P. 1992a. Simulation of spatially evolving compressible turbulence and the applicability of Taylor's hypothesis. *Phys. Fluids A* **4**, 1521-1530.
85. Lee, S., Moin, P. and Lele, S.K. 1992b. Interaction of isotropic turbulence with a shock wave. *Report TF-52*, Department of Mechanical Engineering, Stanford University.
86. Lee, S., Lele, S.K. and Moin, P. 1993. Direct numerical simulation of isotropic turbulence interacting with a weak shock wave. *J. Fluid Mech.*, **251**, 533-562: corrigendum in *J. Fluid Mech.* **264**, 373- 374.
87. Lee, S., Lele, S. K. and Moin, P. 1997. Interaction of isotropic turbulence with shock waves: effect of shock strength. *J. Fluid Mech.* **340**, 225-247.
88. Lees, L. and Lin, C. C. 1946. Investigation of the stability of the laminar boundary layer in a compressible fluid. NACA TN 1115.
89. Lessen, M., Fox., J. A. and Zien, H. M. 1965. On the inviscid stability of the laminar mixing of two parallel streams of a compressible fluid. *J. Fluid Mech.* **23**, 355-367.
90. Lele, S. K. 1989. Direct numerical simulation of compressible free shear flows. AIAA Paper 89-0374.
91. Lele, S. K. 1992. Compact finite difference schemes with spectral-like resolution. *J. Comp. Phys.* **103**, 16-42.
92. Lele, S. K. 1994. Compressibility effects on turbulence. *Ann. Rev. Fluid Mech.* **26**, 211-254.
93. Liepmann, H. and Roshko, A. 1957. *Elements of Gas Dynamics.* John Wiley.
94. Lighthill, M. J. 1952. On sound generated aerodynamically: I. General theory. *Proc. Roy. Soc. London*, Ser. A **221** 564-587.
95. Lilley, G. M. 1974. On the noise from jets. *AGARD CP-131.*
96. Lu, G. and Lele, S. K. 1994. On the density ratio effect on the growth rate of a compressible mixing layer. *Phys. Fluids* **6**, 1073-1075.
97. Lund, T., Wu, X. and Squires, K. D. 1998. Generation of turbulent inflow data for spatially-developing boundary layer simulations. *J. Comput. Phys.* **140**, 233-258.
98. Mack, L. M. 1984. Boundary layer linear stability theory. AGARD Report 709.
99. Mahadevan, R. and Loth, E. 1994 High-speed cinematography of compressible mixing layers. *Experiments in Fluids* 17, 179-189.
100. Mahesh, K., Lele, S. K. and Moin, P. 1994. The response of anisotropic turbulence to rapid homogeneous one-dimensional compression. *Phys. Fluids.* **6**, 1052-1062.
101. Mahesh, K., Lee, S., Lele, S. K. and Moin, P. 1995. The interaction of an isotropic field of acoustic waves with a shock wave. *J. Fluid Mech.* **300**, 383-407.
102. Mahesh, K., Lele, S. K. and Moin, P. 1996. The interaction of a shock wave with

a turbulent shear flow. Report TF-69, Thermosciences Division, Mechanical Engineering Department, Stanford University.
103. Mahesh, K., Lele, S. K. and Moin, P. 1997. The influence of entropy fluctuations on the interaction of turbulence with a shock wave. *J. Fluid Mech.* **334**, 353-379.
104. Manning, T. A. and Lele, S. K. 1998. Numerical simulations of shock-vortex interactions in supersonic jet screech. AIAA Paper AIAA-98-0282.
105. Meadows, K.R., Caughey, D.A. and Casper, J. 1993. Computing unsteady shock waves for aeroacoustic applications. *AIAA Paper No.* 93-4329, Long Beach, CA.
106. Miller, M. F., Bowman, C. T. and Mungal, M. G. 1998. An experimental investigation of the effects of compresssibility on a turbulent reacting mixing layer. *J. Fluid Mech.* **356**, 25-64.
107. Mitchell, B. E., Lele, S. K. and Moin, P. 1995. Direct computation of the sound from a compressible co-rotating vortex pair. *J. Fluid Mech.* **285**, 181-202.
108. Mitchell, B. E., Lele, S. K. and Moin, P. 1997. Direct computation of Mach wave radiation in an axisymmetric jet. *AIAA J.* **35**, 1574-1580.
109. Mitchell, B. E., Lele, S. K. and Moin, P. 1999. Direct computation of the sound generated by vortex pairing in an axisymmetric jet. *J. Fluid Mech.* **383**, 113-142.
110. Mohring, W. 1978. On vortex sound at low Mach number. *J. Fluid Mech.* **85**, 685-691.
111. Moin, P. and Mahesh, K. 1998. Direct numerical simulation: A tool in turbulence research. *Ann. Rev. Fluid Mech.* **30**, 539-578.
112. Moore, F. K. 1953. Unsteady oblique interaction of a shock wave with a plane disturbance. *NACA TN-2879.*
113. Moyal, J. E. 1952. The spectra of turbulence in a compressible fluid: eddy turbulence and random noise. *Proc. Cambridge Phil. Soc.* **48**, 329-344.
114. Na, Y. and Moin, P. 1998. Direct numerical simulation of a separated turbulent boundary layer. *J. Fluid Mech.* **374**, 379-405.
115. Papamoschou, D. 1991a. Structure of the compressible turbulent shear layer. *AIAA J.* **29** 680-681.
116. Papamoschou, D. 1991b. Effect of Mach number on communication between regions of a shear layer. *Proc. Eighth Symp. on Turbulent Shear Flows* 21-5-1 to 21-5-6.
117. Papamoschou, D. and Roshko, A. 1988. The compressible turbulent mixing layer: an experimental study. *J. Fluid Mech.* **197**, 453-477.
118. Papamoschou, D. and Lele, S. K. 1993. Vortex-induced disturbance field in a compressible shear layer. *Phys. Fluids A* **5**, 1412-1419.
119. Papamoschou, D. and Bunyajitradulya, A. 1997. Evolution of large eddies in compressible shear layers. *Phys. Fluids* **9**, 756-765.
120. Poinsot, T. and Lele, S. K. 1992. Boundary conditions for direct simulation of compressible viscous flows. *J. Comput. Phys.*, **101**, 104-129.
121. Powell, A. 1964. Theory of vortex sound. *J. Acous. Soc. America* **36**, 177-195.
122. Powell, A. 1960. Aerodynamic noise and a plane boundary. *J. Acous. Soc. America* **32**, 982-990.
123. Rai, M. M. and Moin, P. 1993. Direct numerical simulation of transition and turbulence in a spatially evolving boundary layer. *J. Comput. Phys.* **109**, 169-192.
124. Ragab, S. A. and Wu, J. L. 1989. Linear instability in two-dimensional compressible mixing layers. *Phys. Fluids A* **1**, 957-966.
125. Ribner, H.S. 1953. Convection of a pattern of vorticity through a shock wave. *NACA TN-2864.*
126. Ribner, H.S. 1954. Shock-turbulence interaction and the generation of noise. *NACA TN-3255.*
127. Ribner, H. S. 1985. Cylindrical sound wave generation by shock-vortex interaction. *AIAA J.* **23**, 1708-1715.
128. Ribner, H. S. 1987. Spectra of noise and amplified turbulence emanating from shock-turbulence interaction. *AIAA Journal*, **25**, 436-442.
129. Ribner, H. S. 1998. Comment on 'Experimental study of a normal

shock/homogeneous turbulence interaction'. *AIAA J.* **36**, 494-495.
130. Ristorcelli, J. R. 1997. A pseudo-sound constitutive relationship for the dilatational covariances in compressible turbulence. *J. Fluid Mech.* **347**, 37-70.
131. Ristorcelli, J. R. and Blaisdell, G. A. 1997. Consistent initial conditions for the DNS of compressible turbulence. *Phys. Fluids* **9**, 4-6.
132. Sandham, N. D. and Reynolds, W. C. 1990. Compressible mixing layer: linear theory and direct simulation. *AIAA J.* **28**, 618-624.
133. Sarkar, S., Erlebacher, G., Hussaini, M. Y. and Kreiss, H. O. 1991. The analysis and modeling of dilatational terms in compressible turbulence. *J. Fluid Mech.* **227**, 473-493.
134. Sarkar, S. 1995. The stabilizing effect of compressibility in turbulent shear flow. *J. Fluid Mech.* **282**, 163-186.
135. Schilichting, H. 1955. *Boundary Layer Theory*, McGraw Hill.
136. Shariff, K. and Wang, M. 1996. A numerical experiment on the role of surface shear stress in the generation of sound. *Bull. American Physical Soc.* **41**, 1735.
137. Shapiro, A. 1953. *The Dynamics and Thermodynamics of Compressible Flow* Vol. 1. John Wiley.
138. Shu, C.-W. 1990. Numerical experimentation on the accuracy of ENO and modified ENO schemes. *J. Sci. Comp.* **5**, 127-149.
139. Shu, C.-W. and Osher, S. 1989. Efficient implementation of essentially non-oscillatory shock-capturing schemes, II. *J. Comp. Phys.* **83**, 32-78.
140. Seiner, J. M. 1984. Advances in high speed jet aeroacoustics. AIAA Paper 84-2275.
141. Simone, A., Coleman, G. N. and Cambon, C. 1997. The effect of compressibility on turbulent shear flow: a rapid-distortion-theory and direct-numerical-simulation study. *J. Fluid Mech.* **330**, 307-338.
142. Spina, E. F., Smits, A. J. and Robinson, S. K. 1994. The physics of supersonic turbulent boundary layers. *Ann. Rev. Fluid Mech.* **26**, 287-319.
143. Spalart, P. 1988. Direct numerical simulation of a turbulent boundary layer up to $R_\theta = 1410$. *J. Fluid Mech.* **187**, 61-98.
144. Staquet, C. 1995. Two-dimensional secondary instabilities in a strongly stratified shear layer. *J. Fluid Mech.* **296**, 73-126.
145. Streett, C. L. and Macaraeg, M. G. 1989. Spectral multidomain for large-scale fluid dynamic simulations. *Appl. Numer. Math.* **6**, 123-139.
146. Stromberg, J. L., McLaughlin, D. K. and Trout, T. R. 1980. Flow field and acoustic properties of Mach number 0.9 jet at a low Reynolds number. *J. Sound Vib.* **72**, 159-176.
147. Taasan, S. and Nark, D. M. 1995. An absorbing buffer zone technique of acoustic wave propagation. AIAA Paper 95-0164.
148. Tam, C. K. W, and Morris, P. J. 1980. The radiation of sound by the instability waves of a compressible plane turbuleny shear layer. *J. Fluid Mech.* **98**, 349-381.
149. Tam, C. K. W. and Burton, D. E. 1984. Sound generated by instability waves of supersonic flow. Part 2: axisymmetric jets. *J. Fluid Mech.* **138**, 273-295.
150. Tam, C. K. W. and Hu, F. Q. 1989a. On three families of instability waves of high speed jets. *J. Fluid Mech.* **201**, 447-483.
151. Tam, C. K. W. and Hu, F. Q. 1989b. The instability and acoustic wave modes of supersonic mixing layers inside a rectangular channel. *J. Fluid Mech.* **203**, 51-76.
152. Tam, C. K. W. and Webb, J. C. 1993. Dispersion-relation-preserving finite difference schemes for computational acoustics. *J. Comput. Phys.* **107**, 262-281.
153. Tam, C. K. W. and Webb, J. C. 1994. Radiation boundary condition and anisotropy correction for finite difference solutions of the Helmholtz equation. *J. Comput. Phys.* **113**, 122-133.
154. Tam, C. K. W. and Dong, Z. 1994. Wall boundary conditions for high-order finite difference schemes in computational aeroacoustics. *Theor. and Comput. Fluid Dyn.* **6**, 303-322.
155. Tam, C. K. W. 1995. Computational aerocoustics: issues and methods. *AIAA J.*

33, 1788-1796.
156. Thompson, P.A. 1985 *Compressible-Fluid Dynamics.* Maple Press.
157. Thompson, K. W. 1990. Time dependent boundary conditions for hyperbolic systems II. *J. Comput. Phys.* **89**, 439-461.
158. Tennekes, H. and Lumley, J. L. 1974. *A First Course in Turbulence.* MIT Press.
159. Trefethen, L. N. 1982. Group velocity in finite difference schemes. *SIAM Review* **24**, 113-136.
160. Urban, W. D. and Mungal, M. G. 1998. A PIV study of compressible shear layers. in *9th International Symposium on Applications of Laser Techniques to Fluid Mechanics*, Lisbon, Portugal, July 13-16, 1998.
161. Vichnevetsky, R. and Bowles, J. B. 1982. *Fourier analysis of numerical approximations of hyperbolic equations.* SIAM.
162. von Mises, R. 1958. *Mathematical theory of compressible fluid flow.* Academic Press.
163. Vreman, A. W., Sandham, N. D. and Luo, K. H. 1996. Compressible mixing layer growth rate and turbulence characteristics. *J. Fluid Mech.* **320**, 235-258.
164. Wang, M., Lele, S. K. and Moin, P. 1996a. Sound radiation during local laminar breakdown in a low-Mach-number boundary layer. *J. Fluid Mech.* **319**, 197-218.
165. Wang, M., Lele, S. K. and Moin, P. 1996b. Computation of quadrupole noise using acoustic analogy. *AIAA J.* **34**, 2247-2254.
166. Wang, M. 1997. Progress in large-eddy simulation of trailing-edge turbulence and aeroacoustics. Annual Research Briefs, 1997, 37-49, Center for Turbulence Research, Stanford University.
167. Wray, A.A. 1986. Very low storage time-advancement schemes. Internal Report, NASA-Ames Research Center.
168. Yates, J. E. 1978. Application of the Bernoulli enthalpy concept to the study of vortex noise and jet impingement noise. NASA Contractor Rep. 2987.
169. Zank, G. P. and Matthaeus, W. H. 1991. The equations of nearly incompressible fluids. I. Hydrodynamics, turbulence, and waves. *Phys. Fluids A* **3**, 69-82.
170. Zeman, O. 1990. Dilatation dissipation: the concept and application in modeling compressible mixing layers. *Phys. Fluids A* **3**, 951-955.

Chapter 8

TURBULENT COMBUSTION MODELLING

JAMES J. RILEY
Dept. of Mechanical Engineering, University of Washington
Seattle, WA 98195, USA

8.1. Introduction

8.1.1. GENERAL FEATURES

Turbulent combustion, or, more generally, turbulent reacting flows, is a phenomenon of importance in both nature and technology. For example, in nature, forest fires are an important aspect of the evolution of a forest ecosystem (see figure 8.1). In another example, the dispersion of man-made pollutants into the atmosphere and oceans, and subsequent chemical reaction with ambient chemical species, often have important consequences, for example in the generation of greenhouse gases or in the reduction of ozone in the polar stratospheres. In technology, turbulent combustion is important in a wide range of processes, ranging from the behavior of various propulsion systems, to energy generation and materials processing.

The understanding and prediction of the behavior of turbulent, reacting flows is generally much more difficult than that for nonreacting flows. First of all the flow is usually compressible, so that, in addition to the conservation of mass and momentum and constitutive relationships, also the conservation of energy as well as gas laws must be considered. Furthermore, conservation equations for chemical species must be included, often for many chemical species and numerous chemical reactions. The reactions can result in large heat release, resulting, for example, in large temperature and density variations. Turbulent combustion often involves multiple phases, for example liquid fuel droplets in a gaseous flow. These fuel droplets can be evaporating, and possibly agglomerating as well. The large temperatures resulting from the chemical reactions can lead to significant

Figure 8.1. Aerial view of large forest fire

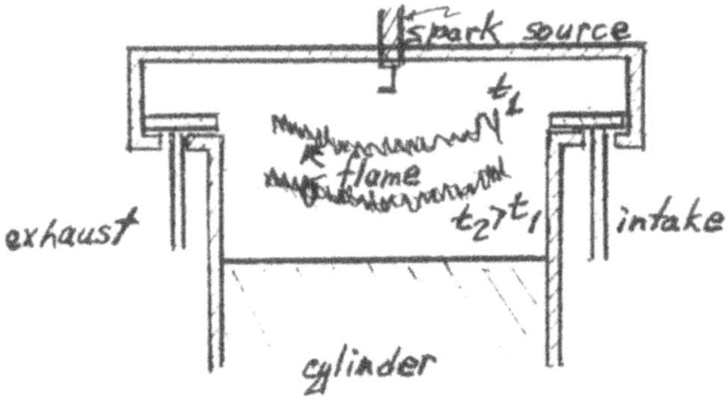

Figure 8.2. Sketch of premixed flame propagation

effects of electromagnetic radiation, especially in the presence of particles, *e.g.*, soot. Finally, some combustion problems contain locally supersonic flow speeds and shock waves, resulting from either very strong heat release (detonations) or large ambient flow speeds.

To conceptually simplify the problem of turbulent combustion, especially in the development and utilization of models, two classifications of reactions are often used: premixed and nonpremixed. In premixed reactions, *e.g.*, in an internal combustion engine (see figure 8.2), the premixed fuel and oxidant are ignited locally, for example by a spark. The flame then propagates through the mixture, leaving hot products in its wake. In nonpremixed combustion, for example a fuel jet issuing into air (see figure 8.3), the fuel and air are introduced in two separate streams. Chemical reaction

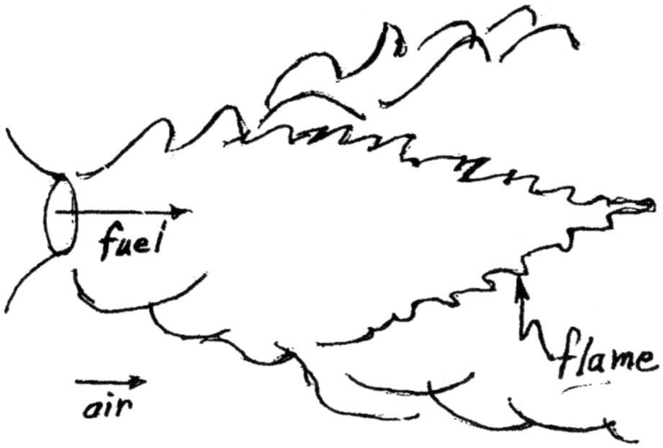

Figure 8.3. Sketch of nonpremixed flame

can occur as the fuel and air mix locally at the molecular level. Therefore the flame can exist where the fuel and air are mixed locally, so that the chemical reaction is, at least to some extent, controlled by the mixing rate. Finally, some reactions occur with partially premixed species. For example, in the problem of a fuel jet into air, it is possible for the fuel and air to mix somewhat before ignition occurs.

8.1.2. PREDICTIVE APPROACHES

The various modelling approaches applied to nonreacting flows have been extended to treat reacting flows. As explained below, however, the difficulties in treating terms involving the averaged reaction rate have led to the development of several other different approaches.

The Reynolds averaging approach

Consider the dynamic quantity $f(\mathbf{x}, t)$, which could represent, *e.g.*, temperature, the concentration of a chemical species, or a component of velocity. Here the vector $\mathbf{x}$ gives the spatial coordinates and t is time. The Reynolds average of f, denoted by $\bar{f}$, can be defined as an ensemble, spatial, and/or temporal average, depending on the problem. It is often convenient to use Reynolds' decomposition,

$$f = \bar{f} + f',$$

where f' is the fluctuation about the average. For variable density flows it is usually helpful to introduce the density-weighted, or Favre, average, denoted by $\tilde{f}$ and defined by

$$\tilde{f} = \overline{\rho f}/\bar{\rho},$$

where $\rho(\mathbf{x}, t)$ is the fluid density. In terms of the Favre average, the decomposition of f is written as

$$f = \tilde{f} + f'' ,$$

where f'' is the fluctuation about the density-weighted average.

To understand the basic difficulty of applying averaging methods to reacting flows, consider the approximately constant density, one-step Arrhenius reaction,

$$F + r\,O \rightarrow (1+r)P + \text{heat} .$$

Here, one unit of mass of fuel (F) combines with r units of mass of oxidant (O) to produce $(1+r)$ units of mass of product (P) in addition to heat; the fuel reaction rate is approximately given by [22]

$$\dot{\omega}_f = -\rho^2 B e^{-T_a/T} Y_f Y_o .$$

Here B is a rate constant, T_a is the activation temperature, T is the fluid temperature, and Y_f and Y_o are the mass fractions of fuel and oxidant, respectively. The conservation equation for fuel is given by:

$$\frac{\partial Y_f}{\partial t} + \mathbf{v} \cdot \nabla Y_f = -\rho B e^{-T_a/T} Y_f Y_o + \mathcal{D} \nabla^2 Y_f . \tag{8.1}$$

Here $\mathbf{v}$ is the fluid velocity, and $\mathcal{D}$ is the diffusion coefficient of Y_f, assumed here for simplicity to be a constant. Averaging Equation 8.1 gives:

$$\frac{\partial \bar{Y}_f}{\partial t} + \bar{\mathbf{v}} \cdot \nabla \bar{Y}_f = -\overline{\rho B e^{-T_a/T} Y_f Y_o} + \mathcal{D} \nabla^2 \bar{Y}_f - \nabla \cdot \left(\overline{\mathbf{v}' Y_f'} \right) .$$

Two unclosed terms appear due to the averaging procedure. One, $\nabla \cdot \left(\overline{\mathbf{v}' Y_f'} \right)$, representing the transport of $\bar{Y}_f$ by the turbulence, is analogous to the Reynolds stress term arising in the averaged momentum equation, and can be modeled by standard methods.

The second term, $\overline{\rho B e^{-T_a/T} Y_f Y_o}$, is the average of the reaction rate term, and presents the essential difficulties. Consider first the case of approximately constant temperature T, *e.g.*, for liquids. Then, with

$$k = \rho^2 B e^{-T_a/T} \approx \text{const.} ,$$

$$\bar{\dot{\omega}}_f = -k \left(\bar{Y}_f \bar{Y}_o + \overline{Y_f' Y_o'} \right) .$$

Often, for example in nonpremixed reactions, the flame zone is very thin. In this case the fluid consists of regions of either fuel or oxidant, with very little overlap (where the flame can occur). The species are highly segregated, and so

$$\left| \frac{\bar{\dot{\omega}}_f}{k} \right| , \overline{Y_f Y_o} \ll \bar{Y}_f \bar{Y}_o , |\overline{Y_f' Y_o'}| .$$

In this case, $\overline{Y_f'Y_o'} \sim -\bar{Y}_f\bar{Y}_o$, and great care must be taken in its modelling, for the small difference between $\bar{Y}_f\bar{Y}_o$ and the model for $\overline{Y_f'Y_o'}$ gives the average reaction rate.

Note that this averaging approach, for the case with $T \approx$ const., is often used in problems in chemical engineering (see, *e.g.*, [17], [17]).

If $T \neq$ const., then, expanding the exponent in a Taylor series, and for simplicity ignoring the temperature dependence of ρ gives (*e.g.*, [1], [1]):

$$\begin{aligned}\overline{\dot{\omega}_f} &= -\rho B\overline{\left(1 - \frac{T_a}{T} + \frac{1}{2}(\frac{T_a}{T})^2 + \ldots\right)(Y_f Y_o)} \\ &= -\rho B\left(\overline{Y_f Y_o} - T_a\overline{(\frac{Y_f Y_o}{T})} + \frac{1}{2}\overline{(\frac{Y_f Y_o}{T})^2} + \ldots\right).\end{aligned}$$

Clearly, even for small temperature fluctuations, this will be very difficult to model, as various higher order correlations of $1/T$ and the species mass fractions will need to be modeled accurately to produce reliable predictions. This is generally not possible, so that other avenues of approach have been developed.

Phenomenological models

Because of the difficulties inherent in modelling the averaged reaction rate, a number of phenomenological models have been developed, which are designed to include some of the physics of the coupled turbulent mixing and reaction process. Two prominent, but distinctly different, models are the eddy-breakup model [31] and the linear eddy model [18].

The eddy-breakup model relies on the fact that the average reaction rate depends on the average mixing rate, especially for nonpremixed reactions. Therefore the average reaction rate, for example for fuel, is modeled as

$$\overline{\dot{\omega}_f} \sim -\frac{\bar{Y}_f}{T_m},$$

where T_m is a "mixing time". The problem is then to obtain a reasonable estimate for T_m, of which several have been proposed. Usually it is argued that the rate of mixing is controlled by the large eddies, and hence T_m is related to the large-eddy time, *i.e.*,

$$T_m \sim \frac{k}{\epsilon},$$

where $k = \frac{1}{2}\overline{v_i'^2}$ is the turbulent kinetic energy per unit mass, and ϵ is the energy dissipation rate.

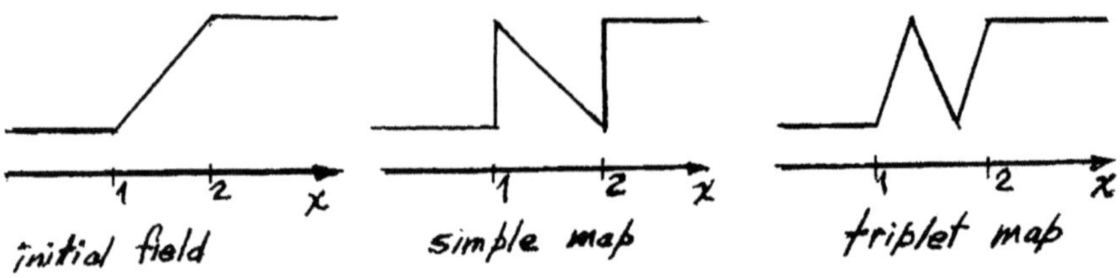

Figure 8.4. Mappings used in linear eddy model

When combined with the Reynolds-Averaged Navier-Stokes (RANS) equations as a model for the average reaction rate, the eddy-breakup model in its various forms has been utilized extensively in applications.

In the linear eddy model (LEM), it is realized that there is no hope of a full numerical simulation of high Reynolds number, three-dimensional turbulent flows, due to the numerical resolution required (see Section 3.1 below). There can be enough resolution in a one-dimensional simulation, however, assuming that the effects of three-dimensionality and the large-scale turbulence can be suitably modeled. In the LEM approach, the problem is treated on a very high resolution, one-dimensional grid. The effects of the larger-scale turbulence are treated with a randomly applied mapping of the scalar field (see figure 8.4), while the molecular diffusion and reaction are computed directly by solving the one-dimensional conservation equations. The statistics of the mapping process, in particular its length and time scales, are chosen for consistency with various turbulence statistics. To implement the LEM, pseudo-random sequences of mappings are generated numerically, enabling realizations of the phenomena of interest to be computed. Statistical results are obtained by ensemble, space or time averaging as appropriate.

The LEM has been found to be useful in testing theoretical ideas, *e.g.*, the effects of differential diffusion of reacting species [19], and more recently it has been shown to have potential as a subgrid-scale model in large-eddy simulations [23]. And very recently, a dynamic version of this model, consistent with a momentum balance and labeled one-dimensional turbulence (ODT), has been formulated.

Probability density function (pdf) models

Standard techniques exist for deriving the evolutionary equation for the joint probability density of the chemical species concentrations (and sometimes the temperature, pressure and fluid velocity; see, *e.g.*, [29], [29]). This equation has the important advantage that the reaction rate terms are closed, *i.e.*, they require no modelling. On the other hand new terms arise requiring closure, most notably the turbulent mixing term. Another difficulty with this approach is that the dimensionality of the equations increases with the number of chemical reactants (and other quantities included in the joint probability density). Therefore, for any number of reactants more than about one or two, this equation is probably impossible to directly integrate numerically.

A number of models have been proposed for the turbulent mixing term, leading to reasonable results, although weaknesses have been pointed out in each case. To numerically integrate the equations, Monte-Carlo methods are appropriate. In analogy with the Monte-Carlo solution of the Fokker-Planck equation by solving the Langevin equation [32], Langevin-type models can be constructed that are consistent with the modeled pdf evolution equation. This approach has the advantage that the required computer resources increase only linearly with the number of chemical species, making the simulations feasible.

Several researchers, most notably Pope and his collaborators, have led in the development and application of this methodology, and it is currently receiving considerable attention in applications. Also, a version of it has been recently developed as a subgrid-scale model for large eddy simulations [7].

Mixture fraction based models

For nonpremixed chemical reactions, it can be very useful to work with conserved scalars, most importantly the mixture fraction, written here as Z. Assuming equal molecular diffusivities for all chemical species, the mixture fraction can be defined as the local mass fraction of material originating in the fuel stream. For infinite rate or equilibrium chemistry, the concentrations of the various chemical species can be functionally related to the mixture fraction, so that average concentrations can be expressed in terms of these relationships and the probability density of the mixture fraction. For finite-rate chemistry, theories have been developed relating the average concentrations to the joint probability density of both Z and its dissipation rate χ.

The mixture-fraction approach, when combined with the RANS equations, has proven to be very useful and will be discussed in Section 2; its use as a subgrid-scale model will be addressed in Section 3.

Numerical simulation

As computers have become more powerful, and numerical methods more accurate, it has become possible to simulate the detailed, unsteady, three-dimensional structures of turbulent flows. In direct numerical simulation

(DNS), all of the relevant spatial scales are resolved on the computational grid. This approach has the advantage that the closure problem introduced by averaging is circumvented, so that no modelling is required. Due to computer memory and speed limitations, however, the method is limited in the range of Reynolds and Damköhler numbers that can be treated. At the present time this approach is mainly limited to a research tool.

In order to avoid the limitations of DNS, the technique of large-eddy simulation (LES) has been developed. In this approach the governing equations are spatially filtered prior to numerical solution, and only length scales resolvable on the computational mesh are retained. Modelling the effects of the filtered, subgrid-scales is then required. If sufficient resolution can be applied to such flows, then very little energy is removed by the filtering operation, so that the resulting flow may be somewhat insensitive to the model for the subgrid scales. Furthermore, the smallest scales of turbulent flow tend to be universal, so that a successful model for one case may be applicable to many other flows. Finally, the larger-scale motions, which control the behavior of the flow, are computed directly, so that the approach has the potential to be very reliable.

This approach also has significant computer requirements compared to RANS; however, it has great potential in applications.

Both DNS and LES applied to chemically-reacting flows will be discussed in detail in Section 3 on large-eddy simulations.

8.1.3. MATHEMATICAL PROBLEM

In the presentation to follow, a number of simplifying assumptions will be employed in order to ease the discussion. The assumptions are:

1. low Mach number (the combustion approximation)
2. single phase (gaseous) flow, usually of an ideal gas
3. neglect of electromagnetic radiation
4. simplified chemistry: 1-step, Arrhenius (mostly)
5. Newtonian fluid; Fourier's and Fick's laws for the diffusion of heat and mass
6. nonpremixed (mostly)

All of these assumptions but the first, low Mach number, are self evident. The assumption of low Mach number implies that the ambient flow speeds are low, and that heat release is not strong, *i.e.*, that there are no detonation waves. This implies that acoustic waves are not dynamically important, and enables much larger time steps to be taken in any numerical simulation. In this case, density variations are due to differences in the density of the species and heat release resulting from the chemical reactions.

Consider the problem of the low speed flow of a fuel jet of velocity U_J, density ρ_J, and viscosity μ_J issuing into air at pressure p_E. The jet nozzle diameter is D. Using U_J, ρ_J, and D to nondimensionalize the governing equations, the momentum equation in particular becomes:

$$\frac{\partial}{\partial t}\rho v_i + \frac{\partial}{\partial x_j}\rho v_i v_j = -\gamma M^2 \frac{\partial p}{\partial x_i} + \frac{1}{Re}\frac{\partial}{\partial x_j}\tau_{ij}$$

Here γ is the ratio of specific heats for the fuel, M is the Mach number defined by

$$M^2 = \frac{U_J^2}{\gamma p_E/\rho_J},$$

and Re is the Reynolds number,

$$Re = \frac{\rho_J U_J D}{\mu_J}.$$

Also, p is the pressure, and τ_{ij} is the viscous stress tensor.

Assuming the Mach number is small, specifically $\gamma M^2 \ll 1$, the dependent variables are expanded in power series in γM^2, *e.g.*,

$$p(\mathbf{x}, t) = p_0(\mathbf{x}, t) + \gamma M^2 p_1(\mathbf{x}, t) + (\gamma M^2)^2 p_2(\mathbf{x}, t) + \ldots$$

Plugging in such expansions and equating coefficients of γM^2 to zero gives:

lowest order: $\quad 0 = \dfrac{\partial p_0}{\partial x_i}$

first order:

$$\frac{\partial}{\partial t}\rho_0 v_{0i} + \frac{\partial}{\partial x_j}\rho_0 v_{0i} v_{0j} = -\frac{\partial p_1}{\partial x_i} + \frac{1}{Re}\frac{\partial}{\partial x_j}\tau_{0ij} \tag{8.2}$$

The lowest order equation implies that p_0, interpreted as the thermodynamic pressure, is uniform in space, consistent with the assumption of low Mach number (very large sound speed). The pressure p_1 in the first order equation is interpreted as the dynamic pressure.

With the assumptions given above, the mathematical problem becomes, dropping most of the subscripts:

- conservation of mass:

$$\frac{\partial \rho}{\partial t} + \frac{\partial}{\partial x_i}\rho v_i = 0 \tag{8.3}$$

- conservation of momentum:

$$\frac{\partial}{\partial t}\rho v_i + \frac{\partial}{\partial x_j}\rho v_i v_j = -\frac{\partial p_1}{\partial x_i} + \frac{1}{Re}\frac{\partial}{\partial x_j}\tau_{ij} \tag{8.4}$$

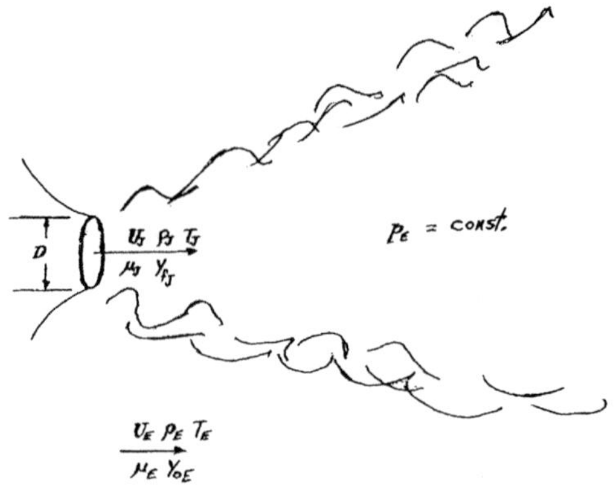

Figure 8.5. Mathematical problem: jet into coflowing stream

- conservation of energy (h is the mixture enthalpy; h_α is that of specie α)

$$\frac{\partial}{\partial t}\rho h + \frac{\partial}{\partial x_k}\rho h v_k = \frac{\partial}{\partial x_k}\left[\frac{\mu}{\sigma}\frac{\partial h}{\partial x_k} + \mu\sum_{\alpha=1}^{N}\left(\frac{1}{Sc_\alpha} - \frac{1}{\sigma}\right)h_\alpha\frac{\partial Y_\alpha}{\partial x_k}\right] \tag{8.5}$$

- equation of state

$$p_0 = \frac{\rho R_0 T}{W} = \rho R_0 T\sum_{\alpha=1}^{N}\frac{Y_\alpha}{W_\alpha} = \text{uniform in space} \tag{8.6}$$

- conservation of species

$$\frac{\partial}{\partial t}\rho Y_\alpha + \frac{\partial}{\partial x_k}\rho Y_\alpha v_k = \dot{\omega}_\alpha + \frac{\partial}{\partial x_k}(\rho \mathcal{D}_\alpha\frac{\partial}{\partial x_k}) \tag{8.7}$$

Here, for example, $\alpha = F, O, P$, representing the fuel, oxidant and product, μ is the mixture viscosity, $\sigma = \frac{c_p\mu}{\kappa}$ is the Prandtl number, with c_p the specific heat at constant pressure and κ the heat diffusivity, $Sc_\alpha = \frac{\mu}{\rho\mathcal{D}_\alpha}$, h_α, W_α and $\mathcal{D}_\alpha$ are, respectively, the Schmidt number, the heat of formation, the molecular weight, and the molecular diffusivity of the species α.

These equations describe a range of mathematical problems. For example, consider the problem of a fuel jet with parameters U_J, Y_{fJ}, T_J, and ρ_J, issuing into a coflow stream of air with the parameters U_E, Y_{oE}, T_E, and p_E (see figure 8.5). These quantities can all be used in upstream boundary conditions when considering an initial value, boundary value problem. The thermodynamic pressure p_E is considered to be a constant. Far from the axis of the jet, the various dependent variables should asymptote to the coflow values, *e.g.*,

$$\text{as} \quad r \to \infty, \quad \mathbf{v} \to U_E\mathbf{i}, \quad Y_f \to 0, \text{etc.}, \tag{8.8}$$

where $\mathbf{i}$ is a unit vector in the streamwise direction.

8.1.4. IMPORTANT PARAMETERS

In problems in fluid mechanics, and especially combustion, it is very important to identify the important dimensional and nondimensional parameters. Especially the nondimensional parameters give significant insight regarding the character and flow regime for a particular problem at hand. For the jet flow given in the previous subsection, the (external) dimensional parameters which define the problem are:

$$\mathcal{D}\,,\, U_J\,,\, U_E$$

$$p_E\,,\, \rho_J\,,\, T_J(\text{related})\,,\, c_J = \sqrt{\gamma R T_J}$$

$$Y_{f\,J}\,,\, Y_{oE}$$

$$\mu_J\,,\, \kappa_J\,,\, \mathcal{D}_J$$

$$E\,,\, R_0\,,\, T_a = E/R_0$$

Parameters (internal) which characterize the flow field, and depend on $\mathbf{x}, t$ are:

$$\tilde{v}_i\,,\, v''\,,\, L\,(\text{integral scale of } v_i)\,,\, \tau_T = L/v''$$

$$\delta \quad \text{flame thickness}$$

$$\tau_c = (Be^{-T_a/T})^{-1} \quad \text{a reaction time scale}$$

$$\ell_k = (\nu^3/\epsilon)^{1/4} \quad \text{Kolmorogov (dissipation) scale}$$

$$\ell_{mf} = (\mathcal{D}^3/\epsilon)^{1/4} \quad \text{Corrsin-Obukov scalar dissipation scale}$$

A number of nondimensional parameters can be defined from these dimensional ones. Most important among these are:

$$Re = \frac{\rho_J U_J D}{\mu_J}\,\text{(Reynolds number)} \quad M_J = \frac{U_J}{c_J}\,\text{(Mach number)}$$

$$R_t = \frac{\rho v'' L}{\mu} \quad \text{indicating the strength of the turbulence}$$

$$\frac{\ell_k}{L} = R_t^{-3/4} \qquad \text{if } R_t \gg 1\,,\text{then } \frac{\ell_k}{L} \ll 1$$

$$\frac{\ell_{mf}}{L} = (\frac{\nu_J}{\mathcal{D}_J})^{1/2} R_t^{-3/4} = Sc_J R_t^{-3/4}$$

$$Da = \frac{\tau_T}{\tau_c} \quad \text{(Damköhler number)}$$

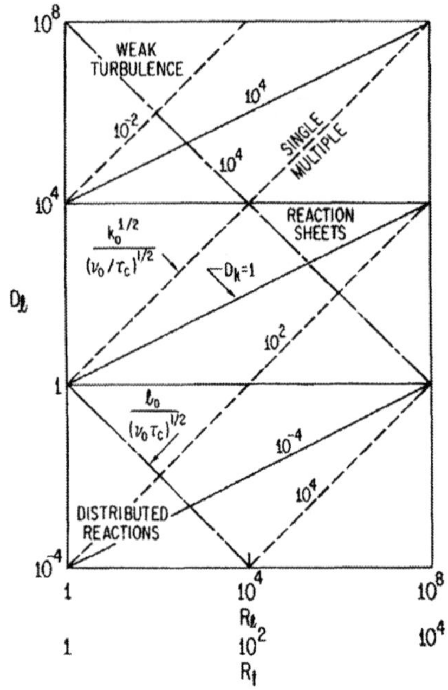

Figure 8.6. Regimes of turbulent combustion (Libby and Williams, 1994)

Generally speaking, if $Da \ll 1$, the chemistry is considered slow, while if $Da \gg 1$, it is considered fast. If $\delta/\ell_k \ll 1$, the flames are considered thin (flamelets), while if $\delta/L \approx 1$, the flames are considered distributed.

These nondimensional parameters can be put together to describe typical regimes for turbulent reacting flows, an example of which is given in figure 8.6 [22].

8.2. Mixture Fraction Based Theories

Several of the most effective theoretical approaches to treating nonpremixed reactions are based upon the concept of a mixture fraction. To understand the basic ideas of this approach consider again the reacting, coflowing jet with fuel of mass fraction $Y_{f\,J}$ issuing from the jet, and oxidant in the coflow with mass fraction Y_{oE}. Assume a one-step, irreversible reaction:

$$F + r\,O \rightarrow (1+r)P + \text{heat}\,,$$

which implies that

$$\dot{\omega}_f = \frac{1}{r}\,\dot{\omega}_o = -\frac{1}{1+r}\,\dot{\omega}_p\,.$$

The following argument can be easily generalized to multi-step reactions.

A major assumption in this approach is that the molecular diffusivities of all three species are equal, and furthermore their diffusivities equal the molecular diffusivity of heat, *i.e.*,

$$\mathcal{D}_f = \mathcal{D}_o = \mathcal{D}_p = \frac{\kappa}{\rho c_p} \equiv \mathcal{D}.$$

This assumption is justified for certain species, *e.g.*, for atomic and molecular oxygen, $\mathcal{D}_O \approx \mathcal{D}_{O_2}$. Furthermore, for high Reynolds number turbulent flows, it is expected that the overall reaction rates become independent of these diffusivities, a fact that has some experimental, theoretical, and numerical support [19]. It must be realized, however, that the molecular diffusivity of some species is much different from that of others, *e.g.*, in comparing molecular hydrogen and oxygen $\mathcal{D}_{H_2} \gg \mathcal{D}_{O_2}$. In addition, this assumption may break down in regions of gaseous flows where the temperature T is high, and hence the viscosity μ is high and the Reynolds number is lower. Finally, it is important to note that recent theoretical work has suggested methods of relaxing this assumption (*e.g.*, [26], [26]; [25], [25]; [28], [28]).

The concept of a mixture fraction is based upon conserved scalars. Examples of this are elementary mass fractions, *e.g.*, those for oxygen and hydrogen, Z_O and Z_H. But, more generally, consider the conservation equations for the fuel and oxidant, assuming equal diffusivities:

$$\frac{\partial}{\partial t}\rho Y_f + \frac{\partial}{\partial x_k}\rho Y_f v_k = \dot{\omega}_f + \frac{\partial}{\partial x_k}\left(\rho\mathcal{D}\frac{\partial Y_f}{\partial x_k}\right) \tag{8.9}$$

$$\frac{\partial}{\partial t}\rho Y_o + \frac{\partial}{\partial x_k}\rho Y_o v_k = \underbrace{\dot{\omega}_o}_{r\dot{\omega}_f} + \frac{\partial}{\partial x_k}\left(\rho\mathcal{D}\frac{\partial Y_o}{\partial x_k}\right) \tag{8.10}$$

Multiplying the first equation by r, and subtracting the second gives:

$$\frac{\partial}{\partial t}\rho(rY_f - Y_o) + \frac{\partial}{\partial x_k}\rho(rY_f - Y_o)v_k = \frac{\partial}{\partial x_k}\left(\rho\mathcal{D}\frac{\partial(rY_f - Y_o)}{\partial x_k}\right),$$

so that $(rY_f - Y_o)$ is a conserved quantity. In fact, for this problem, there are three conserved scalar functions, call Zhvab-Zeldovich functions, obtained by manipulating the species conservation equations. They are:

$$\begin{aligned}
\beta_{fo} &= rY_f - Y_o \\
\beta_{fp} &= (1+r)Y_f + Y_p \\
\beta_{op} &= \frac{(1+r)}{r}Y_o + Y_p
\end{aligned}$$

It is convenient to normalize these so that

$$\begin{aligned} \beta &\rightarrow 1 \quad \text{in the fuel stream} \, (Y_f = Y_{fJ}, Y_o = 0) \\ \beta &\rightarrow 0 \quad \text{in the oxidant stream} \, (Y_f = 0, Y_o = Y_{oE}) \end{aligned}$$

This normalization gives the mixture fraction:

$$Z = \frac{rY_f - Y_o + Y_{oE}}{rY_{fJ} - Y_{oE}},$$

with the properties that it is conserved, and satisfies the limiting conditions:

$$Z \rightarrow \begin{cases} 1 & \text{in the fuel stream} \\ 0 & \text{in the oxidant stream} \end{cases}$$

An interpretation of Z is that it is the local mass fraction of material originating in the fuel stream, while $(1 - Z)$ is the local mass fraction of material originating in the oxidant stream.

An important value of Z is its stoichiometric value, Z_{st}, attained when F and O are in stoichiometric proportions, *i.e.*, $Y_f = Y_o/r$. Plugging this into the definition equation for Z gives

$$Z_{st} = \frac{Y_{oE}}{rY_{fJ} - Y_{oE}}.$$

It will be shown below that, for large Damköhler numbers, it is expected that the chemical reactions (the flame) occur near the stoichiometric surfaces.

Note that the Favre-averaged equations for Z and $Z'' = Z - \tilde{Z}$ are:

$$\frac{\partial}{\partial t}\bar{\rho}\tilde{Z} + \frac{\partial}{\partial x_i}\bar{\rho}\tilde{v}_i\tilde{Z} = \frac{\partial}{\partial x_i}\left(\overline{\rho\mathcal{D}\frac{\partial Z}{\partial x_i}} - \overline{\rho v_i'' Z''}\right), \text{ and}$$

$$\frac{\partial}{\partial t}\bar{\rho}\widetilde{Z''^2} + \frac{\partial}{\partial x_i}\bar{\rho}\widetilde{Z''^2}\tilde{v}_i = 2\overline{\rho v_i'' Z''}\frac{\partial \tilde{Z}}{\partial x_i} - \frac{\partial}{\partial x_i}\overline{\rho v_i'' Z''^2} - \bar{\rho}\tilde{\epsilon}_z$$

The interpretation of the various terms in these equations are discussed elsewhere in this volume. Sketches of a typical stoichiometric surface and surfaces of constant $\tilde{Z}$ are given in figure 8.7.

8.2.1. FAST CHEMISTRY LIMIT

As the reaction rate becomes very large, F and O cannot coexist in space, *i.e.*, the flame becomes very thin (see figure 8.8). In the fuel region, where

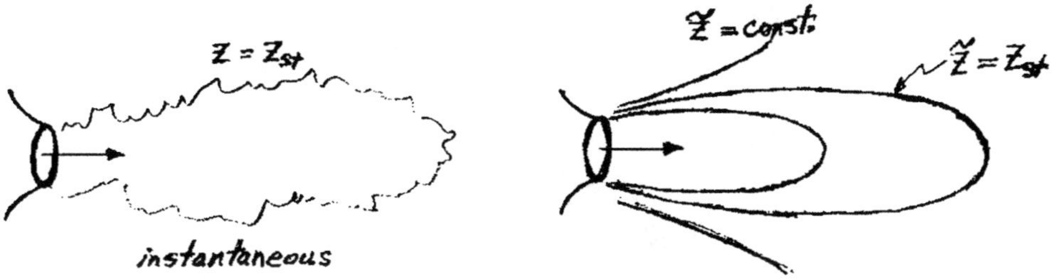

Figure 8.7. Sketches of surfaces of $Z = Z_{st}$ and $\tilde{Z} = \text{const.}$ for a jet

Figure 8.8. Thin flame in fast chemistry limit

$Z > Z_{st}$, $Y_f \neq 0$, but $Y_o = 0$, and hence, from the definition of Z,

$$Z = \frac{rY_f + Y_{oE}}{rY_{f\,J} - Y_{oE}} = \frac{rY_f}{rY_{f\,J} - Y_{oE}} + Z_{st}\,,$$

or

$$\begin{aligned} Y_f &- \frac{1}{r}(rY_{f\,J} - Y_{oE})(Z - Z_{st}) \\ Y_o &= 0 \end{aligned}$$

Similarly, in the oxidant region,

$$\begin{aligned} Y_f &= 0 \\ Y_o &= -(rY_{f\,J} - Y_{oE})(Z - Z_{st})\,. \end{aligned}$$

The infinitely thin reaction occurs at the stoichiometric surface $Z = Z_{st}$.

From above, then, both Y_f and Y_o can be expressed solely in terms of Z, *i.e.*,

$$Y_f = Y_f(Z) = \begin{cases} \frac{1}{r}(rY_{f\,J} - Y_{oE})(Z - Z_{st}) & \text{for} \quad Z > Z_{st} \\ 0 \quad \text{for} \quad Z < Z_{st} & \end{cases}$$

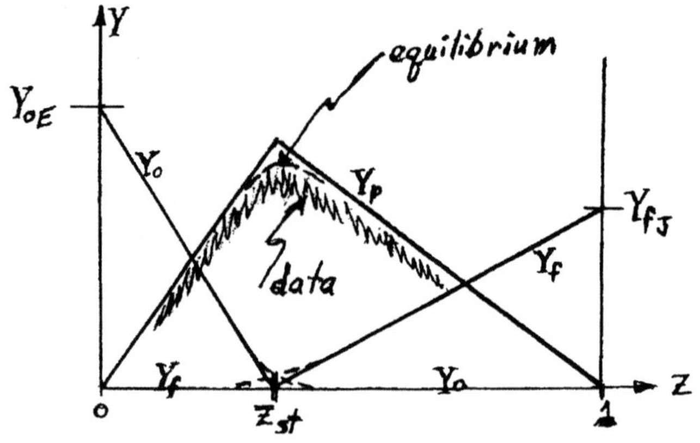

Figure 8.9. Sketch of Y_α versus Z for equilibrium case

$$Y_o = Y_o(Z) = \begin{cases} 0 \quad \text{for} \quad Z > Z_{st} \\ \\ -(rY_{f_J} - Y_{oE})(Z - Z_{st}) \quad \text{for} \quad Z < Z_{st} \end{cases}$$

with a similar expression for Y_p. Therefore, determining Y_f and Y_o reduces to a problem of determining Z, *i.e.*, the problem is really one of pure mixing. A sketch of these curves is given in figure 8.9. For equilibrium chemistry, allowing for reverse reactions to occur, the flame is spread in space to a region of finite extent, as indicated by the dashed lines in the figure. Furthermore, in the low Mach number approximation, equilibrium relations can also be obtained for temperature, density and enthalpy, *i.e.*,

$$\begin{aligned} Y_\alpha &= Y_\alpha^e(Z) \text{ for } \alpha = F, O, P \\ T &= T^e(Z) \\ \rho &= \rho^e(Z) \\ h &= h^e(Z) \end{aligned}$$

Henceforth, the term 'equilibrium chemistry' will be used to include both the equilibrium case, with reversible reactions, and the case with irreversible reactions discussed above. Given equilibrium (or infinite rate) chemistry, the average value for, say Y_f, is given as

$$\bar{Y}_f(\mathbf{x}, t) = \int_0^1 Y_f^e(Z) p(Z; \mathbf{x}, t) dZ\,,$$

where $p(Z; \mathbf{x}, t)$ is the probability density (pdf) of Z. Similar expressions are obtained for $\bar{Y}_o$, $\bar{Y}_p$, $\bar{T}$, $\bar{\rho}$, and $\bar{h}$. The problem thus reduces to determining the pdf of Z. Qualitatively, p takes on forms as shown in the sketch in figure 8.10.

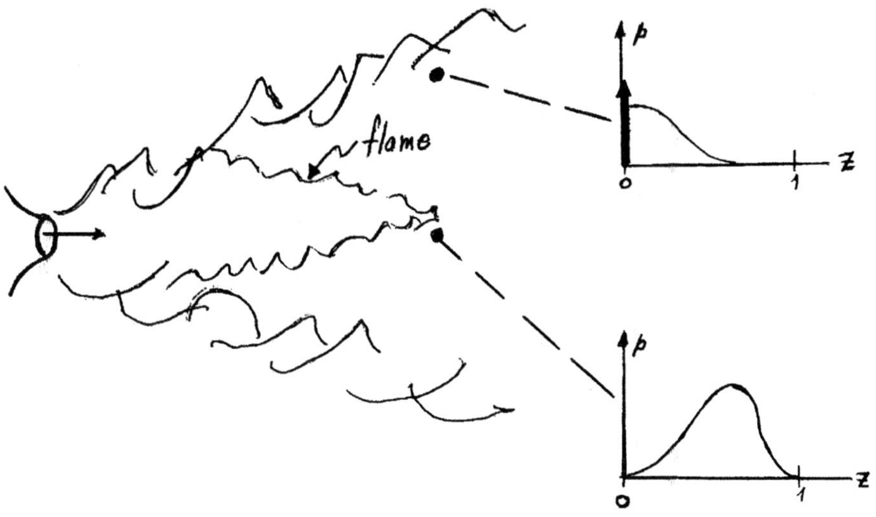

Figure 8.10. Examples of pdf's of Z for a jet

This result suggests a possible approach for obtaining the average values, called the assumed pdf approach. The pdf of Z is assumed to have a particular function form, for example that of a β-function, which generally depends on two parameters and is determined by $\bar{Z}$ and $Z' = \overline{(Z-\bar{Z})^2}^{1/2}$ (or the corresponding Favre-averaged quantities). Standard RANS closures can be used to obtain $\bar{\mathbf{v}}$, k, ϵ, ..., $\bar{Z}$, and Z'. The mixture fractions and thermodynamic variables can then be determined as, *e.g.*,

$$\bar{Y}_f \simeq \int_0^1 Y_f^e(Z) p_a(Z; \bar{Z}, Z') dZ \,,$$

where p_a is the assumed form for the pdf. Results will be presented below to show that, for near equilibrium chemistry, this is a very effective approach.

It is interesting to examine the reacting rates, *e.g.*, $\dot{\omega}_f$, for this equilibrium case. From the equation for Y_f (Equation 8.9), assuming, for simplicity, constant ρ and $\mathcal{D}$:

$$\dot{\omega}_f = \rho(\frac{\partial Y_f}{\partial t} + \mathbf{v} \cdot \nabla Y_f) - \rho \mathcal{D} \nabla^2 Y_f \,.$$

But with equilibrium chemistry, $Y_f(\mathbf{x}, t) = Y_f^e[Z(\mathbf{x}, t)]$. Plugging this into Equation 8.2.1 and using the chain rule gives:

$$\dot{\omega}_f = \underbrace{\rho[\frac{\partial Z}{\partial t} + \mathbf{v} \cdot \nabla Z - \rho \mathcal{D} \nabla^2 Z]}_{=0} - \rho \mathcal{D} |\nabla Z|^2 \frac{d^2 Y_f^e}{dZ^2}$$

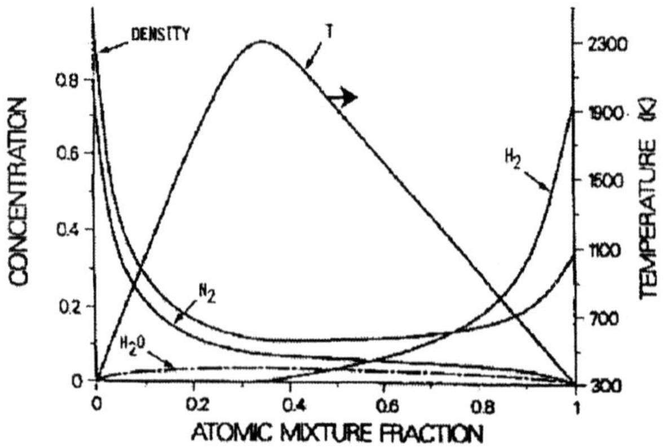

Figure 8.11. Equilibrium relationships for hydrogen jet (Chen and Kollmann, 1994)

$$= -\rho \mathcal{D} |\nabla Z|^2 \frac{d^2 Y_f^e}{dZ^2}$$

So $\dot{\omega}_f = -\frac{1}{2}\chi \frac{d^2 Y_f^e}{dZ^2}$, where $\chi = 2\mathcal{D}|\nabla Z|^2$, the scalar dissipation rate. The reaction rate thus depends directly on χ, which carries information about molecular mixing. Furthermore, $\overline{\dot{\omega}_f}$ depends on the joint probability density of Z and χ.

Equilibrium chemistry has been used extensively in the past in computing averages for major species when they are close to equilibrium. Consider the laboratory experiments of [13] ([13]), employing a hydrogen/air jet. One of their cases utilized a jet of 78% hydrogen diluted with 22% argon, issuing at 150 m/s from a nozzle with diameter $D = 5.2$ mm. The coflow air had a speed of 9.2 m/s, and the jet Reynolds number was approximately 18,000.

[6] ([6]) performed RANS calculations for this case, using the equilibrium assumption to treat the major species, temperature and density. Figure 8.11 gives plots of their equilibrium relationships obtained for this case. Figure 8.12 gives a comparison of model predictions and laboratory data, at a typical downstream location, for the radial profile of the Favre-averaged axial velocity, while figure 8.13 gives the corresponding turbulence intensity. The agreement for these quantities, which are affected by the chemical reaction, is very good. Figure 8.14 gives radial profiles of the average density at three downstream locations. Again the agreement is very good for this thermodynamic quantity. Similar agreement is obtained for other velocity components, and for the major species. Various minor species are not in chemical equilibrium, however, and must be treated with nonequilibrium methods, one of which is discussed next.

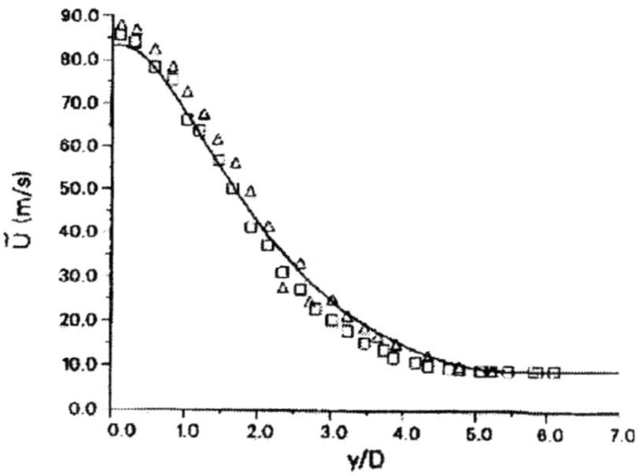

Figure 8.12. Radial profiles of $\tilde{u}$ for hydrogen jet (Chen and Kollmann, 1994)

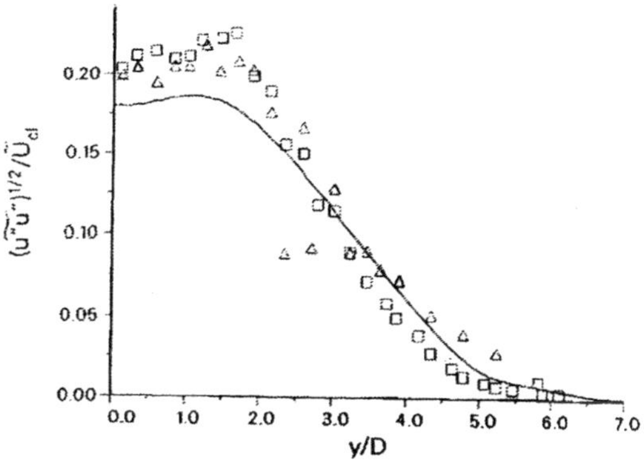

Figure 8.13. Radial profiles of axial turbulence intensity for hydrogen jet (Chen and Kollman, 1994)

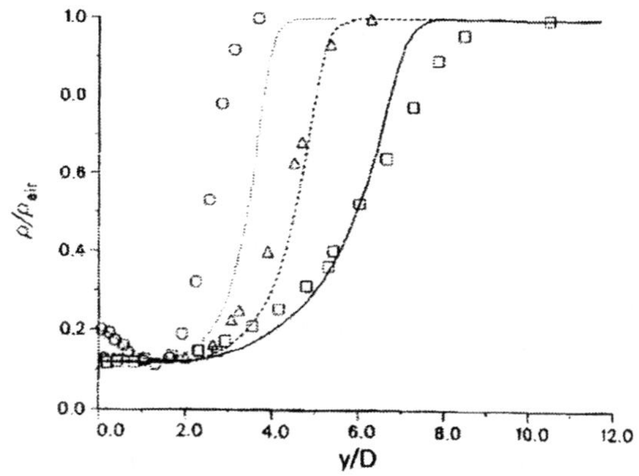

Figure 8.14. Radial profiles of average density for hydrogen jet (Chen and Kollmann, 1994)

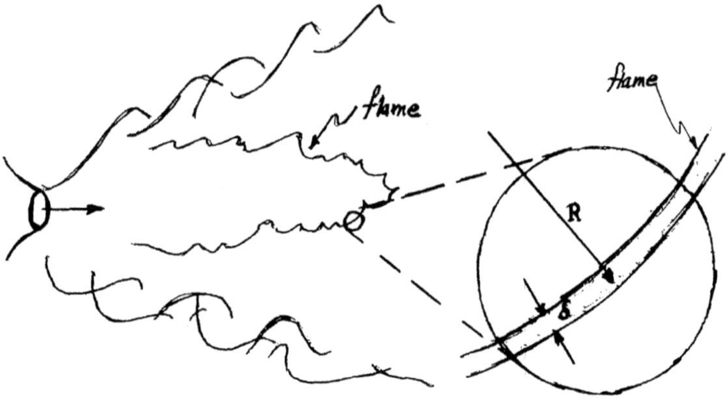

Figure 8.15. Sketch of jet flame showing δ and R

8.2.2. FINITE-RATE CHEMISTRY

For finite rate chemistry, two related approaches utilizing the mixture fraction have been shown to lead to accurate predictions, the (laminar) flamelet model [27] and the Conditional Moment Closure (CMC) approach [3]. Here the flamelet approach will be emphasized, partially because of its recent extension to large-eddy simulations, which will be treated in the following chapter.

In the flamelet model, the Damköhler number ($Da = \tau_T/\tau_c$) is assumed to be large enough that the flame regions are very 'thin'. Here thin will be taken to mean that the flame thickness δ is much smaller than either of the local principal radii of curvature R of the flame itself (see the sketch in figure 8.15), *i.e.*, $\delta/R \ll 1$. There is some argument that, in order for this inequality to hold, in addition to the assumption of large Damköhler number, the flame thickness must be small compared to the local Kolmogorov scale $\ell_k = (\nu^3/\epsilon)^{1/4}$ [2], although this requirement is in dispute.

At any point in the flow field a local orthogonal coordinate system is set up, with one unit vector $\mathbf{n}$ in the direction of ∇Z, *i.e.*,

$$\mathbf{n} = \frac{\nabla Z}{|\nabla Z|}.$$

The equations are transformed as $(x, y, z) \to (n, t_1, t_2)$ where $n = |\mathbf{n}|$ and t_1 and t_2 are local coordinates perpendicular to $\mathbf{n}$. Then, since Z is locally monotonic in the direction of $\mathbf{n}$, n can be replaced by Z as an independent variable. The result is, for example, for the conservation equation for the fuel mass fraction Y_f, neglecting the spatial and temporal variations in t_1

and t_2 [27]

$$\begin{aligned}\rho\frac{\partial Y_f}{\partial t} &= \dot{\omega}_f + \rho\mathcal{D}|\nabla Z|^2\frac{\partial^2 Y_f}{\partial Z^2} \\ &+ \rho\Big(\underbrace{-\mathbf{v}_t\cdot\nabla_t Y_f}_{\sim\frac{\delta}{R}} + \underbrace{\mathcal{D}\nabla_t^2 Y_f}_{\sim(\frac{\delta}{R})^2} + \underbrace{2\mathcal{D}\nabla_t(\frac{\partial Y_f}{\partial Z})\cdot\nabla_t Z}_{\sim\frac{\delta}{R}}\Big) + \underbrace{\nabla_t(\rho\mathcal{D})\cdot\nabla_t Y_f}_{\sim(\frac{\delta}{R})^2}\end{aligned}$$

Scaling arguments give the order of magnitude of the last three terms on the RHS as δ/R, $(\delta/R)^2$, δ/R, and $(\delta/R)^2$ as shown. Therefore, with the assumption that $\delta/R \ll 1$, or, for example, $\frac{\partial^2}{\partial t_1^2}(\cdot) \ll \frac{\partial^2}{\partial Z^2}(\cdot)$, the conservation equation reduces to

$$\rho\frac{\partial Y_f}{\partial t} = \dot{\omega}_f + \rho\mathbf{D}|\nabla Z|^2\frac{\partial^2 Y_f}{\partial Z^2},$$

which is called the flamelet equation. For large Damkóhler numbers, which implies that the chemical time scale is much smaller than the turbulence time scale, it is often argued that, furthermore,

$$\frac{\partial Y_f}{\partial t} \approx 0,$$

so that the flamelet equation reduces to

$$\begin{aligned}\dot{\omega}_f &= -\rho\mathcal{D}|\nabla Z|^2\frac{\partial^2 Y_f}{\partial Z^2} \\ &= -\frac{1}{2}\chi\frac{\partial^2 Y_f}{\partial Z^2},\end{aligned}$$

the steady flamelet equation. Note that this is a simplified form of the conservation equation for the fuel, assumed to hold at each point in the flow. Similar equations can be derived for the mass fractions of oxidant and product, and for the enthalpy.

Note that, as in the equilibrium case, the scalar dissipation rate χ enters into the theory, carrying information about the local scalar mixing. The solution of the steady flamelet equation might then be written as

$$Y_\alpha = Y_\alpha^{fl}(Z,\chi) \quad \alpha = F, O, P,$$

an obvious extension of the equilibrium result where Y_α only depends upon Z. The average value for Y_α can then be written as

$$\bar{Y}_\alpha = \int_0^\infty\int_0^1 Y_\alpha^{fl}(Z,\chi)\,p(Z,\chi)dZd\chi,$$

where $p(Z, \chi)$ is the joint probability density of Z and χ.

To develop the model further, assumptions are needed regarding χ and its relationship to Z. One approach is to assume that Z and χ are statistically independent, since Z depends on the large scale motions while χ depends on the small scales and, for high Reynolds number flows, the motions at these two scales are expected to be independent. Furthermore it has been suggested that $\bar{Y}_i$ depends mainly on $\bar{\chi}$, and only weakly on χ' [20], so that

$$\bar{Y}_\alpha = \int_0^1 Y_\alpha^{fl}(Z, \bar{\chi})\, p(Z) dZ \,.$$

An alternative approach for χ is to assume that the flow in the vicinity of the flame is a local counterflow, implying $\delta < \ell_k$ and locally uniform strain. The quasi-steady solution to the equation for Z [27] then implies that

$$\chi = \chi_0 F(Z)\,, \quad \text{where} \tag{8.11}$$

$$F(Z) = \exp\{-2[\mathrm{erf}^{-1}(2Z)]^2\}\,. \tag{8.12}$$

Here χ_0 is the local peak of χ, and erf^{-1} is the inverse error function. Laboratory data have indicated that this expression holds locally [4].

Assuming, analogous to the above, the $\bar{Y}_\alpha$ depends mainly on $\bar{\chi}_0$, but only weakly on χ_0', and the statistical independence of Z and χ_0, then

$$\bar{Y}_\alpha = \int_0^1 Y_\alpha^{fl}(Z, \bar{\chi}_0)\, p(Z) dZ \,.$$

Furthermore, averaging Equation 8.11 gives

$$\bar{\chi} = \bar{\chi}_0 \int_0^1 F(Z)\, p(Z) dZ \,.$$

Therefore $\bar{\chi}$ and $\bar{\chi}_0$ are directly related.

It is again convenient to assume a β-distribution for Z, as in the equilibrium case. Then

$$\bar{\chi} = \bar{\chi}_0 \int_0^1 F(Z)\, p_a(Z; \bar{Z}, Z')\, dZ = \bar{\chi}_0\, g(\bar{Z}, Z')\,, \text{and}$$

$$\bar{Y}_\alpha = \int_0^1 Y_\alpha^{fl}(Z, \bar{\chi})\, p_a(Z; \bar{Z}, Z')\, dZ = \bar{Y}_\alpha(\bar{\chi}, \bar{Z}, Z') \quad \alpha = F, O, P$$

In theory, this equation can be tabulated ahead of time, suggesting the following algorithm to treat finite rate, non-premixed chemistry:

– from a RANS model, determine $\bar{\mathbf{v}}$, $\overline{v_i v_j}$, ϵ, $\bar{Z}$, Z', and $\bar{\chi}$

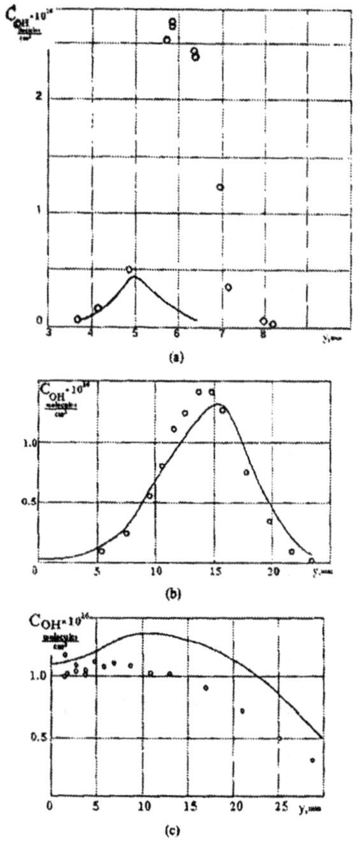

Figure 8.16. Radial profiles of OH concentration in hydrogen jet (Buriko *et al.*, 1994)

– use the flamelet tables to determine $\bar{Y}_\alpha = \bar{Y}_\alpha(\bar{Z}, Z', \bar{\chi})$ locally (similarly for $\bar{T}$, $\bar{h}$, and $\bar{\rho}$)

As implied, the methodology can be easily extended to the variable density, low Mach number case.

Several researchers have compared the results of the flamelet model with laboratory data, and with significant success. For example, [5] ([5]) have compared the predictions of the flamelet model with the laboratory data of [14] ([14]) for the case of an H_2 jet into air. A typical comparison is shown in figure 8.16, giving the model predictions and laboratory data for the mass fraction of OH, Y_{OH}, a quantity that is not in equilibrium in this case. The results show that the model predicts the experimental results rather well, especially away from the near field, where the flow is not expected to be well-developed turbulence. Similar results have been obtained by [21] ([21]) for a syngas jet into air, and by [28] ([28]) for an H_2 jet into air. In the latter case good predictions were obtained for all species except NO, which

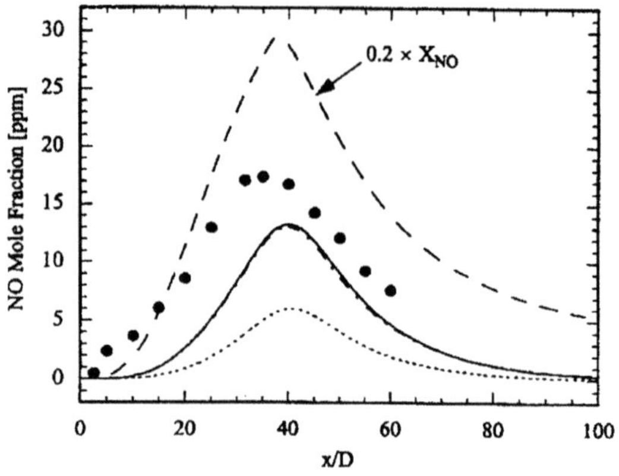

Figure 8.17. Axial behavior of NO in hydrogen jet (Pitsch and Peters, 1998)

was significantly over-predicted. The authors showed that, by including the unsteady term in the flamelet equation, the agreement could be significantly improved (see figure 8.17). Other possible cases of discrepancies between model predictions and data have been suggested to be:

- uncertainties in the prediction of $\tilde{\chi}$, although it is found that the results are somewhat insensitive to the values of $\tilde{\chi}$
- neglect of thermal radiation, which is estimated to decrease Y_{NO} by about a factor of 2 for this case, and
- the differential diffusion of chemical species and also heat, since, *e.g.*, the molecular diffusivity of H_2 is much larger than the corresponding quantities for the other species and temperature.

8.3. Large-Eddy Simulations

8.3.1. INTRODUCTION

As computer memory and speed continue to significantly increase, and numerical methods are improved, it becomes increasingly possible to avoid some of the modelling required in treating the Reynolds-Averaged Navier Stokes equations, and to directly integrate the governing equations. Two closely related approaches have been developed for this purpose: direct numerical simulation (DNS) and large-eddy simulation (LES).

Direct numerical simulation involves solving the three-dimensional, time-dependent Navier Stokes equations and species conservation equations. No Reynolds or Favre averaging is employed; all of the relevant length and time scales are resolved numerically. Since (i) the size of the computational domain must be at least of the order of the integral scale of the turbulence, say L_T, (ii) the flow must be resolved down to at least the Kolmogorov scale, ℓ_k, and (iii) the simulations must be three-dimensional, the number of computational grid points must go like

$$N_g \sim \left(\frac{L_T}{\ell_k}\right)^3 \sim Re^{9/4}\,.$$

Furthermore, due to numerical stability and accuracy considerations, the time step size must decrease in proportion to the grid spacing, so that the number of time steps must scale like

$$N_t \sim Re^{3/4}\,.$$

These conditions are very restrictive, and limit the range of Reynolds and Damköhler numbers that can be simulated. For example, at the present time flows with Reynolds numbers, based upon the Taylor microscale, of about 100 can be simulated (see, *e.g.*, [12], [12]).

The approach of direct numerical simulation has the distinct advantage, when compared to the RANS approach, of avoiding the closure problem. Furthermore, as compared to laboratory experiments, all of the dependent variables, *e.g.*, $\mathbf{v}$, p, ρ, Y_f, are known at each point in space and time. This implies that most statistics of interest can be studied, and flow structures can be analyzed in detail. Finally, within the limitations mentioned above, compared to laboratory experiments parameters such as the Reynolds and Damköhler numbers can be easily varied, and flow conditions are easily controllable. This approach can be considered analogous and complementary to laboratory experiments; at the present time it is mainly useful as a research tool.

It is important to note the distinctions between direct numerical simulations and RANS simulations. To see this, consider a mixing layer down stream from a splitter plate (see figure 8.18). In direct numerical simulations (or in large-eddy simulations to be discussed below) of this flow, the instantaneous, unsteady, three-dimensional structures in this flow are directly computed (figure 8.18a). In a RANS simulation, however, only the averaged fields are computed, so that no direct information is obtained about the flow structures (figure 8.18b).

Because of the limitations inherent in the DNS approach, the technique of large-eddy simulations of turbulence has been developed. In the approach of LES, prior to performing the simulations, the governing equations are spatially filtered to remove motions at length scales not resolvable on the computational mesh. For example, consider the velocity field $\mathbf{v}(\mathbf{x}, t)$; it can be spatially filtered as

$$\hat{\mathbf{v}}(\mathbf{x}, t) = \int G(\mathbf{x} - \mathbf{x}')\,\mathbf{v}(\mathbf{x}', t)\,d\mathbf{x}'\,,$$

(a)

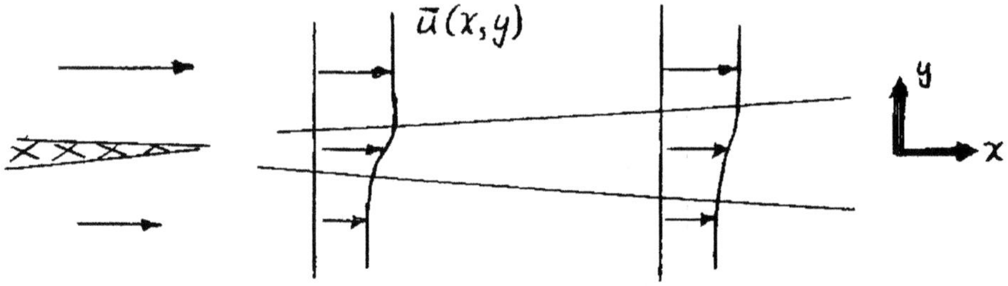

(b)

Figure 8.18. Sketch distinguishing DNS (or LES) from RANS for a mixing layer

where the filter function G, usually a top-hat or a Gaussian function, is chosen with the following properties:

$$\begin{aligned} G(\mathbf{x}) &= G(-\mathbf{x}) \\ G(\mathbf{x}) &\geq 0 \\ \int G(\mathbf{x}) d\mathbf{x} &= 1 \end{aligned}$$

To understand the effect of this filtering operation, consider filtering the momentum equation for constant density flows:

$$\frac{\partial}{\partial t}\hat{v}_i + \hat{v}_j \frac{\partial}{\partial x_j}\hat{v}_i = -\frac{1}{\rho}\frac{\partial}{\partial x_i}\hat{p} + \nu \frac{\partial^2}{\partial x_k^2}\hat{v}_i + \underbrace{\frac{\partial}{\partial x_k}\hat{\tau}_{ik}}_{sgs} .$$

The equation is identical to the momentum equation prior to filtering, except for the addition of the final term on the RHS. The quantity $\hat{\tau}_{ik}$, defined as

$$\hat{\tau}_{ik} = -(\widehat{v_i v_k} - \hat{v}_i \hat{v}_k) ,$$

is called the subgrid-scale stress tensor, and represents the effects of the subgrid-scale motions (which have been filtered out) on the grid-scale motions (computed in the simulations). This stress tensor requires modelling. (Subgrid-scale modelling is discussed elsewhere in this text.) Present models estimate the effect of the subgrid scales using information about the

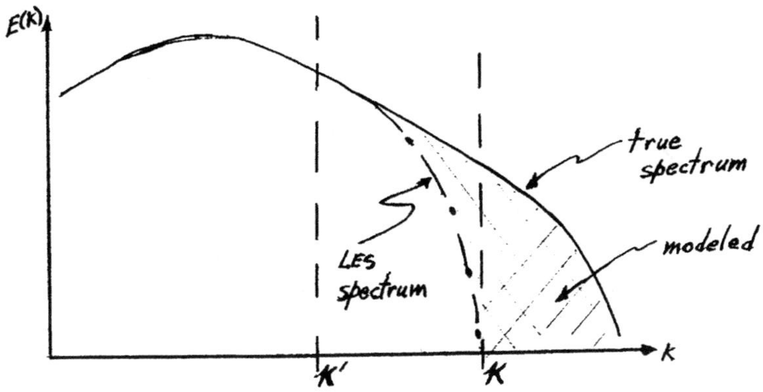

Figure 8.19. Sketch of energy spectrum in an LES

motions close to the scale of the computational cutoff (see figure 8.19). For example, in one modelling approach, due to [30], the subgrid-scale stress tensor is modeled analogous to the Reynolds stress in mixing length theory as

$$\hat{\tau}_{ij} = -C\Delta|\hat{S}|\hat{S}_{ij}|\,,$$

where $\hat{S}_{ij}$ is the filtered strain-rate tensor, and Δ is the computational grid spacing. The parameter C is often chosen by a 'dynamic' method (*e.g.*, [16], [16]), taking into account the behavior of the velocity field at the smallest resolvable length scales.

The approach of LES has distinct advantages and disadvantages. When compared to DNS, it has the potential advantage of treating large Reynolds number and large Damköhler number problems. Closure models must be introduced, however, to treat the effects of the subgrid-scale motions. When compared to RANS, LES has the disadvantage of being a very large calculation, being three-dimensional and time-dependent, and with as many length scales resolved as possible. It has the advantages, however, that the large-scale motions are treated directly, so that only the small-scale motions require modelling. For turbulent flows the large-scale motions tend to control the overall behavior of the flow, so their direct computation is a distinct advantage. Furthermore, the small, filtered scales of motion contain little energy, so that the computed results are expected to be somewhat insensitive to the model used. Finally, the small scale motions of a turbulent flow tend to be more universal, especially when compared to the larger-scale motions, so that a proven model should have broader applicability.

At the present time the approach of LES is very useful as a research tool, and is being used in a limited number of applications. There is some optimism that, as computers continue to develop, and subgrid-scale mod-

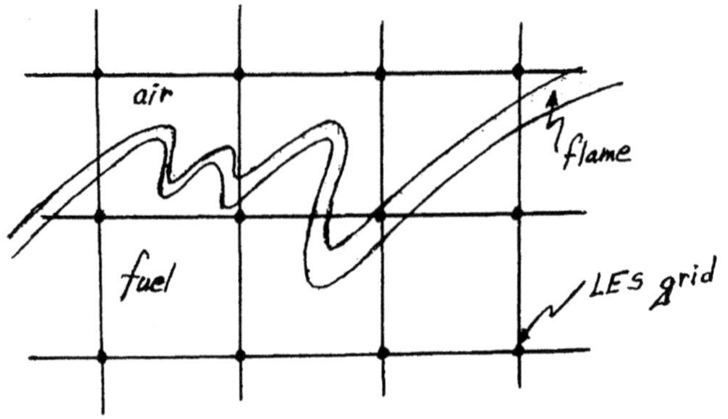

Figure 8.20. Sketch of reaction zone in an LES

elling continues to improve, this approach will prove more and more useful in applications in the future.

8.3.2. LES OF CHEMICALLY-REACTING FLOWS

In order to understand the potential problems involved in large-eddy simulations of turbulent, reacting flows, consider again a constant density, single-step, irreversible, non-premixed chemical reaction occurring on a turbulent flow field. Filtering the conservation equation for fuel (Equation 8.9) for this case gives:

$$\frac{\partial}{\partial t}\hat{Y}_f + \hat{v}_i\frac{\partial}{\partial x_i}\hat{Y}_f = \frac{1}{Pe}\frac{\partial^2}{\partial x_i^2}\hat{Y}_f + \widehat{\dot{\omega}_f} + \frac{\partial}{\partial x_i}\hat{\sigma}_i\,.$$

Here $\hat{\sigma}_i = -(\widehat{v_i Y_f} - \hat{v}_i\hat{Y}_f)$, representing the flux of Y_f due to the subgrid-scale motions, is of the same form as a term arising in the equation for the mixture fraction Z, and can be treated, *e.g.*, by Smagorinsky's model with a dynamically computed coefficient. The filtered reaction rate term is given by

$$\widehat{\dot{\omega}_f} = Da\widehat{Y_f Y_o}\,,$$

and the new issue is how to model this term.

For moderate to large Damköhler numbers, the flame will be very thin, and the chemical reaction will occur almost entirely at the subgrid scale (see figure 8.20). This implies that the spectrum for the reaction rate term $\dot{\omega}_f$ lies generally past the cut-off wave number in an LES, as shown qualitatively in figure 8.21. This appears to have discouraging implications, and implies that modelling procedures for this term must be very different from those for the flux term $\hat{\sigma}_i$.

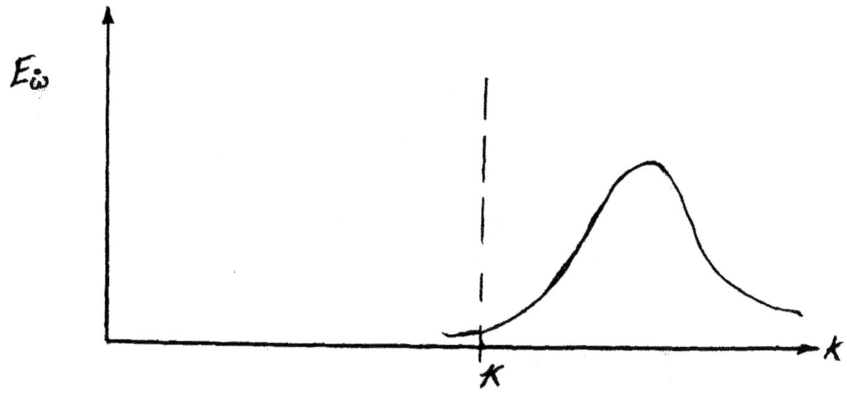

Figure 8.21. Sketch of spectrum of reaction rate

Of course, any models must be based upon grid-scale quantities. What gives hope that a reasonable model exists is that the overall reaction rate depends on the mixing rate, and the mixing rate depends on the behavior of the larger-scale eddies. The mixture fraction provides a natural link between the larger-scale eddies and the molecular mixing rate, and hence the chemical reaction rate.

Equilibrium chemistry

First consider the case of equilibrium chemistry. Assuming equal diffusivities, then, as discussed in Section 2.1, the mass fractions of the various chemical species can be written as known functions of the mixture fraction, *e.g.*,

$$Y_f = Y_f^e(Z)\,.$$

Consider a small, subgrid-scale volume element where possibly a flame exists (see figure 8.22). Let $\hat{p}(Z)$ be the frequency distribution of Z within that volume element. The function $\hat{p}$ is known as the subgrid-scale pdf [15] or the filtered-density function [7]. Then

$$\hat{Y}_f = \int_0^1 Y_f^e(Z)\,\hat{p}(Z)\,dZ\,.$$

Therefore, if $\hat{p}$ can be determined, then so can $\hat{Y}_f$ and the other filtered mass fractions.

In analogy with the assumed pdf approach discussed in Section 2.1 for the Reynolds-averaged equations, one model involves assuming a functional form for $\hat{p}$, in particular that of a β-function. Then $\hat{p}$ can be obtained from

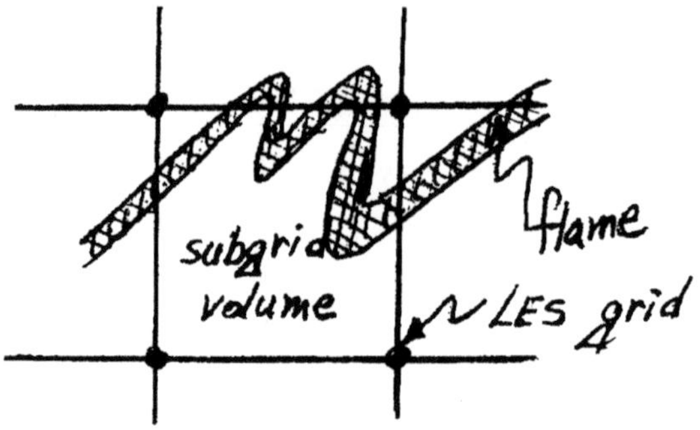

Figure 8.22. Sketch of subgrid-scale volume element

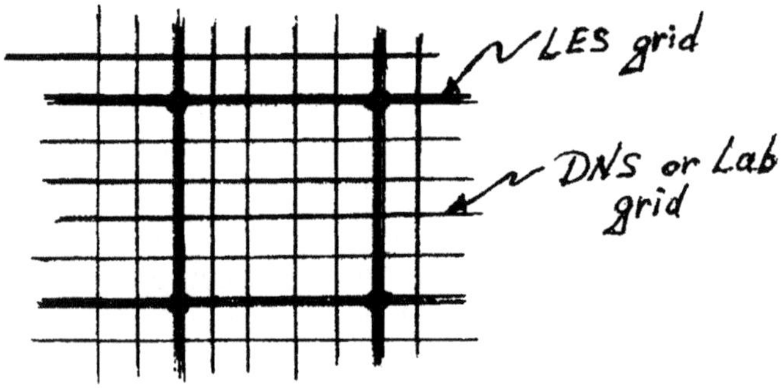

Figure 8.23. Sketch of DNS and LES grids

the first two filtered moments of Z, $\hat{Z}$ and $Z_v = (\widehat{Z^2} - \hat{Z}^2)^{1/2}$. The problem then reduces to developing accurate procedures to predict both $\hat{Z}$ and Z_v.

This approach can be readily tested in an *a priori* fashion using data for Z from either high-resolution DNS or highly-resolved scalar fields measured in laboratory experiments. The values of the mixture fraction (or some other conserved scalar) are assumed to be given at each point on a high-resolution, three-dimensional grid, as shown in figure 8.23. Assuming chemical equilibrium, the values of the mass fractions for each of the chemical species are also known at each point on the high-resolution grid. A lower resolution, LES grid can be defined, as also shown in the figure. If enough grid points are contained in each LES volume element (typically, in the examples given below, these volume elements contain 16^3 to 32^3 grid points), then reliable filtered values for all quantities can be obtained in each volume element by averaging over the higher-resolution grid points contained within that volume element. Therefore, *e.g.*, $\hat{Z}$, Z_v, $\hat{Y}_f$, $\hat{Y}_o$, and $\hat{Y}_p$ can be obtained in each volume element on the LES grid.

Also, $\hat{Y}_f$, $\hat{Y}_o$, and $\hat{Y}_p$ can be obtained from the model mentioned above, assuming a β-function form for $\hat{p}$, and using the directly computed values

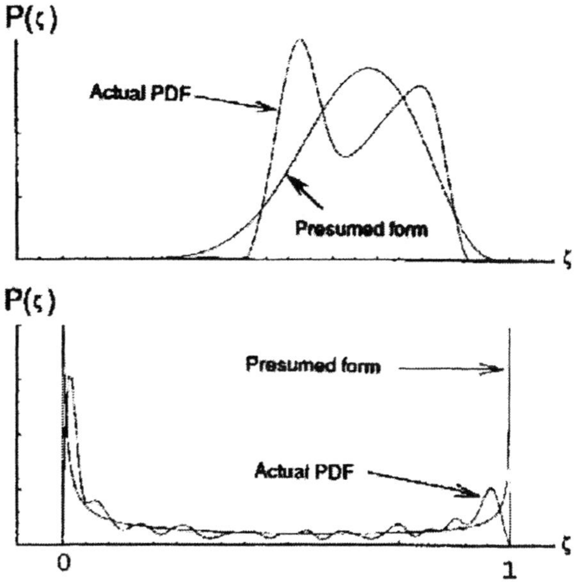

Figure 8.24. Plots of actual and modeled pdf's (Cook and Riley, 1994)

for $\hat{Z}$ and Z_v. These model values will be referred to as $\hat{Y}_{fm}$, $\hat{Y}_{om}$, and $\hat{Y}_{pm}$. The model can therefore be tested by comparing $\hat{Y}_f$ and $\hat{Y}_{fm}$, for example, for each volume element on the LES grid. Note that this is mainly testing the model for $\hat{p}$, determining whether it is well-represented as a β-function.

Two high resolution data sets have been used to test this model [8]. The first is taken from high-resolution DNS of homogeneous, turbulence decay carried out on a 128^3-point computational grid by [24]. A mixture fraction was initialized for this simulation, and allowed to evolve in time. For this case, $Sc = 1$, the initial Reynolds number based upon the Taylor microscale was about 50, the stoichiometric value of Z was taken to be 0.25, and the data sets were examined at the times $\tau = tv_0'/L_0 = 0.54, 1.1, 2.0$, and 4.3, where v_0' is the initial rms velocity and L_0 the initial velocity integral scale.

Figure 8.24 gives typical plots of the computed frequency distribution $\hat{p}$ compared to the β-function model. Locally computed values for $\hat{Z}$ and Z_v were used in obtaining the model curves. It is seen that the β-function is a reasonable representation of the actual distribution, although by no means highly accurate. Figure 8.25 gives a typical plot of the values of $\hat{Y}_f$ plotted against those for $\hat{Y}_{fm}$. Each point on the plot corresponds to one particular subvolume on the LES grid. If the model were exact, the points would lie exactly on a line at 45°. It is seen that the points are very close to this line, indicating the high degree of accuracy of the model. Apparently the errors in the model for $\hat{p}$ displayed in figure 8.24 are significantly reduced

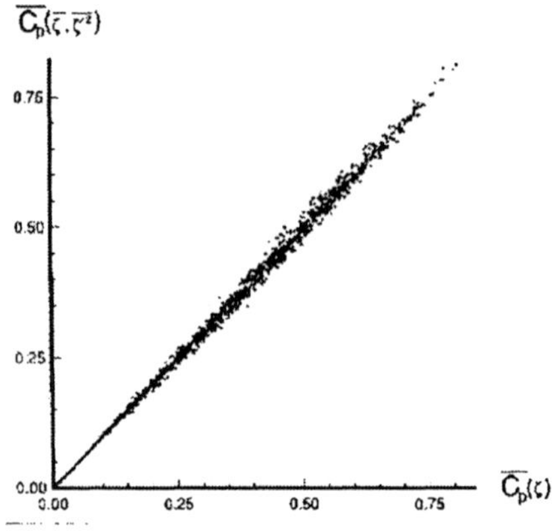

Figure 8.25. Plots of $\hat{Y}_{fm}$ versus $\hat{Y}_f$ (Cook and Riley, 1994)

upon integration. Similar results were obtained for the mass fractions of the oxidant and product, and for all the times of decay addressed.

Stoichiometric values of the mixture fraction are much lower than 0.25 for common reactions. This can potentially lead to difficulties in the model, since the integral then depends mainly on the shape of $\hat{p}$ between 0 and Z_{st}. Figure 8.26 gives plots of $\hat{Y}_p$ versus $\hat{Y}_{pm}$ for several different values of Z_{st} [10]. It is seen that, for values typical of methane/air reactions ($Z_{st} \simeq 0.0476$) and hydrogen/air reactions ($Z_{st} \simeq 0.0244$), the model is still fairly accurate. For the value of $Z_{st} = 0.005$, however, it has clearly become inaccurate.

The model was also tested using the laboratory data of [4] ([4]). The data were taken in a jet of water into water; the jet was seeded with fluorescine dye. The Reynolds number based upon the exit velocity and the nozzle diameter was 3,700. Scanned, planar laser-induced fluorescence was used to obtain 'instantaneous' values of the passive dye concentration field on a grid of 256×256×9 points. With this grid the scalar dissipation range could be effectively resolved. The scalar data fields were processed in the same *a priori* manner as were the DNS fields. Again the comparisons of the directly computed values of, *e.g.*, $\hat{Y}_f$ with the modeled values, *e.g.*, $\hat{Y}_{fm}$, were very good.

Finite-rate chemistry

To treat subgrid-scale effects for finite-rate chemistry, a version of the flamelet model introduced in Section 2.2 has been employed ([9], [9]; [10], [10]). Assuming equal molecular diffusivities of all chemical species, and that the diffusivity of heat equals these molecular diffusivities, *i.e.*, all the Lewis numbers are 1, and also assuming high Damköhler numbers, it was

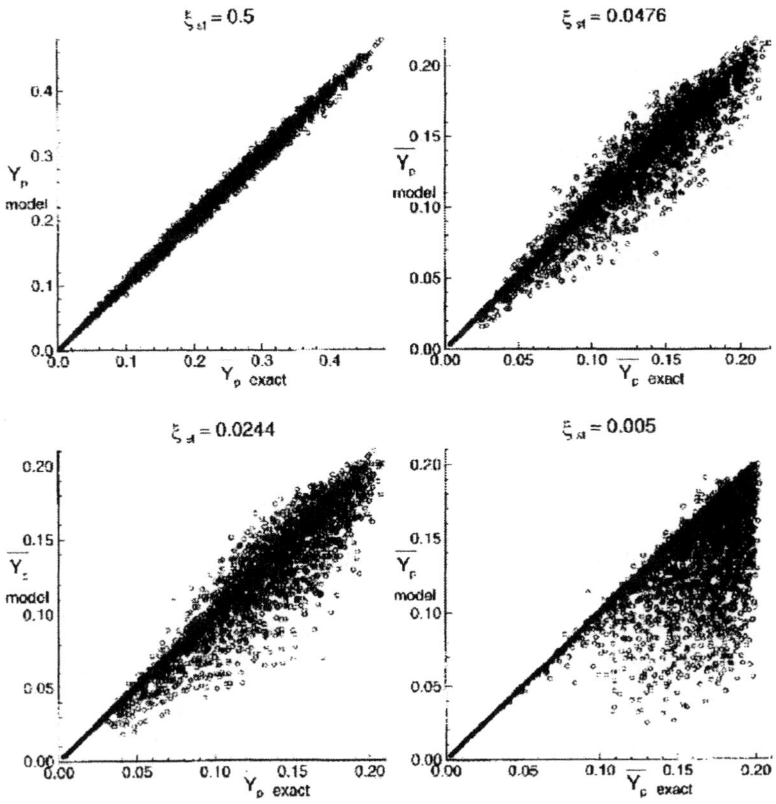

Figure 8.26. Correlation of exact and modeled results for various stoichiometries (Cook and Riley, 1998)

shown that the equation for each specie and for enthalpy reduces locally to the flamelet equation. For example, for the one-step reaction of fuel and oxidant,

$$\dot{\omega}_\alpha = -\frac{1}{2}\chi\frac{d^2Y_\alpha}{dZ^2}, \quad \alpha = F, O, P,$$

with the boundary conditions, *e.g.*,

$$\begin{aligned} Y_f &\to 0 \quad \text{as} \quad Z \to 0 \\ Y_f &\to 1 \quad \text{as} \quad Z \to 1 \end{aligned}$$

The counterflow model for χ is employed, so that the solution to the flamelet equation can be written as, *e.g.*,

$$Y_f = Y_f^{fl}(Z, \chi_0),$$

where χ_0 is the local maximum of χ. Therefore, with $\hat{p}(Z, \chi_0)$ representing the joint frequency distribution of Z and χ_0 in a subgrid-scale volume element, then

$$\hat{Y}_f = \int_{-\infty}^{\infty} \int_0^1 Y_f^{fl}(Z, \chi_0)\, \hat{p}(Z, \chi_0)\, dZ d\chi_0 \,.$$

Assumptions are now required to determine $\hat{p}(Z, \chi_0)$. Again, following the discussion in Section 2.2, it is assumed that Z and χ_0 are statistically independent inside each volume element, and that $\hat{Y}_f$ depends mainly on $\hat{\chi}_0$, and only weakly on $\chi_0' = (\widehat{\chi_0^2} - \hat{\chi}_0^2)^{1/2}$, which results in:

$$\hat{Y}_f = \int_0^1 Y_f^{fl}(Z, \hat{\chi}_0)\, \hat{p}(Z)\, dZ \,.$$

Again, with the counterflow model for χ, and the statistical independence of Z and χ_0,

$$\hat{\chi} = \hat{\chi}_0 \int_0^1 F(Z)\, \hat{p}(Z)\, dZ \,,$$

where F is the inverse error function, defined in Section 2.2. All that remains is to determine $\hat{p}$.

From the results for the equilibrium case, it is clear that the β-function model provides an adequate approximation for $\hat{p}(Z)$, given accurate values for $\hat{Z}$ and Z_v. Denoting this assumed form as $\hat{p}_a(Z; \hat{Z}, Z_v)$, then the model result for $\hat{Y}_f$ for finite-rate chemistry is:

$$\begin{aligned} \hat{Y}_f &= \int_0^1 Y_f^{fl}(Z, \hat{\chi}_0)\, p_a(Z; \hat{Z}, Z_v)\, dZ && (8.13) \\ &= \hat{Y}_f(\hat{Z}, Z_v, \hat{\chi}) \,, && (8.14) \end{aligned}$$

where $\hat{\chi}$ is related to $\hat{\chi}_0$ as above.

These results suggest the following methodology to determine the filtered values of, say, the fuel mass fraction in an LES of non-premixed chemical reactions. Prior to the simulations, tables for $\hat{Y}_f(\hat{Z}, Z_v, \hat{\chi})$ can be set up. This is accomplished by first numerically solving the flamelet equations for an appropriate range of $(Z, \hat{\chi}_0)$, giving $\hat{Y}_f^{fl}(Z, \hat{\chi}_0)$. Then this solution can be used in numerically integrating Equation 8.13 for an appropriate range of $\hat{Z}$, Z_v and $\hat{\chi}_0$ to obtain $\hat{Y}_f(\hat{Z}, Z_v, \hat{\chi})$. Once these tables are set up, then the LES is performed; in it $\hat{Z}$, Z_v and $\hat{\chi}$ would be needed to be computed locally on a point-wise basis; the values for $\hat{Y}_f$ would then be obtained by table look-up. Note that one can also determine $\hat{Y}_o$, $\hat{Y}_p$ and $\hat{\omega}_f$ as:

$$\begin{aligned} \hat{Y}_o &= Y_{oE}(1 - Z) + r(\hat{Y}_f - \hat{Z}\, Y_{fJ}) \\ \hat{Y}_p &= (r + 1)(\hat{Z}\, Y_{fJ} - \hat{Y}_f) \\ \hat{\omega}_f &= Da \int_0^1 Y_f^{fl}(Z, \hat{\chi}_0)\, Y_o^{fl}(Z, \hat{\chi}_0)\, \hat{p}(Z)\, dZ \end{aligned}$$

Also, this approach can be readily extended to realistic chemistry with exothermicity [10].

This model has also been tested in an *a priori* fashion, using results from high-resolution DNS. For example, [9] utilized the DNS data fields computed by [24] for isothermal, one-step, irreversible reactions occurring on decaying, homogeneous turbulence. The DNS fields were computed on 128^3-point grids, while the LES mesh was chosen to be a 16^3-point grid, so that subgrid-scale volume averages over $8^3 = 512$ points could be taken. For Case A of [24], the Schmidt number was 1.0, the Damköhler number $Da = k/v_0' L_0$ was 2.1, and the Reynolds number $Re = v_0' L_0/\nu$ was 148. Here k is the constant coefficient of the fuel reaction rate term. The stoichiometric value of Z was 0.25. In figure 8.27 is compared the actual versus modeled values for fuel, oxidant and product mass fractions. Also included are the equilibrium and frozen ($k = 0$) model predictions.

It is seen that the model predictions for this lower Damköhler number case, although more accurate than the equilibrium prediction, are in significant error. The reaction rate predicted by the model is too fast, resulting in, *e.g.*, too much product formation.

When the Damköhler number is increased to 8.4 (see figure 8.28), the model predictions are significantly improved. The predictions for all of the mass fractions are very good, and significantly better than those for the equilibrium model.

These results have been extended to lower stoichiometric values for Z ($Z_{st} = 0.111$), and to cases with Arrhenius chemistry [10]. The results are very similar to those given above. It appears that, for larger Damköhler numbers, and given accurate values of $\hat{Z}$, Z_v and $\hat{\chi}$, the flamelet model provides very good predictions of the filtered mass fractions of the chemical species.

Recently the flamelet model has been utilized directly in LES of non-premixed reactions [11]. Accurate models for $\hat{Z}$, Z_v and $\hat{\chi}$ are required. Computing the conservation equation for $\hat{Z}$ directly, and using scaling models for both Z_v and $\hat{\chi}$, it is found that these models are adequate to again provide accurate predictions of the filtered mass fractions. These results are aided, to some extent, by the weak sensitivity of the filtered mass fractions to $\hat{\chi}$. The accuracy of this and other approaches to the LES of chemically reacting, turbulent flows is presently the subject of intense research.

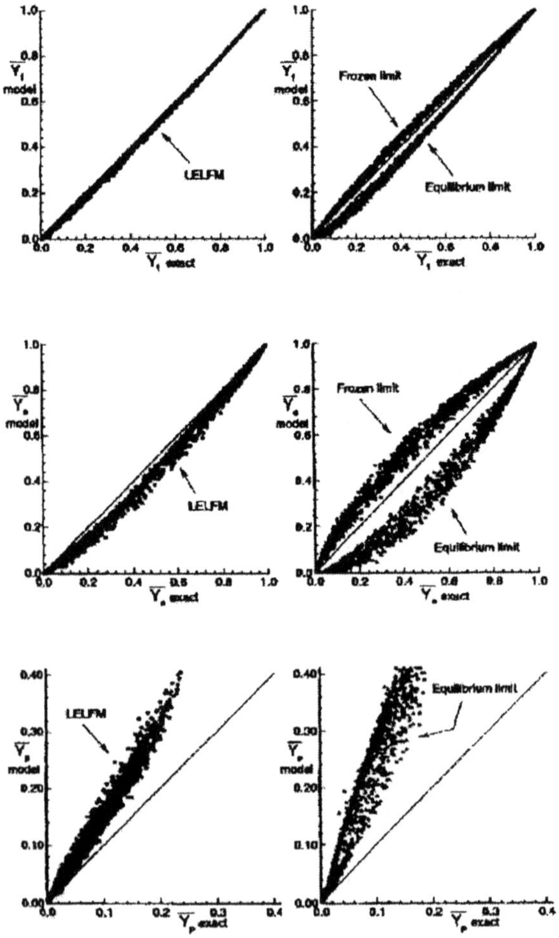

Figure 8.27. Correlation of exact and modeled results, $Da = 2.1$ (Cook *et al.*, 1997)

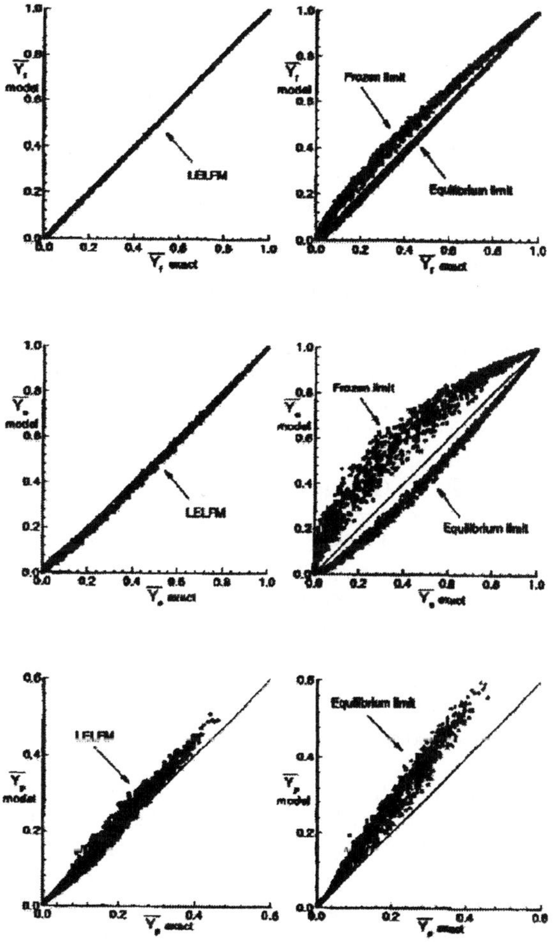

Figure 8.28. Correlation of exact and modeled results, $Da = 8.4$ (Cook *et al.*, 1997)

Acknowledgment

The author would especially like to thank George Kosály, and also Andy Cook and Steve deBruynKops, for many illuminating discussions on this subject. His research on turbulent reacting flows has been recently supported by the U.S. National Science Foundation and the U.S. Air Force Office of Scientific Research.

References

1. Bilger, R. W. 1976. Reaction zone thickness and formation of nitric oxide in turbulent diffusion flames. *Prog. Energy Comb. Sci.*, **1**, 87.
2. Bilger, R. W. 1988. The structure of turbulent nonpremixed flames. *Twenty-second Symposium (International) on Combustion (The Combustion Institute)*, 475.
3. Bilger, R. W. 1993. Conditional moment closure for reacting flows. *Phys. Fluids A*, **5**, 436.
4. Bish, E. S., and Dahm, W. J. A. 1995. Strained dissipation and reaction layer analyses of nonequilibrium chemistry in turbulent reacting flows. *Combust. Flame*, **100**, 457.
5. Buriko, Y. Y., Kuznetsov, V. R., Volkov, D. V. and Zaitsev, S. A. 1990. A test of a flamelet model for turbulent nonpremixed combustion. *Combust. Flame*, **96**, 104.
6. Chen, J.-Y., and Kollman, W. 1994. Comparison of prediction and measurement in nonpremixed turbulent flames. in *Turbulent Reacting Flows*, edited by Libby and Williams, Academic Press, 211.
7. Colucci, P. J., Jaberi, F. A., Givi, P. and Pope, S. B. 1998. Filtered density function for large eddy simulation of turbulent reacting flows.*Phys. Fluids*, **10**, 499.
8. Cook, A. W., and Riley, J. J. 1994. A subgrid model for equilibrium chemistry in turbulent flows. *Phys. Fluids*, **6**, 2868.
9. Cook, A. W., Riley, J. J. and Kosály, G. 1997. A laminar flamelet approach to subgrid-scale chemistry in turbulent flows. *Combust. Flame*, **109**, 332.
10. Cook, A. W., and Riley, J. J. 1998. Subgrid-scale modeling for turbulent reacting flows. *Combust. Flame*, **112**, 593.
11. de Bruyn Kops, S. M., Riley, J. J., Kosály, G. and Cook, A. W. 1998. Investigation of modeling for nonpremixed turbulent combustion. *Flow, Turb. and Comb.*, **60**(1), 105.
12. de Bruyn Kops, S. M., and Riley, J. J. 1998. Direct numerical simulation of laboratory experiments in isotropic turbulence. *Phys. Fluids*, **10**, 2125.
13. Dibble, R. W., Schefer, R. W., Chen, J.-Y., Hartmann, V. and Kollman, W. 1986. Combustion Institute Western States Meeting, Paper WSS/CI 86-65.
14. Drake, M. C., Pitz, R. W. and Lapp, M. 1986. Laser measurements on nonpremixed H_2-air flames for assessment of turbulent combustion models. *AIAA J*, **24**, 905.
15. Gao, F., and O'Brien, E. E. 1993. A large-eddy simulation scheme for turbulent reacting flows. *Phys. Fluids A*, **5**, 1282.
16. Germano, M., Piomelli, U., Moin, P. and Cabot, W. H. 1991. A dynamic subgrid-scale eddy viscosity model. *Phys. Fluids A*, **3**, 1760.
17. Hill, J. C. 1976. Homogeneous turbulent mixing with chemical reaction. *Ann. Rev. Fluid Mech.*, **8**, 135.
18. Kerstein, A. R. 1988. A linear-eddy model of turbulent scalar transport and mixing. *Combust. Sci. Tech.*, **60**, 391.
19. Kerstein, A. R. 1990. Linear-eddy modeling of turbulent transport. Part 3. Mixing and differential molecular diffusion in round jets. *J. Fluid Mech.*, **216**, 411.
20. Kuznetsov, V. R., and Sabel'nikov, V. A. 1990. *Combustion: An International Series*, N. Chigier, ed., Hemisphere Publishing Corp.

21. Lentini, D. 1994. Assessment of the stretched laminar flamelet approach for non-premixed turbulent combustion. *Combust. Sci. Techn.*, **100**, 95.
22. Libby, P. A., and Williams, F. A. 1994. Fundamental aspects and a review. in *Turbulent Reacting Flows*, edited by Libby and Williams, Academic Press, 1.
23. McMurtry, P. A., Menon, S. and Kerstein, A. R. 1992. A linear eddy subgrid model for turbulent reacting flows: application to hydrogen-air combustion. *Twenty-fourth Symposium (International) on Combustion (The Combustion Institute)*, 1990.
24. Mell, W. E., Nilsen, V., Kosály, G. and Riley, J. J. 1994. Investigation of closure models for turbulent reacting flows. *Phys. Fluids A*, **6**, 1331.
25. Nilsen, V. 1998. *Investigation of Differentially Diffusing Scalars in Isotropic Turbulence*, Ph.D. Thesis, University of Washington.
26. Nilsen, V., and Kosály, G. 1997. Differentially diffusing scalars in turbulence. *Phys. Fluids*, **9**, 3385.
27. Peters, N. 1984. Laminar diffusion flamelet models in nonpremixed turbulent combustion. *Prog. Energy Combust. Sci.*, **10.**, 319.
28. Pitsch, H., Chen, M. and Peters, N. 1998 Unsteady flamelet modeling of turbulent hydrogen/air diffusion flames. submitted to the *Twenty-seventh International Symposium on Combustion.*
29. Pope, S. B. 1990 Computations of turbulent combustion: progress and challenges. *Twenty-third Symposium (International) on Combustion (The Combustion Institute)*, 591.
30. Smagorinsky, J. 1963. General circulation experiments with the primitive equations. I. The basic experiment*Mon. Weather Rev.*, **91**, 99.
31. Spalding, D. B. 1967. The spread of turbulent flames confined in ducts. *Eleventh Symposium (International) on Combustion (The Combustion Institute)*, 807.
32. van Kampen, N. B. 1981. *Stochastic Processes in Physics and Chemistry*, North Holland.

ERCOFTAC SERIES

1. A. Gyr and F.-S. Rys (eds.): *Diffusion and Transport of Pollutants in Atmospheric Mesoscale Flow Fields.* 1995 ISBN 0-7923-3260-1
2. M. Hallbäck, D.S. Henningson, A.V. Johansson and P.H. Alfredsson (eds.): *Turbulence and Transition Modelling.* Lecture Notes from the ERCOFTAC/IUTAM Summerschool held in Stockholm. 1996 ISBN 0-7923-4060-4
3. P. Wesseling (ed.): *High Performance Computing in Fluid Dynamics.* Proceedings of the Summerschool held in Delft, The Netherlands. 1996 ISBN 0-7923-4063-9
4. Th. Dracos (ed.): *Three-Dimensional Velocity and Vorticity Measuring and Image Analysis Techniques.* Lecture Notes from the Short Course held in Z~rich, Switzerland. 1996 ISBN 0-7923-4256-9
5. J.-P. Chollet, P.R. Voke and L. Kleiser (eds.): *Direct and Large-Eddy Simulation II.* Proceedings of the ERCOFTAC Workshop held in Grenoble, France. 1997 ISBN 0-7923-4687-4

KLUWER ACADEMIC PUBLISHERS – DORDRECHT / BOSTON / LONDON

ICASE/LaRC Interdisciplinary Series in Science and Engineering

1. J. Buckmaster, T.L. Jackson and A. Kumar (eds.): *Combustion in High-Speed Flows.* 1994 ISBN 0-7923-2086-X
2. M.Y. Hussaini, T.B. Gatski and T.L. Jackson (eds.): *Transition, Turbulence and Combustion.* Volume I: Transition. 1994 ISBN 0-7923-3084-6; set 0-7923-3086-2
3. M.Y. Hussaini, T.B. Gatski and T.L. Jackson (eds.): *Transition, Turbulence and Combustion.* Volume II: Turbulence and Combustion. 1994 ISBN 0-7923-3085-4; set 0-7923-3086-2
4. D.E. Keyes, A. Sameh and V. Venkatakrishnan (eds): *Parallel Numerical Algorithms.* 1997 ISBN 0-7923-4282-8
5. T.G. Campbell, R.A. Nicolaides and M.D. Salas (eds.): *Computational Electromagnetics and Its Applications.* 1997 ISBN 0-7923-4733-1
6. V. Venkatakrishnan, M.D. Salas and S.R. Chakravarthy (eds.): *Barriers and Challenges in Computational Fluid Dynamics.* 1998 ISBN 0-7923-4855-9

KLUWER ACADEMIC PUBLISHERS – DORDRECHT / BOSTON / LONDON

FLUID MECHANICS AND ITS APPLICATIONS

Series Editor: R. Moreau

Aims and Scope of the Series

The purpose of this series is to focus on subjects in which fluid mechanics plays a fundamental role. As well as the more traditional applications of aeronautics, hydraulics, heat and mass transfer etc., books will be published dealing with topics which are currently in a state of rapid development, such as turbulence, suspensions and multiphase fluids, super and hypersonic flows and numerical modelling techniques. It is a widely held view that it is the interdisciplinary subjects that will receive intense scientific attention, bringing them to the forefront of technological advancement. Fluids have the ability to transport matter and its properties as well as transmit force, therefore fluid mechanics is a subject that is particularly open to cross fertilisation with other sciences and disciplines of engineering. The subject of fluid mechanics will be highly relevant in domains such as chemical, metallurgical, biological and ecological engineering. This series is particularly open to such new multidisciplinary domains.

1. M. Lesieur: *Turbulence in Fluids.* 2nd rev. ed., 1990 ISBN 0-7923-0645-7
2. O. Métais and M. Lesieur (eds.): *Turbulence and Coherent Structures.* 1991 ISBN 0-7923-0646-5
3. R. Moreau: *Magnetohydrodynamics.* 1990 ISBN 0-7923-0937-5
4. E. Coustols (ed.): *Turbulence Control by Passive Means.* 1990 ISBN 0-7923-1020-9
5. A.A. Borissov (ed.): *Dynamic Structure of Detonation in Gaseous and Dispersed Media.* 1991 ISBN 0-7923-1340-2
6. K.-S. Choi (ed.): *Recent Developments in Turbulence Management.* 1991 ISBN 0-7923-1477-8
7. E.P. Evans and B. Coulbeck (eds.): *Pipeline Systems.* 1992 ISBN 0-7923-1668-1
8. B. Nau (ed.): *Fluid Sealing.* 1992 ISBN 0-7923-1669-X
9. T.K.S. Murthy (ed.): *Computational Methods in Hypersonic Aerodynamics.* 1992 ISBN 0-7923-1673-8
10. R. King (ed.): *Fluid Mechanics of Mixing.* Modelling, Operations and Experimental Techniques. 1992 ISBN 0-7923-1720-3
11. Z. Han and X. Yin: *Shock Dynamics.* 1993 ISBN 0-7923-1746-7
12. L. Svarovsky and M.T. Thew (eds.): *Hydroclones.* Analysis and Applications. 1992 ISBN 0-7923-1876-5
13. A. Lichtarowicz (ed.): *Jet Cutting Technology.* 1992 ISBN 0-7923-1979-6
14. F.T.M. Nieuwstadt (ed.): *Flow Visualization and Image Analysis.* 1993 ISBN 0-7923-1994-X
15. A.J. Saul (ed.). *Floods and Flood Management.* 1992 ISBN 0-7923-2078-6
16. D.E. Ashpis, T.B. Gatski and R. Hirsh (eds.): *Instabilities and Turbulence in Engineering Flows.* 1993 ISBN 0-7923-2161-8
17. R.S. Azad: *The Atmospheric Boundary Layer for Engineers.* 1993 ISBN 0-7923-2187-1
18. F.T.M. Nieuwstadt (ed.): *Advances in Turbulence IV.* 1993 ISBN 0-7923-2282-7
19. K.K. Prasad (ed.): *Further Developments in Turbulence Management.* 1993 ISBN 0-7923-2291-6
20. Y.A. Tatarchenko: *Shaped Crystal Growth.* 1993 ISBN 0-7923-2419-6

Kluwer Academic Publishers – Dordrecht / Boston / London

Mechanics

FLUID MECHANICS AND ITS APPLICATIONS

Series Editor: R. Moreau

21. J.P. Bonnet and M.N. Glauser (eds.): *Eddy Structure Identification in Free Turbulent Shear Flows.* 1993 ISBN 0-7923-2449-8
22. R.S. Srivastava: *Interaction of Shock Waves.* 1994 ISBN 0-7923-2920-1
23. J.R. Blake, J.M. Boulton-Stone and N.H. Thomas (eds.): *Bubble Dynamics and Interface Phenomena.* 1994 ISBN 0-7923-3008-0
24. R. Benzi (ed.): *Advances in Turbulence V.* 1995 ISBN 0-7923-3032-3
25. B.I. Rabinovich, V.G. Lebedev and A.I. Mytarev: *Vortex Processes and Solid Body Dynamics.* The Dynamic Problems of Spacecrafts and Magnetic Levitation Systems. 1994 ISBN 0-7923-3092-7
26. P.R. Voke, L. Kleiser and J.-P. Chollet (eds.): *Direct and Large-Eddy Simulation I.* Selected papers from the First ERCOFTAC Workshop on Direct and Large-Eddy Simulation. 1994 ISBN 0-7923-3106-0
27. J.A. Sparenberg: *Hydrodynamic Propulsion and its Optimization.* Analytic Theory. 1995 ISBN 0-7923-3201-6
28. J.F. Dijksman and G.D.C. Kuiken (eds.): *IUTAM Symposium on Numerical Simulation of Non-Isothermal Flow of Viscoelastic Liquids.* Proceedings of an IUTAM Symposium held in Kerkrade, The Netherlands. 1995 ISBN 0-7923-3262-8
29. B.M. Boubnov and G.S. Golitsyn: *Convection in Rotating Fluids.* 1995 ISBN 0-7923-3371-3
30. S.I. Green (ed.): *Fluid Vortices.* 1995 ISBN 0-7923-3376-4
31. S. Morioka and L. van Wijngaarden (eds.): *IUTAM Symposium on Waves in Liquid/Gas and Liquid/Vapour Two-Phase Systems.* 1995 ISBN 0-7923-3424-8
32. A. Gyr and H.-W. Bewersdorff: *Drag Reduction of Turbulent Flows by Additives.* 1995 ISBN 0-7923-3485-X
33. Y.P. Golovachov: *Numerical Simulation of Viscous Shock Layer Flows.* 1995 ISBN 0-7923-3626-7
34. J. Grue, B. Gjevik and J.E. Weber (eds.): *Waves and Nonlinear Processes in Hydrodynamics.* 1996 ISBN 0-7923-4031-0
35. P.W. Duck and P. Hall (eds.): *IUTAM Symposium on Nonlinear Instability and Transition in Three-Dimensional Boundary Layers.* 1996 ISBN 0-7923-4079-5
36. S. Gavrilakis, L. Machiels and P.A. Monkewitz (eds.): *Advances in Turbulence VI.* Proceedings of the 6th European Turbulence Conference. 1996 ISBN 0-7923-4132-5
37. K. Gersten (ed.): *IUTAM Symposium on Asymptotic Methods for Turbulent Shear Flows at High Reynolds Numbers.* Proceedings of the IUTAM Symposium held in Bochum, Germany. 1996 ISBN 0-7923-4138-4
38. J. Verhás: *Thermodynamics and Rheology.* 1997 ISBN 0-7923-4251-8
39. M. Champion and B. Deshaies (eds.): *IUTAM Symposium on Combustion in Supersonic Flows.* Proceedings of the IUTAM Symposium held in Poitiers, France. 1997 ISBN 0-7923-4313-1
40. M. Lesieur: *Turbulence in Fluids.* Third Revised and Enlarged Edition. 1997 ISBN 0-7923-4415-4; Pb: 0-7923-4416-2

Kluwer Academic Publishers – Dordrecht / Boston / London

GPSR Compliance

The European Union's (EU) General Product Safety Regulation (GPSR) is a set of rules that requires consumer products to be safe and our obligations to ensure this.

If you have any concerns about our products, you can contact us on ProductSafety@springernature.com

In case Publisher is established outside the EU, the EU authorized representative is:

Springer Nature Customer Service Center GmbH
Europaplatz 3
69115 Heidelberg, Germany

Batch number: 10755248

Printed by Printforce, the Netherlands